VOLUME ONE HUNDRED AND FOUR

# CURRENT TOPICS IN DEVELOPMENTAL BIOLOGY

## Epigenetics and Development

# CURRENT TOPICS IN DEVELOPMENTAL BIOLOGY

*"A meeting-ground for critical review and discussion of developmental processes"*
*A.A. Moscona and Alberto Monroy (Volume 1, 1966)*

VOLUME ONE HUNDRED AND FOUR

# Current Topics in DEVELOPMENTAL BIOLOGY

## Epigenetics and Development

Edited by

**EDITH HEARD**

*Mammalian Developmental Epigenetics Group*
*Unité de Génétique et Biologie du Développement*
*INSERM U934 / CNRS UMR3215*
*Institut Curie - Centre de Recherche*
*Pôle de Biologie du Développement et Cancer*
*Paris, France*

ELSEVIER

AMSTERDAM • BOSTON • HEIDELBERG • LONDON
NEW YORK • OXFORD • PARIS • SAN DIEGO
SAN FRANCISCO • SINGAPORE • SYDNEY • TOKYO
Academic Press is an imprint of Elsevier

Academic Press is an imprint of Elsevier
525 B Street, Suite 1800, San Diego, CA 92101-4495, USA
225 Wyman Street, Waltham, MA 02451, USA
The Boulevard, Langford Lane, Kidlington, Oxford, OX5 1GB, UK
32 Jamestown Road, London NW1 7BY, UK
Radarweg 29, PO Box 211, 1000 AE Amsterdam, The Netherlands

First edition 2013



ISBN: 978-0-12-416027-9
ISSN: 0070-2153

Printed and bound by CPI Group (UK) Ltd, Croydon, CR0 4YY
Transfered to Digital Printing, 2013

# CONTENTS

# CONTRIBUTORS

**Håvard Aanes**
BasAM, Norwegian School of Veterinary Science, Oslo, Norway

**Peter Aleström**
BasAM, Norwegian School of Veterinary Science, Oslo, Norway

**Ingrid S. Andersen**
Stem Cell Epigenetics Laboratory, Institute of Basic Medical Sciences, Faculty of Medicine, and Norwegian Center for Stem Cell Research, University of Oslo, Oslo, Norway

**Déborah Bourc'his**
Unité Génétique Biologie du Développement, Institut Curie, UMR3215/INSERMU394, Paris, France

**Philippe Collas**
Stem Cell Epigenetics Laboratory, Institute of Basic Medical Sciences, Faculty of Medicine, and Norwegian Center for Stem Cell Research, University of Oslo, Oslo, Norway

**Denis Duboule**
National Research Centre "Frontiers in Genetics"- School of Life Sciences, Ecole Polytechnique Fédérale, Lausanne, and Department of Genetics and Evolution, University of Geneva, Geneva, Switzerland

**Rachel Duffié**
Unité Génétique Biologie du Développement, Institut Curie, UMR3215/INSERMU394, Paris, France

**André Eid**
Institut de Génétique et de Biologie Moléculaire et Cellulaire, CNRS/INSERM U964, Université de Strasbourg, Illkirch, France

**Anas Fadloun**
Institut de Génétique et de Biologie Moléculaire et Cellulaire, CNRS/INSERM U964, Université de Strasbourg, Illkirch, France

**Amanda G. Fisher**
MRC Clinical Sciences Centre, Imperial College London, London, United Kingdom

**Daniel Grimanelli**
Institut de Recherche pour le Développement, UMR 232, Université de Montpellier II, Montpellier, France

**Sylvain Guibert**
Biotechnologie et signalisation cellulaire, Université de Strasbourg, CNRS/UMR7242, Illkirch, France

**Harry G. Leitch**
Wellcome Trust/Cancer Research UK Gurdon Institute of Cancer and Developmental Biology; Department of Physiology, Development, and Neuroscience, and Wellcome Trust-Medical Research Council Cambridge Stem Cell Institute, University of Cambridge, Cambridge, United Kingdom

**Leif C. Lindeman**
Stem Cell Epigenetics Laboratory, Institute of Basic Medical Sciences, Faculty of Medicine, and Norwegian Center for Stem Cell Research, University of Oslo, Oslo, Norway

**Peter Nestorov**
Friedrich Miescher Institute for Biomedical Research, and Faculty of Sciences, University of Basel, Basel, Switzerland

**Daan Noordermeer**
National Research Centre "Frontiers in Genetics", School of Life Sciences, Ecole Polytechnique Fédérale, Lausanne, Switzerland

**Antoine H.F.M. Peters**
Friedrich Miescher Institute for Biomedical Research, and Faculty of Sciences, University of Basel, Basel, Switzerland

**Andrew H. Reiner**
Stem Cell Epigenetics Laboratory, Institute of Basic Medical Sciences, Faculty of Medicine, and Norwegian Center for Stem Cell Research, University of Oslo, Oslo, Norway

**François Roudier**
Institut de Biologie de l'Ecole Normale Supérieure, CNRS-UMR 8197/INSERM U1024, Paris, France

**Olga Østrup**
Stem Cell Epigenetics Laboratory, Institute of Basic Medical Sciences, Faculty of Medicine, and Norwegian Center for Stem Cell Research, University of Oslo, Oslo, Norway

**M. Azim Surani**
Wellcome Trust/Cancer Research UK Gurdon Institute of Cancer and Developmental Biology; Department of Physiology, Development, and Neuroscience, and Wellcome Trust-Medical Research Council Cambridge Stem Cell Institute, University of Cambridge, Cambridge, United Kingdom

**Walfred W.C. Tang**
Wellcome Trust/Cancer Research UK Gurdon Institute of Cancer and Developmental Biology, and Department of Physiology, Development, and Neuroscience, University of Cambridge, Cambridge, United Kingdom

**Mathieu Tardat**
Friedrich Miescher Institute for Biomedical Research, Basel, Switzerland

**Maria-Elena Torres-Padilla**
Institut de Génétique et de Biologie Moléculaire et Cellulaire, CNRS/INSERM U964, Université de Strasbourg, Illkirch, France

**Tomomi Tsubouchi**
MRC Genome Damage and Stability Centre, University of Sussex, Falmer, United Kingdom

**Michael Weber**
Biotechnologie et signalisation cellulaire, Université de Strasbourg, CNRS/UMR7242, Illkirch, France

# PREFACE

The question of how a single cell, the fertilized egg, gives rise to hundreds of different cell types, all with specialized roles during the development of an organism, even though the DNA information within each of these cells is identical, continues to fascinate scientists. The highly complex question of developmental regulation can be broken down into a series of simpler questions: how are some genes expressed and others not, in different cell types and lineages? How does a particular cell type not only know what genes to express but remember to keep expressing them, sometimes for years, or over hundreds of cell divisions? Furthermore, once differentiation has occurred, how can this process sometimes be reversed? For example at fertilization, when the highly differentiated sperm and egg cells come together, *tabula rasa* is somehow re-established to begin life. How does this occur?

Development is regulated by differential gene activity, the activation and repression of specific transcriptional programs. Transcription factors clearly play a key role in promoting expression of developmental genes; however maintenance of specific states throughout cell division needs additional mechanisms. These mechanisms often rely on the fact that DNA is associated with chemical modifications, proteins, and RNA partners, all of which can modulate its readability—and hence the expression of genes—as well as the heritability of expression states through cell division. This, in very general terms, is what we mean by "Epigenetics": from the Greek "epi"—"beyond" or "above"—"genetics." The nature of the epigenetic changes that come into play during development and the mechanisms that regulate them have only recently started to be unraveled. Antibody-based technologies are now available enabling us to detect epigenetic marks associated with chromatin, both in single cells using immunofluorescence and across the genome using chromatin immunoprecipitation. Although heterochromatin and euchromatin were identified cytologically more than a century ago, these different types of chromatin and the parts of the genome they represent can now be analyzed in exquisite detail. Combined with molecular techniques enabling the analysis of DNA methylation, as well as the detection of coding and non-coding RNAs in a genome-wide fashion, this has led to tremendous advances in our understanding of epigenetics during development.

Different organisms exploit epigenetic mechanisms to different extents and in a variety of ways during development. This volume of *Current Topics in Developmental Biology* includes nine chapters by experienced investigators who discuss the dynamics of epigenetic changes in a range of organisms, as well as the role of epigenetic mechanisms during development and in the germ line.

In Chapter 1, Maria-Elena Torres-Padilla and colleagues review the changes in chromatin states and nuclear organization that occur during preimplantation mouse development. They focus in particular on the mechanisms that drive heterochromatin formation after fertilization in the early embryo and the role of RNA as a regulator of heterochromatic loci.

In Chapter 2, Michael Weber and Sylvain Guibert review the roles of different DNA epigenetic marks in development, epigenetic reprogramming, and pluripotency in mammals. The genomic patterns of 5-methylcytosine (5 mC) are extensively reprogrammed during early embryonic development as well as in the germ cell lineage. More recently, 5-hydroxymethylcytosine (5 hmC) was discovered as another prominent cytosine modification, catalyzed by the enzymes of the TET family and abundant in certain cell types. 5 hmC may serve as an intermediate in the reaction of DNA demethylation or act as a signal for chromatin factors.

In Chapter 3, Philippe Collas and colleagues explore epigenetic marking in zebrafish embryos, in particular on the transitions in DNA methylation and posttranslational histone modifications occurring during early zebrafish development, as the embryo acquires transcriptional competence and initiates its own gene expression program. A characteristic of anamniote development is a long period of embryonic cell divisions in the absence of on-going transcription. In zebrafish, this lasts for 10 cell cycles, after which zygotic genome activation (ZGA) takes place. In this chapter, data supporting the emerging concept of epigenetic prepatterning of the embryonic gene expression program prior to the onset of ZGA are discussed.

In Chapter 4, Daan Noordermeer and Denis Duboule explore the fascinating question of collinearity, whereby *Hox* genes are sequentially expressed based on their relative positions within their gene clusters. The underlying molecular mechanisms have remained mysterious but recently important insights have been obtained on the dynamic changes in chromatin organization that accompany collinearity. The authors propose a testable framework for collinearity, based on the changes in chromatin and chromosomal organizational features over time.

Chapter 5, by Surani and colleagues, explores the highly dynamic epigenetic reprogramming that occurs in primordial germ cells (PGCs) which are the embryonic precursors of the gametes and founder cells of the germline. They present an overview of PGC development in the mouse and compare this with the limited information available for other mammalian species. A particular focus is placed on the mechanisms that ensure that the genome emerges intact and carrying the appropriate epigenetic information during its passage through the germline.

The role of epigenetics in plant development and growth is discussed in Chapter 6 by Daniel Grimanelli and François Roudier. They review chromatin-based mechanisms involved in the regulation of key aspects of plant development—pointing out conservation between plants and animals, but also several plant-specific innovations that enable chromatin dynamics linked to some of the specific requirements of plants.

In Chapter 7, Tomomi Tsubouchi and Amanda G. Fisher provide a review of the remarkable properties of pluripotent cells, both embryonic stem cells and reprogrammed pluripotent cells. Such cells are immortal, have a rather unusual and rapid cell cycle, and retain the capacity to form multiple tissue types upon differentiation. It is still not clear to what extent epigenetic mechanisms involving chromatin are required for self-renewal or reprogramming. The authors propose that regulation of the cell cycle may be a critical determinant of pluripotent reprogramming.

In Chapter 8, Antoine Peters and colleagues explore two key epigenetic silencing pathways, H3K9/HP1 and Polycomb-mediated repression. They review data in the mouse that have exploited both genetic tools combined with immunofluorescence and *in situ* analyses, to investigate the manner in which these pathways are dynamically regulated and how they contribute to very early embryogenesis.

Finally, Chapter 9 by Rachel Duffié and Déborah Bourc'his reviews recent findings on strategies employed, particularly in mammals, to achieve epigenetic gametic inheritance. Epigenetic marking in the germ line is important for processes such as genomic imprinting and parent-of-origin expression.

This volume therefore gathers together nine timely and exciting reviews on the rapidly moving topic of epigenetic mechanisms and development in different systems.

EDITH HEARD

CHAPTER ONE

# Mechanisms and Dynamics of Heterochromatin Formation During Mammalian Development: Closed Paths and Open Questions

**Anas Fadloun[1], André Eid[1], Maria-Elena Torres-Padilla[2]**
Institut de Génétique et de Biologie Moléculaire et Cellulaire, CNRS/INSERM U964, Université de Strasbourg, Illkirch, France
[1]These authors contributed equally to this work.
[2]Corresponding author: e-mail address: metp@igbmc.fr

## Contents



## Abstract

Early embryonic development in mammals is characterized by major changes in the components of the chromatin and its remodeling. The embryonic chromatin and the nuclear organization in the mouse preimplantation embryo display particular features that are dramatically different from somatic cells. These include the highly specific organization of the pericentromeric heterochromatin within the nucleus and the suggested lack of conventional heterochromatin. We postulate that the plasticity of

*Current Topics in Developmental Biology*, Volume 104
ISSN 0070-2153
http://dx.doi.org/10.1016/B978-0-12-416027-9.00001-2

the cells in the early embryo relies on the distinctive heterochromatin features that prevail during early embryogenesis. Here, we review some of these features and discuss recent findings on the mechanisms driving heterochromatin formation after fertilization, in particular, the emerging role of RNA as a regulator of heterochromatic loci also in mammals. Finally, we believe that there are at least three major avenues that should be addressed in the coming years: (i) Is heterochromatin a driving force in development? (ii) Does it have a role in lineage allocation? (iii) How can heterochromatin "regulate" epigenetic reprogramming?

## 1. INTRODUCTION

Embryonic development is a specificity of metazoans. It starts with the fertilization of the oocyte by a sperm. Following fertilization, the gametes undergo intense chromatin remodeling and epigenetic reprogramming, which is necessary to revert into a totipotent state, essential to start a new developmental program. Importantly, in its natural context, such reprogramming should occur with 100% efficiency in order to sustain development.

In mammals, fertilization is followed by a series of successive divisions during which the embryo generates a higher cell number but maintains the same size overall until the blastocyst stage. The early blastocyst is composed of two lineages: the pluripotent cells of the inner cell mass (ICM), which will give rise to the embryo proper, and the trophectoderm, which will give rise to the placenta and is considered the first differentiated tissue in the embryo. A third lineage, the primitive endoderm, which is extraembryonic, emerges by the late blastocyst stage. After implantation of the blastocyst into the uterine wall, the embryo undergoes gastrulation, during which the three germ layers of the embryo are established and will be complemented by subsequent somitogenesis and organogenesis. During all these developmental processes, the structure of the chromatin is expected to be largely remodeled and modified. The changes to the structure of the chromatin are thought to have a direct effect on gene expression and therefore a key role in the control of developmental gene expression or repression.

In eukaryotic cells, the DNA is organized into chromatin, which regulates the accessibility of the genetic information. The building block of the chromatin is the nucleosome, which consists of two copies of each of the core histones H2A, H2B, H3, and H4 wrapped with ~146 bp of DNA (Luger, Mader, Richmond, Sargent, & Richmond, 1997). The histones are subject to an increasing number of covalent modifications such as methylation and acetylation, which have been shown to regulate chromatin-mediated processes in multiple ways (Kouzarides, 2007). Variation in what is referred to as the primary structure of the chromatin can be thus achieved as a result of incorporation of an increasing number of histone variants and their accompanying post-translational modifications (Luger, Dechassa, & Tremethick, 2012). This variation has the potential to move the equilibrium to different chromatin states and therefore to impact on chromatin function. Globally, the chromatin is organized in two main functional and structural types or states: euchromatin and heterochromatin (Grewal & Elgin, 2007). The former is considered to be an open structure favorable for transcription and is gene rich, whereas the latter is considered to be in a closed structure that tends to be refractory for transcription and is gene poor. The heterochromatin can be further subdivided into two different types, facultative and constitutive. Facultative heterochromatin is characterized mainly by high levels of trimethylation of the lysine 27 of histone H3 (H3K27me3), a modification which is established by the polycomb repressive complex 2 (PRC2) and which also plays a role in the repression of developmental genes (Cao et al., 2002). However, the constitutive heterochromatin is characterized by strong enrichment of H3K9me3, H4K20me3, and H3K64me3, as well as of high levels of DNA methylation (Daujat et al., 2009; Peters et al., 2001; Schotta et al., 2004). The constitutive heterochromatin assembles mainly on centromeric, pericentromeric, and telomeric regions that are known to harbor repeated sequences such as the major and minor satellites. Constitutive heterochromatin is also present at imprinted genes in an allele-specific fashion and is considered to be a heritable trait that can be passed on to daughter cells and maintained (Regha et al., 2007). Several questions arise on how heterochromatin is established at these specific genomic regions after fertilization and naturally on how they are maintained and propagated through the cell cycle. Another key question that has so far been under-investigated is whether there is any role for heterochromatin as such, in regulating embryonic development and the restriction of cell fate and plasticity. Indeed, as we propose below, the earliest stages of mammalian embryogenesis are characterized by a lack of a "conventional" heterochromatin.

The period that follows fertilization is particularly interesting in terms of chromatin remodeling due mainly to the genome-wide epigenetic reprogramming that the parental genomes are subject to (reviewed in Burton & Torres-Padilla, 2010). Erasure of most of the epigenetic information carried by the two highly differentiated gametes is thought to be necessary to restore developmental plasticity in the newly formed organism. The formation of the newly fertilized zygote constitutes therefore the climax of totipotency because of the resulting zygote's inherent ability to produce all cell types in a new organism. However, this amazing capacity of the cells in the mouse embryo to generate all cell types seems to be transient as transplantation experiments have shown that cells from the ICM of a late blastocyst no longer have the potential to form trophectoderm derivatives (Rossant & Lis, 1979). Indeed, the ability of the early embryo to reprogram somatic nuclei decreases as development proceeds (Eckardt, Leu, Kurosaka, & McLaughlin, 2005), suggesting that the capacity to reprogram to totipotency also decreases during time.

What makes the cells in the early embryo capable of supporting such a large degree of plasticity? How is this plastic state achieved after fertilization and how is this state maintained in the early embryo? These questions have remained largely unanswered and are central for our understanding of cell plasticity, development, and reprogramming. We propose that the basis of such plasticity relies—at least partially—on the distinctive heterochromatin features that prevail during early embryogenesis. Here, we will review some of these major features and discuss recent findings that have shed some light on the mechanisms driving heterochromatin formation after fertilization, in particular, in light of what is known in other model systems where we have a more in depth mechanistic knowledge on the formation and maintenance of heterochromatin. Finally, we will put forward some of the questions that in our view represent the major key challenges to address in the field for the coming years.

## 2. HETEROCHROMATIN IN THE EARLY EMBRYO: A RATHER PARTICULAR ENVIRONMENT

Addressing the mechanisms behind the establishment of heterochromatin during the development of preimplantation mouse embryos is particularly relevant because of the huge changes on the chromatin that occur after fertilization. Indeed, most chromatin signatures have to be established *de novo* at the start of development. Because of this, the mouse embryo is one of the few model systems that offer the possibility of dissecting the mechanisms that underlie the establishment of heterochromatin in mammals, as opposed to studying heterochromatin maintenance and/or spreading in somatic cells, where heterochromatin only has to be maintained. Perhaps most interestingly, the question of whether a global rearrangement of heterochromatin and its structure impact on cell potency and development is a very attractive one.

Histones can be posttranslationally modified, and this provides an important level of functionality to the chromatin (Kouzarides, 2007; Strahl & Allis,

2000). The state of the chromatin can control the transcriptional state of a cell and the accessibility to its DNA. As pointed earlier, the two major chromatin states, the euchromatin and the heterochromatin, fluctuate dynamically and it is thought that they are only stabilized once the cells reach their "final" differentiation status. In this sense, the global state of the chromatin could for instance be considered to define the state of potency and plasticity of a cell, through the control of the accessibility to specific sequences of the DNA. In order to dynamically alter nucleosome spacing and/or accessibility, several mechanisms have been described that might in some instances work in a cooperative basis. These include (i) ATP-dependent chromatin remodelers, (ii) DNA methylation (and hydroxymethylation), (iii) histone modifications, (iv) incorporation of specific histone variants, and (v) the nuclear architecture, including the position that a given genomic region occupies within the nucleus.

The constitutive heterochromatin is marked by posttranslational modifications both on the N-terminus of histone tails and in the nucleosomal core region, in particular, by the enrichment of H3K9me3, H3K64me3, and H4K20me3 (Daujat et al., 2009; Peters et al., 2001; Schotta et al., 2004). This state of chromatin is localized to centromeric and pericentromeric regions as well as telomeric regions (the latter only containing H3K9me3) and is also known to be deacetylated (Grewal & Elgin, 2007). Repetitive elements in mammalian genomes are also known to be silenced through the acquisition of a constitutive heterochromatic signature (see Sections 3.5 and 3.6). While the constitutive heterochromatin is considered to assemble mainly at gene poor regions and repetitive elements, the facultative heterochromatin can be present in gene-rich regions that can switch their state between euchromatin and heterochromatin depending on multiple factors (Bernstein et al., 2006). The facultative heterochromatin is characterized by a strong enrichment of H3K27me3 and the related PRC1-catalyzed ubiquitination of H2AK119 (Grewal & Elgin, 2007).

Most posttranslational modifications of histones described to date, including phosphorylation, acetylation, and ubiquitination, are monolayered, which means that they have only one level of modification. In contrast, methylation of lysine residues has been described to occur at three layers: mono-, di-, and trimethylation, each of them with apparent different downstream effects. Indeed, this gives rise to different chromatin landscapes and/or recognition motifs at each layer of modification that can be read or modified by different types of readers and remodelers (Bonasio, Tu, & Reinberg, 2010; Kouzarides, 2007; Santos-Rosa et al., 2002). This has been very clearly demonstrated through genome-wide analysis of a number of histone modifications using chromatin immunoprecipitation (ChIP)-Seq on native chromatin from resting human CD4 cells (Barski et al., 2007). The results from this analysis showed a different genomic distribution between the monomethylation states of H3K27, H3K9, and H4K20, which are mostly enriched at actively transcribed promoters, compared to the di- and trimethylation states of the three corresponding lysines, which are characteristic of silenced promoters. Thus, we will focus in our review on the function and analysis of the di- and trimethylation of these residues. It is of course unclear whether the results of these analyses, performed in somatic cells, can be applied to ES cells and embryos, where the chromatin seems to be in a slightly atypical configuration and gene expression is considered more dynamic. Therefore, an analysis in mouse embryos is necessary to establish a genome-wide correlation between gene expression and histone modifications or histone variants.

## 2.1. Dynamics of establishment of histone modifications during development and function of their histone methyltransferases

As an immediate response to fertilization by the sperm, the oocyte undergoes its second meiotic division. A totipotent zygote is formed. It is the only cell that contains two separate haploid nuclei within the same cytoplasm, which are therefore referred to as pronuclei. The paternal and maternal genomes remain physically distinct until at least the 2-cell stage and are marked by different chromatin modifications (Mayer, Niveleau, Walter, Fundele, & Haaf, 2000). At the zygote stage, the maternal pronucleus is characterized by the presence of histone trimethylation marks that are specific of constitutive heterochromatin, whereas the paternal genome generally lacks such heterochromatic marks. In particular, H3K9me3, H4K20me3, and H3K64me3 are found to localize to the pericentromeric chromatin in the maternal chromatin, but not in the paternal one (Arney, Bao, Bannister, Kouzarides, & Surani, 2002; Daujat et al., 2009; Kourmouli et al., 2004; Santos, Peters, Otte, Reik, & Dean, 2005). Perhaps the only known exception of a histone mark exclusively associated with the maternal chromatin that is not heterochromatic is H3K36me3, which is transmitted to the embryo on the maternal chromatin and then rapidly remodeled to become undetectable at the 2-cell stage (Boskovic et al., 2012). At around 12–14 h after fertilization, the zygote undergoes the first mitotic division and two daughter cells are formed, each containing a copy of the parental chromosomes. H3K9me3 can be clearly seen in two-cell nuclei, but it marks asymmetrically half of each nucleus, which corresponds to the maternal chromatin that remains physically segregated from the paternal chromatin within the nucleus (Liu, Kim, & Aoki, 2004; Santenard et al., 2010) (Fig. 1.1). The levels of H4K20me3 are significantly reduced at the 2-cell stage to almost undetectable levels (Kourmouli et al., 2004; van der Heijden et al., 2005; Wongtawan, Taylor, Lawson, Wilmut, & Pennings, 2011), and H3K64me3 is no longer detected at the late 2-cell stage (Daujat et al., 2009). The 4-cell stage is thus characterized by the absence of H3K64me3, H4K20me3, and a marked decrease in H3K9me3 levels. H4K20me3 is not reestablished until after implantation, but the detailed spatiotemporal pattern of *de novo* methylation of H4K20me3 has not been studied. Global levels of H3K9me3 on the other hand continue to be reduced during the 8-cell stage judged from immunofluorescence analysis, and it would appear that H3K9me3 levels start to increase at the 16-cell stage and through the blastocyst stage, and are maintained thereafter (Puschendorf et al., 2008 and our unpublished observations). It should be noted though that most of these analyses are based on immunofluorescence approaches that have not been thoroughly or precisely quantified. It is worth mentioning that reacquisition of H3K9me3 globally seems to coincide with the time when the blastomeres undergo what can be described as the first differentiation step

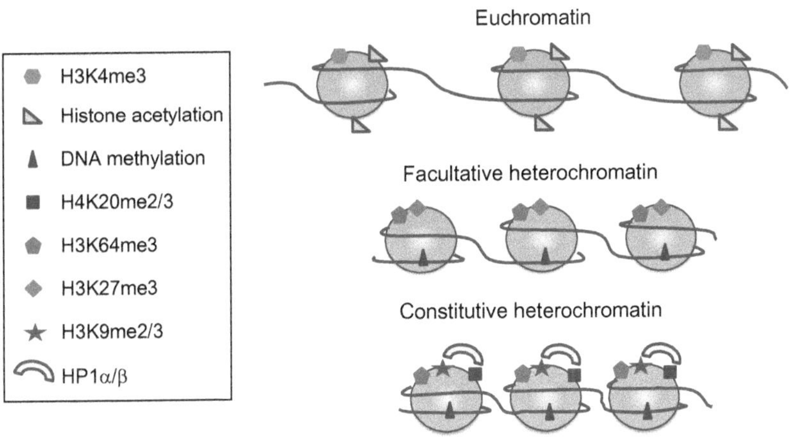


**Figure 1.1** Histone posttranslational modifications and the two chromatin states. Euchromatin is globally marked by H3K4me3 and acetylation of histone tails. However, facultative heterochromatin is marked by H3K27me2/3 and DNA methylation. The latter together with H3K64me3, H3K9me2/3, and H4K20me2/3 are typical of constitutive heterochromatin. (For color version of this figure, the reader is referred to the online version of this chapter.)

in the embryo, which is referred to as compaction and is achieved through the establishment of E-cadherin junctions at the end of the 8-cell stage.

Regarding the histone methyltransferases that catalyze the methylation of H3K9 and H4K20, many of them have been zygotically knocked-out in the mouse and some of them have early embryonic phenotypes (Table 1.1). In contrast, the methyltransferase(s) responsible for H3K64me3 have not been identified so far. At least eight proteins (Suv39h1, Suv39h2, G9a/Ehmt2, Eset/Setdb1, EuHMTase/GLP, Kmt1d, CLL8, SpClr4) have been described to methylate H3K9 (Kouzarides, 2007), and the list has been recently expanded to 10, as monomethyltransferase activity toward H3K9 has been observed for Prdm3 and Prdm16 (Pinheiro et al., 2012). The first five enzymes are active in mammalian cells (Dodge et al., 2004; Kouzarides, 2007). G9a is responsible for the mono- and dimethylation of H3K9 in euchromatic regions (Rice et al., 2003; Tachibana et al., 2002, 2005). Eset is responsible for the trimethylation of H3K9me3 in euchromatin (Wang et al., 2003; Yang et al., 2002), whereas EuHMTase promotes H3K9 dimethylation of the same regions. In contrast, Suv39h1/h2 can di- and trimethylate H3K9 in pericentric, centromeric, and telomeric regions (Peters et al., 2001; Rea et al., 2000). A key downstream "reader" of H3K9me3 is the heterochromatin protein 1 (HP1) (Bannister et al., 2001; Lachner, O'Carroll, Rea, Mechtler, & Jenuwein, 2001), which specifically recognizes H3K9me3 through its chromodomain. HP1 has been shown in turn to recruit the H4K20 methyltransferase Suv4-20 via its chromoshadow domain (Nielsen et al., 2001; Schotta et al., 2004). Suv4-20 would subsequently establish di- and trimethylation of H4K20me1, the latter being catalyzed by PR-Set7 (Nishioka et al., 2002; Schotta et al., 2004, 2008). These steps have been shown to operate in somatic cells to establish a constitutive heterochromatic configuration over the specific genomic regions (Rea et al., 2000; Schotta et al., 2004, 2008). The presence of H3K9me3 and H4K20me3 is characteristic of maternal centromeric and pericentromeric regions in the zygote, which are organized in a ring-like shape around the nucleolar-like bodies (NLBs). It has been shown recently that another mark is present on these regions: H3K64me3 (Daujat et al., 2009). However, the mechanism of establishment of this modification seems to be independent from the one that we described earlier (Lange et al., submitted for publication). Therefore, more work should be done in order to identify the role and effect of this histone mark during development.

A recent study has shown that a novel methyltransferase, Smyd5, can trimethylate H4K20. It was shown that the trimethylation established by this

**Table 1.1** Embryonic phenotypes of the knock-outs of proteins and complexes involved in heterochromatin establishment and/or maintenance

| Methyltransferase | Domain(s) | Function(s) | Location | Knockout | Phenotype | References |
|---|---|---|---|---|---|---|
| G9a | SET (*Suv*3-9, *E*(z)h, *Trx*) | Catalyzes H3K9me1/2 | Euchromatin | Embryo (full KO) | Lethal between E9.5 and E12.5. Developmental abnormalities. Reduction in H3K9me1/2 | Tachibana et al. (2002) |
| | ANK repeats | Binds H3K9me1/2 | | Germ line (conditional KO) | Defects in meiotic prophase progression and gametogenesis. | Tachibana, Nozaki, Takeda, and Shinkai (2007) |
| | | | | ES (full KO) | Altered DNA methylation in the Prader-Willi imprinted region. | |
| ESET | SET | Catalyzes H3K9me3 | Euchromatin | Embryo (full KO) | KO of the SET domain: lethality at peri-implantation. Defects in ICM formation | Dodge, Kang, Beppu, Lei, and Li (2004) |
| | MBD (methyl-CpG-binding domain) | Binds methylated DNA | | | | |
| | Tudor | Recognizes arginine, H4K20 methylation | | | | |

*Continued*

**Table 1.1** Embryonic phenotypes of the knock-outs of proteins and complexes involved in heterochromatin establishment and/or maintenance—cont'd

| Methyltransferase | Domain(s) | Function(s) | Location | Knockout | Phenotype | References |
|---|---|---|---|---|---|---|
| GLP | SET | Catalyzes H3K9me3 | Euchromatin | Embryo (full KO) | Lethal between 9.5 and 12.5 dpc. Developmental abnormalities | Tachibana et al. (2005) |
| | | Catalyzes H3K27me *in virto* | | | Reduction not only in H3K9me3 but also in H3K9me1 and H3K9me2 | |
| | ANK repeats | Binds to H3K9me1 and H3K9me2 | | | Mislocalization of HP1 | |
| Suv39H1 | SET | Catalyzes H3K9me3 | Centromeric, pericentromeric, and telomeric heterochromatin | Embryo (full KO) | No developmental effect | Peters et al. (2001) |
| | Chromo | Recognizes methylated lysines (H3K9me3) | | | | |
| Suv39H1 | SET | Catalyzes H3K9me3 | Centromeric, pericentromeric, and telomeric heterochromatin | Embryo (full KO) | No developmental effect | Peters et al. (2001) |
| | Chromo | Recognizes methylated sites (H3K9me3) | | | | |

| | | | | | | |
|---|---|---|---|---|---|---|
| Suv39H1/2 | SET | Catalyzes H3K9me3 | Centromeric, pericentromeric, and telomeric heterochromatin | Embryo (full KO) | Prenatal lethality | Peters et al. (2001) |
| | Chromo | Recognizes methylated lysines (e.g., H3K9me3) | | | Replacement of H3K9me3 in heterochromatin with H3K27me3 | |
| | | | | Germ line (conditional KO) | Infertile male mice | |
| | | | | | Increased defects following DNA damage and missegregation during meiosis | Puschendorf et al. (2008) |
| HP1α | Chromo | Recognizes methylated sites (H3K9me3) | Heterochromatin | Embryo (full KO) | No developmental phenotype reported | Aucott et al. (2008) |
| | Chromoshadow | Protein–protein interactions | | | | Brown et al. (2010) |
| | Hinge | RNA/DNA interactions | | | | |

*Continued*

**Table 1.1** Embryonic phenotypes of the knock-outs of proteins and complexes involved in heterochromatin establishment and/or maintenance—cont'd

| Methyltransferase | Domain(s) | Function(s) | Location | Knockout | Phenotype | References |
|---|---|---|---|---|---|---|
| HP1β | Chromo | Recognizes methylated lysines (H3K9me3) | Heterochromatin | Embryo (full KO) | Perinatal lethality | Aucott et al. (2008) |
| | Chromoshadow | Protein–protein interactions | | | Aberrant cerebral cortex development | |
| | Hinge | RNA/DNA interactions | | | | |
| HP1γ | Chromo | Recognize methylated lysines (H3K9me3) | Euchromatin | Embryo (full KO) | Developmental effect? | Takada et al. (2011) |
| | Chromoshadow | Protein–protein interactions | | | Sterility of male mice | Brown et al. (2010) |
| | Hinge | RNA/DNA interactions | | | | Naruse et al. (2007) |
| Suv4-20H1 | SET | Catalyzes H4K20me2/3 | Heterochromatin | Embryo (full KO) | No developmental phenotype reported | Schotta et al. (2008) |
| Suv4-20H2 | SET | Catalyzes H4K20me2/3 | Heterochromatin | Embryo (full KO) | No developmental phenotype reported | Schotta et al. (2008) |

| | | | | | | |
|---|---|---|---|---|---|---|
| Suv4-20H1/2 | SET | Catalyzes H4K20me2/3 | Heterochromatin | Embryo (full KO) | Perinatal lethality | Schotta et al. (2004) |
| | | | | MEFs (conditional KO) | Increase in DNA damage sensitivity | Schotta et al. (2008) |
| | | | | | Cell cycle defects and chromosomal aberrations | |
| EED | WD repeats | Protein–protein interactions and assembly | | Embryo (full KO) | Problems with gastrulation leading to lethality | Shumacher, Faust, and Magnuson (1996) |
| | | | | | Loss of maintenance of the inactive state of the X chromosome in trophoblast cells | Wang et al. (2001) |
| E(z)h2 | SET | Catalyzes H3K27me3 | | Embryo (full KO) | Embryonic lethality caused by problems by implantation, gastrulation and growth defects | O'Carroll et al. (2001) |
| | | | | Germ line (conditional KO) | No effect due to a rescue of the phenotype probably caused by the paternal allele | Erhardt et al. (2003) |
| Suz12 | Zinc finger domain | | | Embryo (full KO) | Lethality around E8.5 with similar phenotype to Eed and E(z)h −/− | Pasini, Bracken, Jensen, Lazzerini Denchi, and Helin (2004) |

enzyme has a repressive effect on gene function, but its function *in vivo* or whether Smyd5 integrates in the heterochromatin feedback loop described earlier has not been addressed (Stender et al., 2012).

Loss-of-function strategies of the corresponding methyltransferases have been used in order to study the function of histone methylation of H3K9 and H4K20. However, in most cases, they have not been deleted specifically in the maternal germ line and therefore the embryos analyzed carry maternal contribution of these methyltransferases, since their mRNAs are present in the oocyte (Hamatani, Carter, Sharov, & Ko, 2004; Wang et al., 2004; Zeng & Schultz, 2005). A role for H3K9 and H4K20 methylation during the earliest stages of preimplantation development has therefore not been addressed directly.

Double null embryos for G9a show high lethality between embryonic day (E) 9.5 and E12.5, with many morphological defects and loss of weight compared to heterozygous littermates. This result showed that G9a is necessary for proper embryonic development and thus that methylation of H3K9 in euchromatic regions is functionally essential for mid-gestation (Tachibana, Sugimoto, Fukushima, & Shinkai, 2001; Tachibana et al., 2002). Furthermore, conditional KO of G9a in the mouse germ line showed that it is essential for meiotic prophase progression and is involved in gametogenesis (Tachibana et al., 2007). Finally, conditional KO of G9a in CD4(+) T cells (Lehnertz et al., 2010) or adult neuronal cells (Schaefer et al., 2009) in adult mice showed a functional role for the reduction of H3K9me2 in immune and cognitive abnormalities. Interestingly, studies have demonstrated that KO of GLP has the same phenotype as $G9a^{-/-}$ during embryogenesis. This is somewhat expected since GLP and G9a can form heterodimeric structures and might positively regulate each other's catalytic activity: indeed, KO of GLP causes a reduction not only in H3K9me3 but also in H3K9me1 and H3K9me2 in euchromatic regions, resulting in relocalization of HP1 (Tachibana et al., 2005). Deletion of the SET domain of ESET results in embryonic death around peri-implantation with defects in ICM formation and inability to derive ES cell lines (Dodge et al., 2004). RNAi-mediated silencing of ESET in murine ES cells causes the differentiation of these cells into trophoblastic lineage, thus demonstrating a potential role in restricting the fate of ICM cells during development (Yuan et al., 2009).

Single KO of either Suv39h1 or Suv39h2 seems not to have an effect on development, but double null mice for Suv39h1 and Suv39h2 show increased prenatal lethality than wild-type mice as well as genome instability upon DNA damage and missegregation during meiosis (Peters et al., 2001).

However, no early embryonic defects have been described. It has also been observed in *dn* Mouse Embryonic Fibroblasts (MEFs) mutant for Suv39h1/2 that the loss of H3K9me3 was accompanied by an increase of H3K27me3 in, for example, pericentromeric repeats (Martens et al., 2005), which seemed to compensate for the lack of Suv3-9 activity on these regions. In the mouse zygote, the Peters lab proposed that there is a hierarchy between the constitutive and facultative heterochromatin at the pericentric and centromeric regions. In case of the incorrect establishment of H3K9me3 mark in the maternal pronucleus, H3K27me3 is established as a fail–safe mechanism to silence these repetitive sequences (Puschendorf et al., 2008). However, the sensing or targeting mechanism for the ability of the PRC2 to deposit H3K27me3 in this context has not been established yet. It was also not reported whether the loss of H3K9me3 on the maternal pericentric chromatin had any effect on the recruitment of Suv420H1/2 and thus if there was any accompanying reduction in H4K20me2/3.

In mammals, there are two genes that code for Suv4-20, *Suv4-20h1* and *Suv4-20h2*, and they that been genetically inactivated both in isolation and in combination. The double KO of Suv4-20h1 and Suv4-20h2 mice also display perinatal lethality, as well as an increase in DNA damage sensitivity, cell cycle defects and chromosomal aberrations (Schotta et al., 2004, 2008). A potential role for *Suv4-20h1/h2* in preimplantation development has not been reported.

The other main component of constitutive heterochromatin is HP1. Mammals have three different HP1 isoforms: HP1$\alpha$, $\beta$, and $\gamma$ in humans (CBX5, 1 and 3, respectively, in mice). The three HP1s contain the same modular structure: a chromoshadow domain that is involved in protein–protein interactions (Nielsen et al., 2001), a hinge domain that binds ssDNA and RNA (Muchardt et al., 2002), and a chromodomain that recognizes and binds to H3K9me2/me3 (Lachner et al., 2001). An HP1$\alpha$ KO has been reported to have no apparent effect on embryonic development (Aucott et al., 2008; Brown et al., 2010). There seems to be some degree of redundancy among HP1 proteins. Indeed, the HP1$\gamma$ KO did not show any developmental phenotype, and the authors suggested that the other paralogues of HP1 might compensate for its loss. The absence of HP1$\gamma$ does result, however, in major defects in chromosome segregation in spermatocytes and leads to severe male sterility (Naruse et al., 2007; Takada et al., 2011). Perhaps surprisingly, in contrast to the KO model, a hypomorphic mutant for HP1$\gamma$ shows high lethality after birth with only 1% of mice reaching adulthood but with similar effect on spermatogenesis (Brown et al., 2010). Of the three HP1 isoforms, only the HP1$\beta$ KO mice show a very strong phenotype, with perinatal lethality and aberrant cerebral cortex development, suggesting a role for HP1β in late stages of development (Aucott et al., 2008). As mentioned earlier, it will be important to probe the function of the HP1 proteins after fertilization by performing, for example, conditional KOs in the maternal germ line. This is particularly relevant in light of the fact that HP1α seems to be absent from embryonic chromatin until the blastocyst stage, whereas HP1β is tethered onto embryonic pericentromeric chromatin as early as in the zygote stage (Arney et al., 2002; Santenard et al., 2010; Santos et al., 2005; Wongtawan et al., 2011). This mirroring temporal expression pattern might suggest different functions of these two HP1 isoforms during epigenetic reprogramming *in vivo*.

Although not strictly a constitutive heterochromatic mark, H3K27me3 has probably a "dual" role during early development. Indeed, it seems to be able to replace functionally the absence of H3K9me3/H4K20me3 on the paternal pericentric heterochromatin (Puschendorf et al., 2008; Santenard et al., 2010). The KOs of several of the components of the PRC2 do show early developmental phenotypes. The catalytic subunit of PRC2, Ezh2, is required for implantation, and mutant embryos for Ezh2 show defective growth, poor implantation rates, and gastrulation defects (O'Carroll et al., 2001). The maternal deletion of Ezh2 has a growth retardation effect on born pups, but the phenotype seems to be rescued by the expression of the paternal allele (Erhardt et al., 2003). The second PRC2 core subunit, Eed, is necessary for gastrulation (Faust, Lawson, Schork, Thiel, & Magnuson, 1998; Shumacher et al., 1996) and for the maintenance of the inactive state of the X chromosome in trophoblast cells (Wang et al., 2001). The KO of the third and last core subunit of PRC2, Suz12, shows a similar phenotype to the *Ezh2* and *Eed*$^{-/-}$ mice with lethality occurring at E8.5 (Pasini et al., 2004). All these phenotypes are in line with the observation that all three subunits are absolutely required for H3K27me3. Deletion of other Polycomb proteins such as YY1 results in preimplantation lethality and in defective ICM growth (Donohoe et al., 1999). It is therefore very well established that Polycomb proteins play key and essential developmental roles during preimplantation development.

These loss-of-function strategies have established that heterochromatin (through the di- and trimethylation of H4K20, H3K9, and H3K27) is important for developmental processes but do not establish the mechanism by which these modifications confer a cellular effect. Furthermore, several groups have studied the KO of some of these genes in the germ cells and showed an effect on spermatogenesis, but only a few/none of them have probed the effects on the oocyte development and the maternal contribution to the embryo. Many mRNAs as well as most of the cellular machinery (ribosomes, mitochondria, etc.) are inherited from the oocyte (Poznanski & Calarco, 1991; Smith & Alcivar, 1993; Wagner, 1972). Furthermore, the cytoplasmic components of the oocyte have all the necessary material to reprogram a somatic nucleus (Chan, Smith, Egli, Regev, & Meissner, 2012). Therefore, it would be highly important to directly address the effects of the lack of these methyltransferases from the earliest stages of development. Although with some caveats, the most straightforward way to probe this is probably through the elaboration of conditional KO in the oocyte.

In summary, histone modifications typical of constitutive heterochromatin in somatic cells are exclusively detected on the maternal chromatin at the very beginning of development only. They either become undetectable at the 2-cell stage or are "diluted" upon division until methylation starts to occur *de novo* globally at least three cell divisions later depending on the mark. Indeed, the kinetics of reacquisition of these histone methylation marks seems to be slightly different depending on the histone modification in question (Fig. 1.2). This lack of conventional heterochromatin during the first divisions following fertilization presumably creates a chromatin environment that is permissive for epigenetic reprogramming and a "window of opportunity" for transcription of heterochromatic regions that are normally not transcribed such as retrotransposons (see Section 3.5).

## 2.2. DNA methylation and hydroxymethylation

DNA methylation of cytosines functions in gene silencing (Goll & Bestor, 2005). In spite of being a highly differentiated cell, the sperm shows a promoter DNA methylation landscape that resembles globally that of pluripotent ES cells with a few key exceptions (Farthing et al., 2008). This feature has been suggested to reflect the epigenetic reprogramming of the germ line prior to fertilization and to be important in the transmission of pluripotency to the embryo. The oocyte shows an overall global hypomethylation status and does contribute with a number of specific methylated regions, including differentially methylated regions to the embryo (Smallwood et al., 2011; Smith et al., 2012). Upon fertilization, there is a reduction in global DNA methylation levels that is specific to the paternal genome (Mayer et al., 2000; Rougier et al., 1998). These conclusions were made mainly by using a 5-methylcytosine (5mC) antibody in immunostaining, which clearly reveals a loss of the epitope in the paternal pronucleus. A current emerging notion is, however, that the extent of this demethylation is lesser

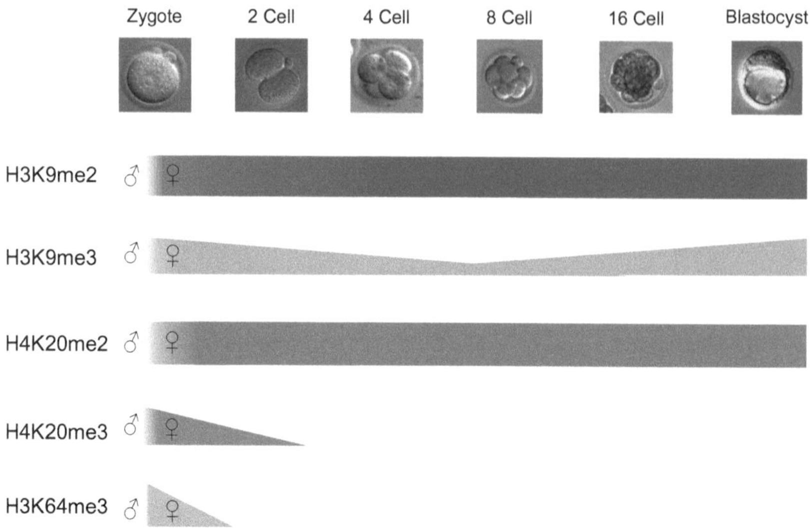


**Figure 1.2** Dynamics of the main constitutive heterochromatic marks during preimplantation development. All heterochromatin marks analysed to date are asymmetrically localized on the maternal chromatin during the zygote stage. The levels of H3K9me2 and H4K20me2 are maintained relatively stable during the early stages of development, whereas global H3K9me3 levels decrease between the 4-cell and the 16-cell stage. H4K20me3 is undetectable on embryonic chromatin by the end of the 2-cell stage and similar kinetics have been described for H3K64me3, which is only present during the zygote stage and is absent afterward. (See Color Insert.)

than what was originally thought and that DNA methylation in the zygote occurs both actively (e.g., before the first round of replication) and passively (concomitant with replication). DNA demethylation has been suggested to be, at least partially, the consequence of the conversion of cytosine methylation to hydroxymethylation by the Tet proteins (Gu et al., 2011; Inoue & Zhang, 2011; Tahiliani et al., 2009; Wossidlo et al., 2011). Together, this results in an asymmetry in DNA modifications between the paternal genome (enriched in DNA hydroxymethylation) and the maternal genome (enriched in DNA methylation) at the zygote stage (Gu et al., 2011; Inoue & Zhang, 2011; Iqbal, Jin, Pfeifer, & Szabo, 2011; Wossidlo et al., 2011). As development progresses, the differences in DNA methylation and hydroxymethylation disappear (Inoue & Zhang, 2011; Iqbal et al., 2011). The overall DNA methylation levels in the embryo decrease during early development and then increase in the epiblast at the blastocyst stage (Mayer et al., 2000; Santos, Hendrich, Reik, & Dean, 2002). On the imprinted genes, however, DNA methylation is not completely lost and is maintained through development until the formation of the primordial germ cells (PGCs) (Goll & Bestor, 2005). Nevertheless, when looking at single copy genes in detail, the levels ofDNA methylation do not seem to obey to a simple kinetic behavior, but there seems to be different waves of re/demethylation in specific regions throughout preimplantation development and even after implantation. It seems more likely that an equilibrium between loss of methylation and *de novo* methylation at different stages of development results in the final methylation landscape of somatic cells (Borgel et al., 2010; Proudhon et al., 2012). Three enzymes catalyze DNA methylation: Dnmt1, Dnmt3A, and Dnmt3B. Dnmt3A and Dnmt3B require for their function another related protein, Dnmt3L (Bourc'his, Xu, Lin, Bollman, & Bestor, 2001; Hata, Okano, Lei, & Li, 2002). The KO of Dnmt1 is lethal at mid-gestation with growth defects, but Dnmt1 is not required for ES cell maintenance (Li, Beard, & Jaenisch, 1993). The KO of either Dnmt3A or Dnmt3B is also lethal during development, with Dnmt3B specifically required for methylation of centromeric minor satellite repeats (Okano, Bell, Haber, & Li, 1999). Both of these genes are required for *de novo* methylation during development and in ES cells (Okano et al., 1999). Loss of Dnmt3L had the same phenotype as Dnmt3A with loss of imprinting at almost all of the same sites, but not in pericentric regions (Bourc'his & Bestor, 2004; Kaneda et al., 2004). Altogether, these studies would argue that *de novo* DNA methylation *per se* is not required prior to implantation.

Three Tet family enzymes (Tet1/2/3) convert 5mC to 5-hydroxymethylcytosine (5hmC) (reviewed in Tan & Shi, 2012 and Wu & Zhang, 2011). The KO of Tet1 has no effect on development or on pluripotency of ES cells (Dawlaty et al., 2011), although there are conflicting reports concerning the effect of Tet1 silencing using small heterochromatic RNAs (shRNAs) in ES cells (Ito et al., 2010). KO of Tet2 seems also to be dispensable for development (Quivoron et al., 2011). The conditional KO of Tet3 in the oocyte results in loss of conversion of 5hmC to 5mC on the paternal genome in the mouse zygote and in increased incidence of developmental failure as well as reduced fertility (Gu et al., 2011). It has been shown recently that H3K9me2 can be recognized by PGC7/Stella in order to prevent the conversion of 5mC to 5hmC, thus protecting DNA methylation (Nakamura et al., 2012). This might be a mechanism whereby the asymmetry between the paternal and maternal genomes in 5mC and 5hmC is generated, since as we mentioned earlier, H3K9me2 is exclusively present in the maternal chromatin.

## 2.3. Histone variants in heterochromatin in the embryo

The two gametes have a very different landscape of histone variants and chromatin compaction. For instance, the sperm is almost devoid of histones and is instead packed by protamines, a histone-like protein (Nonchev & Tsanev, 1990). This replacement is important physically and physiologically because it will help to package the paternal genome to fit into the sperm head for subsequent transportation through the uterus and ovidcut to reach the oocyte. The DNA bound to protamines reaches 6–20 times more compact packaging compared to the nucleosomal organization (Jenkins & Carrell, 2012; Ward & Coffey, 1991). The protamination of the sperm DNA is tightly regulated and is important for fertility (Aoki, Liu, & Carrell, 2005; Balhorn, Reed, & Tanphaichitr, 1988), but it was also considered to result in the loss of any potential epigenetic information carried through the histones. However, the replacement of histones is not complete and between 5% and 15% of histones depending on the species remains bound to DNA (Jenkins & Carrell, 2012; Wykes & Krawetz, 2003). This retention seems not to be stochastic since developmental gene promoters that are important for embryonic development seem to retain histones specifically (Brykczynska et al., 2010; Hammoud et al., 2009). More interestingly, such promoters, which are silent during preimplantation development, retain H3K27me3 in the sperm (Hammoud et al., 2009). This result suggests that the nucleosomes retained in the sperm might act as transmitters of silencing information for some genes during early development. Additionally, testis-specific histone variants exist (Moss, Challoner, & Groudine, 1989; Trostle-Weige, Meistrich, Brock, & Nishioka, 1984; Witt, Albig, & Doenecke, 1996) and have been described to be posttranslationally modified (Lu et al., 2009), although the function of these modifications has not been addressed in the context of the zygotic reprogramming. Most interestingly, constitutive heterochromatic regions display a distinctive organization in the sperm, with both telomeres and centromeres retaining histones and most likely some of their modifications (Govin et al., 2007; Wykes & Krawetz, 2003). In particular, testis-specific variants of H2A, H2AL1, H2AL2, and TH2B are found in the pericentric chromatin to achieve a DNA packaging structure that protects a ~60 bp DNA fragment (Govin et al., 2007).

The oocyte also contains some specific histone variants but most likely retains the traditional nucleosome packaging of the DNA (Chang et al., 2005). The oocyte is thought to further contain a pool of histones that are to be inherited by the zygote in order to replace the protamines of the paternal genome upon fertilization. The mechanism by which the replacement takes place is not well understood. The histone H3.3 has been shown to localize preferentially to the paternal chromatin during the zygote stage (Loppin et al., 2005; Santenard et al., 2010; Torres-Padilla, Bannister, Hurd, Kouzarides, & Zernicka-Goetz, 2006; van der Heijden et al., 2005). We have shown in the lab that the mutation of the lysine K27 to alanine of H3.3 results in a missegregation of chromosomes as well as developmental arrest and mislocalization of HP1. The same mutation in H3.1 does not seem to have any effect on either HP1 localization or developmental progression (Santenard et al., 2010). The same work suggested that incorporation of H3.3 in the paternal pericentric heterochromatin is important for the initial establishment of pericentromeric heterochromatin through its lysine 27. It therefore seems that H3.3 has acquired a role in remodeling heterochromatin after fertilization. This is in line with the preferential heterochromatic localization of H3.3 during the first embryonic divisions (Akiyama, Suzuki, Matsuda, & Aoki, 2011; Santenard et al., 2010).

## 2.4. Nuclear architecture in the mammalian embryo

The most astonishing observation that can be made immediately after fertilization is probably the difference in the size of the two pronuclei and the distinctive organization of the heterochromatin in both of them, which is radically different to that of somatic cells (Fig. 1.3).

The paternal pronucleus is almost 25% bigger than the maternal one. This spatial organization reflects the decondensation of the paternal genome that starts with fertilization. This process, which is accompanied by cycles of recondensation and further decondensation (Bouniol-Baly, Nguyen, Besombes, & Debey, 1997; Martin et al., 2006; Probst et al., 2010; Probst, Santos, Reik, Almouzni, & Dean, 2007), might be important for the subsequent establishment of the nuclear compartments and/or characteristic embryonic nuclear architecture. No heterochromatic marks can be detected on the paternal chromatin at this time, but H3K9me1 and H3K27me1 have been observed on the paternal pronucleus as early as during the first decondensation (Santos et al., 2005). This coincidence might reflect some spatiotemporal requirement that needs to be put in place before heterochromatic marking (e.g., H3K27me3) can be added a couple of hours later. Indeed, the changes in decondensation coincide with the emergence of the NLBs. These are ring-like structures around which the centromeric and pericentromeric sequences localize (Probst et al., 2007). The organization of the pericentromeric domains around NLBs is maintained at least until

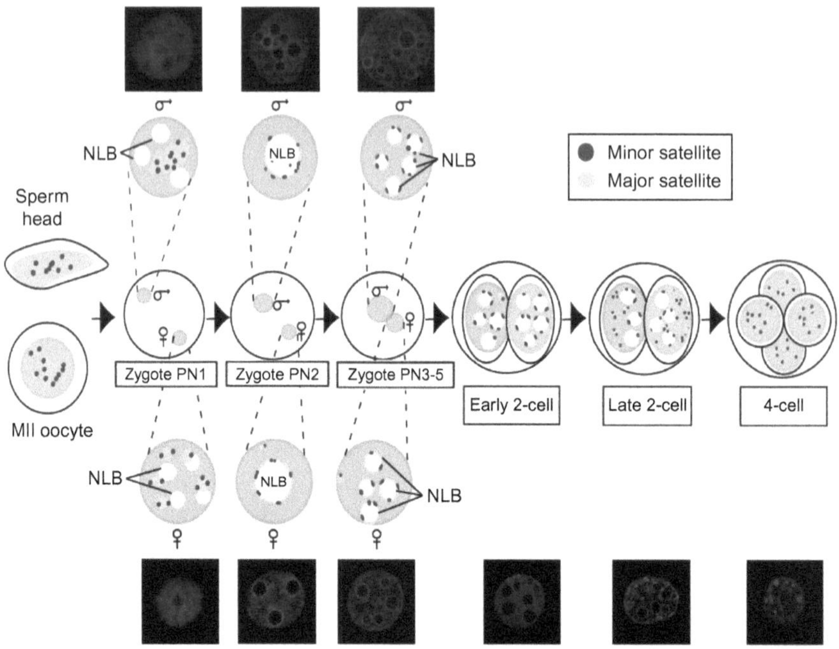


**Figure 1.3** Nuclear architecture changes during preimplantation development in the mouse. The nuclear organization changes dynamically after fertilization. The nucleolar-like bodies (NLBs) are established after fertilization at the pronuclear stage (PN) 1 in the zygote. At the PN2 stage, the minor and major satellites (MMS) reorganize to form a ring-like structure around the NLBs. Between the PN3 and PN5 an increase in the number of NLBs occurs, but the position of the MMS remains the same. At the early 2-cell stage (E2C), the nuclear organization starts to change and by the end of the 2-cell stage (L2C) many MMS start to form new chromocenters independently of the NLBs structure. By the 4-cell stage, most of the NLBs have disappeared and instead the majority of MMS are clustered into "somatic"-like chromocenters. Note that nuclei are not at the same scale. (See Color Insert.)

the 4-cell stage in some nuclei (Martin et al., 2006; Probst et al., 2010, 2007). As development progresses, the NLBs are replaced with chromocenters, which are the clusters of centromeric and pericentromeric regions that appear typically brightly stained by 4′,6-diamidino-2-phenylindole (DAPI) in most somatic cells. The chromocenters are a defining feature of somatic nuclei organization. The disappearance of any reminiscent NLBs in the mouse embryo seems to coincide with the 8-cell stage, which is the stage when cells undertake their first differentiation step (polarization and compaction). It is interesting to note that at this stage, the levels of H3K9me3 seem to be lowest (Fig. 1.2), and H4K20me3 is also absent from embryonic chromatin, suggesting that a loss of heterochromatin might be important for this structural change in the conformation of pericentromeric chromatin. This is even more important if we take into account the reports, indicating that the establishment of NLBs-like structure after somatic nuclear transfer might increase the probability of success in reprogramming the somatic nucleus (Maalouf et al., 2009; Martin et al., 2006). Functionally, whether this particular spatial nuclear positioning of the heterochromatic and/or its evolution toward a chromocenter configuration is important for development has not been addressed.

From all the above, it is clear that global heterochromatin dynamics seem to be particularly relevant during the earliest stages of development. They most likely impact not only on inheritance but also on the establishment of the embryonic epigenome and the embryo's subsequent differentiation capacities. The dramatically different configuration of embryonic heterochromatin compared to somatic cells is probably linked functionally to the plasticity of the early embryo and to the reprogramming process that occurs during this period. Our understanding of the molecular mechanisms that govern *de novo* establishment of heterochromatin domains after fertilization is, however, still very poor. It is therefore essential to uncover the mechanisms driving heterochromatin formation in mammals in order to fully understand the regulation of epigenetic reprogramming and establishment of pluripotency and plasticity.

Because the paternal genome, due to its packing into protamines, has to acquire a nucleosomal configuration and all the subsequent chromatin signatures for the first time after fertilization, the early mouse embryo constitutes a unique system to address the mechanisms of heterochromatin formation in mammals. However, it is not known how chromatin domains and their epigenetic signatures are established *de novo* in the zygote nor it is the extent to which these domains contribute to the regulation of cell potency. Most of our knowledge on heterochromatin formation and in particular on its establishment comes from work in *Schizosaccharomyces pombe* and *Arabidopsis* (Grewal & Elgin, 2007; Martienssen, Kloc, Slotkin, & Tanurdzic, 2008), but little is known on how heterochromatin is initially established in mammalian cells. This is mainly because in most cells heterochromatin only needs to be maintained as opposed to established *de novo*.

In the next section, we will first discuss extensively the main known mechanisms of establishment and inheritance of heterochromatin in other model systems, in particular, in fission yeast. We will then devote a section to draw parallels and open questions in the mammalian embryo and will discuss how other heterochromatic regions such as retrotransposons could be regulated during early mammalian embryogenesis.

# 3. MECHANISMS OF HETEROCHROMATIN ESTABLISHMENT AND INHERITANCE

## 3.1. Heterochromatin establishment and centromeric chromatin

The histone modification marks that are characteristic of constitutive heterochromatin along with the enzymes that produce these modifications and the proteins that recognize them are highly conserved from fission yeast to human (Kouzarides, 2007). From fission yeast to mammals, methylation of H3K9 is considered to be crucial for heterochromatin assembly (Bannister et al., 2001; Lachner et al., 2001; Rea et al., 2000). As pointed out earlier, in mammals several proteins can methylate H3K9, whereas in fission yeast, Clr4, a single histone H3K9 methyltransferase, directs all methylation of K9 on histone H3 (Yamada, Fischle, Sugiyama, Allis, & Grewal, 2005). Fission yeast lacks the enzymatic machinery for methylation of H3K27 and is also devoid of DNA methylation.

The centromere is a well-known landmark of silent chromatin and a paradigm for epigenetic inheritance. Ultrastructurally, it takes the form of a distinct primary constriction on the condensed metaphase chromosome of higher eukaryotes. The constricted region comprises a different chromatin structure consisting of DNA and protein complexes (the kinetochores) to which microtubules bind to effect proper chromosome movements (Folco, Pidoux, Urano, & Allshire, 2008; Pidoux & Allshire, 2005). The DNA sequence in the centromere is not conserved between organisms, yet the centromere displays similar features across evolution such as the presence of repetitive elements that include the alpha satellite in humans, the minor satellite in mice, the AATAT and TTCTC satellites in *Drosophila* (Cleveland, Mao, & Sullivan, 2003). Consequently, the sequence requirements, if any, for a functional centromere are not established.

In fission yeast, the constitutive heterochromatin regions are linked to centromeres, telomeres, and the mating-type locus. Although there are some variations, the global mechanisms of heterochromatin assembly on all these regions are similar. The centromeres in yeast range from 35 to 110 kb in length, with a central domain on which the kinetochore assembles flanked by outer repeat (*otr*) sequences (consisting of *dg* and *dh* repeats) coated in heterochromatin that resemble the pericentomeric heterochromatin in mammals (Steiner, Hahnenberger, & Clarke, 1993; Wood et al., 2002). Experiments performed with minichromosomes have demonstrated that DNA sequences from both the *otrs* and the central domain are required for full centromere function (Folco et al., 2008; Wood et al., 2002). Interestingly, the role of the *otr* sequences seems to be purely to provide a functional platform for heterochromatin assembly, since centromeres lacking *otr* sequences become functional when an enzyme that drives heterochromatin assembly is tethered adjacent to the central domain sequences (Kagansky et al., 2009).

It was originally thought that fission yeast centromeres were transcriptionally inert, as a marker gene inserted within centromeric sequence exhibited classical position effect variegation (Allshire, Javerzat, Redhead, & Cranston, 1994; Allshire, Nimmo, Ekwall, Javerzat, & Cranston, 1995). This silencing was thought to reflect spreading of heterochromatin over the inserted gene, thereby blocking access of RNA polymerase II (Pol II). However, it has been demonstrated that centromeres and *otrs* are transcribed in both fission yeast and mammals, and that transcript abundance in yeast is regulated by the RNAi machinery (Lehnertz et al., 2003; Volpe et al., 2002). Most importantly, the transcription of these repeats that triggers the RNAi machinery is an essential part of the heterochromatin assembly pathway. Centromeric transcription occurs during S-phase, during a window of time in which the repressive histone marks at centromeres become diluted upon DNA replication, allowing Pol II access (Chen et al., 2008; Kloc, Zaratiegui, Nora, & Martienssen, 2008).

## 3.2. Establishment and assembly of constitutive heterochromatin

Heterochromatin assembly is a multistep process. Studies from diverse systems suggest that a common set of structural components contribute to the construction of the heterochromatin platform. Initial targeting of heterochromatin to nucleation sites seems to be distinct from the subsequent heterochromatic spreading and maintenance steps. The strategies that are used by the cell to target heterochromatin differ depending on the chromosomal context. Local *cis*-acting sequences can promote the establishment of facultative heterochromatin, as exemplified by retinoblastoma protein and KRAB–KAP1-mediated recruitment of HP1 and SUV39 proteins (Nielsen et al., 2001; Schultz, Ayyanathan, Negorev, Maul, & Rauscher, 2002). The establishment of constitutive heterochromatin is however most often related to the presence of repetitive DNA elements. It is rather the repetitive nature of the genomic regions where heterochromatin assembles and not the DNA sequence itself, which functions as a trigger for heterochromatin formation (Luff, Pawlowski, & Bender, 1999). These repeats serve as template for the production of noncoding RNA (ncRNA) from both DNA strands, which is often associated with the generation of dsRNA. Although heterochromatin is a silent chromatin structure that blocks transcription, RNA Pol II transcribes these repeats in an exquisitely regulated temporal fashion at particular stages of the cell cycle (Chen et al., 2008; Kloc et al., 2008). Therefore, there seems to be an important and conserved role for both ncRNA and RNAi in this context (Matzke & Birchler, 2005).

The RNAi machinery was originally defined as a regulator of posttranscriptional silencing (Muller et al., 2002), but it is now clear that the RNAi machinery also alters chromatin structure and effects silencing at the transcriptional level. Most importantly, RNAi is central for initiating heterochromatin assembly not only at repetitive DNA in the centromeres and mating-type loci of fission yeast but also at retrotransposons in the *Arabidopsis* germ line (Grewal & Elgin, 2007). In fission yeast, siRNAs derived from heterochromatic repeats are present within the cell and are loaded into the RNA-induced transcriptional silencing (RITS) complex, which is composed of Ago1, Tas3, and the chromodomain-containing protein Chp1(Verdel et al., 2004). RITS is thought to bind to heterochromatic ncRNA using siRNA as a guiding molecule. It further participates in "amplifying" the response since it recruits the RNA-dependent RNA polymerase complex (RDRC), which consists of Rdp1, a poly(A) polymerase (Cid12), and a putative helicase (Hrr1), most likely via physical interactions (Verdel et al., 2004). The RDRC enhances the generation of siRNA by synthesizing dsRNAs from centromeric transcripts as substrates for Dcr1 (Colmenares, Buker, Buhler, Dlakic, & Moazed, 2007; Sugiyama, Cam, Verdel, Moazed, & Grewal, 2005). RITS also recruits Clr4 through the LIM domain protein, Stc1, such that the heterochromatin spreads onto the *dg* and *dh* repeats (Zhang, Mosch, Fischle, & Grewal, 2008). Stc1 associates with RITS on centromeric transcripts and recruits the Clr4-containing complex (CLRC), thereby coupling RNAi to chromatin modification (Bayne et al., 2010). Methylation of H3K9me3 is also recognized by Swi6/HP1, which further reinforces the silencing environment by mediating targeting of HDACs and is also responsible of recruiting the JmjC domain-containing antisilencing factor Epe1 that facilitates the transcription of heterochromatic repeats (see below) (Zofall & Grewal, 2006).

In the RNAi-mediated heterochromatin assembly system, siRNA generation and heterochromatin formation are interdependent, forming a "self-reinforcing loop." The self-reinforcing as well as the *cis*-acting nature of siRNA during RNAi-mediated heterochromatin formation suggests that the whole process is coupled to the chromatin. This is supported by ChIP experiments, which show the physical association of Ago1 and Rdp1 with chromatin (Cam et al., 2005; Volpe et al., 2002). Recently, DNA adenine methyltransferase identification methods to identify DNA binding sites *in vivo* were used to show that Dcr1 also associates with heterochromatin (Woolcock, Gaidatzis, Punga, & Buhler, 2011). This RNAi-mediated heterochromatin model also applies to the centromere-like repeats in the mating locus and to the subtelomeric regions (Cam et al., 2005).

Whether a similar or related RNAi-dependent mechanism exists in mammals is debatable. For example, there is no direct homolog of Chp1 identified so far in mammalian cells, although it is reasonable to believe that a chromodomain-containing member, like those related to the CBX family, could potentially fulfill a similar role. Furthermore, although the major and minor tandem satellite repeats in the mammalian genome can potentially generate dsRNA (Martens et al., 2005), how and whether these dsRNA molecules are processed has not been addressed.

## 3.3. Maintenance of heterochromatin throughout the cell cycle

During DNA replication, the chromatin is assumed to be drastically perturbed by the passage of the DNA replication machinery. With this, two main questions arise: (i) How is the heterochromatin propagated during S-phase? (ii) How can such a close structure be amenable to remodeling, DNA synthesis, and recondensation? Indeed, constitutive heterochromatin remains silent throughout most of the cell cycle, thanks to the recruitment of a myriad of factors that confer transcriptional silencing RNA Pol II occupancy at heterochromatin is restricted for most of the cell cycle (Chen et al., 2008).

Fission yeast spends most of its time in G2, during which HP1 and Chp2 help in the recruitment and spreading of chromatin modifiers such as SHREC and mediate the assembly of repressive chromatin refractory to RNA Pol II transcription (Grewal & Jia, 2007; Sugiyama et al., 2007). During mitosis, HP1 is lost from H3K9-methylated heterochromatin through the phosphorylation of the neighboring serine 10, which creates a "methyl switch" output (Kloc et al., 2008). H3S10P-mediated decrease in the chromatin association of HP1 proteins during mitosis serves two functions. The first one is to allow for the recruitment of the condensin complex, which is essential for chromosome segregation. The second one is to allow heterochromatic transcripts to accumulate in S-phase by creating a short window of opportunity in which heterochromatin is relatively accessible to RNA Pol II for transcription of the underlying repeat sequences (Chen et al., 2008; Kloc et al., 2008) (Fig. 1.4). This increased transcription during S-phase coincides temporally with the recruitment of the factors involved in heterochromatin assembly at the repeats. The upregulation of heterochromatin transcripts occurs preferentially on the antisense strand, and these transcripts are believed to attract heterochromatin assembly factors to the nascent transcripts of the repeats in *cis*. Indeed, proteins like Rik1 (a ClrC subunit) and Ago1 (a RITS complex subunit) are preferentially enriched at heterochromatic repeats during S-phase (Chen et al., 2008). At this time, heterochromatic repeats harbor paradoxically marks of active transcription, including H3K36 methylation. Again H3K36 is thought to mediate two functions, one is to promote transcription and the second one is to subsequently recruit HDAC silencing activities to heterochromatic repeats for further heterochromatin reconstitution during G2 to coordinately silence heterochromatic sequences with *cis*-acting posttranscriptional gene silencing by RNAi (Grewal & Jia, 2007; Sugiyama et al., 2007).

Finally, H3K9 methylation stabilizes the chromatin association of RNAi factors and also engages additional ClrC complexes through the Clr4 chromodomain, which is thought in turn to help methylating neighboring, newly incorporated histones. Altogether, this enables the reacquisition of an equivalent parental heterochromatic pattern on the newly formed chromatin after S-phase (Zhang et al., 2008). Importantly, histone chaperones themselves might play their part in this process as CAF1 has been shown to associate with heterochromatin factors (Quivy et al., 2004).

## 3.4. Targetting heterochromatin

A central and prevailing question in chromatin biology is how does a genomic region "know" that it has to be silenced, or in other words, how are chromatin modifiers targeted to the correct region. In the case of heterochromatin formation, this question remains largely unresolved. Recently, it has been shown that heterochromatic ncRNAs can be associated with chromatin through DNA:RNA hybrid formation (Nakama, Kawakami, Kajitani, Urano, & Murakami, 2012), which provides a target for the RITS complex, suggesting that heterochromatic ncRNAs are retained on chromatin via the formation of DNA–RNA hybrids and provide a platform for RNAi-directed heterochromatin assembly. This further suggests that DNA–RNA hybrid formation plays a role in chromatin-related ncRNA functions (Nakama et al., 2012).

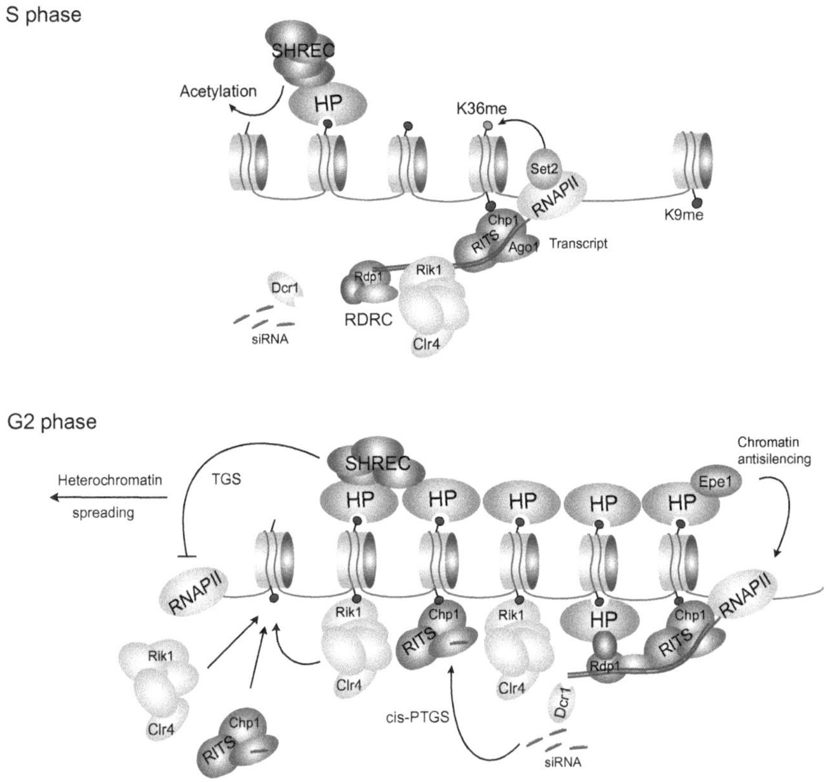


**Figure 1.4** Heterochromatin assembly in fission yeast requires the coordinated action of histone-modifying enzymes and RNAi. During S-phase (top), RNA Pol II activity at centromeric repeats occurs. This, in turn, stimulates the recruitment of heterochromatin assembly factors such as the CIRC subunit Rik1 and the RITS subunit Argonaute 1 (Ago1), as well as histone H3 lysine 36 methylation by the Set2 methyltransferase. Interaction between ClrC and RITS stabilizes their binding to chromatin and facilitates the processing of centromeric repeat RNAs to siRNAs. Recruitment of ClrC may also be mediated by downstream siRNA products such as double-stranded RNAs. Methylation of lysine 9 on histone H3 (H3K9me) by the Clr4 subunit of ClrC not only recruits HP1 proteins but also establishes a positive feedback loop by stabilizing the chromatin association of ClrC (via Clr4 chromodomain) and RNAi components such as RITS (via Chp1 chromodomain). In G2 phase (bottom), HP1 proteins bound to H3K9me recruit not only silencing factors such as the HDAC complex SHREC but also an antisilencing factor, Epe1, that promotes Pol II transcription. Spreading of HP1 proteins and H3K9me from the original nucleation sites allows heterochromatin to serve as a recruiting platform to reinforce its nature, including RITS components. *Adapted from Cam, Chen, & Grewal (2009).* (For color version of this figure, the reader is referred to the online version of this chapter.)

Studies in *Saccharomyces cerevisiae* and *S. pombe* have provided important insights into the process of assembly. A common theme that has emerged is that heterochromatin assembly is nucleated at specific regulatory sites and then spreads to nearby sequences. This spread typically requires some physical coupling of chromatin modifiers and more structural proteins such as Sir3 and Sir4, and Swi6/HP1 (Grewal & Elgin, 2002; Moazed, 2001).

In *S. cerevisiae*, site-specific DNA-binding proteins bind to nucleation sites (silencers) and then recruit to DNA the Sir2/Sir4 complex (Sir-depending spreading) (Huang, 2002). *S. cerevisiae* lacks H3K9me3, and therefore, the silencing occurs through different mechanisms than in fission yeast. Telomeres and the silent mating-type loci *HML* and *HMR* are packaged in silent chromatin that contains the silent information regulator (SIR) proteins Sir2, Sir3, and Sir4. SIR proteins silence nearby genes at telomeres, a phenomenon known as the telomere position effect. Silent telomeric chromatin is nucleated by the binding of Rap1 (a repressor/activator protein) to the telomeric CAAA repeats. Rap1 then recruits the SIR protein Sir4 (Luo, Vega-Palas, & Grunstein, 2002) which in turn recruits Sir2 and Sir3 (Hoppe et al., 2002; Rusche, Kirchmaier, & Rine, 2002). Silencing at the silent mating-type loci differs from telomere position effect primarily at this nucleation step: silencers contain binding sites for Rap1, the related Abf1 (an ARS binding protein), and Orc1 (an origin recognition complex protein). Sir1 seems to bind Orc1 together with a nearby Sir4 to stabilize nucleation of the SIR complex (Bose et al., 2004). The extending of chromatin, away from the nucleation site, comprises Sir2, Sir3, Sir4, and nucleosomes (Strahl-Bolsinger, Hecht, Luo, & Grunstein, 1997). Sir4 forms a complex with Sir2, a NAD-dependent histone deacetylase that deacetylates lysine 16 (K16) on histone H4 (Liou, Tanny, Kruger, Walz, & Moazed, 2005). Deacetylation of H4K16 allows Sir3 and Sir4 to bind to the H3 and H4 tails (Carmen, Milne, & Grunstein, 2002; Hecht, Laroche, Strahl-Bolsinger, Gasser, & Grunstein, 1995). These observations led to a model in which recurrent rounds of deacetylation by the Sir2–Sir4 complex and subsequent binding of Sir3 and Sir2–Sir4 to deacetylated histone tails propagate silencing (Hoppe et al., 2002; Luo et al., 2002; Rusche et al., 2002).

The SIR complex in budding yeast seems to be unique: of the SIR proteins, only Sir2 has clear orthologues. In most other eukaryotes, silencing relies on a common set of related proteins that make up heterochromatin. As introduced earlier, in *S. pombe*, specialized repetitive sequences and RNAi cooperate to initiate heterochromatin formation (stepwise assembly model). In this case, the shRNAs, generated by RNAi-mediated processing, provide the specificity for targeting histone-modifying activities to the corresponding genomic location in which they are generated.

In mammals, it has recently been suggested that some transcription factors could act as repressors of heterochromatic pericentromeric repeats and thereby regulate heterochromatic signatures at those repeats (Bulut-Karslioglu et al., 2012). However, it is not yet clear whether this is a generalized mechanism and/or whether the establishment of heterochromatin after fertilization obeys a potential nucleation step that depends upon a transcription factor. In any case, the general agreement suggests that it is the transcripts themselves that are generated from heterochromatic-to-be loci that function as initial nuleators of heterochromatin formation.

## 3.5. Establishment of pericentric heterochromatin in mammals

In mammalian cells, transcription of major satellite repeats occurs during S-phase, and their transcription seems to take place before replication of the heterochromatic sequences (Lu & Gilbert, 2007). Their transcription is induced in response to cell proliferation and may involve H3S10P or dissociation of HP1 proteins triggered by specific signals during the cell cycle. But it has also been reported that gamma-satellite sequences (e.g., major satellite sequences) are repressed upon retinoic acid treatment (Rudert, Bronner, Garnier, & Dolle, 1995). Inducing muscle differentiation using C2C12 cells has also revealed that differentiation is accompanied by a spatial reorganization of constitutive pericentromeric repeats, that is, associated with elevated major and minor satellite transcripts (Terranova, Sauer, Merkenschlager, & Fisher, 2005). Contrasting with these results, ES cell differentiation induced by retinoic acid has been shown to result in increased accumulation or major satellite transcripts (Martens et al., 2005). Thus, although there are a few reports documenting major satellite transcription in mammalian cells, the physiological and differentiation contexts of such reports are varied and it seems difficult to reconcile a potential unique function to those observations. In particular, it is not known whether S-phase transcription facilitates heterochromatin assembly in a manner similar to the process of transcription-coupled establishment of silenced chromatin in fission yeast that we described earlier.

Recent analyses in early mouse embryos have attempted to investigate whether transcription of pericentromeric repeats is functionally linked to heterochromatin formation in mammals. Work by Almouzni and colleagues documented a very precise temporal regulation of major satellite transcription following fertilization (Probst et al., 2010). Transcription from both forward and reverse strand was detected, but at different degrees and with different temporal dynamics. Interestingly, the paternal pronucleus shows a significantly higher transcriptional rate of major satellites compared to the maternal one (Probst et al., 2010; Puschendorf et al., 2008; Santenard et al., 2010). A burst of transcription of the forward strand correlated temporally with the formation of chromocenters. Importantly, by blocking the effect of major satellite transcription using LNA gapmers, it was shown that there is a strand-specific requirement for the remodeling of heterochromatin into a somatic chromocenter configuration (Probst et al., 2010). These data argue strongly that transcription from the major satellites after fertilization is required for heterochromatin remodeling and developmental progression.

Largely inspired by the work in *S. pombe* and *Arabidopsis*, work from our lab has shown that the transcription of major satellites in the zygote is associated with tethering of HP1β and also with the spatial nuclear localization typical of pericentromeric chromatin around the NLBs in the embryo (Santenard et al., 2010). This further suggests that there might be a functional link between spatial localization and silencing of pericentric domains after fertilization. Although the mechanisms behind are still to be determined, our data suggested that dsRNA from the major satellites can lead to the localization of HP1β to the highly condensed, DAPI-rich regions around the NLBs in the mouse embryo. Interestingly, and in contrast to the *S. pombe* model, it is the hinge region of HP1β rather than its chromodomain that seems to be the main determinant for HP1β localization in embryonic heterochromatin, at least in the zygote and at the 2-cell stage. In agreement with this, mutation of the chromodomain of HP1β does not seem to elicit a major defect in its subnuclear localization (Santenard et al., 2010). This is in line with the fact that the paternal chromatin lacks detectable H3K9me3. In the embryo, silencing of pericentric heterochromatin occurs therefore in the absence of any detectable levels of H3K9me3, and it is instead H3K27me2/3 that seems to be a major player in this process (Puschendorf et al., 2008; Santenard et al., 2010).

Thus, it would seem that transcription of pericentromeric repeats also lies at the heart of the formation of heterochromatic signatures *de novo* in mammals. A number of questions remain, however, unanswered, mainly to determine whether the downstream effectors of such transcription are the components of the RNAi machinery or any related protein(s) and whether the maintenance mechanism throughout the cell cycle depends also on this transcription.

## 3.6. Heterochromatin assembly: A more general RNA-related mechanism also in mammals?

More generally, it is reasonable to say that ncRNA molecules of various sizes appear to play a major role in the regulation of silent chromatin biology. For example, RNAs play an important role in chromosome-specific localization of the activities of chromatin modifiers required for dosage compensation in *Drosophila* and mammals (Park & Kuroda, 2001) and also in some cases of genomic imprinting (Sleutels, Zwart, & Barlow, 2002). In mammals, *Xist* RNA originating from the X-inactivation center is required for initiation but not for the subsequent inheritance of X-inactivation (Avner & Heard, 2001), and silencing is also regulated by *Tsix*, an *Xist* antisense transcript (Cohen & Lee, 2002). Interestingly, it has been recently shown that long interspersed repeated elements (LINEs) that are enriched on the X chromosome compared with autosomes (Boyle, Ballard, & Ward, 1990) participate in creating a silent nuclear compartment into which X-linked genes become recruited. These LINE repeats contain a subset of "young" LINE-1 elements that are therefore relatively active and expressed during X-inactivation, rather than being silenced. Expression of these LINEs requires the specific heterochromatic state induced by *Xist*. These LINEs often lie within escape-prone regions of the X chromosome, but close to the genes that are subject to X-inactivation, and are associated with putative endo-siRNAs. LINEs may thus facilitate X-inactivation at different levels, with silent LINEs participating in assembly of a heterochromatic nuclear compartment induced by *Xist*, and active LINEs participating in local propagation of X-inactivation into regions that would otherwise be prone to escape (Chow et al., 2010).

It is possible that other repetitive elements in the mammalian genome, in particular, retrotransposons, can function as nucleators of silent compartments and/or as spreading or antispreading mechanisms for heterochromatin. In most differentiated cells, such repetitive elements are fully silenced to avoid phenomena such as retrotransposition and recombination that would otherwise compromise DNA integrity (Maksakova, Mager, & Reiss, 2008). As mentioned earlier, the lack of conventional heterochromatic modifications during the earliest stages of embryogenesis has been suggested to provide a window of opportunity for the reactivation of nondesired 'guests' in the genome like retrotransposons. Indeed, transcripts derived from repetitive elements are found in the early embryo (Bachvarova, 1988; Efroni et al., 2008; Evsikov et al., 2004; Packer, Manova, & Bachvarova, 1993; Peaston et al., 2004). We have recently found that the transcriptional activation of these elements decreases at the 8-cell stage after a peak of reactivation at the 2-cell stage (Fadloun et al, 2013). It is therefore intriguing how retrotransposons are silenced during preimplantation development, since at the time when their transcriptional activity decreases, H4K20me3 and H3K64me3 are undetectable from the embryonic chromatin, and H3K9me3 and DNA methylation seem to decrease, both globally and in LINEs (Santos et al., 2002; Smith et al., 2012). In this sense, the retrotransposons provide actually an excellent model to address how heterochromatic regions other than pericentric domains are silenced during embryogenesis and it will be important to determine how such regions are silenced to start with. LINEs and long terminal repeat retrotransposons (LTRs) are known to be demethylated immediately after fertilization (Lane et al., 2003; Smith et al., 2012; Wossidlo et al., 2010). Most of the data available in the literature, including recent reduced representation bisulfite sequence, suggest that levels of DNA methylation on LTRs and on LINEs do not increase prior to implantation and therefore alternative mechanisms are probably in place to silence retrotransposons in the early embryo (Smith et al., 2012). This is in line with elegant experiments in which deletion of KAP1—which is known to silence endogenous retroviruses through an H3K9me3/DNA methylation pathway—at several developmental times revealed that KAP1-directed DNA methylation is dispensable for the silencing of endogenous retroviruses before E3.5 (Rowe et al., 2010). Although it is well known that retrotransposons are silenced through a piRNA mechanism in the germ line, the piRNA/Dnmt3L pathway seems not to be active in the early embryo (Aravin, Sachidanandam, Girard, Fejes-Toth, & Hannon, 2007; Bourc'his & Bestor, 2004; Carmell et al., 2007; Zamudio & Bourc'his, 2010). It is therefore possible that other RNA-mediated mechanisms also regulate transcription and/or silencing of repetitive loci during early embryogenesis, similar to what it has been shown for pericentromeric repeats. Recent analyses performed in our lab indicating that very short RNAs smaller than 18 nt can regulate expression of LINE elements in the mouse zygote support such a scenario (Fadloun et al., 2013). Thus, it could be that RNA has a more generalized role in regulating transcriptional activity of heterochromatic loci during development.

## 4. CONCLUDING REMARKS

It is clear from all the stated earlier that the organization of the heterochromatin itself as well as its potential role in regulating gene expression will have a key role in imparting epigenetic decisions during early development.

We believe that there are at least three different aspects where future research will have to provide mechanistic insights to these important questions.

The first major question is whether heterochromatin can act as a driving force in development. Although there is a suggestion that plasticity correlates with a more open chromatin configuration in ES cells (Meshorer et al., 2006), a direct causal link between the two has not been established and it is unclear whether the more open configuration is a consequence of pluripotency or the other way around. Second, it will be important to determine whether heterochromatin has a potential role in lineage allocation: Is heterochromatin different in the cells that are destined to become trophectoderm than in the cells that will form the ICM? In support to this hypothesis, it has been shown by electron microscopy that global chromatin architecture differs between the two lineages of the early blastocyst (Ahmed et al., 2010). It will have to be established whether such differences emerge *prior* to lineage segregation and if so, whether they could have an instructive role in cell fate determination. It is known that the dynamics of OCT4 differ between cells of embryos at both 4- and 8-cell stage and that these dynamics are predictive of lineage allocation in the blastocyst (Plachta, Bollenbach, Pease, Fraser, & Pantazis, 2011). The dynamics of OCT4 were suggested to reflect accessibility of OCT4 to its binding sites, and it is therefore possible that a global hetero- versus euchromatin organization impacts on such accessibility. And third, could heterochromatin or the absence of conventional heterochromatin be a regulator of epigenetic reprogramming? Overall, the data we have described earlier are so far correlative, mostly because of the technical difficulties and limited approaches that can be applied to early embryos. Nevertheless, it is clear that there are major chromatin remodeling events after fertilization, in particular, of heterochromatic marks. Importantly, this is paralleled in the germ line during the phase of epigenetic reprogramming (Surani, Hayashi, & Hajkova, 2007), strongly suggesting that absence of heterochromatic marks is necessary for reprogramming. However, there is no formal demonstration that this is indeed the case and that heterochromatin needs to be remodeled in order for reprogramming to take place. One could also envisage another provocative scenario: Could it be that reprogramming is a cause of losing heterochromatin? Or is it that reprogramming can occur because the lack of conventional heterochromatin provides a window of opportunity for it to take place? These are all exciting questions that await further investigation.

## ACKNOWLEDGMENTS

Work in the Torres-Padilla lab is supported by grants from the ANR, ANR-09-Blanc-0114, Epigenesys NoE and ERC-2011-StG 280840 "NuclearPotency." A. F. was supported by a postdoctoral fellowship from the Fondation pour la Recherche Medicale and A. E. is supported by a fellowship from the Ministère de la Recherche et de la Technologie. We thank A. Santenard for providing the template for Fig. 1.3.

## REFERENCES

Ahmed, K., Dehghani, H., Rugg-Gunn, P., Fussner, E., Rossant, J., & Bazett-Jones, D. P. (2010). Global chromatin architecture reflects pluripotency and lineage commitment in the early mouse embryo. *PLoS One*, *5*, e10531.

Akiyama, T., Suzuki, O., Matsuda, J., & Aoki, F. (2011). Dynamic replacement of histone H3 variants reprograms epigenetic marks in early mouse embryos. *PLoS Genetics*, 7, e1002279.

Allshire, R. C., Javerzat, J. P., Redhead, N. J., & Cranston, G. (1994). Position effect variegation at fission yeast centromeres. *Cell*, *76*, 157–169.

Allshire, R. C., Nimmo, E. R., Ekwall, K., Javerzat, J. P., & Cranston, G. (1995). Mutations derepressing silent centromeric domains in fission yeast disrupt chromosome segregation. *Genes & Development*, *9*, 218–233.

Aoki, V. W., Liu, L., & Carrell, D. T. (2005). Identification and evaluation of a novel sperm protamine abnormality in a population of infertile males. *Human Reproduction (Oxford, England)*, *20*, 1298–1306.

Aravin, A. A., Sachidanandam, R., Girard, A., Fejes-Toth, K., & Hannon, G. J. (2007). Developmentally regulated piRNA clusters implicate MILI in transposon control. *Science*, *316*, 744–747.

Arney, K. L., Bao, S., Bannister, A. J., Kouzarides, T., & Surani, M. A. (2002). Histone methylation defines epigenetic asymmetry in the mouse zygote. *The International Journal of Developmental Biology*, *46*, 317–320.

Aucott, R., Bullwinkel, J., Yu, Y., Shi, W., Billur, M., Brown, J., et al. (2008). HP1-beta is required for development of the cerebral neocortex and neuromuscular junctions. *The Journal of Cell Biology*, *183*, 597–606.

Avner, P., & Heard, E. (2001). X-chromosome inactivation: Counting, choice and initiation. *Nature Reviews. Genetics*, *2*, 59–67.

Bachvarova, R. (1988). Small B2 RNAs in mouse oocytes, embryos, and somatic tissues. *Developmental Biology*, *130*, 513–523.

Balhorn, R., Reed, S., & Tanphaichitr, N. (1988). Aberrant protamine 1/protamine 2 ratios in sperm of infertile human males. *Experientia*, *44*, 52–55.

Bannister, A. J., Zegerman, P., Partridge, J. F., Miska, E. A., Thomas, J. O., Allshire, R. C., et al. (2001). Selective recognition of methylated lysine 9 on histone H3 by the HP1 chromo domain. *Nature*, *410*, 120–124.

Barski, A., Cuddapah, S., Cui, K., Roh, T. Y., Schones, D. E., Wang, Z., et al. (2007). High-resolution profiling of histone methylations in the human genome. *Cell*, *129*, 823–837.

Bayne, E. H., White, S. A., Kagansky, A., Bijos, D. A., Sanchez-Pulido, L., Hoe, K. L., et al. (2010). Stc1: A critical link between RNAi and chromatin modification required for heterochromatin integrity. *Cell*, *140*, 666–677.

Bernstein, B. E., Mikkelsen, T. S., Xie, X., Kamal, M., Huebert, D. J., Cuff, J., et al. (2006). A bivalent chromatin structure marks key developmental genes in embryonic stem cells. *Cell*, *125*, 315–326.

Bonasio, R., Tu, S., & Reinberg, D. (2010). Molecular signals of epigenetic states. *Science*, *330*, 612–616.

Borgel, J., Guibert, S., Li, Y., Chiba, H., Schubeler, D., Sasaki, H., et al. (2010). Targets and dynamics of promoter DNA methylation during early mouse development. *Nature Genetics*, *42*, 1093–1100.

Bose, M. E., McConnell, K. H., Gardner-Aukema, K. A., Muller, U., Weinreich, M., Keck, J. L., et al. (2004). The origin recognition complex and Sir4 protein recruit Sir1p to yeast silent chromatin through independent interactions requiring a common Sir1p domain. *Molecular and Cellular Biology*, *24*, 774–786.

Boskovic, A., Bender, A., Gall, L., Ziegler-Birling, C., Beaujean, N., & Torres-Padilla, M. E. (2012). Analysis of active chromatin modifications in early mammalian embryos reveals uncoupling of H2A.Z acetylation and H3K36 trimethylation from embryonic genome activation. *Epigenetics*, *7*, 747–757.

Bouniol-Baly, C., Nguyen, E., Besombes, D., & Debey, P. (1997). Dynamic organization of DNA replication in one-cell mouse embryos: Relationship to transcriptional activation. *Experimental Cell Research*, *236*, 201–211.

Bourc'his, D., & Bestor, T. H. (2004). Meiotic catastrophe and retrotransposon reactivation in male germ cells lacking Dnmt3L. *Nature*, *431*, 96–99.

Bourc'his, D., Xu, G. L., Lin, C. S., Bollman, B., & Bestor, T. H. (2001). Dnmt3L and the establishment of maternal genomic imprints. *Science*, *294*, 2536–2539.

Boyle, A. L., Ballard, S. G., & Ward, D. C. (1990). Differential distribution of long and short interspersed element sequences in the mouse genome: Chromosome karyotyping by fluorescence in situ hybridization. *Proceedings of the National Academy of Sciences of the United States of America*, *87*, 7757–7761.

Brown, J., Bullwinkel, J., Baron-Lühr, B., Billur, M., Schneider, P., Winking, H., et al. (2010). HP1gamma function is required for male germ cell survival and spermatogenesis. *Epigenetics & chromatin*, *3*, 9.

Bulut-Karslioglu, A., Perrera, V., Scaranaro, M., de la Rosa-Velazquez, I. A., van de Nobelen, S., Shukeir, N., et al. (2012). A transcription factor-based mechanism for mouse heterochromatin formation. *Nature Structural & Molecular Biology*, *19*, 1023–1030.

Burton, A., & Torres-Padilla, M. E. (2010). Epigenetic reprogramming and development: A unique heterochromatin organization in the preimplantation mouse embryo. *Briefings in Functional Genomics*, *9*, 444–454.

Cam, H. P., Sugiyama, T., Chen, E. S., Chen, X., FitzGerald, P. C., & Grewal, S. I. (2005). Comprehensive analysis of heterochromatin- and RNAi-mediated epigenetic control of the fission yeast genome. *Nature Genetics*, *37*, 809–819.

Cam, H. P., Chen, E. S., & Grewal, S. I. (2009). Transcriptional scaffolds for heterochromatin assembly. *Cell*, *136*(4), 610–614. http://dx.doi.org/10.1016/j.cell.2009.02.004.

Cao, R., Wang, L., Wang, H., Xia, L., Erdjument-Bromage, H., Tempst, P., et al. (2002). Role of histone H3 lysine 27 methylation in Polycomb-group silencing. *Science*, *298*, 1039–1043.

Carmell, M. A., Girard, A., van de Kant, H. J., Bourc'his, D., Bestor, T. H., de Rooij, D. G., et al. (2007). MIWI2 is essential for spermatogenesis and repression of transposons in the mouse male germline. *Developmental Cell*, *12*, 503–514.

Carmen, A. A., Milne, L., & Grunstein, M. (2002). Acetylation of the yeast histone H4 N terminus regulates its binding to heterochromatin protein SIR3. *The Journal of Biological Chemistry*, *277*, 4778–4781.

Chan, M. M., Smith, Z. D., Egli, D., Regev, A., & Meissner, A. (2012). Mouse ooplasm confers context-specific reprogramming capacity. *Nature Genetics*, *44*, 978–980.

Chang, C. C., Ma, Y., Jacobs, S., Tian, X. C., Yang, X., & Rasmussen, T. P. (2005). A maternal store of macroH2A is removed from pronuclei prior to onset of somatic macroH2A expression in preimplantation embryos. *Developmental Biology*, *278*, 367–380.

Chen, E. S., Zhang, K., Nicolas, E., Cam, H. P., Zofall, M., & Grewal, S. I. (2008). Cell cycle control of centromeric repeat transcription and heterochromatin assembly. *Nature, 451*, 734–737.

Chow, J. C., Ciaudo, C., Fazzari, M. J., Mise, N., Servant, N., Glass, J. L., et al. (2010). LINE-1 activity in facultative heterochromatin formation during X chromosome inactivation. *Cell, 141*, 956–969.

Cleveland, D. W., Mao, Y., & Sullivan, K. F. (2003). Centromeres and kinetochores: From epigenetics to mitotic checkpoint signaling. *Cell, 112*, 407–421.

Cohen, D. E., & Lee, J. T. (2002). X-chromosome inactivation and the search for chromosome-wide silencers. *Current Opinion in Genetics & Development, 12*, 219–224.

Colmenares, S. U., Buker, S. M., Buhler, M., Dlakic, M., & Moazed, D. (2007). Coupling of double-stranded RNA synthesis and siRNA generation in fission yeast RNAi. *Molecular Cell, 27*, 449–461.

Daujat, S., Weiss, T., Mohn, F., Lange, U. C., Ziegler-Birling, C., Zeissler, U., et al. (2009). H3K64 trimethylation marks heterochromatin and is dynamically remodeled during developmental reprogramming. *Nature Structural & Molecular Biology, 16*, 777–781.

Dawlaty, M. M., Ganz, K., Powell, B. E., Hu, Y. C., Markoulaki, S., Cheng, A. W., et al. (2011). Tet1 is dispensable for maintaining pluripotency and its loss is compatible with embryonic and postnatal development. *Cell Stem Cell, 9*, 166–175.

Dodge, J., Kang, Y.-K., Beppu, H., Lei, H., & Li, E. (2004). Histone H3-K9 methyltransferase ESET is essential for early development. *Molecular and Cellular Biology, 24*, 2478–2486.

Donohoe, M. E., Zhang, X., McGinnis, L., Biggers, J., Li, E., & Shi, Y. (1999). Targeted disruption of mouse Yin Yang 1 transcription factor results in peri-implantation lethality. *Molecular and Cellular Biology, 19*, 7237–7244.

Eckardt, S., Leu, N. A., Kurosaka, S., & McLaughlin, K. J. (2005). Differential reprogramming of somatic cell nuclei after transfer into mouse cleavage stage blastomeres. *Reproduction, 129*, 547–556.

Efroni, S., Duttagupta, R., Cheng, J., Dehghani, H., Hoeppner, D. J., Dash, C., et al. (2008). Global transcription in pluripotent embryonic stem cells. *Cell Stem Cell, 2*, 437–447.

Erhardt, S., Su, I. H., Schneider, R., Barton, S., Bannister, A., Perez-Burgos, L., et al. (2003). Consequences of the depletion of zygotic and embryonic enhancer of zeste 2 during preimplantation mouse development. *Development (Cambridge, England), 130*, 4235–4248.

Evsikov, A. V., de Vries, W. N., Peaston, A. E., Radford, E. E., Fancher, K. S., Chen, F. H., et al. (2004). Systems biology of the 2-cell mouse embryo. *Cytogenetic and Genome Research, 105*, 240–250.

Fadloun, A., Le Gras, S., Jost, B., Ziegler-Birling, C., Takahashi, H., Gorab, E., et al. (2013). Chromatin signatures and retrotransposon profiling in mouse embryos reveal regulation of LINE-1 by RNA. *Nature Structural & Molecular Biology*, http://dx.doi.org/10.1038/nsmb.2495. [Epub ahead of print].

Farthing, C. R., Ficz, G., Ng, R. K., Chan, C. F., Andrews, S., Dean, W., et al. (2008). Global mapping of DNA methylation in mouse promoters reveals epigenetic reprogramming of pluripotency genes. *PLoS Genetics, 4*, e1000116.

Faust, C., Lawson, K. A., Schork, N. J., Thiel, B., & Magnuson, T. (1998). The Polycomb-group gene eed is required for normal morphogenetic movements during gastrulation in the mouse embryo. *Development (Cambridge, England), 125*, 4495–4506.

Folco, H. D., Pidoux, A. L., Urano, T., & Allshire, R. C. (2008). Heterochromatin and RNAi are required to establish CENP-A chromatin at centromeres. *Science, 319*, 94–97.

Goll, M. G., & Bestor, T. H. (2005). Eukaryotic cytosine methyltransferases. *Annual Review of Biochemistry, 74*, 481–514.

Govin, J., Escoffier, E., Rousseaux, S., Kuhn, L., Ferro, M., Thevenon, J., et al. (2007). Pericentric heterochromatin reprogramming by new histone variants during mouse spermiogenesis. *The Journal of Cell Biology*, *176*, 283–294.

Grewal, S. I., & Elgin, S. C. (2002). Heterochromatin: New possibilities for the inheritance of structure. *Current Opinion in Genetics & Development*, *12*, 178–187.

Grewal, S. I., & Elgin, S. C. (2007). Transcription and RNA interference in the formation of heterochromatin. *Nature*, *447*, 399–406.

Grewal, S. I., & Jia, S. (2007). Heterochromatin revisited. *Nature Reviews. Genetics*, *8*, 35–46.

Gu, T. P., Guo, F., Yang, H., Wu, H. P., Xu, G. F., Liu, W., et al. (2011). The role of Tet3 DNA dioxygenase in epigenetic reprogramming by oocytes. *Nature*, *477*, 606–610.

Hamatani, T., Carter, M. G., Sharov, A. A., & Ko, M. S. (2004). Dynamics of global gene expression changes during mouse preimplantation development. *Developmental Cell*, *6*, 117–131.

Hammoud, S. S., Nix, D. A., Zhang, H., Purwar, J., Carrell, D. T., & Cairns, B. R. (2009). Distinctive chromatin in human sperm packages genes for embryo development. *Nature*, *460*, 473–478.

Hata, K., Okano, M., Lei, H., & Li, E. (2002). Dnmt3L cooperates with the Dnmt3 family of de novo DNA methyltransferases to establish maternal imprints in mice. *Development (Cambridge, England)*, *129*, 1983–1993.

Hecht, A., Laroche, T., Strahl-Bolsinger, S., Gasser, S. M., & Grunstein, M. (1995). Histone H3 and H4 N-termini interact with SIR3 and SIR4 proteins: A molecular model for the formation of heterochromatin in yeast. *Cell*, *80*, 583–592.

Hoppe, G. J., Tanny, J. C., Rudner, A. D., Gerber, S. A., Danaie, S., Gygi, S. P., et al. (2002). Steps in assembly of silent chromatin in yeast: Sir3-independent binding of a Sir2/Sir4 complex to silencers and role for Sir2-dependent deacetylation. *Molecular and Cellular Biology*, *22*, 4167–4180.

Huang, Y. (2002). Transcriptional silencing in Saccharomyces cerevisiae and Schizosaccharomyces pombe. *Nucleic Acids Research*, *30*, 1465–1482.

Inoue, A., & Zhang, Y. (2011). Replication-dependent loss of 5-hydroxymethylcytosine in mouse preimplantation embryos. *Science*, *334*, 194.

Iqbal, K., Jin, S. G., Pfeifer, G. P., & Szabo, P. E. (2011). Reprogramming of the paternal genome upon fertilization involves genome-wide oxidation of 5-methylcytosine. *Proceedings of the National Academy of Sciences of the United States of America*, *108*, 3642–3647.

Ito, S., D'Alessio, A. C., Taranova, O. V., Hong, K., Sowers, L. C., & Zhang, Y. (2010). Role of Tet proteins in 5mC to 5hmC conversion, ES-cell self-renewal and inner cell mass specification. *Nature*, *466*, 1129–1133.

Jenkins, T. G., & Carrell, D. T. (2012). The sperm epigenome and potential implications for the developing embryo. *Reproduction*, *143*, 727–734.

Kagansky, A., Folco, H. D., Almeida, R., Pidoux, A. L., Boukaba, A., Simmer, F., et al. (2009). Synthetic heterochromatin bypasses RNAi and centromeric repeats to establish functional centromeres. *Science*, *324*, 1716–1719.

Kaneda, M., Okano, M., Hata, K., Sado, T., Tsujimoto, N., Li, E., et al. (2004). Essential role for de novo DNA methyltransferase Dnmt3a in paternal and maternal imprinting. *Nature*, *429*, 900–903.

Kloc, A., Zaratiegui, M., Nora, E., & Martienssen, R. (2008). RNA interference guides histone modification during the S phase of chromosomal replication. *Current Biology: CB*, *18*, 490–495.

Kourmouli, N., Jeppesen, P., Mahadevhaiah, S., Burgoyne, P., Wu, R., Gilbert, D. M., et al. (2004). Heterochromatin and tri-methylated lysine 20 of histone H4 in animals. *Journal of Cell Science*, *117*, 2491–2501.

Kouzarides, T. (2007). Chromatin modifications and their function. *Cell*, *128*, 693–705.

Lachner, M., O'Carroll, D., Rea, S., Mechtler, K., & Jenuwein, T. (2001). Methylation of histone H3 lysine 9 creates a binding site for HP1 proteins. *Nature*, *410*, 116–120.

Lane, N., Dean, W., Erhardt, S., Hajkova, P., Surani, A., Walter, J., et al. (2003). Resistance of IAPs to methylation reprogramming may provide a mechanism for epigenetic inheritance in the mouse. *Genesis*, *35*, 88–93.

Lehnertz, B., Northrop, J. P., Antignano, F., Burrows, K., Hadidi, S., Mullaly, S. C., et al. (2010). Activating and inhibitory functions for the histone lysine methyltransferase G9a in T helper cell differentiation and function. *The Journal of Experimental Medicine*, *207*, 915–922.

Lehnertz, B., Ueda, Y., Derijck, A. A., Braunschweig, U., Perez-Burgos, L., Kubicek, S., et al. (2003). Suv39h-mediated histone H3 lysine 9 methylation directs DNA methylation to major satellite repeats at pericentric heterochromatin. *Current Biology: CB*, *13*, 1192–1200.

Li, E., Beard, C., & Jaenisch, R. (1993). Role for DNA methylation in genomic imprinting. *Nature*, *366*, 362–365.

Liou, G. G., Tanny, J. C., Kruger, R. G., Walz, T., & Moazed, D. (2005). Assembly of the SIR complex and its regulation by O-acetyl-ADP-ribose, a product of NAD-dependent histone deacetylation. *Cell*, *121*, 515–527.

Liu, H., Kim, J.-M., & Aoki, F. (2004). Regulation of histone H3 lysine 9 methylation in oocytes and early pre-implantation embryos. *Development (Cambridge, England)*, *131*, 2269–2280.

Loppin, B., Bonnefoy, E., Anselme, C., Laurencon, A., Karr, T. L., & Couble, P. (2005). The histone H3.3 chaperone HIRA is essential for chromatin assembly in the male pronucleus. *Nature*, *437*, 1386–1390.

Lu, J., & Gilbert, D. M. (2007). Proliferation-dependent and cell cycle regulated transcription of mouse pericentric heterochromatin. *The Journal of Cell Biology*, *179*, 411–421.

Lu, S., Xie, Y. M., Li, X., Luo, J., Shi, X. Q., Hong, X., et al. (2009). Mass spectrometry analysis of dynamic post-translational modifications of TH2B during spermatogenesis. *Molecular Human Reproduction*, *15*, 373–378.

Luff, B., Pawlowski, L., & Bender, J. (1999). An inverted repeat triggers cytosine methylation of identical sequences in Arabidopsis. *Molecular Cell*, *3*, 505–511.

Luger, K., Dechassa, M. L., & Tremethick, D. J. (2012). New insights into nucleosome and chromatin structure: An ordered state or a disordered affair? *Nature Reviews. Molecular Cell Biology*, *13*, 436–447.

Luger, K., Mader, A. W., Richmond, R. K., Sargent, D. F., & Richmond, T. J. (1997). Crystal structure of the nucleosome core particle at 2.8 A resolution. *Nature*, *389*, 251–260.

Luo, K., Vega-Palas, M. A., & Grunstein, M. (2002). Rap1-Sir4 binding independent of other Sir, yKu, or histone interactions initiates the assembly of telomeric heterochromatin in yeast. *Genes & Development*, *16*, 1528–1539.

Maalouf, W. E., Liu, Z., Brochard, V., Renard, J. P., Debey, P., Beaujean, N., et al. (2009). Trichostatin A treatment of cloned mouse embryos improves constitutive heterochromatin remodeling as well as developmental potential to term. *BMC Developmental Biology*, *9*, 11.

Maksakova, I. A., Mager, D. L., & Reiss, D. (2008). Keeping active endogenous retroviral-like elements in check: The epigenetic perspective. *Cellular and Molecular Life Sciences*, *65*, 3329–3347.

Martens, J. H., O'Sullivan, R. J., Braunschweig, U., Opravil, S., Radolf, M., Steinlein, P., et al. (2005). The profile of repeat-associated histone lysine methylation states in the mouse epigenome. *The EMBO Journal*, *24*, 800–812.

Martienssen, R. A., Kloc, A., Slotkin, R. K., & Tanurdzic, M. (2008). Epigenetic inheritance and reprogramming in plants and fission yeast. *Cold Spring Harbor Symposia on Quantitative Biology*, *73*, 265–271.

Martin, C., Beaujean, N., Brochard, V., Audouard, C., Zink, D., & Debey, P. (2006). Genome restructuring in mouse embryos during reprogramming and early development. *Developmental Biology, 292*, 317–332.
Matzke, M. A., & Birchler, J. A. (2005). RNAi-mediated pathways in the nucleus. *Nature Reviews. Genetics, 6*, 24–35.
Mayer, W., Niveleau, A., Walter, J., Fundele, R., & Haaf, T. (2000). Demethylation of the zygotic paternal genome. *Nature, 403*, 501–502.
Meshorer, E., Yellajoshula, D., George, E., Scambler, P. J., Brown, D. T., & Misteli, T. (2006). Hyperdynamic plasticity of chromatin proteins in pluripotent embryonic stem cells. *Developmental Cell, 10*, 105–116.
Moazed, D. (2001). Common themes in mechanisms of gene silencing. *Molecular Cell, 8*, 489–498.
Moss, S. B., Challoner, P. B., & Groudine, M. (1989). Expression of a novel histone 2B during mouse spermiogenesis. *Developmental Biology, 133*, 83–92.
Muchardt, C., Guilleme, M., Seeler, J. S., Trouche, D., Dejean, A., & Yaniv, M. (2002). Coordinated methyl and RNA binding is required for heterochromatin localization of mammalian HP1alpha. *EMBO Reports, 3*, 975–981.
Muller, J., Hart, C. M., Francis, N. J., Vargas, M. L., Sengupta, A., Wild, B., et al. (2002). Histone methyltransferase activity of a Drosophila Polycomb group repressor complex. *Cell, 111*, 197–208.
Nakama, M., Kawakami, K., Kajitani, T., Urano, T., & Murakami, Y. (2012). DNA-RNA hybrid formation mediates RNAi-directed heterochromatin formation. *Genes to Cells: Devoted to Molecular & Cellular Mechanisms, 17*, 218–233.
Nakamura, T., Liu, Y. J., Nakashima, H., Umehara, H., Inoue, K., Matoba, S., et al. (2012). PGC7 binds histone H3K9me2 to protect against conversion of 5mC to 5hmC in early embryos. *Nature, 486*, 415–419.
Naruse, C., Fukusumi, Y., Kakiuchi, D., & Asano, M. (2007). A novel gene trapping for identifying genes expressed under the control of specific transcription factors. *Biochemical and Biophysical Research Communications, 361*, 109–115.
Nielsen, A. L., Oulad-Abdelghani, M., Ortiz, J. A., Remboutsika, E., Chambon, P., & Losson, R. (2001). Heterochromatin formation in mammalian cells: Interaction between histones and HP1 proteins. *Molecular Cell, 7*, 729–739.
Nishioka, K., Rice, J., Sarma, K., Erdjument-Bromage, H., Werner, J., Wang, Y., et al. (2002). PR-Set7 is a nucleosome-specific methyltransferase that modifies lysine 20 of histone H4 and is associated with silent chromatin. *Molecular Cell, 9*, 1201–1213.
Nonchev, S., & Tsanev, R. (1990). Protamine-histone replacement and DNA replication in the male mouse pronucleus. *Molecular Reproduction and Development, 25*, 72–76.
O'Carroll, D., Erhardt, S., Pagani, M., Barton, S., Surani, M., & Jenuwein, T. (2001). The polycomb-group gene Ezh2 is required for early mouse development. *Molecular and Cellular Biology, 21*, 4330–4336.
Okano, M., Bell, D. W., Haber, D. A., & Li, E. (1999). DNA methyltransferases Dnmt3a and Dnmt3b are essential for de novo methylation and mammalian development. *Cell, 99*, 247–257.
Packer, A. I., Manova, K., & Bachvarova, R. F. (1993). A discrete LINE-1 transcript in mouse blastocysts. *Developmental Biology, 157*, 281–283.
Park, Y., & Kuroda, M. I. (2001). Epigenetic aspects of X-chromosome dosage compensation. *Science, 293*, 1083–1085.
Pasini, D., Bracken, A. P., Jensen, M. R., Lazzerini Denchi, E., & Helin, K. (2004). Suz12 is essential for mouse development and for EZH2 histone methyltransferase activity. *The EMBO Journal, 23*, 4061–4071.

Peaston, A. E., Evsikov, A. V., Graber, J. H., de Vries, W. N., Holbrook, A. E., Solter, D., et al. (2004). Retrotransposons regulate host genes in mouse oocytes and preimplantation embryos. *Developmental Cell, 7*, 597–606.

Peters, A. H., O'Carroll, D., Scherthan, H., Mechtler, K., Sauer, S., Schofer, C., et al. (2001). Loss of the Suv39h histone methyltransferases impairs mammalian heterochromatin and genome stability. *Cell, 107*, 323–337.

Pidoux, A. L., & Allshire, R. C. (2005). The role of heterochromatin in centromere function. *Philosophical Transactions of the Royal Society of London. Series B, Biological Sciences, 360*, 569–579.

Pinheiro, I., Margueron, R., Shukeir, N., Eisold, M., Fritzsch, C., Richter, F. M., et al. (2012). Prdm3 and Prdm16 are H3K9me1 methyltransferases required for mammalian heterochromatin integrity. *Cell, 150*, 948–960.

Plachta, N., Bollenbach, T., Pease, S., Fraser, S. E., & Pantazis, P. (2011). Oct4 kinetics predict cell lineage patterning in the early mammalian embryo. *Nature Cell Biology, 13*, 117–123.

Poznanski, A. A., & Calarco, P. G. (1991). The expression of intracisternal A particle genes in the preimplantation mouse embryo. *Developmental Biology, 143*, 271–281.

Probst, A. V., Okamoto, I., Casanova, M., El Marjou, F., Le Baccon, P., & Almouzni, G. (2010). A strand-specific burst in transcription of pericentric satellites is required for chromocenter formation and early mouse development. *Developmental Cell, 19*, 625–638.

Probst, A. V., Santos, F., Reik, W., Almouzni, G., & Dean, W. (2007). Structural differences in centromeric heterochromatin are spatially reconciled on fertilisation in the mouse zygote. *Chromosoma, 116*, 403–415.

Proudhon, C., Duffie, R., Ajjan, S., Cowley, M., Iranzo, J., Carbajosa, G., et al. (2012). Protection against de novo methylation is instrumental in maintaining parent-of-origin methylation inherited from the gametes. *Molecular Cell, 47*, 909–920.

Puschendorf, M., Terranova, R., Boutsma, E., Mao, X., Isono, K., Brykczynska, U., et al. (2008). PRC1 and Suv39h specify parental asymmetry at constitutive heterochromatin in early mouse embryos. *Nature Genetics, 40*, 411–420.

Quivoron, C., Couronne, L., Della Valle, V., Lopez, C. K., Plo, I., Wagner-Ballon, O., et al. (2011). TET2 inactivation results in pleiotropic hematopoietic abnormalities in mouse and is a recurrent event during human lymphomagenesis. *Cancer Cell, 20*, 25–38.

Quivy, J. P., Roche, D., Kirschner, D., Tagami, H., Nakatani, Y., & Almouzni, G. (2004). A CAF-1 dependent pool of HP1 during heterochromatin duplication. *The EMBO Journal, 23*, 3516–3526.

Rea, S., Eisenhaber, F., O'Carroll, D., Strahl, B. D., Sun, Z. W., Schmid, M., et al. (2000). Regulation of chromatin structure by site-specific histone H3 methyltransferases. *Nature, 406*, 593–599.

Regha, K., Sloane, M. A., Huang, R., Pauler, F. M., Warczok, K. E., Melikant, B., et al. (2007). Active and repressive chromatin are interspersed without spreading in an imprinted gene cluster in the mammalian genome. *Molecular Cell, 27*(3), 353–366.

Rice, J., Briggs, S., Ueberheide, B., Barber, C., Shabanowitz, J., Hunt, D., et al. (2003). Histone methyltransferases direct different degrees of methylation to define distinct chromatin domains. *Molecular Cell, 12*, 1591–1598.

Rossant, J., & Lis, W. T. (1979). Potential of isolated mouse inner cell masses to form trophectoderm derivatives in vivo. *Developmental Biology, 70*, 255–261.

Rougier, N., Bourc'his, D., Gomes, D. M., Niveleau, A., Plachot, M., Paldi, A., et al. (1998). Chromosome methylation patterns during mammalian preimplantation development. *Genes & Development, 12*, 2108–2113.

Rowe, H. M., Jakobsson, J., Mesnard, D., Rougemont, J., Reynard, S., Aktas, T., et al. (2010). KAP1 controls endogenous retroviruses in embryonic stem cells. *Nature, 463*, 237–240.

Rudert, F., Bronner, S., Garnier, J. M., & Dolle, P. (1995). Transcripts from opposite strands of gamma satellite DNA are differentially expressed during mouse development. *Mammalian Genome, 6*, 76–83.
Rusche, L. N., Kirchmaier, A. L., & Rine, J. (2002). Ordered nucleation and spreading of silenced chromatin in Saccharomyces cerevisiae. *Molecular Biology of the Cell, 13*, 2207–2222.
Santenard, A., Ziegler-Birling, C., Koch, M., Tora, L., Bannister, A. J., & Torres-Padilla, M. E. (2010). Heterochromatin formation in the mouse embryo requires critical residues of the histone variant H3.3. *Nature Cell Biology, 12*, 853–862.
Santos, F., Hendrich, B., Reik, W., & Dean, W. (2002). Dynamic reprogramming of DNA methylation in the early mouse embryo. *Developmental Biology, 241*, 172–182.
Santos, F., Peters, A. H., Otte, A. P., Reik, W., & Dean, W. (2005). Dynamic chromatin modifications characterise the first cell cycle in mouse embryos. *Developmental Biology, 280*, 225–236.
Santos-Rosa, H., Schneider, R., Bannister, A. J., Sherriff, J., Bernstein, B. E., Emre, N. C., et al. (2002). Active genes are tri-methylated at K4 of histone H3. *Nature, 419*, 407–411.
Schaefer, A., Sampath, S. C., Intrator, A., Min, A., Gertler, T. S., Surmeier, D. J., et al. (2009). Control of cognition and adaptive behavior by the GLP/G9a epigenetic suppressor complex. *Neuron, 64*, 678–691.
Schotta, G., Lachner, M., Sarma, K., Ebert, A., Sengupta, R., Reuter, G., et al. (2004). A silencing pathway to induce H3-K9 and H4-K20 trimethylation at constitutive heterochromatin. *Genes & Development, 18*, 1251–1262.
Schotta, G., Sengupta, R., Kubicek, S., Malin, S., Kauer, M., Callen, E., et al. (2008). A chromatin-wide transition to H4K20 monomethylation impairs genome integrity and programmed DNA rearrangements in the mouse. *Genes & Development, 22*, 2048–2061.
Schultz, D. C., Ayyanathan, K., Negorev, D., Maul, G. G., & Rauscher, F. J., 3rd. (2002). SETDB1: A novel KAP-1-associated histone H3, lysine 9-specific methyltransferase that contributes to HP1-mediated silencing of euchromatic genes by KRAB zinc-finger proteins. *Genes & Development, 16*, 919–932.
Shumacher, A., Faust, C., & Magnuson, T. (1996). Positional cloning of a global regulator of anterior-posterior patterning in mice. *Nature, 383*, 250–253.
Sleutels, F., Zwart, R., & Barlow, D. P. (2002). The non-coding Air RNA is required for silencing autosomal imprinted genes. *Nature, 415*, 810–813.
Smallwood, S. A., Tomizawa, S., Krueger, F., Ruf, N., Carli, N., Segonds-Pichon, A., et al. (2011). Dynamic CpG island methylation landscape in oocytes and preimplantation embryos. *Nature Genetics, 43*, 811–814.
Smith, L. C., & Alcivar, A. A. (1993). Cytoplasmic inheritance and its effects on development and performance. *Journal of Reproduction and Fertility. Supplement, 48*, 31–43.
Smith, Z. D., Chan, M. M., Mikkelsen, T. S., Gu, H., Gnirke, A., Regev, A., et al. (2012). A unique regulatory phase of DNA methylation in the early mammalian embryo. *Nature, 484*, 339–344.
Steiner, N. C., Hahnenberger, K. M., & Clarke, L. (1993). Centromeres of the fission yeast Schizosaccharomyces pombe are highly variable genetic loci. *Molecular and Cellular Biology, 13*, 4578–4587.
Stender, J., Pascual, G., Liu, W., Kaikkonen, M., Do, K., Spann, N., et al. (2012). Control of proinflammatory gene programs by regulated trimethylation and demethylation of histone H4K20. *Molecular Cell, 48*, 28–38.
Strahl, B. D., & Allis, C. D. (2000). The language of covalent histone modifications. *Nature, 403*, 41–45.
Strahl-Bolsinger, S., Hecht, A., Luo, K., & Grunstein, M. (1997). SIR2 and SIR4 interactions differ in core and extended telomeric heterochromatin in yeast. *Genes & Development, 11*, 83–93.

Sugiyama, T., Cam, H. P., Sugiyama, R., Noma, K., Zofall, M., Kobayashi, R., et al. (2007). SHREC, an effector complex for heterochromatic transcriptional silencing. *Cell, 128*, 491–504.

Sugiyama, T., Cam, H., Verdel, A., Moazed, D., & Grewal, S. I. (2005). RNA-dependent RNA polymerase is an essential component of a self-enforcing loop coupling heterochromatin assembly to siRNA production. *Proceedings of the National Academy of Sciences of the United States of America, 102*, 152–157.

Surani, M. A., Hayashi, K., & Hajkova, P. (2007). Genetic and epigenetic regulators of pluripotency. *Cell, 128*, 747–762.

Tachibana, M., Nozaki, M., Takeda, N., & Shinkai, Y. (2007). Functional dynamics of H3K9 methylation during meiotic prophase progression. *The EMBO Journal, 26*, 3346–3359.

Tachibana, M., Sugimoto, K., Fukushima, T., & Shinkai, Y. (2001). Set domain-containing protein, G9a, is a novel lysine-preferring mammalian histone methyltransferase with hyperactivity and specific selectivity to lysines 9 and 27 of histone H3. *The Journal of Biological Chemistry, 276*, 25309–25317.

Tachibana, M., Sugimoto, K., Nozaki, M., Ueda, J., Ohta, T., Ohki, M., et al. (2002). G9a histone methyltransferase plays a dominant role in euchromatic histone H3 lysine 9 methylation and is essential for early embryogenesis. *Genes & Development, 16*, 1779–1791.

Tachibana, M., Ueda, J., Fukuda, M., Takeda, N., Ohta, T., Iwanari, H., et al. (2005). Histone methyltransferases G9a and GLP form heteromeric complexes and are both crucial for methylation of euchromatin at H3-K9. *Genes & Development, 19*, 815–826.

Tahiliani, M., Koh, K. P., Shen, Y., Pastor, W. A., Bandukwala, H., Brudno, Y., et al. (2009). Conversion of 5-methylcytosine to 5-hydroxymethylcytosine in mammalian DNA by MLL partner TET1. *Science, 324*, 930–935.

Takada, Y., Naruse, C., Costa, Y., Shirakawa, T., Tachibana, M., Sharif, J., et al. (2011). HP1γ links histone methylation marks to meiotic synapsis in mice. *Development (Cambridge, England), 138*, 4207–4217.

Tan, L., & Shi, Y. G. (2012). Tet family proteins and 5-hydroxymethylcytosine in development and disease. *Development (Cambridge, England), 139*, 1895–1902.

Terranova, R., Sauer, S., Merkenschlager, M., & Fisher, A. G. (2005). The reorganisation of constitutive heterochromatin in differentiating muscle requires HDAC activity. *Experimental Cell Research, 310*(2), 344–356 [Epub September 21, 2005].

Torres-Padilla, M. E., Bannister, A. J., Hurd, P. J., Kouzarides, T., & Zernicka-Goetz, M. (2006). Dynamic distribution of the replacement histone variant H3.3 in the mouse oocyte and preimplantation embryos. *The International Journal of Developmental Biology, 50*, 455–461.

Trostle-Weige, P. K., Meistrich, M. L., Brock, W. A., & Nishioka, K. (1984). Isolation and characterization of TH3, a germ cell-specific variant of histone 3 in rat testis. *The Journal of Biological Chemistry, 259*, 8769–8776.

van der Heijden, G., Dieker, J., Derijck, A. A. H., Muller, S., Berden, J., Braat, D., et al. (2005). Asymmetry in histone H3 variants and lysine methylation between paternal and maternal chromatin of the early mouse zygote. *Mechanisms of Development, 122*, 1008–1022.

Verdel, A., Jia, S., Gerber, S., Sugiyama, T., Gygi, S., Grewal, S. I., et al. (2004). RNAi-mediated targeting of heterochromatin by the RITS complex. *Science, 303*, 672–676.

Volpe, T. A., Kidner, C., Hall, I. M., Teng, G., Grewal, S. I., & Martienssen, R. A. (2002). Regulation of heterochromatic silencing and histone H3 lysine-9 methylation by RNAi. *Science, 297*, 1833–1837.

Wagner, R. P. (1972). The role of maternal effects in animal breeding. II. Mitochondria and animal inheritance. *Journal of Animal Science, 35*, 1280–1287.

Wang, H., An, W., Cao, R., Xia, L., Erdjument-Bromage, H., Chatton, B., et al. (2003). mAM facilitates conversion by ESET of dimethyl to trimethyl lysine 9 of histone H3 to cause transcriptional repression. *Molecular Cell, 12*, 475–487.

Wang, H., Huang, Z. Q., Xia, L., Feng, Q., Erdjument-Bromage, H., Strahl, B. D., et al. (2001). Methylation of histone H4 at arginine 3 facilitating transcriptional activation by nuclear hormone receptor. *Science, 293*, 853–857.

Wang, Q. T., Piotrowska, K., Ciemerych, M. A., Milenkovic, L., Scott, M. P., Davis, R. W., et al. (2004). A genome-wide study of gene activity reveals developmental signaling pathways in the preimplantation mouse embryo. *Developmental Cell, 6*, 133–144.

Ward, W. S., & Coffey, D. S. (1991). DNA packaging and organization in mammalian spermatozoa: Comparison with somatic cells. *Biology of Reproduction, 44*, 569–574.

Witt, O., Albig, W., & Doenecke, D. (1996). Testis-specific expression of a novel human H3 histone gene. *Experimental Cell Research, 229*, 301–306.

Wongtawan, T., Taylor, J. E., Lawson, K. A., Wilmut, I., & Pennings, S. (2011). Histone H4K20me3 and HP1alpha are late heterochromatin markers in development, but present in undifferentiated embryonic stem cells. *Journal of Cell Science, 124*, 1878–1890.

Wood, V., Gwilliam, R., Rajandream, M. A., Lyne, M., Lyne, R., Stewart, A., et al. (2002). The genome sequence of Schizosaccharomyces pombe. *Nature, 415*, 871–880.

Woolcock, K. J., Gaidatzis, D., Punga, T., & Buhler, M. (2011). Dicer associates with chromatin to repress genome activity in Schizosaccharomyces pombe. *Nature Structural & Molecular Biology, 18*, 94–99.

Wossidlo, M., Arand, J., Sebastiano, V., Lepikhov, K., Boiani, M., Reinhardt, R., et al. (2010). Dynamic link of DNA demethylation, DNA strand breaks and repair in mouse zygotes. *The EMBO Journal, 29*, 1877–1888.

Wossidlo, M., Nakamura, T., Lepikhov, K., Marques, C. J., Zakhartchenko, V., Boiani, M., et al. (2011). 5-Hydroxymethylcytosine in the mammalian zygote is linked with epigenetic reprogramming. *Nature Communications, 2*, 241.

Wu, H., & Zhang, Y. (2011). Mechanisms and functions of Tet protein-mediated 5-methylcytosine oxidation. *Genes & Development, 25*, 2436–2452.

Wykes, S. M., & Krawetz, S. A. (2003). The structural organization of sperm chromatin. *The Journal of Biological Chemistry, 278*, 29471–29477.

Yamada, T., Fischle, W., Sugiyama, T., Allis, C. D., & Grewal, S. I. (2005). The nucleation and maintenance of heterochromatin by a histone deacetylase in fission yeast. *Molecular Cell, 20*, 173–185.

Yang, L., Xia, L., Wu, D., Wang, H., Chansky, H., Schubach, W., et al. (2002). Molecular cloning of ESET, a novel histone H3-specific methyltransferase that interacts with ERG transcription factor. *Oncogene, 21*, 148–152.

Yuan, P., Han, J., Guo, G., Orlov, Y. L., Huss, M., Loh, Y. H., et al. (2009). Eset partners with Oct4 to restrict extraembryonic trophoblast lineage potential in embryonic stem cells. *Genes & Development, 23*, 2507–2520.

Zamudio, N., & Bourc'his, D. (2010). Transposable elements in the mammalian germline: A comfortable niche or a deadly trap? *Heredity, 105*, 92–104.

Zeng, F., & Schultz, R. M. (2005). RNA transcript profiling during zygotic gene activation in the preimplantation mouse embryo. *Developmental Biology, 283*, 40–57.

Zhang, K., Mosch, K., Fischle, W., & Grewal, S. I. (2008). Roles of the Clr4 methyltransferase complex in nucleation, spreading and maintenance of heterochromatin. *Nature Structural & Molecular Biology, 15*, 381–388.

Zofall, M., & Grewal, S. I. (2006). Swi6/HP1 recruits a JmjC domain protein to facilitate transcription of heterochromatic repeats. *Molecular Cell, 22*, 681–692.

CHAPTER TWO

# Functions of DNA Methylation and Hydroxymethylation in Mammalian Development

**Sylvain Guibert, Michael Weber**[1]

Biotechnology and Cell Signaling, University of Strasbourg, CNRS/UMR7242, Illkirch, Strasbourg, France
[1]Corresponding author: e-mail address: michael.weber@unistra.fr

## Contents



## Abstract

DNA methylation occurs at cytosines, predominantly in the CpG dinucleotide context and is a key epigenetic regulator of embryogenesis and stem-cell differentiation in mammals. The genomic patterns of 5-methylcytosine are extensively reprogrammed during early embryonic development as well as in the germ-cell lineage. Thanks to improvements in high-throughput mapping technologies, it is now possible to characterize the dynamics of this epigenetic mark at the genome scale. DNA methylation plays

*Current Topics in Developmental Biology*, Volume 104
ISSN 0070-2153
http://dx.doi.org/10.1016/B978-0-12-416027-9.00002-4

multiple roles during development and serves to establish long-term gene silencing. In 2009, it was revealed that 5-hydroxymethylcytosine (5hmC) is another prominent cytosine modification catalyzed by the enzymes of the TET family and abundant in certain cell types. 5hmC has been thought to serve as an intermediate in the reaction of DNA demethylation or act as a signal for chromatin factors. Here, we review the current knowledge on the roles of these DNA epigenetic marks in development, epigenetic reprogramming, and pluripotency.

## 1. INTRODUCTION

"Epigenetics" defines the processes that ensure the propagation of phenotypes through mitosis or meiosis without irreversible changes in the genetic sequence (Berger, Kouzarides, Shiekhattar, & Shilatifard, 2009). This epigenetic information can reside in posttranslational modifications of histones and cytosine modifications, which modify the accessibility of the DNA and provide recognition motifs for chromatin-binding proteins. When inherited during cell division, these marks provide a memory of gene expression states that is crucial to maintain cellular identity. During development, these epigenetic marks undergo global changes that contribute to the restriction of cellular potential and the formation of distinct cell lineages (Meissner, 2010).

The most abundant epigenetic mark of DNA is methylation of the carbon 5 of cytosines (Fig. 2.1A), which creates 5-methylcytosine (5mC). 5mC is abundant in vertebrates and occurs almost exclusively as a symmetrical mark at CpG dinucleotides (Fig. 2.1B). It is also found in invertebrate species, albeit at lower levels than in vertebrate genomes (Suzuki & Bird, 2008). Cytosine methylation is catalyzed by DNA methyltransferases (DNMTs) that methylate DNA *de novo* during development and propagate DNA methylation during replication (Fig. 2.1B). While initially viewed as a repressive mark, the recent progress in genome-wide mapping sheds new light on the complex relationship between DNA methylation and gene expression.

5-Hydroxymethylcytosine (5hmC), another modified base present in vertebrate DNA (Fig. 2.1A; Penn, Suwalski, O'Riley, Bojanowski, & Yura, 1972), was recently found to be catalyzed by the enzymes of the ten–eleven translocation (TET) family and to be relatively abundant in pluripotent cells and neurons (Kriaucionis & Heintz, 2009; Tahiliani et al., 2009). With the parallel discovery that TET2 is mutated in a large fraction of myeloid malignancies (Abdel-Wahab et al., 2009; Delhommeau et al., 2009; Langemeijer et al., 2009), this groundbreaking discovery led to the

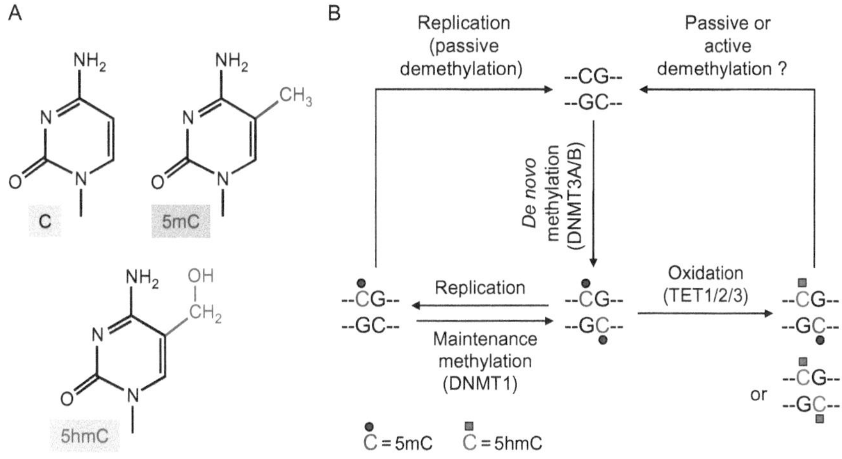


**Figure 2.1** DNA methylation and hydroxymethylation occur on cytosines. (A) Chemical structure of cytosine (C), 5-methylcytosine (5mC), and 5-hydroxymethylcytosine (5hmC). (B) Cytosine methylation occurs symmetrically at CpG dinucleotides. *De novo* methylation is mostly catalyzed by DNMT3A/B enzymes, whereas DNMT1 methylates hemimethylated sites generated after DNA replication. Lack of maintenance methylation results in passive loss of DNA methylation with DNA replication. TET proteins are capable of oxidizing 5mC into 5hmC, which acts as an intermediate in demethylation pathways. (See Color Insert.)

exciting prospect that 5hmC could also play major roles as an epigenetic mark in development and disease, which prompted an impressive number of studies in the past 4 years. In this review, we summarize the current knowledge and discuss recent discoveries on the role of cytosine epigenetic marks during mammalian development.

## 2. ENZYMATIC PLAYERS

### 2.1. DNA methyltransferases

The family of DNMTs comprises four members: DNMT1, DNMT3A, DNMT3B, and DNMT3L (Fig. 2.2A). Another member initially identified by sequence homologies, named DNMT2, was later found to methylate RNA instead of DNA (Goll et al., 2006). DNMT1 is broadly expressed and is responsible for propagating DNA methylation during cell division by applying a methyl group on the newly synthesized DNA strand at hemimethylated CpG sites created by DNA replication (Jones & Liang, 2009; Fig. 2.1B). This function of DNMT1 as a maintenance methyltransferase is supported by the fact that DNMT1 localizes at sites of DNA

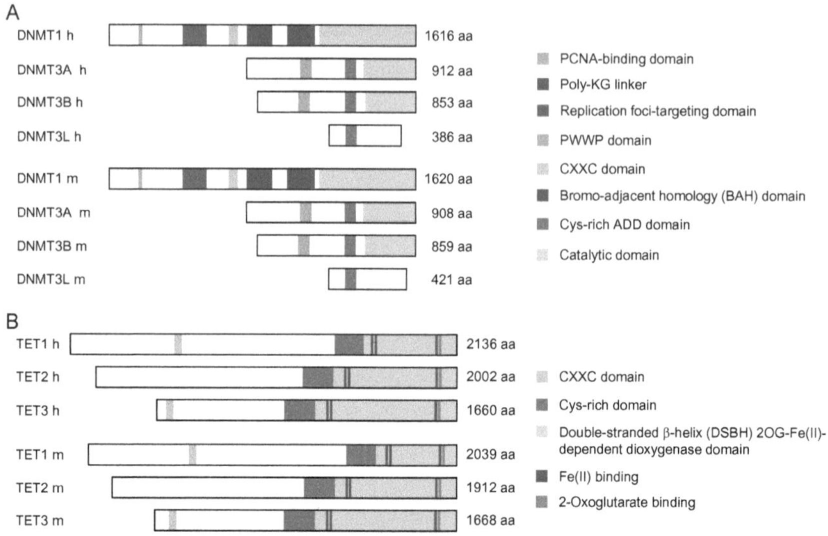


**Figure 2.2** Structure of mammalian DNMT and TET enzymes. (A) DNMTs share a C-terminal catalytic domain, except DNMT3L which does not contain a functional catalytic domain. The N-terminal part of DNMT1 contains interaction domains with PCNA, replication foci, and unmethylated DNA (CXXC domain). DNMT1 also contains two BAH domains. DNMT3A and DNMT3B comprise a PWWP domain recognizing H3K36me3 and an ADD domain mediating interaction with histone H3 unmethylated at H3K4. (B) TET proteins contain in their C terminus a cysteine-rich domain adjacent to a DSBH domain typical of 2-oxoglutarate-Fe(II)-dependent dioxygenases. TET1 and TET3 also contain a CXXC domain. The size of the proteins is indicated by the number of amino acids (aa). (See Color Insert.)

replication during S phase and is preferentially active on hemimethylated CpGs generated after replication (Arand et al., 2012; Leonhardt, Page, Weier, & Bestor, 1992). In mice, the absence of DNMT1 leads to global genome hypomethylation, increase of hemimethylated CpG sites, failure to maintain methylation imprints, and early embryonic death (Arand et al., 2012; Hirasawa et al., 2008; Kurihara et al., 2008; Lei et al., 1996). DNMT1 works as part of a complex as it contains a PCNA-binding domain and also interacts with UHRF1 (also known as NP95), a factor that binds hemimethylated CpGs and is required to recruit DNMT1 at sites of hemimethylation (Bostick et al., 2007; Sharif et al., 2007). The N-terminal part of the protein contains two bromo-adjacent homology domains (BAH1 and BAH2), as well as a CXXC-type zinc-finger domain also found in other chromatin regulators such as MLL, CFP1, MBD1, KDM2A/B, and TET1 (Fig. 2.2A). Similar to MLL, CFP1, or MBD1, the DNMT1 CXXC domain mediates binding to DNA containing unmethylated CpGs (Frauer et al., 2011). This CXXC domain has also been shown to create an autoinhibitory conformational change that prevents interaction of DNA with the catalytic domain when DNMT1 is bound to unmethylated CpGs but not hemimethylated CpGs, therefore providing a mechanism to explain the preference of DNMT1 for hemimethylated over unmethylated DNA (Song, Rechkoblit, Bestor, & Patel, 2011; Song, Teplova, Ishibe-Murakami, & Patel, 2012).

DNMT3A and DNMT3B are responsible for establishing DNA methylation *de novo* during development (Okano, Bell, Haber, & Li, 1999; Fig. 2.1B). DNMT3B is abundant in early embryos and is the main enzyme responsible for the gain of DNA methylation after implantation (Borgel et al., 2010), whereas DNMT3A is more prevalent in differentiated cells. Accordingly, *Dnmt3b* − /− mice die during gestation, whereas *Dnmt3a* − /− mice survive until birth but most of them die around 4 weeks of age (Okano et al., 1999). DNMT3A is also critical for the establishment of DNA methylation imprints in the germ line (Kaneda et al., 2004; Kato et al., 2007). The combined knockout of DNMT3A and DNMT3B leads to a more severe phenotype and all embryos die before E11.5 due to impaired gastrulation, suggesting that both enzymes play synergistic functions during early development (Okano et al., 1999). Besides their activity as *de novo* methyltransferases, DNMT3A and DNMT3B also cooperate with DNMT1 to efficiently maintain DNA methylation during replication at sites that were missed by DNMT1 (Arand et al., 2012; Jones & Liang, 2009). This is illustrated by the fact that DNMT3A/B-deficient embryonic stem (ES) cells show an increased proportion of hemimethylated CpG sites and a progressive loss of methylation upon cell division (Liang et al., 2002). The last member of the family, DNMT3L, lacks crucial amino acids required for a functional catalytic domain (Fig. 2.2A). DNMT3L is highly expressed in germ cells and seems to exert its function mostly in the germ line. *Dnmt3L*− /− mice are viable and proceed normally through development but fail to establish DNA methylation in the male and female germ line, which translates into male sterility (Bourc'his & Bestor, 2004; Bourc'his, Xu, Lin, Bollman, & Bestor, 2001; Kato et al., 2007; Kobayashi et al., 2012; Smallwood et al., 2011). The role of DNMT3L, despite the absence of a functional catalytic domain, is explained by the fact that DNMT3L acts as a cofactor for the DNMT3 enzymes and stimulates their activity (Chedin, Lieber, & Hsieh, 2002; Gowher, Liebert, Hermann, Xu, & Jeltsch, 2005; Jia, Jurkowska, Zhang, Jeltsch, & Cheng, 2007; Ooi et al., 2007). DNMT3A and DNMT3B contain a PWWP domain (Fig. 2.2A) that interacts with histone H3 lysine 36 (H3K36) when it is trimethylated (Dhayalan et al., 2010), suggesting that DNA methylation might be guided to chromatin carrying the H3K36me3 mark. This is compatible with the observation that DNA methylation and H3K36me3 overlap in mammalian cells, in particular, in the body of genes (Ball et al., 2009; Blackledge et al., 2010; Kolasinska-Zwierz et al., 2009). DNMT3A, DNMT3B, and DNMT3L also possess a cysteine-rich ADD (ATRX–DNMT3–DNMT3L) domain (Fig. 2.2A) with affinity to histone H3 when it is unmethylated on lysine 4 (Ooi et al., 2007; Otani et al., 2009; Zhang, Jurkowska, et al., 2010). This suggests that the presence of H3K4 methylation prevents the recruitment of DNMT3 enzymes to chromatin, which provides one possible mechanism for the strong anticorrelation between DNA methylation and H3K4 methylation observed in genome-wide studies (Meissner et al., 2008; Weber et al., 2007).

## 2.2. Proteins of the TET family

The TET proteins TET1–3 are 2-oxoglutarate and Fe(II)-dependent dioxygenases that convert 5mC into 5hmC using α-ketoglutarate as a cosubstrate (Ito et al., 2010; Tahiliani et al., 2009). The founding member TET1 was initially identified as a fusion partner of MLL in acute myeloid leukemia (AML) (Lorsbach et al., 2003; Ono et al., 2002). TET proteins can also further oxidize 5mC into 5-formylcytosine (5fC) and 5-carboxylcytosine (5caC), which is probably of physiological importance because 5fC and 5caC are detectable in DNA from ES cells and early embryos (He et al., 2011; Inoue, Shen, Dai, He, & Zhang, 2011; Ito et al., 2011). The TET proteins share a conserved catalytic domain composed of a double-stranded β-helix (DSBH) region downstream of a cysteine-rich domain (Fig. 2.2B). Another characteristic of the TET proteins is that TET1 and TET3 contain a CXXC domain in their N-terminus (Fig. 2.2B). Unlike the CXXC domains in DNMT1, MLL, MBD1, or CFP1 that bind unmethylated CpGs, the function of the TET1 CXXC domain is less clear. Some studies showed that the human TET1 CXXC domain binds methylated and unmethylated CpG-rich DNA (Xu et al., 2011; Zhang, Zhang, et al., 2010), whereas another study reported that the mouse TET1 CXXC has no DNA-binding activity (Frauer et al., 2011). The three TET proteins have distinct expression patterns, suggesting possible nonoverlapping functions. TET1 and TET2 are highly expressed in ES cells, TET2 is also abundant in hematopoietic cells, and TET3 is present in oocytes. As TET1 is abundant in ES cells, this protein has been the most studied so far. Several groups performed ChIP-Seq in ES cells and observed that TET1 binds preferentially to CpG-rich regions in promoters and within genes, possibly owing to the presence of the CXXC domain (Williams et al., 2011; Wu, D'Alessio, Ito, Xia, et al., 2011; Xu et al., 2011). From this data, it has been proposed that TET1 could contribute to maintain CpG islands in a hypomethylated state by counteracting spurious *de novo* CpG methylation (Branco, Ficz, & Reik, 2011; Williams, Christensen, & Helin, 2012). Yet the exact role of TET1 in ES cells is still debated as several groups reported that TET1 depletion leads to downregulation of pluripotency factors and loss of stem-cell identity (Ficz et al., 2011; Freudenberg et al., 2012; Ito et al., 2010), whereas others did not see major effects on ES cell self-renewal and pluripotency (Dawlaty et al., 2011; Koh et al., 2011; Williams et al., 2011). Another layer of complexity comes from the observation that TET1-bound genes can be both active and inactive, and that downregulation of TET1 in ES cells leads to transcriptional upregulation and downregulation (Ficz et al., 2011; Williams et al., 2011; Wu, D'Alessio, Ito, Xia, et al., 2011; Xu et al., 2011). This suggests that TET1 plays multiple roles in gene regulation, some of which is potentially independent of its catalytic activity (Williams et al., 2011).

Knockouts for the *Tet* genes have now been generated in mice. Surprisingly, *Tet1*-null mice are viable (Dawlaty et al., 2011), indicating that the defects observed in TET1-deficient ES cells do not translate into developmental failure *in vivo*. Nevertheless, *Tet1*−/− mice display a smaller body size at birth, which might reflect a developmental delay. *Tet1*−/− mice are also subfertile, which revealed that TET1 is crucial for the progression through meiosis (Yamaguchi et al., 2012). TET2 plays key roles in hematopoiesis, as first evidenced by the discovery that *TET2* mutations are frequent in human myeloid malignancies such as myeloproliferative neoplasms, chronic myelomonocitic leukemia (CMML), myelodisplastic syndrome, and AML (Abdel-Wahab et al., 2009; Delhommeau et al., 2009; Langemeijer et al., 2009), as well as in B- and T-cell lymphomas (Quivoron et al., 2011). Consistent with TET2 being a key regulator of hematopoiesis, *Tet2* knockout in mice does not affect embryonic development but leads to severe hematopoietic defects (Ko et al., 2011; Li et al., 2011; Moran-Crusio et al., 2011; Quivoron et al., 2011). Mice first show an enlargement of the hematopoietic progenitor cell pool and an increased self-renewal capacity of hematopoietic stem cells, which ultimately leads to CMML-like myeloid malignancies at 4–6 months of age and death. Finally, TET3 is highly expressed in oocytes and has been implicated in epigenetic reprogramming in the zygote. Embryos derived from *Tet3*-deficient oocytes have an increased incidence of developmental failures and fail to properly erase 5mC from the paternal genome in the preimplantation embryos (see Section 5.1; Gu et al., 2011). Even though the levels of 5hmC are the highest in the adult brain, no neurological phenotypes have so far been reported in *Tet*-knockout mice.

# 3. PATTERNS OF DNA METHYLATION AND HYDROXYMETHYLATION DURING DEVELOPMENT

## 3.1. Distribution of 5mC in mammalian genomes

In mammals, cytosine methylation is found almost exclusively in the context of CpG dinucleotides. CpGs occur with a low frequency in vertebrate genomes, that is, on average less than one quarter of the expected frequency in the human and mouse genomes (Fig. 2.3A), because they are progressively lost over evolutionary time by spontaneous deamination of 5mC into thymines in the germ line. In accordance with this model, organisms without DNA methylation (such as *Drosophila melanogaster*) show no signs of CpG depletion (Fig. 2.3A) and methylated sequences in sperm show higher rates of CpG loss during primate evolution (Molaro et al., 2011; Weber et al., 2007). Mammalian genomes contain only few areas with higher CpG density, termed "CpG islands," which are typically 0.5–2-kb long and colocalize with 60–70% of annotated gene promoters, including most housekeeping genes and many tissue-specific genes. These CpG islands do not result from a positive selection for CpGs but rather from a neutral effect reflecting the fact that they are unmethylated in the germ line and therefore not prone to CpG loss during evolution (Cohen, Kenigsberg, & Tanay, 2011).

In order to better define the distribution of CpG methylation, technologies have emerged to map 5mC genome wide with restriction enzymes, affinity purification, or sodium bisulfite treatment in combination with microarrays or high-throughput sequencing (Laird, 2010). The gold-standard approach is now to perform chemical treatment with sodium bisulfite, which converts unmethylated cytosines into uracils while leaving methylated cytosines unchanged, followed by high-throughput sequencing to generate quantitative methylation maps at a single-nucleotide resolution, with the caveat that it does not discriminate between 5mC and 5hmC. The first complete cytosine methylation maps have been generated in mouse and human somatic cells and revealed a bimodal distribution: CpG-poor sequences in exons, introns, and intergenic regions are heavily methylated

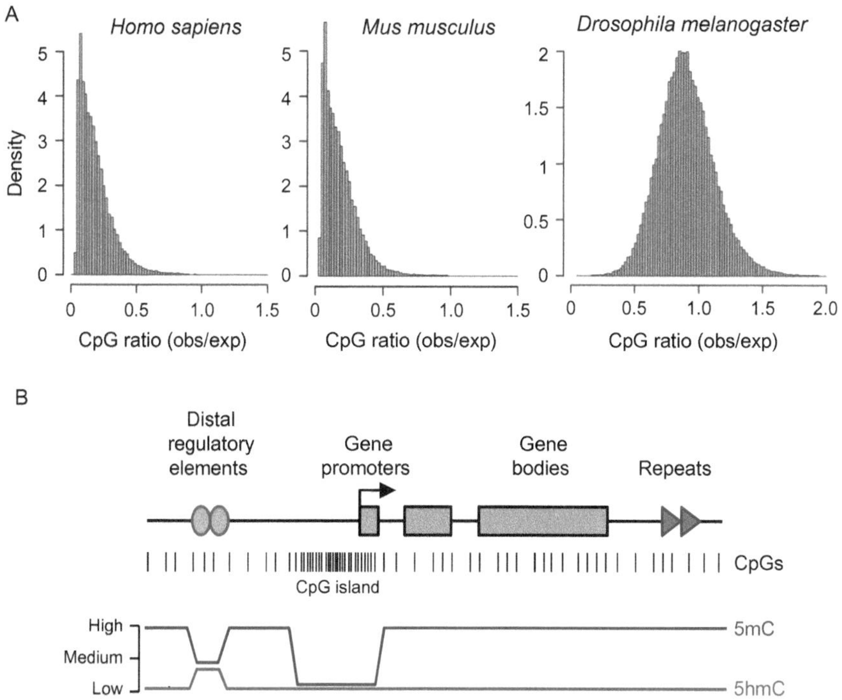


**Figure 2.3** Distribution of CpG methylation and hydroxymethylation in mammalian genomes. (A) Distribution of CpG density in animal genomes. The graphs show the distribution of CpG ratios (observed over expected) measured in 1 kb windows in human, mouse, and *Drosophila* genomes, which illustrates CpG depletion in mammals. The CpG ratio was calculated using the following formula: (*n*CpGs × bp)/(*n*Cs × *n*Gs). (B) Schematic distribution of CpGs, 5mC levels, and 5hmC levels in mouse ES cells. CpGs are locally enriched at CpG islands that are found in many gene promoters. Most CpGs in the genome are hypermethylated except those in CpG islands that are generally unmethylated. CpGs in distal regulatory regions are moderately methylated and show correspondingly elevated levels of 5hmC. (For color version of this figure, the reader is referred to the online version of this chapter.)

(80–90% of CpGs contain a 5mC), whereas CpG islands remain mostly unmethylated (Laurent et al., 2010; Lister et al., 2009; Stadler et al., 2011; Xie et al., 2012; Fig. 2.3B). Cytosine methylation is also abundant in all classes of repeat elements. Advances in methylome profiling also identified low levels of methylation in non-CpG contexts, predominantly in ES cells, oocytes, and brain (Arand et al., 2012; Laurent et al., 2010; Lister et al., 2009; Smith et al., 2012; Stadler et al., 2011; Tomizawa et al., 2011; Xie et al., 2012; Ziller et al., 2011). This suggests that two factors contribute to nonsymmetrical methylation in mammals: either continuous high levels of DNMT3 enzymes or lack of DNA replication (as in oocytes and neurons). It is, however, unknown if this non-CpG methylation plays any functional role. Surprisingly, although initial experiments suggested that intragenic-methylated CpGs inhibit transcriptional elongation (Lorincz, Dickerson, Schmitt, & Groudine, 2004), gene-body methylation positively correlates with gene expression (Ball et al., 2009; Hellman & Chess, 2007; Kobayashi et al., 2012; Laurent et al., 2010). One possible explanation is that intragenic methylation favors elongation by inhibiting transcription from alternative promoters in gene bodies (Illingworth et al., 2010; Maunakea et al., 2010), but this remains to be further studied.

Recently, a class of low-methylated regions (LMRs) was described in mouse ES cells. LMRs have on average 30% CpG methylation, are CpG poor, and correspond to distal regulatory regions such as enhancers, DNase I hypersensitivity sites, or CTCF-binding sites (Fig. 2.3B; Stadler et al., 2011). These sequences undergo demethylation as a consequence of transcription factor binding, which suggests that DNA methylation is not causally involved in their regulation but could potentiate their activity once transcription factors are bound (Serandour et al., 2011; Stadler et al., 2011). Similar to enhancers, there is also evidence that high CpG methylation found at many CpG-poor promoters (Meissner et al., 2008; Weber et al., 2007) does not prevent transcription activation and can be reversed. For example, demethylation at CpG-poor promoters has been observed following transcriptional activation by hormones (Kim et al., 2009; Metivier et al., 2008), and in muscle (Lucarelli, Fuso, Strom, & Scarpa, 2001), liver (Nagae et al., 2011; Waterland et al., 2009), hematopoietic cells (Calvanese et al., 2012), differentiation of monocytes into dendritic cells (Klug et al., 2010), and neurons upon neuronal activation (Guo, Ma, et al., 2011), which does in some but not all genes correlate with gene activation. Interestingly, some of these examples are nondividing cells, which indicates that demethylation must be replication independent.

In contrast to the rest of the genome, CpG islands are mostly unmethylated. This is particularly evident for CpG island promoters of protein-coding genes that are constitutively unmethylated even when the associated gene is not expressed (Meissner et al., 2008; Weber et al., 2007), whereas intergenic and intragenic CpG islands are more susceptible to DNA methylation (Illingworth et al., 2010; Maunakea et al., 2010). It is unclear why CpG islands are refractory to DNA methylation, one possibility being that it has evolved as a way to ensure that promoters of housekeeping genes remain in a transcriptionally competent state. There are only few exceptions of methylated CpG island promoters, for example, at imprinted and pluripotency-related genes, which in this case is associated with long-term transcription silencing (see Section 4.2). What are the mechanisms that exclude DNA methylation from CpG islands? There are probably multiple factors which are not mutually exclusive (Deaton & Bird, 2011). One factor is the underlying DNA sequence, as illustrated by the fact that 1-kb promoter fragments recapitulate the DNA methylation pattern of the endogenous sequence when inserted somewhere else in the genome of mouse ES cells (Lienert et al., 2011). This shows that the local DNA sequence is key in determining hypo- or hypermethylation states. This can be mediated by transcription factors like SP1 that bind to sequence motifs in CpG islands and have a protective effect against DNA methylation (Brandeis et al., 1994; Macleod, Charlton, Mullins, & Bird, 1994), either because they create steric hindrance or because transcription itself inhibits DNA methylation. Another possibility is that the particular epigenetic state of CpG islands is mediated by CXXC proteins that bind unmethylated CpG-rich DNA. One candidate is CXXC finger protein 1 (CFP1), a protein that is part of the mammalian SET1 H3K4 methyltransferase complex. CFP1 binds to unmethylated CpG islands and recruits methylation of the lysine 4 of histone H3 (H3K4) (Thomson et al., 2010), which is found at all CpG islands irrespective of transcription (Meissner et al., 2008; Weber et al., 2007). As methylation of H3K4 has been proposed to inhibit the recruitment of DNMTs to chromatin (Ooi et al., 2007; Zhang, Jurkowska, et al., 2010), it was tempting to speculate that CFP1-dependent H3K4 methylation protects CpG islands from DNA methylation. However, depletion of CFP1 alone is not sufficient to increase DNA methylation at CpG islands in ES cells, suggesting that other factors cooperate with H3K4 methylation (Clouaire et al., 2012). Other CXXC proteins could contribute to the unmethylated state of CpG islands, such as TET1 (as discussed in Section 2.2) or KDM2A. KDM2A binds to CpG islands and removes H3K36me2 (Blackledge et al., 2010), which could contribute to exclude the DNA methylation machinery given that DNMT3A recognizes methylated H3K36 (Dhayalan et al., 2010). Another aspect that deserves attention is the interplay between DNA methylation and nucleosome occupancy (Chodavarapu et al., 2010). In mammalian cells, unmethylated CpG islands are relatively nucleosome free (Fenouil et al., 2012; Kelly et al., 2012), which could render them refractory to the DNA methylation machinery because most of the cellular content of DNMT3A/B is associated with nucleosomes (Gowher, Stockdale, et al., 2005; Jeong et al., 2009). This possibility is supported by the fact that nucleosome insertion precedes DNA methylation at the *OCT4* and *NANOG* loci during cellular differentiation (You et al., 2011).

## 3.2. Distribution of 5hmC in mammalian genomes

In 2009, two groundbreaking studies discovered that TET proteins convert 5mC into 5hmC and that 5hmC is abundant in ES cells and Purkinje neurons (Kriaucionis & Heintz, 2009; Tahiliani et al., 2009). 5hmC was estimated to account for 0.59% and 0.03% of all nucleotides in Purkinje neurons and mouse ES cells, respectively (in comparison 5mC constitutes around 1% of all nucleotides). Other estimates confirmed that 5hmC levels are around 10 times higher in the mouse brain compared to ES cells (Song, Szulwach, et al., 2011). Intriguingly, in contrast to 5mC levels that are relatively equal between cell types, the amount of 5hmC in DNA is variable between cell types and generally below the one observed in ES cells and the nervous system (Globisch et al., 2010; Nestor et al., 2012; Ruzov et al., 2011; Szwagierczak, Bultmann, Schmidt, Spada, & Leonhardt, 2010). The observation that 5hmC is highest in the brain suggests a link with active DNA demethylation because most adult neurons do not divide, and therefore, any demethylation of DNA would require a replication-independent process.

After this discovery, efforts have been concentrated on studying the distribution of 5hmC in the genome. Because standard approaches such as sodium bisulfite sequencing or methylation-sensitive restriction enzymes cannot discriminate 5hmC from 5mC, novel experimental strategies have been developed, which are detailed in some recent reviews (Wu & Zhang, 2011). Using affinity purification, several groups reported that 5hmC in ES cells is present in gene bodies; transcription start sites (TSS), and *cis*-regulatory elements, albeit with no information on the absolute amounts (Ficz et al., 2011; Pastor et al., 2011; Song, Szulwach, et al., 2011; Wu, D'Alessio, Ito, Wang, et al., 2011). 5hmC was also profiled by affinity purification in the brain and found to be enriched in gene bodies but to a lesser extent at TSS (Jin, Wu, Li, & Pfeifer, 2011; Szulwach, Li, Li, Song, Wu, et al., 2011). Recently, two modified bisulfite sequencing protocols have been developed to generate for the first time single-base resolution maps of the absolute abundance of 5hmC in mammalian cells (Booth et al., 2012; Yu et al., 2012). One of these studies generated the first genome-wide maps of 5hmC in mouse and human ES cells, which revealed that 5hmC occurs nearly exclusively in CpGs (Yu et al., 2012). Interestingly, 5hmC is most enriched at CpG-poor distal-regulatory regions corresponding to enhancers, CTCF-binding sites or

DNase I hypersensitive sites (Fig. 2.3B). At these sites, 5hmC occurs at up to 10–20% of all CpGs in the cellular population (Yu et al., 2012). This finding is in agreement with other studies that also point toward a preferential distribution of 5hmC at distal enhancers (Serandour et al., 2012; Stroud, Feng, Morey Kinney, Pradhan, & Jacobsen, 2011; Szulwach, Li, Li, Song, Han, et al., 2011); however, it is intriguing that it does not correlate with the preferential binding of TET1 at CpG-rich sequences in ES cells (discussed in Section 2.2). Interestingly, the same sequences have been identified as LMRs in ES cells and show correspondingly lower levels of 5mC (Stadler et al., 2011; Fig. 2.3B), suggesting that a high proportion of 5hmC is linked with demethylation at transcription factor-binding sites. Future studies should help to clarify whether this is a general feature in all cell types and what mechanisms govern the dynamics of 5mC and 5hmC at distal-regulatory regions.

## 3.3. Reprogramming of DNA methylation during preimplantation development

Patterns of DNA methylation undergo drastic changes during early embryonic development in mammals. Early studies in mice using immunofluorescence and restriction enzymes revealed that DNA methylation inherited from the gametes is globally erased after fertilization from the first cleavage stages up to the blastocyst, before being reestablished after implantation (Monk, Boubelik, & Lehnert, 1987; Santos, Hendrich, Reik, & Dean, 2002; Fig. 2.4). This process occurs via multiple mechanisms: the paternal DNA undergoes rapid loss of 5mC in the zygote before the first cell division, whereas the maternal DNA is demethylated over several cell divisions, probably as a result of lack of maintenance activity (Saitou, Kagiwada, & Kurimoto, 2012). There are still debates on whether the demethylation in the male pronucleus also occurs with similar dynamics and amplitude in other mammals such as sheep or rabbit (Lepikhov et al., 2008; Reis Silva et al., 2011). Several groups have now overcome the difficulty to work with small amounts of cells and applied genome-wide bisulfite sequencing to quantify the dynamics of DNA methylation in mouse early cleavage-stage embryos (Kobayashi et al., 2012; Smallwood et al., 2011; Smith et al., 2012). These studies revealed that oocytes have a relatively hypomethylated genome compared to sperm, and that upon fertilization methylation levels decrease to reach a low point in preimplantation blastocysts (Fig. 2.4). Even though bisulfite sequencing cannot distinguish the kinetics of 5mC and 5hmC, these studies confirm that a major epigenetic reprogramming event occurs after fertilization. After implantation of the embryo, DNA methylation is progressively reestablished and is

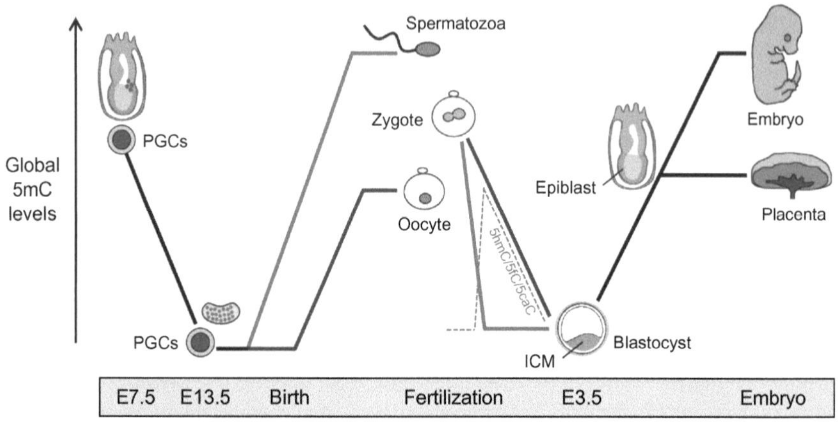


**Figure 2.4** Reprogramming of DNA methylation during mouse development. The graph depicts the variations in global DNA methylation levels during mouse development. Global levels are extrapolated from immunofluorescence and genome-wide studies and do not necessarily reflect the dynamics of all individual sequences. PGCs lose global DNA methylation until E13.5. Methylation is then restored in a sex-dependent manner before birth in spermatozoa and after birth during oocyte growth. After fertilization, the paternal genome (blue line) undergoes rapid erasure of 5mC and the maternal genome (red line) undergoes slow erasure to reach a low point in the blastocyst. Intermediates of demethylation (5hmC, 5fC, 5caC, blue dotted line) appear on the paternal genome and are gradually diluted during preimplantation development. After implantation, DNA methylation levels increase in the embryonic tissues derived from the epiblast and to a mild level in extraembryonic placental structures. (For interpretation of the references to color in this figure legend, the reader is referred to the online version of this chapter.)

recruited by default to CpG-poor regions and to a small set of CpG islands (Fig. 2.4). There are, however, sequences that escape demethylation after fertilization, in particular, imprinted loci and repeats such as intracisternal A particles (IAPs) (Lane et al., 2003), L1Md_A elements (Smith et al., 2012), and LTR ERV1 elements (Guibert, Forne, & Weber, 2012). There is also an unanticipated high number of CpG islands that maintain oocyte-derived DNA methylation in preimplantation embryos (Borgel et al., 2010; Kobayashi et al., 2012; Smallwood et al., 2011; Smith et al., 2012). Many of these regions do not maintain maternal-specific DNA methylation after implantation because they resolve to unmethylation (Smith et al., 2012) or because paternal alleles undergo *de novo* methylation at the time of implantation (Borgel et al., 2010; Proudhon et al., 2012), indicating that gametic DNA methylation rarely translates into imprinted methylation.

### 3.4. Reprogramming of DNA methylation in gametes

In mice, the primordial germ cells (PGCs) are specified from posterior proximal epiblast cells at embryonic day (E) 6.25 and express specific markers such as *Prdm1* (also known as *Blimp1*), *Prdm14*, and *Dppa3* (also known as *PGC7* and *Stella*). PGCs form a cluster of around 40 cells at E7.25 and then colonize the genital ridges by E10.5, where they continue to proliferate until E13.5. Because PGCs are derived from somatic cells that have already acquired a somatic epigenetic profile, they undergo a global resetting leading to repression of somatic genes, changes in histone modifications, exchange of histone variants, and global DNA demethylation (Hackett, Zylicz, & Surani, 2012; Saitou et al., 2012; Fig. 2.4). The genome-wide demethylation in PGCs is evidenced by a global loss of 5mC immunostaining signal from approximately E8.0 (Seki et al., 2005). PGCs also repress *DNMT3A/B* as well as *UHRF1*, a factor required to maintain DNA methylation patterns during replication (Kurimoto et al., 2008; Yabuta, Kurimoto, Ohinata, Seki, & Saitou, 2006). Global mapping by bisulfite sequencing or immunoprecipitation confirmed that E13.5 PGCs have a demethylated genome compared to somatic cells (Guibert et al., 2012; Seisenberger et al., 2012). This includes erasure of methylation at imprinting loci, gene bodies, intergenic regions, mobile elements as well as virtually all CpG island promoters found methylated in early embryos (Guibert et al., 2012; Seisenberger et al., 2012), indicating that the extent of methylation erasure in PGCs is more complete than in preimplantation embryos. The kinetics of demethylation varies depending on the target sequence: certain sequences like ICRs (imprinting control regions) maintain high levels of cytosine methylation until E10.5, whereas most other sequences initiate slow demethylation in early PGC precursors around E8.0 that probably results from a passive lack of maintenance by DNMT1 (Guibert et al., 2012; Seisenberger et al., 2012). These successive waves of demethylation might reflect the existence of multiple mechanisms being used in PGCs to achieve full demethylation (Hackett et al., 2012). As DNA methylation mediates stable silencing of pluripotency-related and meiotic genes (see Section 4.2), this global demethylation in PGCs is necessary to restore an epigenetic state compatible with the expression of the germ-line program. It might also serve to erase deleterious epimutations even though there are rare single-copy and repeated sequences that maintain high levels of methylation in PGCs and could potentially mediate transgenerational epigenetic inheritance (Guibert et al., 2012; Lane et al., 2003; Seisenberger et al., 2012).

### 3.5. Global demethylation in erythropoiesis

Once established in somatic cells, DNA methylation levels were thought to be very stable. This view has been challenged by the surprising discovery that mouse erythropoiesis is associated with global DNA demethylation *in vivo* (Shearstone et al., 2011). The formation of erythrocytes (enucleated red cells) first occurs from erythroid progenitors in the murine fetal liver between E11 and E15 and can be followed by isolating cellular populations at different stages of maturation with surface markers. Using global measurements by ELISA, the authors observed a striking loss of global CpG methylation from 70% to 30% during differentiation in the course of around three cell divisions. Using reduced representation bisulfite sequencing that provides quantitative mapping of 5mC in a large portion of CpG-rich regions, the authors confirmed that there is a progressive decrease in methylation levels from 79% to 55% in the covered regions. Demethylation affects a broad range of sequences including promoters, enhancers, repeats, and even imprinted loci, which is not associated with global changes in gene expression. Further mechanistic studies showed that this demethylation requires cell division and therefore probably reflects a compromised maintenance of DNA methylation during DNA replication (Shearstone et al., 2011). This shows that global variations in DNA methylation not only occur in preimplantation embryos and PGCs but also during somatic cell differentiation.

## 4. ROLE OF CYTOSINE METHYLATION IN GENOME REGULATION

### 4.1. Genomic imprinting and X inactivation

Early experiments showing that mammalian gynogenetic and androgenetic embryos are not viable indicated that the maternal and paternal genomes are not equivalent (McGrath & Solter, 1984). This is attributed to a subset of genes subjected to genomic imprinting, which are expressed only from the paternal or maternal allele. To date, more than 100 imprinted genes have been reported, and these are often organized in small clusters. The imprinted expression of these genes depends on allele-specific DNA methylation in CpG-rich regions called ICRs (imprinted control region). These acquire differential DNA methylation in the gametes and have the unique property to maintain it in the embryo by escaping demethylation on one allele and *de novo* methylation on the other allele. DNA methylation at ICRs is established in the gametes by DNMT3A and DNMT3L (Bourc'his et al.,

2001; Hata, Okano, Lei, & Li, 2002; Kaneda et al., 2004; Kato et al., 2007) and then maintained during early embryogenesis by DNMT1 (Hirasawa et al., 2008). These ICRs sit directly in the promoter of the imprinted gene or correspond to insulators and promoters of noncoding RNAs that regulate imprinted expression in *cis* (Barlow, 2011). The study of imprinted genes was pioneering because it helped to propose novel paradigms; in particular, that DNA methylation is not only repressive but can also stimulate gene expression by regulating *cis*-acting repressors. Imprinting methylation marks are then erased in PGCs of the offspring and set back again according to the sex of the embryo, thereby completing the cycle. The key role of DNA methylation in genomic imprinting is demonstrated by the fact that imprinted expression is perturbed in mice lacking DNMT1 or derived from gametes that failed to establish DNA methylation (Bourc'his et al., 2001; Kaneda et al., 2004; Li, Beard, & Jaenisch, 1993). Genome-wide DNA methylation profiling is now used to discover more imprinted loci and indicates that only few gametic ICRs remain to be discovered (Proudhon et al., 2012; Xie et al., 2012), although the precise inventory will require the systematic investigation of many tissues as imprinting can occur in a tissue-specific manner (Prickett & Oakey, 2012). The key questions are now to elucidate the mechanisms that establish and maintain differential gametic DNA methylation, which involves multiple factors such as DNA sequence, histone modifications, and transcription (Abramowitz & Bartolomei, 2012; Chotalia et al., 2009; Ciccone et al., 2009; Li et al., 2008; Nakamura et al., 2012; Quenneville et al., 2011).

DNA methylation has also been implicated in the inactivation of one X chromosome in female cells. By making one of the two X chromosomes transcriptionally silent in female cells, this process equalizes X-chromosome gene dosage between males and females. In mice, X-chromosome inactivation first occurs on the paternal X chromosome in the female cleavage-stage embryo, and subsequently in a random manner after implantation (Barakat & Gribnau, 2012). Similar to genomic imprinting, X-chromosome inactivation depends on a *cis*-acting X-inactivation center that encodes the Xist noncoding RNA, which coats the inactive X chromosome and initiates a cascade of events that include exclusion of the transcription machinery, recruitment of chromatin-modifying complexes, and DNA methylation of promoter-associated CpG islands (Barakat & Gribnau, 2012). The inactive X chromosome also shows less DNA methylation in gene bodies compared to the active X (Hellman & Chess, 2007), which further reinforces the link between intragenic DNA methylation and transcription (see Section 3.1). Several studies indicate that CpG island DNA methylation at many genes is a late event in the process of X inactivation (Gendrel et al., 2012; Lock, Takagi, & Martin, 1987). In addition, mutant mouse embryos with defective DNA methylation only show modest effects on the silencing of X-linked genes (Blewitt et al., 2008; Sado et al., 2000; Sado, Okano, Li, & Sasaki, 2004). Altogether, this suggests that DNA methylation is not the initiating event in X-chromosome inactivation but might be required to maintain long-term silencing of the inactive X copy.

## 4.2. Regulation of lineage-specific gene expression

The role of DNA methylation in orchestrating gene expression in the embryo has been debated for a long time after initial experiments indicated that CpG island promoters remain constitutively unmethylated irrespective of gene expression. This dogma has been challenged by genome-scale mapping experiments that identified examples of nonimprinted genes with methylated CpG-rich promoters in mouse and human somatic cells. Many of these genes are pluripotency and germ line-specific genes that have constitutively methylated promoters in somatic lineages (Farthing et al., 2008; Mohn et al., 2008; Shen et al., 2007; Weber et al., 2007). These promoters undergo *de novo* methylation at the time of implantation, and reduced methylation at these genes leads to incomplete gene silencing in the embryo (Borgel et al., 2010; Velasco et al., 2010). Using ES cells as an experimental system, it has been possible to dissect the process of *de novo* methylation at pluripotency genes. One well-studied example is *Oct4*, a pluripotency marker that has a promoter with moderate CpG richness. Interestingly DNA methylation in this promoter does not initiate silencing in differentiating ES cells but is a late event that follows histone deacetylation, recruitment of the histone methyltransferase G9a, methylation of H3K9, and binding of HP1 (Feldman et al., 2006). Once DNA methylation has occurred, the *Oct4* gene can no longer be reversed to an active state in culture, suggesting that DNA methylation serves to stabilize the silent state (Fig. 2.5; Epsztejn-Litman et al., 2008; Feldman et al., 2006). From these experiments, one can postulate that DNA methylation has evolved as a secondary safeguarding event that prevents the potentially deleterious reactivation of germ line and pluripotency genes in differentiated somatic cells. An illustration of the role of DNA methylation in repressing pluripotency is that changes in DNA methylation are necessary for the successful reprogramming of differentiated cells into induced pluripotent stem cells. In particular, reprogramming entails the erasure of promoter DNA

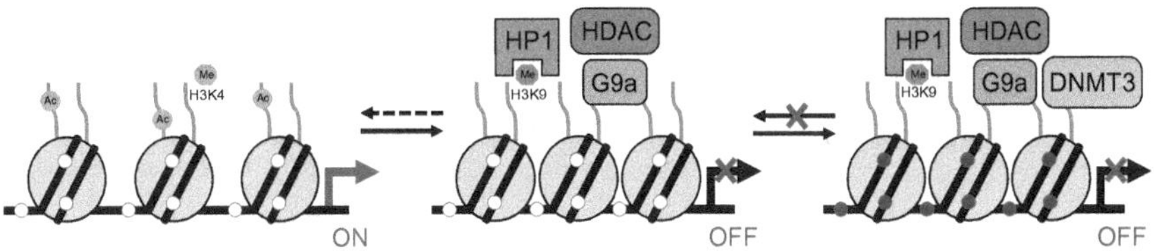


**Figure 2.5** Epigenetic regulation of the *Oct-4* promoter during development. Transcriptionally active *Oct-4* promoter in pluripotent cells harbors acetylated histones, H3K4me, and no CpG methylation (white circles). During differentiation of ES cells, the gene promoter is silenced and undergoes deacetylation by histone deacetylases (HDAC), methylation of H3K9 by methyltransferase G9a, and recruitment of HP1. Silencing is still reversible until CpG methylation (red circles) catalyzed by DNMT3 locks the promoter in an inactive state. (For interpretation of the references to color in this figure legend, the reader is referred to the online version of this chapter.)

methylation at pluripotency genes, which, if incomplete, leads to partial reprogramming (Mikkelsen et al., 2008).

How does DNA methylation maintain promoter silencing? One mechanism is that cytosine methylation inhibits the binding of transcription factors. Many transcription factors recognize sequence motifs that contain CpGs (Deaton & Bird, 2011) and some of them show reduced binding affinity when the recognition motif is methylated (Campanero, Armstrong, & Flemington, 2000; Iguchi-Ariga & Schaffner, 1989; Kim, Kollhoff, Bergmann, & Stubbs, 2003). Another model is that DNA methylation modifies the local chromatin by recruiting 5mC "readers." Proteins that specifically bind methylated DNA have been identified, such as those having a methyl-CpG binding domain (MBD1, MBD2, MBD4, MeCP2) and those recognizing 5mC with zinc-fingers (ZBTB4, ZBTB33, and ZBTB38) (Joulie, Miotto, & Defossez, 2010). Most of these proteins interact with factors that deposit heterochromatin marks, which led to propose that they mediate DNA methylation-dependent gene repression by inducing heterochromatinization at sites of dense DNA methylation. Whereas there is evidence for this model in cancer cells (Lopez-Serra & Esteller, 2012), their role in translating the DNA methylation signal in normal cells still remains to be further clarified because surprisingly little MBD target genes have been identified so far (Joulie et al., 2010). In addition, the several single and combined knockout mice generated so far show no overt phenotype (Martin Caballero, Hansen, Leaford, Pollard, & Hendrich, 2009), which is in clear contrast with the lethality of DNMT-knockout mice.

There are numerous differences in DNA methylation between somatic cell lineages, indicating that further changes in DNA methylation occur in a lineage-specific way after gastrulation and during late fetal and postnatal development. For example, a comprehensive methylome mapping effort in human tissues estimated that 18% of the covered regions show tissue-specific methylation among 16 analyzed tissues (Rakyan et al., 2008). Mapping experiments have also been performed in hematopoietic cells, which offer a convenient experimental system to study cellular differentiation *in vivo*, and revealed widespread changes in DNA methylation associated with hematopoietic cellular commitment (Calvanese et al., 2012; Hodges et al., 2011; Ji et al., 2010). Strikingly, DNA methylation profiles can even be used to segregate hematopoietic cell lineages with a good accuracy (Bock et al., 2012). Some of the tissue-specific DNA methylation occurs at CpG-rich promoters of important developmental genes such as *Hox* or *Pax* genes (Illingworth et al., 2008). Another interesting example is methylation of the *Elf5* gene, which occurs earlier in development. *Elf5* encodes an Ets transcription factor crucial for the development of the trophoblast lineage and its promoter is methylated in embryonic cells but not the trophoblast lineage, which constitutes an epigenetic barrier between these earliest two cell lineages (Ng et al., 2008). These observations suggest that promoter methylation could serve to stabilize lineage decision by inducing stable repression of key developmental genes (Hemberger, Dean, & Reik, 2009). Besides promoters, CpG islands residing outside annotated promoters are more prone to methylation in tissues and could provide additional positive or negative regulation of tissue-specific gene expression in *cis* by regulating insulators or noncoding RNAs (Deaton et al., 2011; Illingworth et al., 2010; Maunakea et al., 2010). Yet the majority of tissue-specific DNA methylation occurs in CpG-poor sequences (Nagae et al., 2011; Rakyan et al., 2008). In this case, because CpG-poor sequences can be demethylated downstream of transcription factor binding (Stadler et al., 2011; see Section 3.1), it cannot be considered a primary cause of lineage-specific gene expression.

## 4.3. Does DNA methylation influence RNA processing?

One exciting recent discovery is that apart from regulating transcription, chromatin epigenetic marks may also influence cotranscriptional RNA processing events (Luco et al., 2010). In line with this idea, sharp transitions in DNA methylation levels were observed at exon–intron boundaries, as well as correlations between exon methylation and alternative splicing, suggesting a possible role of DNA methylation in exon choice (Anastasiadou, Malousi, Maglaveras, & Kouidou, 2011; Chen, Feng, Joo, Jacobsen, & Pellegrini,

2011; Hodges et al., 2009; Laurent et al., 2010; Zhou, Lu, & Tian, 2012). Changes in DNA methylation correlating with alternative splicing have also been reported in insects (Bonasio et al., 2012; Foret et al., 2012; Lyko et al., 2010). Mechanistically, this could be mediated by methylated DNA-binding proteins that regulate splicing via recruitment of splicing factors (Long, Ooi, Yau, & Jones, 2011; Young et al., 2005). Alternatively, factors like CTCF, which binds to DNA in a methylation-dependent manner, could link DNA methylation to splicing by causing local pausing of RNA polymerase (Shukla et al., 2011). Another way for generating transcript diversity is the use of alternative polyadenylation sites. In 2008, a study identified an unusual imprinted gene that utilizes alternative polyadenylation sites in an allele-specific manner (Wood et al., 2008). Intriguingly, this correlates with DNA methylation of a CpG island localized between the two poly(A) sites: hypomethylation of the CpG island correlates with the use of upstream poly(A) sites, whereas hypermethylated alleles preferentially utilize downstream poly(A) sites. In addition, alternative polyadenylation is lost in embryos lacking maternally derived methylation imprints. Another similar case was reported recently (Cowley, Wood, Bohm, Schulz, & Oakey, 2012), and although the mechanisms involved are not clear, this suggests that DNA methylation could also influence polyadenylation.

## 4.4. Maintenance of genome integrity

In addition, another well-documented function of DNA methylation is to repress transcription from transposable elements. These elements, considered as "parasitic," have the ability to move in the genome and therefore need to be controlled to prevent mutagenesis. A big proportion of the methylated CpGs are found in transposable elements and contribute to maintain them in a silent state. In mouse embryos lacking DNMT1, transcription from IAPs, a family of active retrotransposons, is massively increased (Walsh, Chaillet, & Bestor, 1998). Similarly, failure to establish DNA methylation at transposons in male germ cells of DNMT3L-knockout mice leads to transposon reactivation and meiotic failure in spermatocytes (Bourc'his & Bestor, 2004). Because of the increased mutability of methylated CpGs, DNA methylation also serves to permanently silence mobile elements by mutations. Finally, as it is well established that loss of DNA methylation leads to increased genome instability (Dodge et al., 2005; Xu et al., 1999), it is also possible that methylation of interspersed transposons or tandem repeats at centromeric heterochromatin could protect against illegitimate recombination (Bourc'his & Bestor, 2004; De La Fuente et al., 2006). The role of DNA methylation in suppressing mobile elements is also evident in plants and fungi, which had led to speculate that DNA methylation may have primarily evolved for the purpose of repressing mobile elements. This view is, however, challenged by recent data showing that repeats are not methylated in many invertebrate species like the honey bee, silkworm, ants, or sea squirts (Bonasio et al., 2012; Feng et al., 2010; Lyko et al., 2010; Xiang et al., 2010; Zemach, McDaniel, Silva, & Zilberman, 2010).

# 5. EMERGING FUNCTIONS OF CYTOSINE HYDROXYMETHYLATION

## 5.1. Role in DNA methylation erasure

The discovery of the catalytic properties of TET proteins led to the hypothesis that TET-mediated 5mC oxidation could initiate DNA demethylation, both by passive and by active mechanisms. This has been well documented in the context of postfertilization reprogramming. Using specific antibodies against the diverse form of modified cytosines, immunofluorescence studies have revealed that the loss of 5mC signal in the zygote on the paternal genome coincides with the appearance of 5hmC (Iqbal, Jin, Pfeifer, & Szabo, 2011; Wossidlo et al., 2011), as well as 5fC and 5caC (Inoue et al., 2011; Fig. 2.4). The appearance of 5hmC in the paternal pronucleus is also observed in bovine and rabbit zygotes, which suggests an evolutionary conserved process (Wossidlo et al., 2011). In addition, inactivation of TET3 in the zygote by conditional knockout or RNAi impairs the formation of 5hmC and the erasure of DNA methylation measured by immunofluorescence and bisulfite sequencing at LINE1 transposons and genes *Oct4*, *Nanog*, and *Lemd1* (Gu et al., 2011; Wossidlo et al., 2011). This is associated with developmental failures in the embryo, probably due to the persistent DNA methylation that interferes with the activation of early embryonic genes (Gu et al., 2011). Altogether, this reveals a pathway in which TET-mediated conversion of 5mC to 5hmC initiates DNA methylation erasure in preimplantation embryos. TET1 and TET2 are also expressed in PGCs and TET-mediated oxidation occurs in PGCs at the time of global demethylation (Hackett et al., 2013). However, the TET1 knock-out in mice revealed that TET1 is not essential for the genome-wide demethylation in PGCs but participates in the demethylation of meiotic genes (Yamaguchi et al., 2012).

This suggests that TET enzymes have a locus-specific role in demethylation in PGCs or that TET2 compensates for the absence of TET1, which needs to be further investigated. One might also predict that TET2 is involved in DNA demethylation during hematopoietic differentiation, but this has also not been studied.

What are the possible mechanisms for TET-mediated DNA demethylation? One scenario is that the oxidation derivatives lead to passive demethylation upon DNA replication (Fig. 2.6). This is supported by the observation that 5hmC, 5fC, and 5caC signals in preimplantation embryos persist in the cleavage-stage embryos and are gradually diluted with each cell division (Inoue & Zhang, 2011; Inoue et al., 2011). In addition, it has also been shown that DNMT1 is not active on hemihydroxymethylated DNA (Hashimoto et al., 2012). Alternatively, TET-mediated oxidation of 5mC could lead to

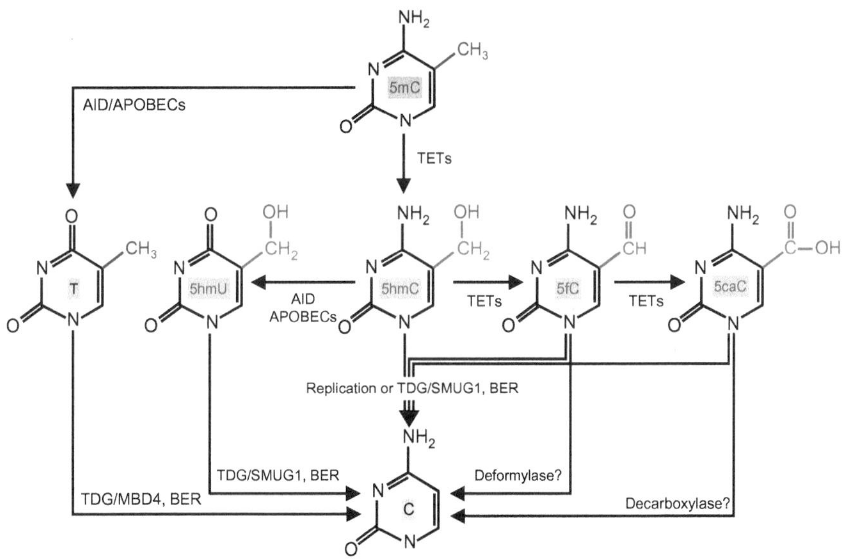


**Figure 2.6** Potential chemical pathways for active cytosine demethylation. One potential mechanism for active replication-independent demethylation is deamination of 5mC into thymine (T) by cytosine deaminases AID or APOBEC, followed by excision of the thymine by DNA glycosylases TDG or MBD4 and base excision repair (BER). 5mC can also be converted into 5hmC by TET enzymes, which can further be oxidized into 5-formylcytosine (5fC) and 5-carboxylcytosine (5caC), or deaminated into 5-hydroxymethyluracil (5hmU). All these modified bases could lead to replication-dependent passive demethylation or active demethylation via excision by DNA glycosylases (TDG, SMUG1) followed by BER. In addition, a putative decarboxylase or deformylase could convert 5caC or 5fC directly back to cytosine, but such enzymes have not yet been discovered. (See Color Insert.)

replication-independent active demethylation involving DNA repair (Fig. 2.6). Recent studies have shown that the DNA glycosylase TDG (thymine DNA glycosylase) can efficiently remove 5caC and 5fC from DNA (He et al., 2011; Maiti & Drohat, 2011), which suggests that 5caC or 5fC could trigger removal by TDG followed by base excision repair (BER). TDG, however, seems to have no activity on 5hmC (Maiti & Drohat, 2011). In support of a role of TDG in epigenetic reprogramming, *Tdg−/−* mice show embryonic lethality and hypermethylation of a number of developmental genes (Cortazar et al., 2011; Cortellino et al., 2011). Another possible route initiated by the TET proteins is deamination of 5hmC by the AID/APOBEC family of deaminases, thereby creating 5hmU that could be removed by BER-mediated DNA repair (Fig. 2.6). This scenario is supported by recent findings showing that DNA glycosylases TDG and SMUG1 (single-strand-selective monofunctional uracil-DNA glycosylase 1) can excise 5hmU from DNA (Cortellino et al., 2011; Guo, Su, Zhong, Ming, & Song, 2011). It has also been shown that AID/APOBEC deaminases use 5hmC as a preferential substrate and promote 5hmC-dependant demethylation of a reporter construct in combination with BER enzymes in HEK293 cultured cells (Guo, Su, et al., 2011). In the same study, the authors also provide experimental evidence that TET1 and AID/APOBEC deaminases participate in neuronal activity-induced demethylation of endogenous targets (Guo, Su, et al., 2011). The physiological relevance of this pathway still needs to be clarified as these deaminases could also act directly on 5mC and trigger demethylation independent of TET activity (Fig. 2.6). In addition, *Aid−/−* mice are viable and only show a moderate defect in DNA methylation erasure in PGCs (Popp et al., 2010), indicating that AID-mediated deamination is only responsible for a small part of demethylation *in vivo*. Finally, one last scenario is that a deformylase or decarboxylase directly converts 5fC or 5caC to C (Fig. 2.6). In summary, experimental evidence point toward a variety of putative TET-dependent demethylation pathways. Additional investigations are needed to determine their relative contribution to physiological demethylation processes, which is currently difficult because the precise kinetics of the modified cytosine forms cannot be determined as traditional bisulfite sequencing does not discriminate between 5mC and 5hmC, or 5caC and C.

## 5.2. Putative roles in gene regulation

In addition to being an intermediate of demethylation, 5hmC might also function as a stable signal that modulates the binding of chromatin effectors and influences gene expression. Because a transient intermediate of demethylation would not be expected to be detected at high levels, this proposition could explain the relatively high steady-state levels of 5hmC measured in various cell types. One possible role is that 5hmC modulates the binding of methyl-CpG binding proteins. Interestingly, several members of this family (MBD1, MBD2, MBD4) do not recognize 5hmC *in vitro* (Jin, Kadam, & Pfeifer, 2010), which suggests that 5mC oxidation could serve to displace MBD binding and relieve transcription silencing. Inversely, MBD3, another member of the MBD family that was long known to have low-binding affinity for 5mC, was proposed to bind to 5hmC-enriched sequences in mouse ES cells (Yildirim et al., 2011). Knockdown of TET1 impairs MBD3 binding to its genomic targets, suggesting that MBD3 binding to chromatin is 5hmC dependent (Yildirim et al., 2011). Another study identified MeCP2 as a major 5hmC-binding protein in brain cells (Mellen, Ayata, Dewell, Kriaucionis, & Heintz, 2012), although earlier studies reported that MeCP2 does not efficiently bind to 5hmC-containing DNA *in vitro* (Valinluck et al., 2004). Finally, quantitative mass-spectrometry-based proteomics recently identified many novel proteins that recognize 5hmC in mouse embryonic and neuronal cells (Spruijt et al., 2013), which further suggests that 5hmC could affect gene expression states by modulating the binding of protein complexes to chromatin. In the future, it will be of great interest to characterize these 5hmC "readers" and study the stability of 5hmC in the cell and potential mechanisms that maintain 5hmC during DNA replication.

## 6. CONCLUSIONS

The research in the past years has led to tremendous progress in the characterization of cytosine methylation patterns in mammalian genomes, as well as the discovery that other forms of modified cytosines play important roles in epigenetic reprogramming and signaling. Whole-genome approaches give us a detailed view of DNA modifications at high resolution and reveal that DNA methylation profiles are more dynamic than previously anticipated and can be regulated upon a variety of stimuli. At CpG islands, DNA methylation is generally not the initiating event in gene silencing but reinforces long-term silencing. Further studies are now needed to understand the role of DNA methylation at other sequences (gene bodies and enhancers), solve the cause/consequence debates, and elucidate the mechanisms that establish or remove methylation marks at specific targets. This is important for a better understanding of not only the molecular mechanisms of development, but also the etiology of many human diseases such as cancer. Cancer cells show both genome-wide hypomethylation and aberrant hypermethylation of CpG islands (Sandoval & Esteller, 2012), as well as abnormal patterns of 5hmC (Jin, Jiang, et al., 2011). The causal role of cytosine modifications in cancer has become even more evident with the discovery that mutations in enzymes that methylate (DNMT3A) and hydroxymethylate (TET2) DNA are frequent in leukemia (Delhommeau et al., 2009; Ley et al., 2010). Therefore, deciphering the role and regulation of DNA methylation and hydroxymethylation in development will also be instrumental in understanding epigenetic deregulation in disease, which can ultimately lead to the development of novel strategies and drugs that target the epigenome.

## ACKNOWLEDGMENTS

Research in our laboratory is supported by the Centre National de la Recherche Scientifique (CNRS), the University of Strasbourg, the ATIP-AVENIR program, the Agence Nationale de la Recherche, the PNRPE program, and the EpiGeneSys network of Excellence.

## REFERENCES

Abdel-Wahab, O., Mullally, A., Hedvat, C., Garcia-Manero, G., Patel, J., Wadleigh, M., et al. (2009). Genetic characterization of TET1, TET2, and TET3 alterations in myeloid malignancies. *Blood*, *114*, 144–147.

Abramowitz, L. K., & Bartolomei, M. S. (2012). Genomic imprinting: Recognition and marking of imprinted loci. *Current Opinion in Genetics and Development*, *22*, 72–78.

Anastasiadou, C., Malousi, A., Maglaveras, N., & Kouidou, S. (2011). Human epigenome data reveal increased CpG methylation in alternatively spliced sites and putative exonic splicing enhancers. *DNA and Cell Biology*, *30*, 267–275.

Arand, J., Spieler, D., Karius, T., Branco, M. R., Meilinger, D., Meissner, A., et al. (2012). In vivo control of CpG and non-CpG DNA methylation by DNA methyltransferases. *PLoS Genetics*, *8*, e1002750.

Ball, M. P., Li, J. B., Gao, Y., Lee, J. H., LeProust, E. M., Park, I. H., et al. (2009). Targeted and genome-scale strategies reveal gene-body methylation signatures in human cells. *Nature Biotechnology*, *27*, 361–368.

Barakat, T. S., & Gribnau, J. (2012). X chromosome inactivation in the cycle of life. *Development*, *139*, 2085–2089.

Barlow, D. P. (2011). Genomic imprinting: A mammalian epigenetic discovery model. *Annual Review of Genetics*, *45*, 379–403.

Berger, S. L., Kouzarides, T., Shiekhattar, R., & Shilatifard, A. (2009). An operational definition of epigenetics. *Genes and Development*, *23*, 781–783.

Blackledge, N. P., Zhou, J. C., Tolstorukov, M. Y., Farcas, A. M., Park, P. J., & Klose, R. J. (2010). CpG islands recruit a histone H3 lysine 36 demethylase. *Molecular Cell*, *38*, 179–190.

Blewitt, M. E., Gendrel, A. V., Pang, Z., Sparrow, D. B., Whitelaw, N., Craig, J. M., et al. (2008). SmcHD1, containing a structural-maintenance-of-chromosomes hinge domain, has a critical role in X inactivation. *Nature Genetics*, *40*, 663–669.

Bock, C., Beerman, I., Lien, W. H., Smith, Z. D., Gu, H., Boyle, P., et al. (2012). DNA methylation dynamics during in vivo differentiation of blood and skin stem cells. *Molecular Cell*, *47*, 633–647.

Bonasio, R., Li, Q., Lian, J., Mutti, N. S., Jin, L., Zhao, H., et al. (2012). Genome-wide and caste-specific DNA methylomes of the ants *Camponotus floridanus* and *Harpegnathos saltator*. *Current Biology*, *22*, 1755–1764.

Booth, M. J., Branco, M. R., Ficz, G., Oxley, D., Krueger, F., Reik, W., et al. (2012). Quantitative sequencing of 5-methylcytosine and 5-hydroxymethylcytosine at single-base resolution. *Science*, *336*, 934–937.

Borgel, J., Guibert, S., Li, Y., Chiba, H., Schubeler, D., Sasaki, H., et al. (2010). Targets and dynamics of promoter DNA methylation during early mouse development. *Nature Genetics*, *42*, 1093–1100.

Bostick, M., Kim, J. K., Esteve, P. O., Clark, A., Pradhan, S., & Jacobsen, S. E. (2007). UHRF1 plays a role in maintaining DNA methylation in mammalian cells. *Science*, *317*, 1760–1764.

Bourc'his, D., & Bestor, T. H. (2004). Meiotic catastrophe and retrotransposon reactivation in male germ cells lacking Dnmt3L. *Nature*, *431*, 96–99.

Bourc'his, D., Xu, G. L., Lin, C. S., Bollman, B., & Bestor, T. H. (2001). Dnmt3L and the establishment of maternal genomic imprints. *Science*, *294*, 2536–2539.

Branco, M. R., Ficz, G., & Reik, W. (2011). Uncovering the role of 5-hydroxymethylcytosine in the epigenome. *Nature Reviews Genetics*, *13*, 7–13.

Brandeis, M., Frank, D., Keshet, I., Siegfried, Z., Mendelsohn, M., Nemes, A., et al. (1994). Sp1 elements protect a CpG island from de novo methylation. *Nature*, *371*, 435–438.

Calvanese, V., Fernandez, A. F., Urdinguio, R. G., Suarez-Alvarez, B., Mangas, C., Perez-Garcia, V., et al. (2012). A promoter DNA demethylation landscape of human hematopoietic differentiation. *Nucleic Acids Research*, *40*, 116–131.

Campanero, M. R., Armstrong, M. I., & Flemington, E. K. (2000). CpG methylation as a mechanism for the regulation of E2F activity. *Proceedings of the National Academy of Sciences of the United States of America*, *97*, 6481–6486.

Chedin, F., Lieber, M. R., & Hsieh, C. L. (2002). The DNA methyltransferase-like protein DNMT3L stimulates de novo methylation by Dnmt3a. *Proceedings of the National Academy of Sciences of the United States of America*, *99*, 16916–16921.

Chen, P. Y., Feng, S., Joo, J. W., Jacobsen, S. E., & Pellegrini, M. (2011). A comparative analysis of DNA methylation across human embryonic stem cell lines. *Genome Biology*, *12*, R62.

Chodavarapu, R. K., Feng, S., Bernatavichute, Y. V., Chen, P. Y., Stroud, H., Yu, Y., et al. (2010). Relationship between nucleosome positioning and DNA methylation. *Nature*, *466*, 388–392.

Chotalia, M., Smallwood, S. A., Ruf, N., Dawson, C., Lucifero, D., Frontera, M., et al. (2009). Transcription is required for establishment of germline methylation marks at imprinted genes. *Genes and Development*, *23*, 105–117.

Ciccone, D. N., Su, H., Hevi, S., Gay, F., Lei, H., Bajko, J., et al. (2009). KDM1B is a histone H3K4 demethylase required to establish maternal genomic imprints. *Nature*, *461*, 415–418.

Clouaire, T., Webb, S., Skene, P., Illingworth, R., Kerr, A., Andrews, R., et al. (2012). Cfp1 integrates both CpG content and gene activity for accurate H3K4me3 deposition in embryonic stem cells. *Genes and Development*, *26*, 1714–1728.

Cohen, N. M., Kenigsberg, E., & Tanay, A. (2011). Primate CpG islands are maintained by heterogeneous evolutionary regimes involving minimal selection. *Cell*, *145*, 773–786.

Cortazar, D., Kunz, C., Selfridge, J., Lettieri, T., Saito, Y., MacDougall, E., et al. (2011). Embryonic lethal phenotype reveals a function of TDG in maintaining epigenetic stability. *Nature*, *470*, 419–423.

Cortellino, S., Xu, J., Sannai, M., Moore, R., Caretti, E., Cigliano, A., et al. (2011). Thymine DNA glycosylase is essential for active DNA demethylation by linked deamination-base excision repair. *Cell*, *146*, 67–79.

Cowley, M., Wood, A. J., Bohm, S., Schulz, R., & Oakey, R. J. (2012). Epigenetic control of alternative mRNA processing at the imprinted Herc3/Nap1l5 locus. *Nucleic Acids Research*, *40*, 8917–8926.

Dawlaty, M. M., Ganz, K., Powell, B. E., Hu, Y. C., Markoulaki, S., Cheng, A. W., et al. (2011). Tet1 is dispensable for maintaining pluripotency and its loss is compatible with embryonic and postnatal development. *Cell Stem Cell*, *9*, 166–175.

Deaton, A. M., & Bird, A. (2011). CpG islands and the regulation of transcription. *Genes and Development*, *25*, 1010–1022.

Deaton, A. M., Webb, S., Kerr, A. R., Illingworth, R. S., Guy, J., Andrews, R., et al. (2011). Cell type-specific DNA methylation at intragenic CpG islands in the immune system. *Genome Research*, *21*, 1074–1086.

De La Fuente, R., Baumann, C., Fan, T., Schmidtmann, A., Dobrinski, I., & Muegge, K. (2006). Lsh is required for meiotic chromosome synapsis and retrotransposon silencing in female germ cells. *Nature Cell Biology*, *8*, 1448–1454.

Delhommeau, F., Dupont, S., Della Valle, V., James, C., Trannoy, S., Masse, A., et al. (2009). Mutation in TET2 in myeloid cancers. *The New England Journal of Medicine*, *360*, 2289–2301.

Dhayalan, A., Rajavelu, A., Rathert, P., Tamas, R., Jurkowska, R. Z., Ragozin, S., et al. (2010). The Dnmt3a PWWP domain reads histone 3 lysine 36 trimethylation and guides DNA methylation. *The Journal of Biological Chemistry*, *285*, 26114–26120.

Dodge, J. E., Okano, M., Dick, F., Tsujimoto, N., Chen, T., Wang, S., et al. (2005). Inactivation of Dnmt3b in mouse embryonic fibroblasts results in DNA hypomethylation, chromosomal instability, and spontaneous immortalization. *The Journal of Biological Chemistry*, *280*, 17986–17991.

Epsztejn-Litman, S., Feldman, N., Abu-Remaileh, M., Shufaro, Y., Gerson, A., Ueda, J., et al. (2008). De novo DNA methylation promoted by G9a prevents reprogramming of embryonically silenced genes. *Nature Structural and Molecular Biology*, *15*, 1176–1183.

Farthing, C. R., Ficz, G., Ng, R. K., Chan, C. F., Andrews, S., Dean, W., et al. (2008). Global mapping of DNA methylation in mouse promoters reveals epigenetic reprogramming of pluripotency genes. *PLoS Genetics*, *4*, e1000116.

Feldman, N., Gerson, A., Fang, J., Li, E., Zhang, Y., Shinkai, Y., et al. (2006). G9a-mediated irreversible epigenetic inactivation of Oct-3/4 during early embryogenesis. *Nature Cell Biology*, *8*, 188–194.

Feng, S., Cokus, S. J., Zhang, X., Chen, P. Y., Bostick, M., Goll, M. G., et al. (2010). Conservation and divergence of methylation patterning in plants and animals. *Proceedings of the National Academy of Sciences of the United States of America*, *107*, 8689–8694.

Fenouil, R., Cauchy, P., Koch, F., Descostes, N., Cabeza, J. Z., Innocenti, C., et al. (2012). CpG islands and GC content dictate nucleosome depletion in a transcription-independent manner at mammalian promoters. *Genome Research*, *22*, 2399–2408.

Ficz, G., Branco, M. R., Seisenberger, S., Santos, F., Krueger, F., Hore, T. A., et al. (2011). Dynamic regulation of 5-hydroxymethylcytosine in mouse ES cells and during differentiation. *Nature*, *473*, 398–402.

Foret, S., Kucharski, R., Pellegrini, M., Feng, S., Jacobsen, S. E., Robinson, G. E., et al. (2012). DNA methylation dynamics, metabolic fluxes, gene splicing, and alternative phenotypes in honey bees. *Proceedings of the National Academy of Sciences of the United States of America*, *109*, 4968–4973.

Frauer, C., Rottach, A., Meilinger, D., Bultmann, S., Fellinger, K., Hasenoder, S., et al. (2011). Different binding properties and function of CXXC zinc finger domains in Dnmt1 and Tet1. *PLoS One*, *6*, e16627.

Freudenberg, J. M., Ghosh, S., Lackford, B. L., Yellaboina, S., Zheng, X., Li, R., et al. (2012). Acute depletion of Tet1-dependent 5-hydroxymethylcytosine levels impairs LIF/Stat3 signaling and results in loss of embryonic stem cell identity. *Nucleic Acids Research*, *40*, 3364–3377.

Gendrel, A. V., Apedaile, A., Coker, H., Termanis, A., Zvetkova, I., Godwin, J., et al. (2012). Smchd1-dependent and -independent pathways determine developmental dynamics of CpG island methylation on the inactive x chromosome. *Developmental Cell*, *23*, 265–279.

Globisch, D., Munzel, M., Muller, M., Michalakis, S., Wagner, M., Koch, S., et al. (2010). Tissue distribution of 5-hydroxymethylcytosine and search for active demethylation intermediates. *PLoS One*, *5*, e15367.

Goll, M. G., Kirpekar, F., Maggert, K. A., Yoder, J. A., Hsieh, C. L., Zhang, X., et al. (2006). Methylation of tRNAAsp by the DNA methyltransferase homolog Dnmt2. *Science*, *311*, 395–398.

Gowher, H., Liebert, K., Hermann, A., Xu, G., & Jeltsch, A. (2005). Mechanism of stimulation of catalytic activity of Dnmt3A and Dnmt3B DNA-(cytosine-C5)-methyltransferases by Dnmt3L. *The Journal of Biological Chemistry*, *280*, 13341–13348.

Gowher, H., Stockdale, C. J., Goyal, R., Ferreira, H., Owen-Hughes, T., & Jeltsch, A. (2005). De novo methylation of nucleosomal DNA by the mammalian Dnmt1 and Dnmt3A DNA methyltransferases. *Biochemistry*, *44*, 9899–9904.

Gu, T. P., Guo, F., Yang, H., Wu, H. P., Xu, G. F., Liu, W., et al. (2011). The role of Tet3 DNA dioxygenase in epigenetic reprogramming by oocytes. *Nature*, *477*, 606–610.

Guibert, S., Forne, T., & Weber, M. (2012). Global profiling of DNA methylation erasure in mouse primordial germ cells. *Genome Research*, *22*, 633–641.

Guo, J. U., Ma, D. K., Mo, H., Ball, M. P., Jang, M. H., Bonaguidi, M. A., et al. (2011). Neuronal activity modifies the DNA methylation landscape in the adult brain. *Nature Neuroscience*, *14*, 1345–1351.

Guo, J. U., Su, Y., Zhong, C., Ming, G. L., & Song, H. (2011). Hydroxylation of 5-methylcytosine by TET1 promotes active DNA demethylation in the adult brain. *Cell*, *145*, 423–434.

Hackett, J. A., Zylicz, J. J., & Surani, M. A. (2012). Parallel mechanisms of epigenetic reprogramming in the germline. *Trends in Genetics*, *28*, 164–174.

Hackett, J. A., Sengupta, R., Zylicz, J. J., Murakami, K., Lee, C., Down, T. A., & Surani, M. A. (2013). Germline DNA demethylation dynamics and imprint erasure through 5-hydroxymethylcytosine. *Science*, *339*(6118), 448–452.

Hashimoto, H., Liu, Y., Upadhyay, A. K., Chang, Y., Howerton, S. B., Vertino, P. M., et al. (2012). Recognition and potential mechanisms for replication and erasure of cytosine hydroxymethylation. *Nucleic Acids Research*, *40*, 4841–4849.

Hata, K., Okano, M., Lei, H., & Li, E. (2002). Dnmt3L cooperates with the Dnmt3 family of de novo DNA methyltransferases to establish maternal imprints in mice. *Development*, *129*, 1983–1993.

He, Y. F., Li, B. Z., Li, Z., Liu, P., Wang, Y., Tang, Q., et al. (2011). Tet-mediated formation of 5-carboxylcytosine and its excision by TDG in mammalian DNA. *Science*, *333*, 1303–1307.

Hellman, A., & Chess, A. (2007). Gene body-specific methylation on the active X chromosome. *Science*, *315*, 1141–1143.

Hemberger, M., Dean, W., & Reik, W. (2009). Epigenetic dynamics of stem cells and cell lineage commitment: Digging Waddington's canal. *Nature Reviews Molecular Cell Biology*, *10*, 526–537.

Hirasawa, R., Chiba, H., Kaneda, M., Tajima, S., Li, E., Jaenisch, R., et al. (2008). Maternal and zygotic Dnmt1 are necessary and sufficient for the maintenance of DNA methylation imprints during preimplantation development. *Genes and Development*, *22*, 1607–1616.

Hodges, E., Molaro, A., Dos Santos, C. O., Thekkat, P., Song, Q., Uren, P. J., et al. (2011). Directional DNA methylation changes and complex intermediate states accompany lineage specificity in the adult hematopoietic compartment. *Molecular Cell*, *44*, 17–28.

Hodges, E., Smith, A. D., Kendall, J., Xuan, Z., Ravi, K., Rooks, M., et al. (2009). High definition profiling of mammalian DNA methylation by array capture and single molecule bisulfite sequencing. *Genome Research*, *19*, 1593–1605.

Iguchi-Ariga, S. M., & Schaffner, W. (1989). CpG methylation of the cAMP-responsive enhancer/promoter sequence TGACGTCA abolishes specific factor binding as well as transcriptional activation. *Genes and Development*, *3*, 612–619.

Illingworth, R. S., Gruenewald-Schneider, U., Webb, S., Kerr, A. R., James, K. D., Turner, D. J., et al. (2010). Orphan CpG islands identify numerous conserved promoters in the mammalian genome. *PLoS Genetics*, *6*, e1001134.

Illingworth, R., Kerr, A., Desousa, D., Jorgensen, H., Ellis, P., Stalker, J., et al. (2008). A novel CpG island set identifies tissue-specific methylation at developmental gene loci. *PLoS Biology*, *6*, e22.

Inoue, A., Shen, L., Dai, Q., He, C., & Zhang, Y. (2011). Generation and replication-dependent dilution of 5fC and 5caC during mouse preimplantation development. *Cell Research*, *21*, 1670–1676.

Inoue, A., & Zhang, Y. (2011). Replication-dependent loss of 5-hydroxymethylcytosine in mouse preimplantation embryos. *Science*, *334*, 194.

Iqbal, K., Jin, S. G., Pfeifer, G. P., & Szabo, P. E. (2011). Reprogramming of the paternal genome upon fertilization involves genome-wide oxidation of 5-methylcytosine. *Proceedings of the National Academy of Sciences of the United States of America*, *108*, 3642–3647.

Ito, S., D'Alessio, A. C., Taranova, O. V., Hong, K., Sowers, L. C., & Zhang, Y. (2010). Role of Tet proteins in 5mC to 5hmC conversion, ES-cell self-renewal and inner cell mass specification. *Nature*, *466*, 1129–1133.

Ito, S., Shen, L., Dai, Q., Wu, S. C., Collins, L. B., Swenberg, J. A., et al. (2011). Tet proteins can convert 5-methylcytosine to 5-formylcytosine and 5-carboxylcytosine. *Science*, *333*, 1300–1303.

Jeong, S., Liang, G., Sharma, S., Lin, J. C., Choi, S. H., Han, H., et al. (2009). Selective anchoring of DNA methyltransferases 3A and 3B to nucleosomes containing methylated DNA. *Molecular and Cellular Biology*, *29*, 5366–5376.

Ji, H., Ehrlich, L. I., Seita, J., Murakami, P., Doi, A., Lindau, P., et al. (2010). Comprehensive methylome map of lineage commitment from haematopoietic progenitors. *Nature*, *467*, 338–342.

Jia, D., Jurkowska, R. Z., Zhang, X., Jeltsch, A., & Cheng, X. (2007). Structure of Dnmt3a bound to Dnmt3L suggests a model for de novo DNA methylation. *Nature*, *449*, 248–251.

Jin, S. G., Jiang, Y., Qiu, R., Rauch, T. A., Wang, Y., Schackert, G., et al. (2011). 5-Hydroxymethylcytosine is strongly depleted in human cancers but its levels do not correlate with IDH1 mutations. *Cancer Research*, *71*, 7360–7365.

Jin, S. G., Kadam, S., & Pfeifer, G. P. (2010). Examination of the specificity of DNA methylation profiling techniques towards 5-methylcytosine and 5-hydroxymethylcytosine. *Nucleic Acids Research*, *38*, e125.

Jin, S. G., Wu, X., Li, A. X., & Pfeifer, G. P. (2011). Genomic mapping of 5-hydroxymethylcytosine in the human brain. *Nucleic Acids Research*, *39*, 5015–5024.

Jones, P. A., & Liang, G. (2009). Rethinking how DNA methylation patterns are maintained. *Nature Reviews Genetics*, *10*, 805–811.

Joulie, M., Miotto, B., & Defossez, P. A. (2010). Mammalian methyl-binding proteins: What might they do? *Bioessays*, *32*, 1025–1032.

Kaneda, M., Okano, M., Hata, K., Sado, T., Tsujimoto, N., Li, E., et al. (2004). Essential role for de novo DNA methyltransferase Dnmt3a in paternal and maternal imprinting. *Nature, 429*, 900–903.

Kato, Y., Kaneda, M., Hata, K., Kumaki, K., Hisano, M., Kohara, Y., et al. (2007). Role of the Dnmt3 family in de novo methylation of imprinted and repetitive sequences during male germ cell development in the mouse. *Human Molecular Genetics, 16*, 2272–2280.

Kelly, T. K., Liu, Y., Lay, F. D., Liang, G., Berman, B. P., & Jones, P. A. (2012). Genome-wide mapping of nucleosome positioning and DNA methylation within individual DNA molecules. *Genome Research, 22*, 2497–2506.

Kim, J., Kollhoff, A., Bergmann, A., & Stubbs, L. (2003). Methylation-sensitive binding of transcription factor YY1 to an insulator sequence within the paternally expressed imprinted gene, Peg3. *Human Molecular Genetics, 12*, 233–245.

Kim, M. S., Kondo, T., Takada, I., Youn, M. Y., Yamamoto, Y., Takahashi, S., et al. (2009). DNA demethylation in hormone-induced transcriptional derepression. *Nature, 461*, 1007–1012.

Klug, M., Heinz, S., Gebhard, C., Schwarzfischer, L., Krause, S. W., Andreesen, R., et al. (2010). Active DNA demethylation in human postmitotic cells correlates with activating histone modifications, but not transcription levels. *Genome Biology, 11*, R63.

Ko, M., Bandukwala, H. S., An, J., Lamperti, E. D., Thompson, E. C., Hastie, R., et al. (2011). Ten-Eleven-Translocation 2 (TET2) negatively regulates homeostasis and differentiation of hematopoietic stem cells in mice. *Proceedings of the National Academy of Sciences of the United States of America, 108*, 14566–14571.

Kobayashi, H., Sakurai, T., Imai, M., Takahashi, N., Fukuda, A., Yayoi, O., et al. (2012). Contribution of intragenic DNA methylation in mouse gametic DNA methylomes to establish oocyte-specific heritable marks. *PLoS Genetics, 8*, e1002440.

Koh, K. P., Yabuuchi, A., Rao, S., Huang, Y., Cunniff, K., Nardone, J., et al. (2011). Tet1 and Tet2 regulate 5-hydroxymethylcytosine production and cell lineage specification in mouse embryonic stem cells. *Cell Stem Cell, 8*, 200–213.

Kolasinska-Zwierz, P., Down, T., Latorre, I., Liu, T., Liu, X. S., & Ahringer, J. (2009). Differential chromatin marking of introns and expressed exons by H3K36me3. *Nature Genetics, 41*, 376–381.

Kriaucionis, S., & Heintz, N. (2009). The nuclear DNA base 5-hydroxymethylcytosine is present in Purkinje neurons and the brain. *Science, 324*, 929–930.

Kurihara, Y., Kawamura, Y., Uchijima, Y., Amamo, T., Kobayashi, H., Asano, T., et al. (2008). Maintenance of genomic methylation patterns during preimplantation development requires the somatic form of DNA methyltransferase 1. *Developmental Biology, 313*, 335–346.

Kurimoto, K., Yabuta, Y., Ohinata, Y., Shigeta, M., Yamanaka, K., & Saitou, M. (2008). Complex genome-wide transcription dynamics orchestrated by Blimp1 for the specification of the germ cell lineage in mice. *Genes and Development, 22*, 1617–1635.

Laird, P. W. (2010). Principles and challenges of genomewide DNA methylation analysis. *Nature Reviews Genetics, 11*, 191–203.

Lane, N., Dean, W., Erhardt, S., Hajkova, P., Surani, A., Walter, J., et al. (2003). Resistance of IAPs to methylation reprogramming may provide a mechanism for epigenetic inheritance in the mouse. *Genesis, 35*, 88–93.

Langemeijer, S. M., Kuiper, R. P., Berends, M., Knops, R., Aslanyan, M. G., Massop, M., et al. (2009). Acquired mutations in TET2 are common in myelodysplastic syndromes. *Nature Genetics, 41*, 838–842.

Laurent, L., Wong, E., Li, G., Huynh, T., Tsirigos, A., Ong, C. T., et al. (2010). Dynamic changes in the human methylome during differentiation. *Genome Research, 20*, 320–331.

Lei, H., Oh, S. P., Okano, M., Juttermann, R., Goss, K. A., Jaenisch, R., et al. (1996). De novo DNA cytosine methyltransferase activities in mouse embryonic stem cells. *Development, 122*, 3195–3205.
Leonhardt, H., Page, A. W., Weier, H. U., & Bestor, T. H. (1992). A targeting sequence directs DNA methyltransferase to sites of DNA replication in mammalian nuclei. *Cell, 71*, 865–873.
Lepikhov, K., Zakhartchenko, V., Hao, R., Yang, F., Wrenzycki, C., Niemann, H., et al. (2008). Evidence for conserved DNA and histone H3 methylation reprogramming in mouse, bovine and rabbit zygotes. *Epigenetics and Chromatin, 1*, 8.
Ley, T. J., Ding, L., Walter, M. J., McLellan, M. D., Lamprecht, T., Larson, D. E., et al. (2010). DNMT3A mutations in acute myeloid leukemia. *The New England Journal of Medicine, 363*, 2424–2433.
Li, E., Beard, C., & Jaenisch, R. (1993). Role for DNA methylation in genomic imprinting. *Nature, 366*, 362–365.
Li, Z., Cai, X., Cai, C. L., Wang, J., Zhang, W., Petersen, B. E., et al. (2011). Deletion of Tet2 in mice leads to dysregulated hematopoietic stem cells and subsequent development of myeloid malignancies. *Blood, 118*, 4509–4518.
Li, X., Ito, M., Zhou, F., Youngson, N., Zuo, X., Leder, P., et al. (2008). A maternal-zygotic effect gene, Zfp57, maintains both maternal and paternal imprints. *Developmental Cell, 15*, 547–557.
Liang, G., Chan, M. F., Tomigahara, Y., Tsai, Y. C., Gonzales, F. A., Li, E., et al. (2002). Cooperativity between DNA methyltransferases in the maintenance methylation of repetitive elements. *Molecular and Cellular Biology, 22*, 480–491.
Lienert, F., Wirbelauer, C., Som, I., Dean, A., Mohn, F., & Schubeler, D. (2011). Identification of genetic elements that autonomously determine DNA methylation states. *Nature Genetics, 43*, 1091–1097.
Lister, R., Pelizzola, M., Dowen, R. H., Hawkins, R. D., Hon, G., Tonti-Filippini, J., et al. (2009). Human DNA methylomes at base resolution show widespread epigenomic differences. *Nature, 462*, 315–322.
Lock, L. F., Takagi, N., & Martin, G. R. (1987). Methylation of the Hprt gene on the inactive X occurs after chromosome inactivation. *Cell, 48*, 39–46.
Long, S. W., Ooi, J. Y., Yau, P. M., & Jones, P. L. (2011). A brain-derived MeCP2 complex supports a role for MeCP2 in RNA processing. *Bioscience Reports, 31*, 333–343.
Lopez-Serra, P., & Esteller, M. (2012). DNA methylation-associated silencing of tumor-suppressor microRNAs in cancer. *Oncogene, 31*, 1609–1622.
Lorincz, M. C., Dickerson, D. R., Schmitt, M., & Groudine, M. (2004). Intragenic DNA methylation alters chromatin structure and elongation efficiency in mammalian cells. *Nature Structural and Molecular Biology, 11*, 1068–1075.
Lorsbach, R. B., Moore, J., Mathew, S., Raimondi, S. C., Mukatira, S. T., & Downing, J. R. (2003). TET1, a member of a novel protein family, is fused to MLL in acute myeloid leukemia containing the t(10;11)(q22;q23). *Leukemia, 17*, 637–641.
Lucarelli, M., Fuso, A., Strom, R., & Scarpa, S. (2001). The dynamics of myogenin site-specific demethylation is strongly correlated with its expression and with muscle differentiation. *The Journal of Biological Chemistry, 276*, 7500–7506.
Luco, R. F., Pan, Q., Tominaga, K., Blencowe, B. J., Pereira-Smith, O. M., & Misteli, T. (2010). Regulation of alternative splicing by histone modifications. *Science, 327*, 996–1000.
Lyko, F., Foret, S., Kucharski, R., Wolf, S., Falckenhayn, C., & Maleszka, R. (2010). The honey bee epigenomes: Differential methylation of brain DNA in queens and workers. *PLoS Biology, 8*, e1000506.
Macleod, D., Charlton, J., Mullins, J., & Bird, A. P. (1994). Sp1 sites in the mouse aprt gene promoter are required to prevent methylation of the CpG island. *Genes and Development, 8*, 2282–2292.

Maiti, A., & Drohat, A. C. (2011). Thymine DNA glycosylase can rapidly excise 5-formylcytosine and 5-carboxylcytosine: Potential implications for active demethylation of CpG sites. *The Journal of Biological Chemistry*, *286*, 35334–35338.

Martin Caballero, I., Hansen, J., Leaford, D., Pollard, S., & Hendrich, B. D. (2009). The methyl-CpG binding proteins Mecp2, Mbd2 and Kaiso are dispensable for mouse embryogenesis, but play a redundant function in neural differentiation. *PLoS One*, *4*, e4315.

Maunakea, A. K., Nagarajan, R. P., Bilenky, M., Ballinger, T. J., D'Souza, C., Fouse, S. D., et al. (2010). Conserved role of intragenic DNA methylation in regulating alternative promoters. *Nature*, *466*, 253–257.

McGrath, J., & Solter, D. (1984). Completion of mouse embryogenesis requires both the maternal and paternal genomes. *Cell*, *37*, 179–183.

Meissner, A. (2010). Epigenetic modifications in pluripotent and differentiated cells. *Nature Biotechnology*, *28*, 1079–1088.

Meissner, A., Mikkelsen, T. S., Gu, H., Wernig, M., Hanna, J., Sivachenko, A., et al. (2008). Genome-scale DNA methylation maps of pluripotent and differentiated cells. *Nature*, *454*, 766–770.

Mellen, M., Ayata, P., Dewell, S., Kriaucionis, S., & Heintz, N. (2012). MeCP2 binds to 5hmC enriched within active genes and accessible chromatin in the nervous system. *Cell*, *151*, 1417–1430.

Metivier, R., Gallais, R., Tiffoche, C., Le Peron, C., Jurkowska, R. Z., Carmouche, R. P., et al. (2008). Cyclical DNA methylation of a transcriptionally active promoter. *Nature*, *452*, 45–50.

Mikkelsen, T. S., Hanna, J., Zhang, X., Ku, M., Wernig, M., Schorderet, P., et al. (2008). Dissecting direct reprogramming through integrative genomic analysis. *Nature*, *454*, 49–55.

Mohn, F., Weber, M., Rebhan, M., Roloff, T. C., Richter, J., Stadler, M. B., et al. (2008). Lineage-specific polycomb targets and de novo DNA methylation define restriction and potential of neuronal progenitors. *Molecular Cell*, *30*, 755–766.

Molaro, A., Hodges, E., Fang, F., Song, Q., McCombie, W. R., Hannon, G. J., et al. (2011). Sperm methylation profiles reveal features of epigenetic inheritance and evolution in primates. *Cell*, *146*, 1029–1041.

Monk, M., Boubelik, M., & Lehnert, S. (1987). Temporal and regional changes in DNA methylation in the embryonic, extraembryonic and germ cell lineages during mouse embryo development. *Development*, *99*, 371–382.

Moran-Crusio, K., Reavie, L., Shih, A., Abdel-Wahab, O., Ndiaye-Lobry, D., Lobry, C., et al. (2011). Tet2 loss leads to increased hematopoietic stem cell self-renewal and myeloid transformation. *Cancer Cell*, *20*, 11–24.

Nagae, G., Isagawa, T., Shiraki, N., Fujita, T., Yamamoto, S., Tsutsumi, S., et al. (2011). Tissue-specific demethylation in CpG-poor promoters during cellular differentiation. *Human Molecular Genetics*, *20*, 2710–2721.

Nakamura, T., Liu, Y. J., Nakashima, H., Umehara, H., Inoue, K., Matoba, S., et al. (2012). PGC7 binds histone H3K9me2 to protect against conversion of 5mC to 5hmC in early embryos. *Nature*, *486*, 415–419.

Nestor, C. E., Ottaviano, R., Reddington, J., Sproul, D., Reinhardt, D., Dunican, D., et al. (2012). Tissue type is a major modifier of the 5-hydroxymethylcytosine content of human genes. *Genome Research*, *22*, 467–477.

Ng, R. K., Dean, W., Dawson, C., Lucifero, D., Madeja, Z., Reik, W., et al. (2008). Epigenetic restriction of embryonic cell lineage fate by methylation of Elf5. *Nature Cell Biology*, *10*, 1280–1290.

Okano, M., Bell, D. W., Haber, D. A., & Li, E. (1999). DNA methyltransferases Dnmt3a and Dnmt3b are essential for de novo methylation and mammalian development. *Cell*, *99*, 247–257.

Ono, R., Taki, T., Taketani, T., Taniwaki, M., Kobayashi, H., & Hayashi, Y. (2002). LCX, leukemia-associated protein with a CXXC domain, is fused to MLL in acute myeloid leukemia with trilineage dysplasia having t(10;11)(q22;q23). *Cancer Research, 62*, 4075–4080.

Ooi, S. K., Qiu, C., Bernstein, E., Li, K., Jia, D., Yang, Z., et al. (2007). DNMT3L connects unmethylated lysine 4 of histone H3 to de novo methylation of DNA. *Nature, 448*, 714–717.

Otani, J., Nankumo, T., Arita, K., Inamoto, S., Ariyoshi, M., & Shirakawa, M. (2009). Structural basis for recognition of H3K4 methylation status by the DNA methyltransferase 3A ATRX-DNMT3-DNMT3L domain. *EMBO Reports, 10*, 1235–1241.

Pastor, W. A., Pape, U. J., Huang, Y., Henderson, H. R., Lister, R., Ko, M., et al. (2011). Genome-wide mapping of 5-hydroxymethylcytosine in embryonic stem cells. *Nature, 473*, 394–397.

Penn, N. W., Suwalski, R., O'Riley, C., Bojanowski, K., & Yura, R. (1972). The presence of 5-hydroxymethylcytosine in animal deoxyribonucleic acid. *The Biochemical Journal, 126*, 781–790.

Popp, C., Dean, W., Feng, S., Cokus, S. J., Andrews, S., Pellegrini, M., et al. (2010). Genome-wide erasure of DNA methylation in mouse primordial germ cells is affected by AID deficiency. *Nature, 463*, 1101–1105.

Prickett, A. R., & Oakey, R. J. (2012). A survey of tissue-specific genomic imprinting in mammals. *Molecular Genetics and Genomics, 287*, 621–630.

Proudhon, C., Duffie, R., Ajjan, S., Cowley, M., Iranzo, J., Carbajosa, G., et al. (2012). Protection against de novo methylation is instrumental in maintaining parent-of-origin methylation inherited from the gametes. *Molecular Cell, 47*, 909–920.

Quenneville, S., Verde, G., Corsinotti, A., Kapopoulou, A., Jakobsson, J., Offner, S., et al. (2011). In embryonic stem cells, ZFP57/KAP1 recognize a methylated hexanucleotide to affect chromatin and DNA methylation of imprinting control regions. *Molecular Cell, 44*, 361–372.

Quivoron, C., Couronne, L., Della Valle, V., Lopez, C. K., Plo, I., Wagner-Ballon, O., et al. (2011). TET2 inactivation results in pleiotropic hematopoietic abnormalities in mouse and is a recurrent event during human lymphomagenesis. *Cancer Cell, 20*, 25–38.

Rakyan, V. K., Down, T. A., Thorne, N. P., Flicek, P., Kulesha, E., Graf, S., et al. (2008). An integrated resource for genome-wide identification and analysis of human tissue-specific differentially methylated regions (tDMRs). *Genome Research, 18*, 1518–1529.

Reis Silva, A. R., Adenot, P., Daniel, N., Archilla, C., Peynot, N., Lucci, C. M., et al. (2011). Dynamics of DNA methylation levels in maternal and paternal rabbit genomes after fertilization. *Epigenetics, 6*, 987–993.

Ruzov, A., Tsenkina, Y., Serio, A., Dudnakova, T., Fletcher, J., Bai, Y., et al. (2011). Lineage-specific distribution of high levels of genomic 5-hydroxymethylcytosine in mammalian development. *Cell Research, 21*, 1332–1342.

Sado, T., Fenner, M. H., Tan, S. S., Tam, P., Shioda, T., & Li, E. (2000). X inactivation in the mouse embryo deficient for Dnmt1: Distinct effect of hypomethylation on imprinted and random X inactivation. *Developmental Biology, 225*, 294–303.

Sado, T., Okano, M., Li, E., & Sasaki, H. (2004). De novo DNA methylation is dispensable for the initiation and propagation of X chromosome inactivation. *Development, 131*, 975–982.

Saitou, M., Kagiwada, S., & Kurimoto, K. (2012). Epigenetic reprogramming in mouse pre-implantation development and primordial germ cells. *Development, 139*, 15–31.

Sandoval, J., & Esteller, M. (2012). Cancer epigenomics: Beyond genomics. *Current Opinion in Genetics and Development, 22*, 50–55.

Santos, F., Hendrich, B., Reik, W., & Dean, W. (2002). Dynamic reprogramming of DNA methylation in the early mouse embryo. *Developmental Biology, 241*, 172–182.

Seki, Y., Hayashi, K., Itoh, K., Mizugaki, M., Saitou, M., & Matsui, Y. (2005). Extensive and orderly reprogramming of genome-wide chromatin modifications associated with specification and early development of germ cells in mice. *Developmental Biology*, *278*, 440–458.

Seisenberger, S., Andrews, S., Krueger, F., Arand, J., Walter, J., Santos, F., et al. (2012). The dynamics of genome-wide DNA methylation reprogramming in mouse primordial germ cells. *Molecular Cell*, *48*(6), 849–862.

Serandour, A. A., Avner, S., Oger, F., Bizot, M., Percevault, F., Lucchetti-Miganeh, C., et al. (2012). Dynamic hydroxymethylation of deoxyribonucleic acid marks differentiation-associated enhancers. *Nucleic Acids Research*, *40*, 8255–8265.

Serandour, A. A., Avner, S., Percevault, F., Demay, F., Bizot, M., Lucchetti-Miganeh, C., et al. (2011). Epigenetic switch involved in activation of pioneer factor FOXA1-dependent enhancers. *Genome Research*, *21*, 555–565.

Sharif, J., Muto, M., Takebayashi, S., Suetake, I., Iwamatsu, A., Endo, T. A., et al. (2007). The SRA protein Np95 mediates epigenetic inheritance by recruiting Dnmt1 to methylated DNA. *Nature*, *450*, 908–912.

Shearstone, J. R., Pop, R., Bock, C., Boyle, P., Meissner, A., & Socolovsky, M. (2011). Global DNA demethylation during mouse erythropoiesis in vivo. *Science*, *334*, 799–802.

Shen, L., Kondo, Y., Guo, Y., Zhang, J., Zhang, L., Ahmed, S., et al. (2007). Genome-wide profiling of DNA methylation reveals a class of normally methylated CpG island promoters. *PLoS Genetics*, *3*, 2023–2036.

Shukla, S., Kavak, E., Gregory, M., Imashimizu, M., Shutinoski, B., Kashlev, M., et al. (2011). CTCF-promoted RNA polymerase II pausing links DNA methylation to splicing. *Nature*, *479*, 74–79.

Smallwood, S. A., Tomizawa, S., Krueger, F., Ruf, N., Carli, N., Segonds-Pichon, A., et al. (2011). Dynamic CpG island methylation landscape in oocytes and preimplantation embryos. *Nature Genetics*, *43*, 811–814.

Smith, Z. D., Chan, M. M., Mikkelsen, T. S., Gu, H., Gnirke, A., Regev, A., et al. (2012). A unique regulatory phase of DNA methylation in the early mammalian embryo. *Nature*, *484*, 339–344.

Song, J., Rechkoblit, O., Bestor, T. H., & Patel, D. J. (2011). Structure of DNMT1-DNA complex reveals a role for autoinhibition in maintenance DNA methylation. *Science*, *331*, 1036–1040.

Song, C. X., Szulwach, K. E., Fu, Y., Dai, Q., Yi, C., Li, X., et al. (2011). Selective chemical labeling reveals the genome-wide distribution of 5-hydroxymethylcytosine. *Nature Biotechnology*, *29*, 68–72.

Song, J., Teplova, M., Ishibe-Murakami, S., & Patel, D. J. (2012). Structure-based mechanistic insights into DNMT1-mediated maintenance DNA methylation. *Science*, *335*, 709–712.

Spruijt, C. G., Gnerlich, F., Smits, A. H., Pfaffeneder, T., Jansen, P. W., Bauer, C., et al. (2013). Dynamic readers for 5-(hydroxy)methylcytosine and its oxidized derivatives. *Cell*, *152*(5), 1146–1159.

Stadler, M. B., Murr, R., Burger, L., Ivanek, R., Lienert, F., Scholer, A., et al. (2011). DNA-binding factors shape the mouse methylome at distal regulatory regions. *Nature*, *480*, 490–495.

Stroud, H., Feng, S., Morey Kinney, S., Pradhan, S., & Jacobsen, S. E. (2011). 5-Hydroxymethylcytosine is associated with enhancers and gene bodies in human embryonic stem cells. *Genome Biology*, *12*, R54.

Suzuki, M. M., & Bird, A. (2008). DNA methylation landscapes: Provocative insights from epigenomics. *Nature Reviews Genetics*, *9*, 465–476.

Szulwach, K. E., Li, X., Li, Y., Song, C. X., Han, J. W., Kim, S., et al. (2011). Integrating 5-hydroxymethylcytosine into the epigenomic landscape of human embryonic stem cells. *PLoS Genetics*, 7, e1002154.

Szulwach, K. E., Li, X., Li, Y., Song, C. X., Wu, H., Dai, Q., et al. (2011). 5-hmC-mediated epigenetic dynamics during postnatal neurodevelopment and aging. *Nature Neuroscience*, *14*, 1607–1616.

Szwagierczak, A., Bultmann, S., Schmidt, C. S., Spada, F., & Leonhardt, H. (2010). Sensitive enzymatic quantification of 5-hydroxymethylcytosine in genomic DNA. *Nucleic Acids Research*, *38*, e181.

Tahiliani, M., Koh, K. P., Shen, Y., Pastor, W. A., Bandukwala, H., Brudno, Y., et al. (2009). Conversion of 5-methylcytosine to 5-hydroxymethylcytosine in mammalian DNA by MLL partner TET1. *Science*, *324*, 930–935.

Thomson, J. P., Skene, P. J., Selfridge, J., Clouaire, T., Guy, J., Webb, S., et al. (2010). CpG islands influence chromatin structure via the CpG-binding protein Cfp1. *Nature*, *464*, 1082–1086.

Tomizawa, S., Kobayashi, H., Watanabe, T., Andrews, S., Hata, K., Kelsey, G., et al. (2011). Dynamic stage-specific changes in imprinted differentially methylated regions during early mammalian development and prevalence of non-CpG methylation in oocytes. *Development*, *138*, 811–820.

Valinluck, V., Tsai, H. H., Rogstad, D. K., Burdzy, A., Bird, A., & Sowers, L. C. (2004). Oxidative damage to methyl-CpG sequences inhibits the binding of the methyl-CpG binding domain (MBD) of methyl-CpG binding protein 2 (MeCP2). *Nucleic Acids Research*, *32*, 4100–4108.

Velasco, G., Hube, F., Rollin, J., Neuillet, D., Philippe, C., Bouzinba-Segard, H., et al. (2010). Dnmt3b recruitment through E2F6 transcriptional repressor mediates germ-line gene silencing in murine somatic tissues. *Proceedings of the National Academy of Sciences of the United States of America*, *107*, 9281–9286.

Walsh, C. P., Chaillet, J. R., & Bestor, T. H. (1998). Transcription of IAP endogenous retroviruses is constrained by cytosine methylation. *Nature Genetics*, *20*, 116–117.

Waterland, R. A., Kellermayer, R., Rached, M. T., Tatevian, N., Gomes, M. V., Zhang, J., et al. (2009). Epigenomic profiling indicates a role for DNA methylation in early postnatal liver development. *Human Molecular Genetics*, *18*, 3026–3038.

Weber, M., Hellmann, I., Stadler, M. B., Ramos, L., Paabo, S., Rebhan, M., et al. (2007). Distribution, silencing potential and evolutionary impact of promoter DNA methylation in the human genome. *Nature Genetics*, *39*, 457–466.

Williams, K., Christensen, J., & Helin, K. (2012). DNA methylation: TET proteins-guardians of CpG islands? *EMBO Reports*, *13*, 28–35.

Williams, K., Christensen, J., Pedersen, M. T., Johansen, J. V., Cloos, P. A., Rappsilber, J., et al. (2011). TET1 and hydroxymethylcytosine in transcription and DNA methylation fidelity. *Nature*, *473*, 343–348.

Wood, A. J., Schulz, R., Woodfine, K., Koltowska, K., Beechey, C. V., Peters, J., et al. (2008). Regulation of alternative polyadenylation by genomic imprinting. *Genes and Development*, *22*, 1141–1146.

Wossidlo, M., Nakamura, T., Lepikhov, K., Marques, C. J., Zakhartchenko, V., Boiani, M., et al. (2011). 5-Hydroxymethylcytosine in the mammalian zygote is linked with epigenetic reprogramming. *Nature Communications*, *2*, 241.

Wu, H., D'Alessio, A. C., Ito, S., Wang, Z., Cui, K., Zhao, K., et al. (2011). Genome-wide analysis of 5-hydroxymethylcytosine distribution reveals its dual function in transcriptional regulation in mouse embryonic stem cells. *Genes and Development*, *25*, 679–684.

Wu, H., D'Alessio, A. C., Ito, S., Xia, K., Wang, Z., Cui, K., et al. (2011). Dual functions of Tet1 in transcriptional regulation in mouse embryonic stem cells. *Nature*, *473*, 389–393.

Wu, H., & Zhang, Y. (2011). Mechanisms and functions of Tet protein-mediated 5-methylcytosine oxidation. *Genes and Development*, *25*, 2436–2452.

Xiang, H., Zhu, J., Chen, Q., Dai, F., Li, X., Li, M., et al. (2010). Single base-resolution methylome of the silkworm reveals a sparse epigenomic map. *Nature Biotechnology, 28*, 516–520.

Xie, W., Barr, C. L., Kim, A., Yue, F., Lee, A. Y., Eubanks, J., et al. (2012). Base-resolution analyses of sequence and parent-of-origin dependent DNA methylation in the mouse genome. *Cell, 148*, 816–831.

Xu, G. L., Bestor, T. H., Bourc'his, D., Hsieh, C. L., Tommerup, N., Bugge, M., et al. (1999). Chromosome instability and immunodeficiency syndrome caused by mutations in a DNA methyltransferase gene. *Nature, 402*, 187–191.

Xu, Y., Wu, F., Tan, L., Kong, L., Xiong, L., Deng, J., et al. (2011). Genome-wide regulation of 5hmC, 5mC, and gene expression by Tet1 hydroxylase in mouse embryonic stem cells. *Molecular Cell, 42*, 451–464.

Yabuta, Y., Kurimoto, K., Ohinata, Y., Seki, Y., & Saitou, M. (2006). Gene expression dynamics during germline specification in mice identified by quantitative single-cell gene expression profiling. *Biology of Reproduction, 75*, 705–716.

Yamaguchi, S., Hong, K., Liu, R., Shen, L., Inoue, A., Diep, D., et al. (2012). Tet1 controls meiosis by regulating meiotic gene expression. *Nature, 492*, 443–447.

Yildirim, O., Li, R., Hung, J. H., Chen, P. B., Dong, X., Ee, L. S., et al. (2011). Mbd3/NURD complex regulates expression of 5-hydroxymethylcytosine marked genes in embryonic stem cells. *Cell, 147*, 1498–1510.

You, J. S., Kelly, T. K., De Carvalho, D. D., Taberlay, P. C., Liang, G., & Jones, P. A. (2011). OCT4 establishes and maintains nucleosome-depleted regions that provide additional layers of epigenetic regulation of its target genes. *Proceedings of the National Academy of Sciences of the United States of America, 108*, 14497–14502.

Young, J. I., Hong, E. P., Castle, J. C., Crespo-Barreto, J., Bowman, A. B., Rose, M. F., et al. (2005). Regulation of RNA splicing by the methylation-dependent transcriptional repressor methyl-CpG binding protein 2. *Proceedings of the National Academy of Sciences of the United States of America, 102*, 17551–17558.

Yu, M., Hon, G. C., Szulwach, K. E., Song, C. X., Zhang, L., Kim, A., et al. (2012). Base-resolution analysis of 5-hydroxymethylcytosine in the mammalian genome. *Cell, 149*, 1368–1380.

Zemach, A., McDaniel, I. E., Silva, P., & Zilberman, D. (2010). Genome-wide evolutionary analysis of eukaryotic DNA methylation. *Science, 328*, 916–919.

Zhang, Y., Jurkowska, R., Soeroes, S., Rajavelu, A., Dhayalan, A., Bock, I., et al. (2010). Chromatin methylation activity of Dnmt3a and Dnmt3a/3L is guided by interaction of the ADD domain with the histone H3 tail. *Nucleic Acids Research, 38*, 4246–4253.

Zhang, H., Zhang, X., Clark, E., Mulcahey, M., Huang, S., & Shi, Y. G. (2010). TET1 is a DNA-binding protein that modulates DNA methylation and gene transcription via hydroxylation of 5-methylcytosine. *Cell Research, 20*, 1390–1393.

Zhou, Y., Lu, Y., & Tian, W. (2012). Epigenetic features are significantly associated with alternative splicing. *BMC Genomics, 13*, 123.

Ziller, M. J., Muller, F., Liao, J., Zhang, Y., Gu, H., Bock, C., et al. (2011). Genomic distribution and inter-sample variation of non-CpG methylation across human cell types. *PLoS Genetics, 7*, e1002389.

CHAPTER THREE

# Epigenetic Marking of the Zebrafish Developmental Program

**Ingrid S. Andersen[*,†], Leif C. Lindeman[*,†], Andrew H. Reiner[*,†], Olga Østrup[*,†], Håvard Aanes[‡], Peter Aleström[‡], Philippe Collas[*,†,1]**

[*]Stem Cell Epigenetics Laboratory, Institute of Basic Medical Sciences, Faculty of Medicine, University of Oslo, Oslo, Norway
[†]Norwegian Center for Stem Cell Research, University of Oslo, Oslo, Norway
[‡]BasAM, Norwegian School of Veterinary Science, Oslo, Norway
[1]Corresponding author: e-mail address: philc@medisin.uio.no

## Contents



## Abstract

A characteristic of anamniote development is a relatively long period of embryonic cell divisions in the absence of on-going transcription. In zebrafish, this period lasts for 10 cell cycles, or ~3-h postfertilization, after which zygotic genome activation (ZGA) takes place during the midblastula transition. How the embryo establishes transcriptional competence and how ZGA is spatially and temporally regulated have not been examined until recently. We review here recent data on the transitions in DNA methylation and posttranslational histone modifications occurring during early zebrafish development, as the embryo

*Current Topics in Developmental Biology*, Volume 104
ISSN 0070-2153
http://dx.doi.org/10.1016/B978-0-12-416027-9.00003-6

acquires transcriptional competence and initiates its own gene expression program. We also address models accounting for the origin of epigenetic states detected in early embryos. From these observations, a concept of epigenetic prepatterning of the embryonic gene expression program prior to the onset of ZGA is emerging. The recent data collectively start shedding light on how ZGA may be programmed and regulated.

## 1. INTRODUCTION: SETTING THE STAGE FOR THE ZEBRAFISH DEVELOPMENTAL GENE EXPRESSION PROGRAM

The zebrafish (*Danio rerio*) has over the years proved to be a powerful animal model for our understanding of the genetics of vertebrate development. Zebrafish are small, hardy, and have a short generation interval; thus, they are relatively easy to maintain in large quantities and at moderate costs compared to rodents. More importantly, zebrafish lay on average more than 100 eggs in a clutch, development is rapid and external, and embryos are optically clear. A reference review of zebrafish developmental stages has been published by Kimmel, Ballard, Kimmel, Ullmann, and Schilling, (1995). Large-scale forward genetic screens, later supplemented with reverse genetics strategies, have led to the identification of genes and pathways essential for vertebrate development (Aleström, Holter, & Nourizadeh-Lillabadi, 2006; Lawson & Wolfe, 2011).

Embryonic development results from a cascade of gene activation and inactivation events triggered after fertilization, which results in the establishment of complex gene regulatory networks (Chan, Longabaugh, et al., 2009). A particularity of anamniotes, including zebrafish, is a relatively long developmental window after fertilization in the absence of on-going transcription. In zebrafish, this window lasts for 3.3 h, during which the zygote undergoes 10 rounds of cell divisions with an unusually rapid cell cycle time of ~15 min (Kane & Kimmel, 1993). In the absence of noticeable G1 and G2 phases, the embryo merely replicates its DNA in each cell cycle during this period. Until the midblastula transition (MBT; 3.3-h postfertilization) at the ~1000-cell stage, zebrafish development is supported by proteins and RNAs stored in the egg cytoplasm (Kane & Kimmel, 1993; Tadros & Lipshitz, 2009). At the MBT, the embryo turns on its own genes, a critical step referred to as the maternal-to-zygotic transition, or MZT (Tadros & Lipshitz, 2009). Zygotic genome activation (ZGA) during the MZT, therefore, ensures that the embryo takes control over its developmental program.

Maternal transcript degradation and onset of ZGA seem to be functionally connected in that zygotic transcription provides proteins and microRNAs (miRNAs) with a feedback effect enhancing the efficiency of maternal RNA clearance (reviewed in Walser & Lipshitz, 2011).

During pre-MBT development, the highly condensed maternal and paternal genomes are remodeled in preparation for embryonic transcription. This remodeling involves removal, exchange, and deposition of DNA- and chromatin-associated proteins on the zygotic genome. It also entails biochemical modifications of DNA, such as DNA demethylation and methylation (Andersen, Reiner, Aanes, Aleström, & Collas, 2012; Mhanni & McGowan, 2004). In addition, recent evidence shows that chromatin remodeling prior to and during the MBT involves the establishment of specific "marks" on the genome, in the form of posttranslationally modifications of histones including methylation and acetylation of specific lysine (K) residues (Aday, Zhu, Lakshmanan, Wang, & Lawson, 2011; Lindeman et al., 2011; Vastenhouw et al., 2010). In this review, we present recent evidence of chromatin changes taking place during zebrafish pre-MBT development, prior to ZGA onset, and as the embryo developments past the MBT. These observations suggest a view of prepatterning of developmental transcription by epigenetic marking of developmentally regulated genes.

## 2. TURNING ON GENE EXPRESSION AT THE MZT

Transcripts loaded in the oocyte during oogenesis (maternal transcripts) have been shown to direct early embryonic development before activation of the embryonic genome. These maternal transcripts have started to be characterized in zebrafish (Abrams & Mullins, 2009; Lindeman & Pelegri, 2010); however, only recently has high-throughput RNA sequencing (RNA-seq) analysis provided a first attempt at estimating the relative abundance of these mRNAs during development through the MBT period (Aanes et al., 2011; Pauli et al., 2012; Vesterlund, Jiao, Unneberg, Hovatta, & Kere, 2011). In addition to highlighting novel transcripts and splice variants, this study identifies five main cohorts of differentially expressed genes during this period (Fig. 3.1).

The first cohort encompasses maternal mRNAs, detected in unfertilized and fertilized eggs, and involved in cell cycle regulation, DNA replication, DNA repair, and protein trafficking (Fig. 3.1, panel i). These maternal transcripts display different kinetics of degradation in the pre-MBT period. Nonetheless, a subset of maternal transcripts subsists and becomes upregulated

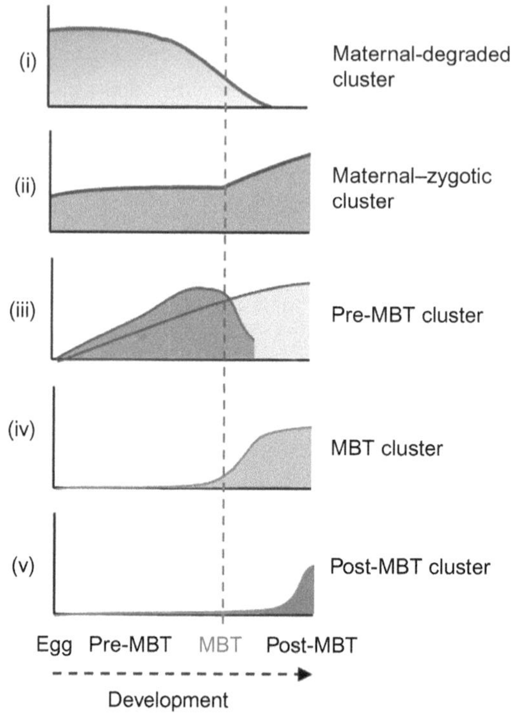


**Figure 3.1** Gene expression cohorts detected during zebrafish development through the MBT period. Profiles are drawn based on our RNA-seq data (Aanes et al., 2011). (For color version of this figure, the reader is referred to the online version of this chapter.)

at the MBT, upon ZGA (Fig. 3.1, panel ii). This "maternal–zygotic" cluster primarily encodes proteins with housekeeping and cellular homeostasis functions, the need for which increases as development proceeds. The third cohort includes a cluster of transcripts detected in pre-MBT stages (pre-MBT cluster), before ZGA onset; this cluster exhibits two distinct profiles: one subcluster is upregulated throughout pre-MBT stages and after the MBT, while the other is only transiently upregulated during pre-MBT development (Fig. 3.1, panel iii). Strikingly, the pre-MBT cluster encodes molecules involved in posttranslational modification, protein processing, transport and degradation, and in gene expression regulation. Interestingly, closer examination of these pre-MBT transcripts reveals a plethora of mRNAs expressed at various levels and encoding histone lysine methyltransferases, DNA methyltransferases (Dnmts), and nonmethylated and methylated CpG-binding proteins (Table 3.1; Fig. 3.2A and B). Detection of these transcripts prior to ZGA onset likely

**Table 3.1** Expression patterns of components of the histone methylation machinery in zebrafish embryos[a]

| | Egg | Pre-MBT | MBT | Post-MBT | HMT target |
|---|---|---|---|---|---|
| **Histone lysine methyltransferases** | | | | | |
| *mll* | 254 | 499 | 774 | 167 | H3K4 |
| *prdm5* | 11 | 88 | 85 | 21 | H3K4 |
| *prdm9* | 144 | 1367 | 1227 | 240 | H3K4 |
| *setd7* | 86 | 122 | 163 | 122 | H3K4 |
| *emht1a* | 858 | 1425 | 1495 | 737 | H3K9 |
| *setdb1a* | 2915 | 935 | 1181 | 239 | H3K9 |
| *setdb1b* | 692 | 1257 | 1410 | 754 | H3K9 |
| *setdb2* | 105 | 179 | 198 | 123 | H3K9 |
| *suv39h1b* | 371 | 1686 | 1970 | 1303 | H3K9 |
| *smyd5* | 26 | 211 | 176 | 74 | H3K27 |
| *ezh2* | 137 | 1269 | 1605 | 1256 | H3K27 |
| *emht2* | 1613 | 1077 | 1038 | 336 | H3K36 |
| *setd3* | 31 | 443 | 547 | 232 | H3K36 |
| *smyd2a* | 94 | 403 | 660 | 101 | H3K36 |
| *smyd2b* | 2 | 0 | 1 | 2 | H4K36 |
| *suv420h1* | 75 | 1239 | 1331 | 680 | H4K20 |
| *suv420h2* | 505 | 555 | 651 | 139 | H4K20 |
| *setd8a* | 2700 | 4658 | 4928 | 4538 | H4K20 |
| *setd8b* | 3114 | 3933 | 3929 | 3139 | H4K20 |
| *setd4* | 164 | 1182 | 1132 | 123 | ? |
| *setd6* | 36 | 153 | 185 | 110 | ? |
| *prdm1a* | 0 | 1 | 2 | 31 | ? |
| *prdm8* | 1 | 5 | 2 | 6 | ? |
| *whsc1* | 2987 | 2765 | 2726 | 1109 | ? |
| **Binding to unmethylated CpGs** | | | | | |
| *cxxc1 (cfp1)* | 259 | 2212 | 2690 | 709 | |

*Continued*

**Table 3.1** Expression patterns of components of the histone methylation machinery in zebrafish embryos—cont'd

| | Egg | Pre-MBT | MBT | Post-MBT | HMT target |
|---|---|---|---|---|---|
| *sp1* | 2211 | 1884 | 2609 | 3041 | |
| *vezf1* | 716 | 863 | 1306 | 1060 | |
| **Binding to methylated CpGs** | | | | | |
| *mbd1* | 627 | 1344 | 1799 | 697 | |
| *mbd2* | 216 | 83 | 148 | 48 | |
| *mbd3a* | 497 | 280 | 330 | 649 | |
| *mbd3b* | 223 | 325 | 571 | 2440 | |
| *mecp2* | 22 | 106 | 153 | 24 | |

[a]Expression levels (length-adjusted RNA-seq reads) are based on data from Aanes et al. (2011). Note the strong variability in expression levels between genes.

reflects a functional requirement for their products during early development, in support of chromatin remodeling events taking place at this time.

Overall, RNA-seq profiles do not exclude the formal possibility that some pre-MBT transcription may occur—as suggested for a handful of genes (Leung, Soll, Arnold, Kemler, & Driever, 2003; Mathavan et al., 2005); yet the detection of mRNAs prior to the MBT has importantly been shown to reflect cytoplasmic polyadenylation of maternal transcripts (Aanes et al., 2011). Inhibiting polyadenylation of these transcripts during pre-MBT stages in zebrafish results in developmental arrest post-MBT, supporting their role in development from the MBT onwards (Aanes et al., 2011).

As expected, ZGA is accompanied by the induction of several additional classes of mRNAs. Indeed, a fourth mRNA cohort includes transcripts virtually undetected pre-MBT but induced from the MBT onwards, at the time of ZGA (Fig. 3.1, panel iv). These genes are linked to signaling pathways involved in development, and encode many developmentally regulated transcription factors. The fifth transcript cohort encompasses mRNAs transcribed post-MBT, after ZGA onset (post-MBT cluster; Fig. 3.1, panel v) and involved in development and differentiation. A striking feature of genes activated upon ZGA is therefore their involvement in developmentally controlled transcriptional regulation processes. Activation of these genes in a spatially and temporally regulated manner requires a fine-tuned mechanism ensuring maintenance of these genes in a temporary repressed yet permissive state, and timely induction when developmental cues are encountered.

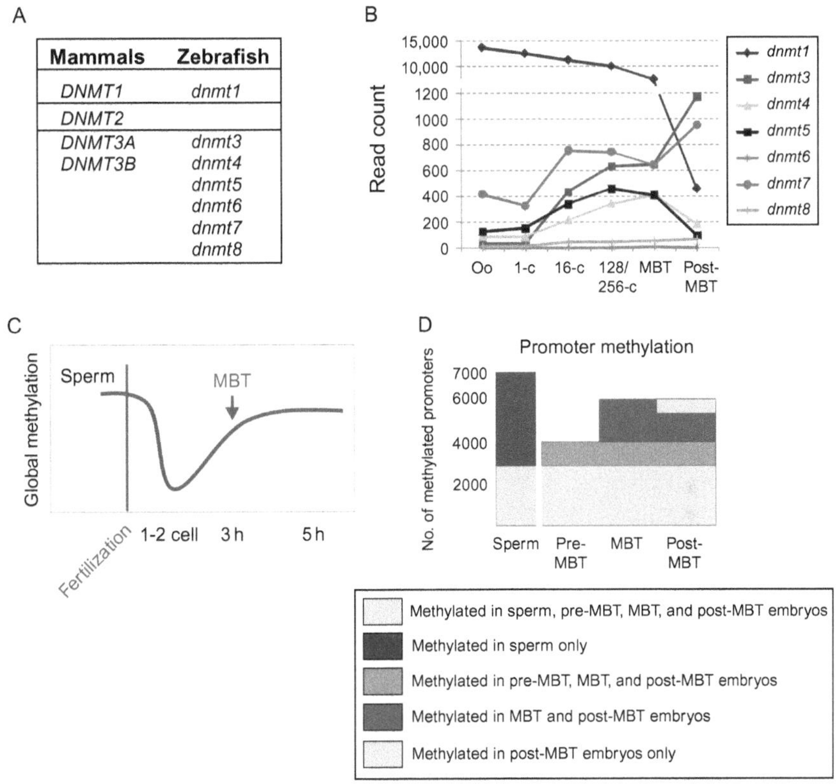


**Figure 3.2** DNA methylation patterns in the zebrafish embryo. (A) *dnmt* gene orthologs in mammals and zebrafish. (B) Expression patterns of zebrafish *dnmt* genes during pre-MBT, MBT, and post-MBT development. Graph depicts mRNA detection levels from our RNA-seq data (Aanes et al., 2011) in unfertilized eggs (Oo), and 1-cell (fertilized), 16-cell, 128-/256-cell, MBT, and post-MBT (4.5-h postfertilization) embryos. (C) Global DNA methylation changes during early development. (D) Changes in promoter DNA methylation levels in sperm and at pre-MBT (256-cell stage), MBT (1- to 2000-cell stage), and post-MBT (4.5-h postfertilization) stages. Distinct shadings indicate developmental stage-specific methylation. (See Color Insert.)

## 3. DNA METHYLATION CHANGES DURING THE MZT

An outstanding question has long been how transcriptional activation is regulated at the time of ZGA. Recent evidence from zebrafish and other species indicates that establishment of the embryo's own gene expression program involves a chromatin remodeling implicating DNA methylation and demethylation, and histone modification mechanisms. How these epigenetic processes contribute to the regulation of gene expression before and at the time of ZGA is the subject of current investigations.

## 3.1. Expression of DNA methyltransferases in zebrafish embryos

DNA methylation is, in general, a negative regulator of gene expression. It organizes long-term gene silencing events during development such as X chromosome inactivation, genomic imprinting, and the silencing of repetitive elements (Reik, 2007). Several mechanisms by which DNA methylation may repress gene expression have been identified (Klose & Bird, 2006): repression may be direct by blocking access of transcriptional activators to promoters, or indirect through the recruitment of methyl-binding proteins such as Kaiso, methyl-CpG-binding domain proteins 1–4, and MeCP2; these in turn may recruit corepressors such as histone deacetylases. Additionally, DNA methyltransferases (DNMTs) themselves may inhibit transcription through interactions with repressive chromatin modifiers (Klose & Bird, 2006; Lee, Kim, Taylor, & Muller, 2010).

In vertebrates, DNA is methylated on cytosines in CpG dinucleotides, by DNMTs, which convert *S*-adenosylmethionine to *S*-adenosylhomocysteine. In mammals, DNA methylation is catalyzed by three DNMTs, among which DNMT1 directs maintenance of methylation after each cell division, and DNMT3a and DNMT3b carry out *de novo* methylation during development and differentiation. DNMT2 enzymes are not thought to have DNA methyltransferase activity but rather seem to methylate RNA.

Zebrafish contains representatives of all mammalian DNMT families, including Dnmt1, and interestingly, six *dnmt* genes (*dnmt3–8*) encoding orthologs to the mammalian DNMT3 family of enzymes (Fig. 3.2A). We and others (Wu, Zhang, Hammoud, et al., 2011) have not found a zebrafish ortholog of the mammalian DNMT3L enzyme which connects unmethylated lysine 4 of histone H3 (H3K4) to DNA methylation (Ooi et al., 2007) and is involved in imprinting. The zebrafish *dnmt3–8* gene products have not been characterized to date; however, transcript profiles for these genes during pre-MBT, MBT, and post-MBT development reveal distinct patterns which may provide hypotheses on how key developmental genes might be maintained in a repressed state before ZGA (Fig. 3.2B).

One can from RNA-seq data (Aanes et al., 2011) identify four classes of *dnmt* expression patterns. (i) Maintenance methylase *dnmt1* mRNAs are highly abundant in the egg, and strikingly, *dnmt1* is downregulated at the MBT. (ii) At the MBT, *dnmt3*, not expressed in eggs, is strongly upregulated; *dnmt3* transcripts are nonetheless already detected before the MBT (Fig. 3.2B), which is indicative of early polyadenylation of transcripts already present in the fertilized egg (Aanes et al., 2011). *dnmt7* expression parallels that of *dnmt3* (Fig. 3.2B), suggesting redundancy between these enzymes. (iii) Remarkably, two additional genes, *dnmt4* and *dnmt5*, display transient expression restricted to the pre-MBT period, increasing levels up to the 256-cell and MBT stages, and marked degradation at the MBT (Fig. 3.2B). It is probable that polyadenylation of these transcripts during pre-MBT development results in the translation of Dnmt enzymes; it would then be interesting to determine what their targets are and whether they mediate (transient?) *de novo* methylation of developmentally important genes, perhaps in a cell type-specific manner within the embryo. (iv) *dnmt8* is apparently not expressed during any of the pre-MBT, MBT, and post-MBT stages examined. Altogether, these findings are consistent with DNA methyltransferase activity in the early zebrafish embryo prior to the MZT onset. They also suggest a switch from maintenance methylation to *de novo* DNA methyltransferase activity around the MZT. Future work is expected to investigate presence, activity, and targets of these putative Dnmts during zebrafish pre-MBT development and whether these play a role in maintaining early developmentally regulated and lineage specification genes in a repressed state prior to ZGA onset.

## 3.2. Dynamics of DNA methylation through the MZT

Recent studies have started to elucidate the impact of DNA methylation on the regulation of gene expression in embryos, notably in mouse (Borgel et al., 2010; Smallwood et al., 2011; Smith et al., 2012; Xie et al., 2012), frog (Bogdanovic et al., 2011; Jones & Takai, 2001; Stancheva, El-Maarri, Walter, Niveleau, & Meehan, 2002; Veenstra & Wolffe, 2001), and zebrafish (Andersen, Reiner, et al., 2012; Mhanni & McGowan, 2004; Rai et al., 2008, 2010, 2006). Two major cycles of genome-wide DNA demethylation and remethylation characterize the mammalian life cycle (Reik, 2007). The first cycle takes place during germ cell formation, when parental imprints are reset by demethylation and differential remethylation of maternal and paternal alleles. This demethylation may also be important for the removal of epimutations that may have arisen during gametogenesis (Reik, Dean, & Walter, 2001). The second occurs after fertilization, when maternal and paternal methylation patterns are erased and reestablished during blastocyst formation. Imprinted genes escape postfertilization reprogramming of DNA methylation, and evidence also indicates that some nonimprinted genes and methylated CpG islands are incompletely demethylated or retain parental methylation patterns (Borgel et al., 2010). *Xenopus* embryos also maintain a high methylation level (Bogdanovic et al., 2011). Therefore, some sperm DNA methylation patterns are retained after fertilization, and the "methylation fate" of specific genes or genomic sequences appears to be diverse in the early embryo.

As in mammals, zebrafish undergoes global demethylation and remethylation after fertilization (Mhanni & McGowan, 2004; Fig. 3.2C). Zebrafish sperm is hypermethylated relative to the oocyte (this does not, however, mean that the entire sperm genome is methylated; see below), and following postfertilization demethylation and remethylation, DNA methylation levels in the embryo are similar to those of somatic tissues (Mhanni & McGowan, 2004; Rai et al., 2008; Wu, Zhang, Hammoud, et al., 2011; Fig. 3.2C). To refine global embryonic methylation patterns, we have recently examined changes in methylation of the promoter of all RefSeq zebrafish genes using a methylated DNA immunoprecipitation (MeDIP) approach coupled to hybridization of the precipitated DNA to promoter arrays (Andersen, Reiner, et al., 2012). Nearly 4000 genes are methylated at the 256-cell stage (i.e., pre-MBT, two cell cycles before ZGA) and an additional 2000 are methylated at the MBT (Fig. 3.2D). The number of methylated genes remains similar after the MBT (4.5-h postfertilization), yet distinct subsets of genes are demethylated and methylated after MBT (Fig. 3.2D). Thus, the transition through the MBT period is accompanied by dynamic changes in promoter DNA methylation, albeit for a relatively minor proportion of genes (2000/12,697 RefSeq genes). For the most part, promoters retain either a methylated or unmethylated state throughout pre-MBT development and through the MBT (Andersen, Reiner, et al., 2012). This suggests that in zebrafish, as in *Xenopus* embryos (Bogdanovic et al., 2011), DNA methylation patterns on upstream regulatory regions seem to be uncoupled from gene expression changes taking place at the time of ZGA.

Several developmental features nevertheless arise from promoter DNA methylation patterns before, during, and after ZGA. In particular, development through the MBT period is associated with differential methylation of low and high CpG content promoters, with preferential methylation of high CpG promoters (Andersen, Reiner, et al., 2012). This suggests that developmentally associated CpG-rich promoters, including CpG islands, are not as strongly protected from *de novo* methylation in embryonic cells as in differentiated cells, in which they are more resilient to *de novo* methylation.

### 3.3. A relationship between DNA methylation and transcriptional repression?

Leaving the aforementioned observations aside, one may wonder whether promoter methylation correlates at all with gene expression in embryos. The overall repressive effect of promoter methylation on transcription—this notably depends on CpG content (Weber et al., 2007)—has previously suggested a view of a DNA methylation-dependent mechanism of pre-ZGA gene silencing (Stancheva et al., 2002). Since DNA methylation loss can lead to the activation of repressed genes, the postfertilization wave of genome-wide DNA demethylation may establish a transcriptionally permissive state favoring gene activation at the time of ZGA (Stancheva et al., 2002). Depletion of the maintenance DNA methyltransferase Dnmt1 in *Xenopus* leads to gene activation before ZGA (Stancheva & Meehan, 2000). This only occurs on genes normally expressed at ZGA, whereas genes scheduled for expression later in development are unaffected. A local gene activation event in the context of overall DNA demethylation would be consistent with a local "prepatterning" of gene expression by some other mechanism—for example, by histone modifications (as discussed below), already established on the cohort of genes scheduled for expression upon ZGA. One might also not exclude an indirect role of DNA methylation on transcriptional repression before ZGA, notably through Kaiso, a methylated CpG-specific repressor important for pre-ZGA gene repression in *Xenopus* and zebrafish (Ruzov et al., 2004, 2009). During organogenesis and terminal differentiation, however, the repressive effect of DNA methylation seems to be restored (Bogdanovic et al., 2011; Rai et al., 2006).

## 4. DEVELOPMENTAL GENE EXPRESSION PREPATTERNING BY DNA HYPOMETHYLATION

The hypomethylated state of promoters appears to be more informative for a patterning of developmental gene expression than the methylated state. In addition to methylated promoters, nearly 50% of RefSeq promoters are hypomethylated; that is, they display a methylation level by MeDIP-chip statistically below genome-average methylation (Andersen, Reiner, et al., 2012; Fig. 3.3A). Hypomethylated promoters are mostly CpG-rich and most remain hypomethylated throughout the MBT period. As for methylation, we find little correlation between expression and DNA hypomethylation during these stages. In fact, cohorts of genes differentially expressed during this period show distinct promoter methylation patterns related to CpG content rather than expression state (Andersen, Reiner, et al., 2012).

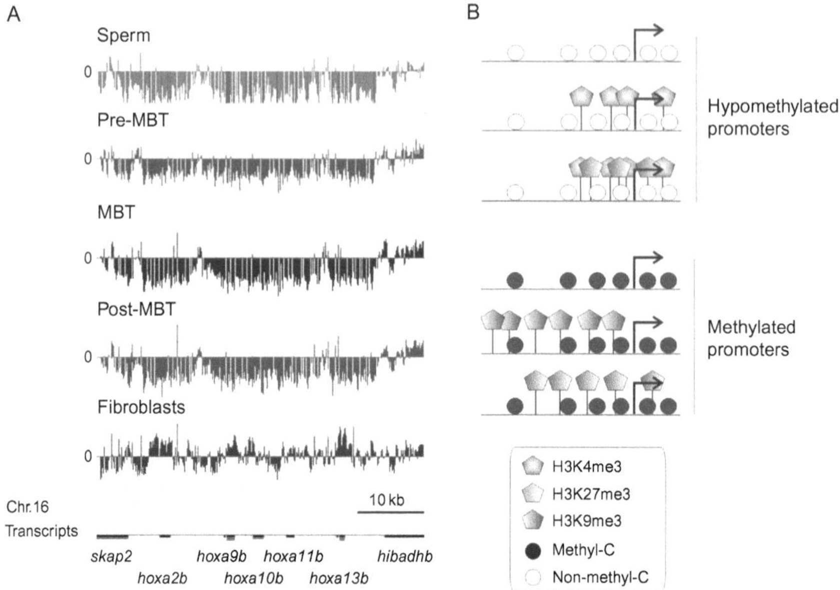


**Figure 3.3** Differential marking of hypomethylated and methylated DNA regions by modified histones in zebrafish embryos. (A) DNA hypomethylation of a developmentally regulated gene cluster. Browser view of DNA methylation profiles over a 54-kb fragment of the zebrafish hoxa locus on chromosome 16. Note the hypomethylated state of the locus in sperm and embryos, and local recurrent methylation in fibroblasts. (B) Differential association of H3K4me3 and H3K9me3 or H3K27me3 with hypomethylated and methylated promoters, respectively, in pre-MBT embryos. (See Color Insert.)

Nevertheless, evidence suggests that the hypomethylated state may create a transcriptionally permissive "platform" for developmental gene expression. Recent work from our laboratory based on differentiation of adipose tissue-derived progenitor cells into adipocytes has shown that whereas methylation of lineage-specific promoters may be predictive of lineage restriction (manifested by inability to differentiate into a specific lineage), the unmethylated state, in contrast, has no predictive value on gene activation and lineage differentiation (Sørensen et al., 2010, 2009). The absence of methylation may nonetheless enable a transcriptionally permissive state for later gene activation, but potential for activation is likely modulated by other mechanisms (such as histone modifications). Similarly, in early zebrafish embryos, hypomethylated regions may also favor a permissive condition. Indeed, the vast majority of genes induced at the time of ZGA or later are either not significantly methylated, or hypomethylated, prior to ZGA (Andersen, Reiner, et al., 2012). Additionally, hypomethylated regions seem to constitute a platform for trimethylation of H3K4 (H3K4me3; Thomson et al., 2010), a transcriptionally permissive modification, prior to ZGA (Andersen, Reiner, et al., 2012; Fig. 3.3B). In contrast, trimethylation of H3K9 (or H3K27me3), associated with transcriptional repression, mainly occurs on methylated promoters (Andersen, Reiner, et al., 2012; Fig. 3.3B). As discussed below, H3K4me3 on hypomethylated developmentally regulated promoters pre-MBT is strongly suggestive of gene expression potential after ZGA onset (Lindeman et al., 2011).

Where does the hypomethylated state come from? A significant fraction of genes hypomethylated in pre-MBT embryos are also hypomethylated in sperm (Andersen, Reiner, et al., 2012; Wu, Zhang, & Cairns, 2011). Indeed, although sperm exhibits DNA methylation levels comparable to those of somatic cells (Wu, Zhang, & Cairns, 2011), both sperm and embryos are marked by regions of methylation and hypomethylation. Hypomethylated genes in sperm and embryos encode for the most part developmentally regulated transcription factors, and hypomethylation occurs either on stand-alone genes or over entire clusters (e.g., *hox* loci; Fig. 3.3A; Andersen, Reiner, et al., 2012; Wu, Zhang, & Cairns, 2011). This suggests that hypomethylation may confer, already in sperm, an instructive role for developmental gene expression. Likewise, the hypomethylated state of maternally expressed genes with cellular homeostatic (housekeeping) functions (Aanes et al., 2011) lends support to the view of an instructive function of promoter hypomethylation also in egg. We recently proposed that one such instructive function may be H3K4 trimethylation (Andersen, Reiner, et al., 2012). Because DNA methylation functionally interacts with histone modifications (Bannister & Kouzarides, 2011; Fuks, 2005; Fuks, Hurd, Deplus, & Kouzarides, 2003; Rai et al., 2006; see also below), it emerges from these studies that a role of DNA methylation in the regulation of ZGA may be the priming of chromatin states, prior to ZGA, for transcriptional competence by association with modified histones.

# 5. PREPATTERNING DEVELOPMENTAL GENE EXPRESSION BY MODIFIED HISTONES BEFORE ZGA

## 5.1. Histone methylation marks of promoters and enhancers

The second class of epigenetic changes that accompany development through the ZGA period involves the posttranslational modifications of histones. These changes occur predominantly in regions of developmental importance, encompassing housekeeping, and developmentally regulated loci. Over 100 posttranslational histone modifications have been reported to date, on the tail and core regions of all four core histones (H2A, H2B, H3, and H4) and of the linker histone H1 (Ruthenburg, Li, Patel, & Allis, 2007). Methylation marks recently examined in the context of embryo development include H3K4 trimethylation (H3K4me3). H3K4me3 marks the transcription start site (TSS) of active and of some inactive genes, and is perceived as a transcriptionally permissive modification. In opposition, H3K9me3 marks inactive promoters and, in contrast to H3K4me3, tends to occupy broader domains over promoter regions. Nevertheless, H3K9me3 can also mark active gene bodies, and a role in the inhibition of spurious transcription from cryptic intragenic start sites has been ascribed to this intragenic enrichment (Carrozza et al., 2005). Another histone modification associated with gene repression, H3K27me3, also occupies promoters where, as H3K4me3, it overlaps with the TSS; its occupancy, however, may often extend upstream of the TSS to cover broader promoter and even enhancer regions. Whether H3K9me3 and H3K27me3 can together form a corepressive module is currently a formal possibility, and evidence suggests that H3K9me3/K27me3 co-occupancy may occur in zebrafish embryos (Andersen, Østrup, et al., 2012; Lindeman et al., 2011) and in human somatic progenitor cells (Sørensen et al., 2010; see below). Trimethylation of H3K36 is found on the bodies of active genes, often in parallel with RNA polymerase II (RNAPII), and it is as such considered as a mark of transcriptional elongation (Kolasinska-Zwierz et al., 2009).

Additional histone marks have been identified on enhancer regions. H3K4 monomethylation (H3K4me1) and H3K27 acetylation (H3K27ac) have been useful in identifying distal regulatory regions (Rada-Iglesias et al., 2011). In embryonic stem cells (ESCs), H3K27ac occupancy on enhancers correlates with transcriptional activity of the nearest gene, whereas enhancer-bound H3K27me3 is associated with enhancer inactivity (Creyghton et al., 2010; Rada-Iglesias et al., 2011). Interestingly, a recent tissue-specific analysis of histone modifications and RNAPII occupancy on enhancers in *Drosophila* embryos has extended our understanding of the spatio-temporal setting of histone marks as they relate to gene expression during development (Bonn et al., 2012). These findings have recently been reviewed elsewhere (Østrup et al., 2012) and are not discussed here. Rather, we focus on promoter-associated changes in chromatin states, defined by combinations of histone modifications, during zebrafish development before, during, and after ZGA.

## 5.2. Dynamic occupancy of the embryonic genome by histone marks during development through the MBT

A picture of the chromatin landscape during zebrafish development through the MBT period starts to emerge from recent chromatin immunoprecipitation (ChIP) data and unveils clues on potential mechanisms regulating embryonic transcription initiated upon ZGA. Initial studies have shown enrichment in H3K4me3 and H3K27me3 from the MBT stage onwards (Aday et al., 2011; Lindeman, Winata, et al., 2010; Vastenhouw et al., 2010). In line with data from ESCs and somatic cells, H3K4me3 predominantly marks housekeeping genes, while H3K27me3 occupies developmentally regulated genes (Vastenhouw et al., 2010). Similar findings are reported in embryos of *Xenopus* (Akkers et al., 2009) and *Drosophila* (Bonn et al., 2012; Schuettengruber et al., 2009). Interestingly, H3K27me3 occupies developmental promoters and enhancers that are temporally and spatially controlled during early development (Akkers et al., 2009; Bonn et al., 2012).

Intriguingly, occupancy of the zebrafish genome by modified histones has initially not been evidenced by ChIP before the MBT (Vastenhouw et al., 2010). This has led to the perception that the "chromatin signature of embryonic pluripotency" is established during embryonic genome activation at the MZT (Vastenhouw et al., 2010). However, one would argue that the rapid cell cycles that precede the MZT in zebrafish make it unlikely that chromatin marks of embryonic development be established in a gene-specific manner concomitantly with ZGA onset. Recent evidence demonstrates occupancy of the zebrafish genome by modified histones prior to the MBT and ZGA onset (Lindeman et al., 2011), suggesting an epigenetic prepatterning of developmental gene expression before the start of embryonic gene expression.

Refinement of the ChIP assay for pre-MBT zebrafish embryos (a technical challenge due to the large amount of yolk relative to cell mass and the low nuclear-cytoplasmic ratio) has recently unveiled the marking of genes by trimethylated H3K4, H3K9, and H3K27 at the 256-cell stage, that is, two cell cycles before ZGA onset (Lindeman et al., 2011). Whether genes are also marked by modified histones at earlier stages is likely but requires further examination (Leif C. Lindeman and Philippe Collas, unpublished data). Using ChIP, we have identified ~200 promoters marked by H3K27me3, ~500 by H3K9me3, and over 1000 promoters marked by H3K4me3 at the 256-cell stage (Fig. 3.4A). Corresponding genes are enriched in cellular homeostasis and developmental functions (H3K4me3-marking), intracellular signaling

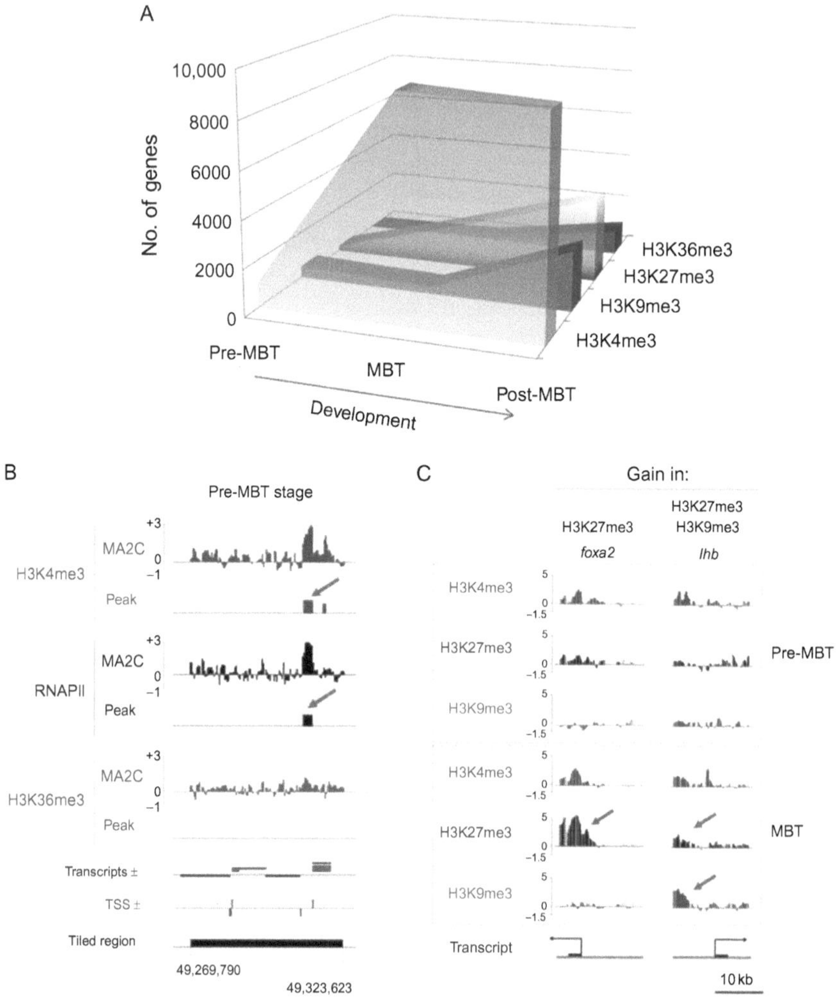


**Figure 3.4** Marking of the zebrafish embryonic genome by posttranslationally modified histones. (A) Developmental enrichment in H3K4me3, H3K9me3, H3K27me3, and H3K36 me3 during pre-MBT, MBT, and post-MBT development. (B) Example of pre-MBT promoter occupancy by H3K4me3 and RNAPII on an inactive locus (ChIP-chip browser shot showing enrichment levels and peaks (arrow)); the tiled region covers nucleotides 49,269,790–49,323,623 of chromosome 1. (C) Examples of gain of pre-MBT H3K4me3-marked genes (upper track) in H3K27me3 (foxa2) and in H3K27me3 and H3K9me3 (lhb). *(B and C) Modified from Lindeman et al. (2011).* (See Color Insert.)

functions (H3K9me3), and transcription regulation and developmental functions (H3K27me3). This pre-MBT enrichment in modified histone has notably been substantiated by (i) ChIP using several antibodies, (ii) native ChIP (ruling out any putative artifacts owing to DNA–protein cross-linking), (iii) immunolabeling of interphase and mitotic chromatin, and (iv) Western blotting (Lindeman et al., 2011). These results together demonstrate the marking of genes of developmental importance by modified histones prior to ZGA onset, challenging the earlier view of developmental chromatin states being established at the MBT.

Nevertheless, most pronounced occupancy of promoters in H3K4me3 is detected at the MBT, when the number of H3K4me3-marked promoters has increased by ninefold relative to the 256-cell stage (Fig. 3.4A). This number remains constant thereafter, suggesting that ZGA is associated with bulk genomic H3K4me3 marking. Genomic occupancy by histone marks is not limited to H3K4me3 through the MBT period, as an increasing number of promoters become enriched in H3K27me3 (Fig. 3.4A). H3K9me3 marking, however, significantly increases only after onset of ZGA, concomitantly with further enrichment in H3K27me3 on apparently the same and on additional promoters. This suggests that overall epigenetic complexity in the embryo increases after ZGA onset, most likely as a reflection of differentiation processes taking place in different blastomeres. In line with transcriptional activation occurring upon ZGA, H3K36me3 is detected on the bodies of genes activated at that time (Fig. 3.4A). It is therefore becoming clear that development through pre-MBT, MBT, and post-MBT stages is accompanied by a reshaping of chromatin by posttranslational histone modifications. The new findings demonstrate the establishment of a chromatin signature on a subset of genes prior to ZGA onset. This epigenetic marking is supported by the storage, in the zebrafish egg, of mature (polyadenylated) transcripts for H3K4, K9, and K27 methyltransferases (Table 3.1).

The discovery that developmentally important genes are marked by H3K4me3 prior to ZGA onset introduces a new concept around H3K4me3 marking. In ESCs, many developmentally regulated genes are marked by H3K4me3 despite their inactivity or low-level expression (Bernstein et al., 2006; Mikkelsen et al., 2007; Ram et al., 2011); however, occupancy is detected in a context of global on-going transcription (ESCs are transcriptionally active). In contrast, transcription is not initiated in the embryo prior to ZGA, so H3K4me3 occupies genes that not only are not transcribed, but also lie in a globally transcriptionally quiescent environment (Lindeman et al., 2011). In fact, only 10% of H3K4me3-marked promoters are occupied by RNAPII which is restricted to the TSS and is therefore not elongating (an example is shown in Fig. 3.4B). Thus, H3K4me3 occupancy of promoters in pre-MBT embryos argues for a role on transcription that is not merely regulatory. Indeed, examination of the post-ZGA expression status of genes premarked by H3K4me3 before ZGA shows that these have a greater propensity to be expressed than all RefSeq genes activated at that time; further, expression level of these genes is higher than that of all RefSeq activated genes (Lindeman et al., 2011). These genes logically retain H3K4me3 at the TSS and acquire both H3K36me3 and RNAPII on gene bodies. These findings strongly suggest that H3K4me3 marking prior to ZGA generates a chromatin state permissive for transcriptional activation after ZGA, and this in the absence of ongoing transcription. Affinity of H3K4me3 for hypomethylated regions (Andersen, Reiner, et al., 2012) reinforces the state of transcriptional permissiveness of this chromatin state. These data raise the view of epigenetic prepatterning of developmental transcription program by selective H3K4me3 marking of housekeeping and developmentally regulated promoters prior to ZGA, and suggest a predictive role of H3K4me3 for developmental gene expression.

The concept of prepatterning of embryonic gene expression is gradually emerging from several studies. Determinant of this prepatterning may be histone modifications such as H3 methylation as described above (Lindeman et al., 2011) and H4 acetylation (Toyama, Rebbert, Dey, Ozato, & Dawid, 2008). Exciting recent findings also suggest that other determinants to developmental transcription may be "stand-by" transcription factors. The binding of factors such as Zelda, STAT92E, or β-catenin to DNA has been shown to be temporally uncoupled from transcription initiation of their target genes in *Drosophila* and *Xenopus* (Blythe, Cha, Tadjuidje, Heasman, & Klein, 2010; Harrison, Li, Kaplan, Botchan, & Eisen, 2011; Tsurumi et al., 2011); they may thus preestablish gene expression potential in a spatio-temporal manner in the embryo. Similarly, in zebrafish, H4 acetylation prior to ZGA mediates binding of the Brd4 transcription factor on genes activated after ZGA onset (Dey, Chitsaz, Abbasi, Misteli, & Ozato, 2003; Toyama et al., 2008); thus, histone marks may premark genes for subsequent transcription factor binding and transcription initiation upon ZGA. The role of transcription factors in the regulation of ZGA and their potential function as determinants of embryonic gene activation has been recently discussed (Østrup et al., 2012). Lastly, DNA motifs directing transcription factor binding may also represent determinants of the site-specific action of these factors. The so-called TAGteam clusters have been shown to be a key determinant in the timing of transcription initiation in *Drosophila melanogaster* and *Aedes aegypti* embryos (Biedler, Hu, Tae, & Tu, 2012; ten Bosch, Benavides, & Cline, 2006; reviewed in Østrup et al., 2012). Whether DNA sequence determinants of embryonic transcription also exist in zebrafish and whether factors such as β-catenin may contribute to spatially and temporally prepattern gene expression during pre-MBT stages are questions of high interest (Chan, Chao, Wang, Yu, & Yuh, 2009; Chan, Longabaugh, et al., 2009).

## 6. DEVELOPMENTAL EPIGENETIC FATE MAP THROUGH THE MBT: A PATH TO MULTIVALENCY?

Despite the potential of pre-MBT H3K4me3-marked genes to be activated after ZGA, a number of these genes have a distinct transcriptional fate shortly after ZGA. An epigenetic fate map of pre-ZGA H3K4me3-marked genes drawn by examining histone modifications associated with these genes during development through the MBT (Lindeman et al., 2011) shows that a significant proportion of these genes, not expressed upon ZGA, becomes enriched in H3K27me3, H3K9me3, or apparently in both marks, while retaining H3K4me3 (Fig. 3.4C).

In zebrafish embryos, H3K4me3/H3K27me3 co-occupancy prevails over H3K4me3/H3K9me3 at the time of ZGA (Andersen, Østrup, et al., 2012). Yet, while most H3K4me3-marked genes are devoid of H3K9me3 or H3K27me3, the latter marks are rarely seen without H3K4me3. Over apparently coenriched regions, H3K4me3 and H3K27me3 may overlap, whereas H3K4me3 and H3K9me3 are mutually exclusive. Intriguingly, however, H3K4me3 and H3K9me3 occasionally coincide, only when H3K27me3 also marks the overlapping area. Furthermore, on metagenes H3K27me3 correlates with local alteration in H3K4me3 density, and H3K9me3 is linked to alterations in the profiles of both H3K4me3 and H3K27me3. Thus, enrichment in trimethylated H3K9 or H3K27 seems to be linked to local remodeling of chromatin manifested by changes in H3K4me3 density (Andersen, Østrup, et al., 2012). This raises the possibility of actual proximity of these marks, supporting a view of bi- or trivalent chromatin domains in the zebrafish embryo, at least at this stage of development.

The coincidence of permissive H3K4me3 with repressive H3K27me3 without or with H3K9me3 on a subset of developmentally regulated promoters raises the question of whether coenrichment of these marks really exists, or whether codetection simply reflects heterogeneity of the cell populations in the embryo. In other words, do developmentally important genes display multivalency in chromatin states or not? In ESCs (Bernstein et al., 2006; Mikkelsen et al., 2007), as well as in somatic stem cells (Cui et al., 2009) and differentiated cells (Barski et al., 2007), including the zebrafish ZF4 fibroblast cell line (Lindeman, Reiner, Mathavan, Aleström, & Collas, 2010), H3K4me3 and H3K27me3 appear to be coenriched on developmentally regulated genes that are poised for later transcription, creating a "bivalent" domain (Bernstein et al., 2006). Claims of H3K4me3/H3K27me3 bivalency have been made based on sequential ChIP experiments, where one modification is ChIPed and the other is in turn ChIPed from the eluate of the first ChIP (Bernstein et al., 2006; Lindeman, Reiner, et al., 2010). Nevertheless, many other studies claim bivalency on the basis of a mere colocalization of two marks on a given locus in the absence of demonstration of coenrichment. Intriguingly, H3K4me3/H3K27me3 coenrichment has been reported exclusively in cultured cell lines and little evidence exists to our knowledge to suggest that bivalency exists *in vivo*.

In embryos, the issue of bi- or multivalency of histone marks remains open (Andersen, Østrup, et al., 2012; Vastenhouw & Schier, 2012). In *Xenopus* gastrula stage embryos, H3K27me3-enriched promoters are devoid of H3K4me3, arguing against bivalency (Akkers et al., 2009). Similarly, in *Drosophila*, there is essentially no evidence of H3K4me3/H3K27me3 bivalency, or if it occurs, its incidence is marginal (Schuettengruber et al., 2009). A distinct finding arises from zebrafish post-MBT stage embryos, however, where rigorous sequential ChIP experiments suggest coenrichment of developmentally regulated promoters by H3K27me3 and H3K4me3 (Vastenhouw et al., 2010). We have no such evidence of coenrichment in zebrafish embryos in our hands, although this may be due to technical difficulties associated with the low amount of material recovered after sequential ChIP (Leif C. Lindeman and Philippe Collas, unpublished observations). The discrepancy between findings in *Xenopus*, *Drosophila*, and zebrafish may be due to species differences, developmental stages examined, and stringency and sensitivity of the assays used to question bivalency. Conceptually, bi- or multivalency with coenrichment of permissive and repressive marks creating an "idling" chromatin state constitutes an attractive mechanism whereby in H3K27me3 may keep lineage-specific genes in a temporarily repressed (halted) state preventing unscheduled premature activation, while H3K4me3 provides permissiveness enabling activation upon development and differentiation.

Domains with coinciding H3K4me3 and H3K27me3 are also clearly detectable in sperm (Wu, Zhang, & Cairns, 2011). These relatively large chromatin "blocks" predominantly contain developmental genes, and even entire clusters of such genes (e.g., *hox* loci) (Wu, Zhang, & Cairns, 2011). Because sperm cells are presumably similar or identical in their chromatin states (this remains to be demonstrated), one can speculate that these domains are bivalent. In addition, an interesting modality of sperm chromatin is the combinatorial association of histone modifications, which, to our knowledge, has not been reported in other cell types. A striking example is H3K36me3, a mark commonly detected on the coding regions of expressed genes, yet found on inactive developmental promoters in zebrafish sperm, often overlapping with blocks of H3K27me3 (Wu, Zhang, & Cairns, 2011). Whether chromatin domains enriched in H3K27me3 and H3K36me3 exist in embryos is uncertain, as only in rare occasions do we observe overlapping domains enriched in these marks. To what extent this and other cases of bi- or multivalency may also be the result of the partial tetraploidy of zebrafish, with different alleles or duplicated regions in a distinct transcriptional state, has to our knowledge not be taken into consideration and is worthy of close attention.

## 7. TRANSGENERATIONAL INHERITANCE OR *DE NOVO* DEPOSITION OF EMBRYONIC EPIGENETIC MARKS? TWO MODELS

The presence of histone marks on the genome of pre-MBT embryos suggests two alternatives as to the origin of these marks (Lindeman et al., 2011). A first model puts forward a direct inheritance of modified histones from gametes. To support this view, a significant proportion of genes in pre-MBT embryos harbors the same histone modifications in sperm (Andersen, Reiner, et al., 2012; Wu, Zhang, & Cairns, 2011). Furthermore, in mouse (Brykczynska et al., 2010; Puschendorf et al., 2008) and *C. elegans* (Arico, Katz, van der Vlag, & Kelly, 2011), male and female genomes are marked by modified histones at the 1-cell stage, consistent with a transmission of marks through fertilization. An implication of these findings is that developmental instructions may already be predetermined by epigenetic states in gametes, an idea supported by findings that high levels of H3K4me3 and H3K14ac (a mark of transcriptional activity) in sperm correlate well with transcriptional activation at or after ZGA (Wu, Zhang, & Cairns, 2011). These marks may also reflect transcriptional activity in the germline of the previous generation, as proposed in *C. elegans* (Arico et al., 2011). An exciting idea emerging from these observations is that inheritance of modified histones through fertilization may constitute a mechanism of "bookmarking" gene expression patterns between generations.

A second model involves the removal of modified histones or modifications themselves from sperm chromatin and restoration of these modifications after fertilization, on the same target genes as those premarked in sperm. We and others have observed that genes enriched in modified histones in zebrafish sperm are not all marked in pre-MBT embryos (Lindeman et al., 2011; Wu, Zhang, & Cairns, 2011), suggesting removal of histone marks from parental gametes. This is also the case for DNA methylation, as genes methylated in sperm are not necessarily methylated in the embryo (Andersen, Reiner, et al., 2012; Wu, Zhang, & Cairns, 2011). Conversely, a significant number of genes marked by modified histones in pre-MBT embryos are not marked in sperm (Lindeman et al., 2011; Wu, Zhang, & Cairns, 2011), arguing in favor of postfertilization *de novo* embryonic epigenetic marking.

## 8. CONCLUSIONS AND PERSPECTIVES

A prepatterning of developmental gene expression by post-translationally modified histones is consistent with an instructive function of histone modifications on gene expression patterns. Maintenance of transcriptional repression before ZGA by H3K27me3 (or H3K9me3) may ensure temporal and spatial modulation of embryonic gene activation to pattern development. From recent studies, a working model of epigenetic control of developmental gene expression in zebrafish emerges (Fig. 3.5). The model puts forward a prepatterning of developmental transcription during pre-MBT development, followed by a priming of transcription initiation or repression at the MBT, when the embryonic genome is turned on. This priming leads to distinct chromatin states on genes pending on their transcriptional fate.

The jury remains open on which of the transgenerational inheritance or removal/redeposition mechanisms prevails to epigenetically prepare the embryonic genome for transcriptional activation. The models proposed are not mutually exclusive, as while some genes may remain epigenetically marked in sperm and embryos, others clearly lose or gain modifications. Even in a context of removal and reestablishment of histone marks, a challenging question is how this remarking is directed to previously marked sites. This may involve instructions "encoded" in the sperm chromatin or DNA,

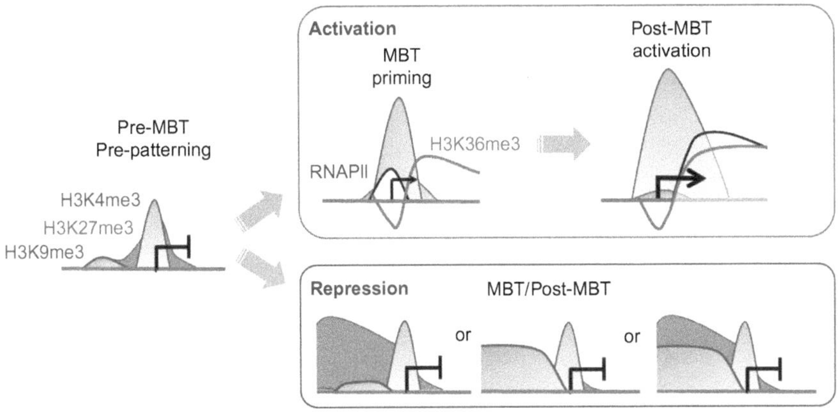


**Figure 3.5** Model of prepatterning and priming of embryonic gene expression before and after the MBT. A "prepatterning" of expression profiles is set by histone modifications on a subset of housekeeping and developmentally regulated genes during pre-MBT development, before ZGA. These modifications may occur alone (e.g., H3K4me3 only) or in combinations (e.g., H3K4me3 together with H3K27me3 or H3K9me3). At the MBT, on genes that are to be activated (right, upper panel), "priming" for activation takes place with enhanced H3K4me3, RNAPII loading at the TSS and moderate H3K36 occupancy in the 5′ end of the gene. Full activation post-MBT, that is, after ZGA onset, is accompanied by spreading of RNAPII and H3K36me3 along the coding region. On genes that are to be repressed, or kept in a repressed state at and after the MBT (right, lower panel), various chromatin states are detected, with enrichment in H3K27me3, H3K9me3, or both marks. Note that some genes may also lose the pre-MBT marks altogether. (For color version of this figure, the reader is referred to the online version of this chapter.)

speculatively (at present) in the form of DNA methylation, transposons, noncoding RNAs, or even sequence determinants (Lindeman et al., 2011). Several lines of evidence support this possibility (Aravin & Bourc'his, 2008; Bogdanovic et al., 2011; Branciamore, Chen, Riggs, & Rodin, 2010; Illingworth et al., 2010; Lienert et al., 2011; Thomson et al., 2010; Weber et al., 2007; Wu, Zhang, & Cairns, 2011), although a formal demonstration remains pending.

## REFERENCES


Aanes, H., Winata, C. L., Lin, C. H., Chen, J. P., Srinivasan, K. G., Lee, S. G., et al. (2011). Zebrafish mRNA sequencing decifers novelties in transcriptome dynamics during maternal to zygotic transition. *Genome Research*, *21*, 1328–1338.

Abrams, E. W., & Mullins, M. C. (2009). Early zebrafish development: It's in the maternal genes. *Current Opinion in Genetics and Development*, *19*, 396–403.

Aday, A. W., Zhu, L. J., Lakshmanan, A., Wang, J., & Lawson, N. D. (2011). Identification of cis regulatory features in the embryonic zebrafish genome through large-scale profiling of H3K4me1 and H3K4me3 binding sites. *Developmental Biology*, *357*, 450–462.

Akkers, R. C., van Heeringen, S. J., Jacobi, U. G., Janssen-Megens, E. M., Francoijs, K. J., Stunnenberg, H. G., et al. (2009). A hierarchy of H3K4me3 and H3K27me3 acquisition in spatial gene regulation in Xenopus embryos. *Developmental Cell, 17*, 425–434.

Aleström, P., Holter, J. L., & Nourizadeh-Lillabadi, R. (2006). Zebrafish in functional genomics and aquatic biomedicine. *Trends in Biotechnology, 24*, 15–21.

Andersen, I. S., Østrup, O., Lindeman, L. C., Aanes, H., Reiner, A. H., Mathavan, S., et al. (2012). Epigenetic complexity during the zebrafish mid-blastula transition. *Biochemical and Biophysical Research Communications, 417*, 1139–1144.

Andersen, I. S., Reiner, A. H., Aanes, H., Aleström, P., & Collas, P. (2012). Developmental features of DNA methylation during activation of the embryonic zebrafish genome. *Genome Biology, 13*, R65.

Aravin, A. A., & Bourc'his, D. (2008). Small RNA guides for de novo DNA methylation in mammalian germ cells. *Genes & Development, 22*, 970–975.

Arico, J. K., Katz, D. J., van der Vlag, J., & Kelly, W. G. (2011). Epigenetic patterns maintained in early Caenorhabditis elegans embryos can be established by gene activity in the parental germ cells. *PLoS Genetics*, 7, 1–15.

Bannister, A. J., & Kouzarides, T. (2011). Regulation of chromatin by histone modifications. *Cell Research, 21*, 381–395.

Barski, A., Cuddapah, S., Cui, K., Roh, T. Y., Schones, D. E., Wang, Z., et al. (2007). High-resolution profiling of histone methylations in the human genome. *Cell, 129*, 823–837.

Bernstein, B. E., Mikkelsen, T. S., Xie, X., Kamal, M., Huebert, D. J., Cuff, J., et al. (2006). A bivalent chromatin structure marks key developmental genes in embryonic stem cells. *Cell, 125*, 315–326.

Biedler, J. K., Hu, W., Tae, H., & Tu, Z. (2012). Identification of early zygotic genes in the yellow fever mosquito Aedes aegypti and discovery of a motif involved in early zygotic genome activation. *PLoS One*, 7, e33933.

Blythe, S. A., Cha, S. W., Tadjuidje, E., Heasman, J., & Klein, P. S. (2010). Beta-Catenin primes organizer gene expression by recruiting a histone H3 arginine 8 methyltransferase, Prmt2. *Developmental Cell, 19*, 220–231.

Bogdanovic, O., Long, S. W., van Heeringen, S. J., Brinkman, A. B., Gomez-Skarmeta, J. L., Stunnenberg, H. G., et al. (2011). Temporal uncoupling of the DNA methylome and transcriptional repression during embryogenesis. *Genome Research, 21*, 1313–1327.

Bonn, S., Zinzen, R. P., Girardot, C., Gustafson, E. H., Perez-Gonzalez, A., Delhomme, N., et al. (2012). Tissue-specific analysis of chromatin state identifies temporal signatures of enhancer activity during embryonic development. *Nature Genetics, 44*, 148–156.

Borgel, J., Guibert, S., Li, Y., Chiba, H., Schubeler, D., Sasaki, H., et al. (2010). Targets and dynamics of promoter DNA methylation during early mouse development. *Nature Genetics, 42*, 1093–1100.

Branciamore, S., Chen, Z. X., Riggs, A. D., & Rodin, S. N. (2010). CpG island clusters and pro-epigenetic selection for CpGs in protein-coding exons of HOX and other transcription factors. *Proceedings of the National Academy of Sciences of the United States of America, 107*, 15485–15490.

Brykczynska, U., Hisano, M., Erkek, S., Ramos, L., Oakeley, E. J., Roloff, T. C., et al. (2010). Repressive and active histone methylation mark distinct promoters in human and mouse spermatozoa. *Nature Structural and Molecular Biology, 17*, 679–687.

Carrozza, M. J., Li, B., Florens, L., Suganuma, T., Swanson, S. K., Lee, K. K., et al. (2005). Histone H3 methylation by Set2 directs deacetylation of coding regions by Rpd3S to suppress spurious intragenic transcription. *Cell, 123*, 581–592.

Chan, T. M., Chao, C. H., Wang, H. D., Yu, Y. J., & Yuh, C. H. (2009). Functional analysis of the evolutionarily conserved cis-regulatory elements on the sox17 gene in zebrafish. *Developmental Biology, 326*, 456–470.

Chan, T. M., Longabaugh, W., Bolouri, H., Chen, H. L., Tseng, W. F., Chao, C. H., et al. (2009). Developmental gene regulatory networks in the zebrafish embryo. *Biochimica et Biophysica Acta, 1789*, 279–298.

Creyghton, M. P., Cheng, A. W., Welstead, G. G., Kooistra, T., Carey, B. W., Steine, E. J., et al. (2010). Histone H3K27ac separates active from poised enhancers and predicts developmental state. *Proceedings of the National Academy of Sciences of the United States of America, 107*, 21931–21936.

Cui, K., Zang, C., Roh, T. Y., Schones, D. E., Childs, R. W., Peng, W., et al. (2009). Chromatin signatures in multipotent human hematopoietic stem cells indicate the fate of bivalent genes during differentiation. *Cell Stem Cell, 4*, 80–93.

Dey, A., Chitsaz, F., Abbasi, A., Misteli, T., & Ozato, K. (2003). The double bromodomain protein Brd4 binds to acetylated chromatin during interphase and mitosis. *Proceedings of the National Academy of Sciences of the United States of America, 100*, 8758–8763.

Fuks, F. (2005). DNA methylation and histone modifications: Teaming up to silence genes. *Current Opinion in Genetics and Development, 15*, 490–495.

Fuks, F., Hurd, P. J., Deplus, R., & Kouzarides, T. (2003). The DNA methyltransferases associate with HP1 and the SUV39H1 histone methyltransferase. *Nucleic Acids Research, 31*, 2305–2312.

Harrison, M. M., Li, X. Y., Kaplan, T., Botchan, M. R., & Eisen, M. B. (2011). Zelda binding in the early Drosophila melanogaster embryo marks regions subsequently activated at the maternal-to-zygotic transition. *PLoS Genetics, 7*, e1002266.

Illingworth, R. S., Gruenewald-Schneider, U., Webb, S., Kerr, A. R., James, K. D., Turner, D. J., et al. (2010). Orphan CpG islands identify numerous conserved promoters in the mammalian genome. *PLoS Genetics, 6*, e1001134.

Jones, P. A., & Takai, D. (2001). The role of DNA methylation in mammalian epigenetics. *Science, 293*, 1068–1070.

Kane, D. A., & Kimmel, C. B. (1993). The zebrafish midblastula transition. *Development, 119*, 447–456.

Kimmel, C. B., Ballard, W. W., Kimmel, S. R., Ullmann, B., & Schilling, T. F. (1995). Stages of embryonic development of the zebrafish. *Developmental Dynamics, 203*, 253–310.

Klose, R. J., & Bird, A. P. (2006). Genomic DNA methylation: The mark and its mediators. *Trends in Biochemical Sciences, 31*, 89–97.

Kolasinska-Zwierz, P., Down, T., Latorre, I., Liu, T., Liu, X. S., & Ahringer, J. (2009). Differential chromatin marking of introns and expressed exons by H3K36me3. *Nature Genetics, 41*, 376–381.

Lawson, N. D., & Wolfe, S. A. (2011). Forward and reverse genetic approaches for the analysis of vertebrate development in the zebrafish. *Developmental Cell, 21*, 48–64.

Lee, G. E., Kim, J. H., Taylor, M., & Muller, M. T. (2010). DNA methyltransferase 1-associated protein (DMAP1) is a co-repressor that stimulates DNA methylation globally and locally at sites of double strand break repair. *Journal of Biological Chemistry, 285*, 37630–37640.

Leung, T., Soll, I., Arnold, S. J., Kemler, R., & Driever, W. (2003). Direct binding of Lef1 to sites in the boz promoter may mediate pre-midblastula-transition activation of boz expression. *Developmental Dynamics, 228*, 424–432.

Lienert, F., Wirbelauer, C., Som, I., Dean, A., Mohn, F., & Schubeler, D. (2011). Identification of genetic elements that autonomously determine DNA methylation states. *Nature Genetics, 43*, 1091–1097.

Lindeman, L. C., Andersen, I. S., Reiner, A. H., Li, N., Aanes, H., Østrup, O., et al. (2011). Pre-patterning of developmental gene expression by modified histones before zygotic genome activation. *Developmental Cell, 21*, 993–1004.
Lindeman, R. E., & Pelegri, F. (2010). Vertebrate maternal-effect genes: Insights into fertilization, early cleavage divisions, and germ cell determinant localization from studies in the zebrafish. *Molecular Reproduction and Development, 77*, 299–313.
Lindeman, L. C., Reiner, A. H., Mathavan, S., Aleström, P., & Collas, P. (2010). Tiling histone H3 lysine 4 and 27 methylation in zebrafish using high-density microarrays. *PLoS One, 5*, e15651.
Lindeman, L. C., Winata, C. L., Aanes, H., Mathavan, S., Aleström, P., & Collas, P. (2010). Chromatin states of developmentally-regulated genes revealed by DNA and histone methylation patterns in zebrafish embryos. *International Journal of Developmental Biology, 54*, 803–813.
Mathavan, S., Lee, S. G., Mak, A., Miller, L. D., Murthy, K. R., Govindarajan, K. R., et al. (2005). Transcriptome analysis of zebrafish embryogenesis using microarrays. *PLoS Genetics, 1*, 260–276.
Mhanni, A. A., & McGowan, R. A. (2004). Global changes in genomic methylation levels during early development of the zebrafish embryo. *Development Genes and Evolution, 214*, 412–417.
Mikkelsen, T. S., Ku, M., Jaffe, D. B., Issac, B., Lieberman, E., Giannoukos, G., et al. (2007). Genome-wide maps of chromatin state in pluripotent and lineage-committed cells. *Nature, 448*, 553–560.
Ooi, S. K., Qiu, C., Bernstein, E., Li, K., Jia, D., Yang, Z., et al. (2007). DNMT3L connects unmethylated lysine 4 of histone H3 to de novo methylation of DNA. *Nature, 448*, 714–717.
Østrup, O., Andersen, I. S., & Collas, P. (2012). Chromatin-linked determinants of zygotic genome activation. *Cellular and Molecular Life Sciences*, Sep. 11 [Epub ahead of print].
Pauli, A., Valen, E., Lin, M. F., Garber, M., Vastenhouw, N. L., Levin, J. Z., et al. (2012). Systematic identification of long noncoding RNAs expressed during zebrafish embryogenesis. *Genome Research, 22*, 577–591.
Puschendorf, M., Terranova, R., Boutsma, E., Mao, X., Isono, K., Brykczynska, U., et al. (2008). PRC1 and Suv39h specify parental asymmetry at constitutive heterochromatin in early mouse embryos. *Nature Genetics, 40*, 411–420.
Rada-Iglesias, A., Bajpai, R., Swigut, T., Brugmann, S. A., Flynn, R. A., & Wysocka, J. (2011). A unique chromatin signature uncovers early developmental enhancers in humans. *Nature, 470*, 279–283.
Rai, K., Huggins, I. J., James, S. R., Karpf, A. R., Jones, D. A., & Cairns, B. R. (2008). DNA demethylation in zebrafish involves the coupling of a deaminase, a glycosylase, and gadd45. *Cell, 135*, 1201–1212.
Rai, K., Jafri, I. F., Chidester, S., James, S. R., Karpf, A. R., Cairns, B. R., et al. (2010). Dnmt3 and G9a cooperate for tissue-specific development in zebrafish. *Journal of Biological Chemistry, 285*, 4110–4121.
Rai, K., Nadauld, L. D., Chidester, S., Manos, E. J., James, S. R., Karpf, A. R., et al. (2006). Zebra fish Dnmt1 and Suv39h1 regulate organ-specific terminal differentiation during development. *Molecular and Cellular Biology, 26*, 7077–7085.
Ram, O., Goren, A., Amit, I., Shoresh, N., Yosef, N., Ernst, J., et al. (2011). Combinatorial patterning of chromatin regulators uncovered by genome-wide location analysis in human cells. *Cell, 147*, 1628–1639.
Reik, W. (2007). Stability and flexibility of epigenetic gene regulation in mammalian development. *Nature, 447*, 425–432.
Reik, W., Dean, W., & Walter, J. (2001). Epigenetic reprogramming in mammalian development. *Science, 293*, 1089–1093.

Ruthenburg, A. J., Li, H., Patel, D. J., & Allis, C. D. (2007). Multivalent engagement of chromatin modifications by linked binding modules. *Nature Reviews. Molecular Cell Biology, 8*, 983–994.

Ruzov, A., Dunican, D. S., Prokhortchouk, A., Pennings, S., Stancheva, I., Prokhortchouk, E., et al. (2004). Kaiso is a genome-wide repressor of transcription that is essential for amphibian development. *Development, 131*, 6185–6194.

Ruzov, A., Savitskaya, E., Hackett, J. A., Reddington, J. P., Prokhortchouk, A., Madej, M. J., et al. (2009). The non-methylated DNA-binding function of Kaiso is not required in early Xenopus laevis development. *Development, 136*, 729–738.

Schuettengruber, B., Ganapathi, M., Leblanc, B., Portoso, M., Jaschek, R., Tolhuis, B., et al. (2009). Functional anatomy of polycomb and trithorax chromatin landscapes in Drosophila embryos. *PLoS Biology, 7*, e13.

Smallwood, S. A., Tomizawa, S., Krueger, F., Ruf, N., Carli, N., Segonds-Pichon, A., et al. (2011). Dynamic CpG island methylation landscape in oocytes and preimplantation embryos. *Nature Genetics, 43*, 811–814.

Smith, Z. D., Chan, M. M., Mikkelsen, T. S., Gu, H., Gnirke, A., Regev, A., et al. (2012). A unique regulatory phase of DNA methylation in the early mammalian embryo. *Nature, 484*, 339–344.

Sørensen, A. L., Jacobsen, B. M., Reiner, A. H., Andersen, I. S., & Collas, P. (2010). Promoter DNA methylation patterns of differentiated cells are largely programmed at the progenitor stage. *Molecular Biology of the Cell, 21*, 2066–2077.

Sørensen, A. L., Timoskainen, S., West, F. D., Vekterud, K., Boquest, A. C., Ährlund-Richter, L., et al. (2009). Lineage-specific promoter DNA methylation patterns segregate adult progenitor cell types. *Stem Cells and Development, 19*, 1257–1266.

Stancheva, I., El-Maarri, O., Walter, J., Niveleau, A., & Meehan, R. R. (2002). DNA methylation at promoter regions regulates the timing of gene activation in Xenopus laevis embryos. *Developmental Biology, 243*, 155–165.

Stancheva, I., & Meehan, R. R. (2000). Transient depletion of xDnmt1 leads to premature gene activation in Xenopus embryos. *Genes & Development, 14*, 313–327.

Tadros, W., & Lipshitz, H. D. (2009). The maternal-to-zygotic transition: A play in two acts. *Development, 136*, 3033–3042.

ten Bosch, J. R., Benavides, J. A., & Cline, T. W. (2006). The TAGteam DNA motif controls the timing of Drosophila pre-blastoderm transcription. *Development, 133*, 1967–1977.

Thomson, J. P., Skene, P. J., Selfridge, J., Clouaire, T., Guy, J., Webb, S., et al. (2010). CpG islands influence chromatin structure via the CpG-binding protein Cfp1. *Nature, 464*, 1082–1086.

Toyama, R., Rebbert, M. L., Dey, A., Ozato, K., & Dawid, I. B. (2008). Brd4 associates with mitotic chromosomes throughout early zebrafish embryogenesis. *Developmental Dynamics, 237*, 1636–1644.

Tsurumi, A., Xia, F., Li, J., Larson, K., LaFrance, R., & Li, W. X. (2011). STAT is an essential activator of the zygotic genome in the early Drosophila embryo. *PLoS Genetics, 7*, e1002086.

Vastenhouw, N. L., & Schier, A. F. (2012). Bivalent histone modifications in early embryogenesis. *Current Opinion in Cell Biology, 24*, 374–386.

Vastenhouw, N. L., Zhang, Y., Woods, I. G., Imam, F., Regev, A., Liu, X. S., et al. (2010). Chromatin signature of embryonic pluripotency is established during genome activation. *Nature, 464*, 922–926.

Veenstra, G. J., & Wolffe, A. P. (2001). Constitutive genomic methylation during embryonic development of Xenopus. *Biochimica et Biophysica Acta, 1521*, 39–44.

Vesterlund, L., Jiao, H., Unneberg, P., Hovatta, O., & Kere, J. (2011). The zebrafish transcriptome during early development. *BMC Developmental Biology, 11*, 30.

Walser, C. B., & Lipshitz, H. D. (2011). Transcript clearance during the maternal-to-zygotic transition. *Current Opinion in Genetics and Development, 21*, 431–443.
Weber, M., Hellmann, I., Stadler, M. B., Ramos, L., Paabo, S., Rebhan, M., et al. (2007). Distribution, silencing potential and evolutionary impact of promoter DNA methylation in the human genome. *Nature Genetics, 39*, 457–466.
Wu, S. F., Zhang, H., & Cairns, B. R. (2011). Genes for embryo development are packaged in blocks of multivalent chromatin in zebrafish sperm. *Genome Research, 21*, 578–589.
Wu, S. F., Zhang, H., Hammoud, S. S., Potok, M., Nix, D. A., Jones, D. A., et al. (2011). DNA methylation profiling in zebrafish. *Methods in Cell Biology, 104*, 327–339.
Xie, W., Barr, C. L., Kim, A., Yue, F., Lee, A. Y., Eubanks, J., et al. (2012). Base-resolution analyses of sequence and parent-of-origin dependent DNA methylation in the mouse genome. *Cell, 148*, 816–831.

CHAPTER FOUR

# Chromatin Architectures and *Hox* Gene Collinearity

**Daan Noordermeer**[*], **Denis Duboule**[*,†,1]

[*]National Research Centre "Frontiers in Genetics", School of Life Sciences, Ecole Polytechnique Fédérale, Lausanne, Switzerland
[†]National Research Centre "Frontiers in Genetics", Department of Genetics and Evolution, University of Geneva, Geneva, Switzerland
[1]Corresponding author: e-mail address: denis.duboule@epfl.ch

## Contents



## Abstract

Ever since the observation that collinearity, that is, the sequential activity of *Hox* genes based on their relative positions within their gene clusters, is conserved throughout most of the animal kingdom, the question has been raised as to what are the underlying molecular mechanisms. In recent years, technological advances have allowed to uncover changes in chromatin organization that accompany collinearity at *Hox* gene clusters. Here, we discuss insights in the dynamics of histone modifications and 3D organization in *Drosophila* and mammals and relate these findings to genomic organization of *Hox* gene clusters. Using these findings, we propose a framework for collinearity, based on five components: clustering, coating, compaction, compartmentalization, and contacts. We argue that these five components may be sufficient to provide a mechanistic ground for the readout of collinearity in *Drosophila* and vertebrates.

*Current Topics in Developmental Biology*, Volume 104
ISSN 0070-2153
http://dx.doi.org/10.1016/B978-0-12-416027-9.00004-8

## 1. INTRODUCTION

The formation of an embryo from a single fertilized cell is one the most fundamental processes in nature. A diverse array of morphogenetic strategies has evolved to bring together various differentiated cell types and tissues, which will eventually produce a coherent organism. These strategies generally require, at some point, the determination of global embryonic axes, such as the anterior to posterior (AP) axis from the head to the caudal part of the embryo. The subsequent deployment of various structures along this major body axis can be achieved in different ways, for example, with or without segmentation. Development can thus rely exclusively upon cell lineage instructions, like in *Caenorhabditis elegans*, or upon the reiteration of segments that at later stages will acquire their identities. In the case of segmentation, two programs can be followed: (1) segmentation can be sequential in time and progress from anterior to posterior, as in all vertebrates and many invertebrates or (2) segmentation can be simultaneous, like in *Drosophila* and other long germ band insects. Importantly, in all these scenarios, the identification of structures along this AP axis depends on the activity of the homeobox-containing *Hox* gene family (de Rosa et al., 1999; Garcia-Fernandez, 2005; McGinnis & Krumlauf, 1992).

Various combinations of HOX proteins along the AP axis will trigger the morphogenesis of different structures, whereas incorrect combinations will ill-specify body parts, a phenomenon referred to as homeotic transformations (Lewis, 1978). The correct distribution of protein members from the HOX family to the AP axis is therefore of utmost importance. The precision achieved in this spatial distribution is largely due to a mechanism that translates the genomic position of any given *Hox* gene into both its order of activation and its maintained activity (Figs. 4.1 and 4.2). This "spatial collinearity", first recognized by Lewis when working out the genetics of the Bithorax complex (BC-X) in *Drosophila melanogaster* mutants (Lewis, 1978), was observed in vertebrates to regulate this highly conserved *Hox* gene family as well (Duboule & Dolle, 1989; Gaunt, Sharpe, & Duboule, 1988; Graham, Papalopulu, & Krumlauf, 1989; Kmita & Duboule, 2003; Krumlauf, 1994). In animals that develop in an anterior to posterior temporal sequence, *Hox* genes are further activated in a temporal sequence that reflects the genes' positions along the *Hox* clusters, a mechanism referred to as temporal collinearity (Izpisua-Belmonte, Falkenstein, Dolle, Renucci, & Duboule, 1991). Over the years, this progressive gene activation in time and space, following the genomic topography, has proved to be a difficult question to solve. However, recent

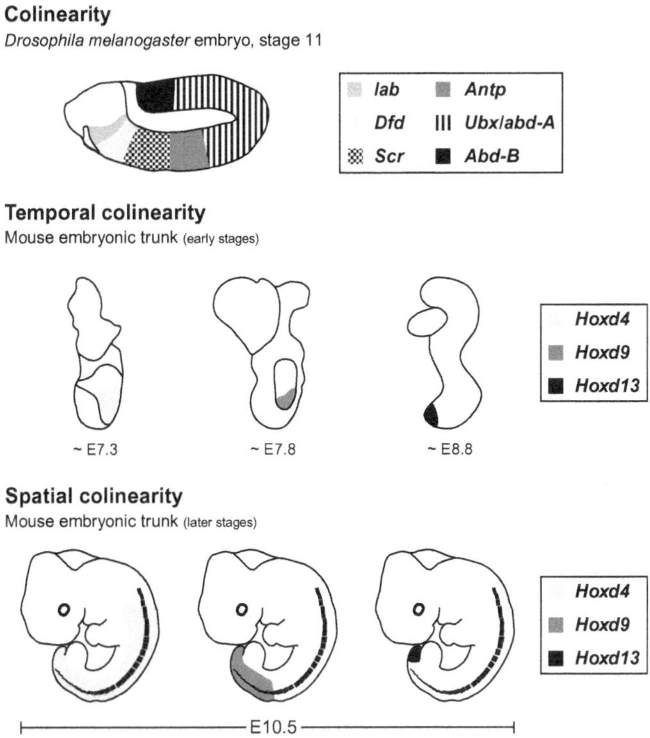


**Figure 4.1** Schematics of *Hox* gene expression in *Drosophila* and mouse. Top: Collinear expression domains of *Hox* genes in stage 11 *Drosophila* embryo. Middle: Temporal collinear expression domains of *Hoxd* genes in the early embryonic mouse trunk, at different stages of embryonic development. Bottom: Overlapping spatial collinear expression domains of *Hoxd* genes in the late embryonic mouse trunk. *Top panel data is taken from Kosman et al. (2004).*

studies focusing on the epigenetic status and the three-dimensional (3D) organization of *Hox* clusters, combined with both reverse genetics and the increased numbers of available animal genomes, have started to reveal some of the regulatory rules associated with these mechanisms. In this chapter, we compare recent insights into these processes obtained from both *Drosophila* and mammals and try to relate these findings to the function and evolution of *Hox* genes.

## 2. *HOX* GENE FUNCTION AND GENOMIC ORGANIZATION

Mutations in *Hox* genes of *Drosophila* were originally identified because of their ability to switch imaginal disk identity during embryonic development, which resulted in dramatic changes in adult body structures.

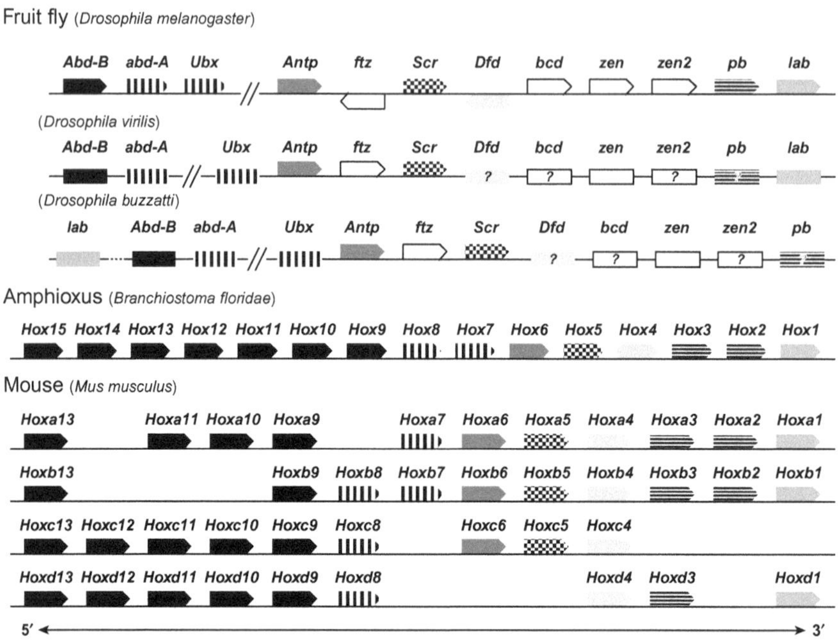


**Figure 4.2** Schematic organization of *Hox* clusters in *Drosophila*, Amphioxus, and mouse. Genes with a known transcriptional orientation are indicated with arrows and genes with an unknown orientation as boxes. Orthologous genes are depicted with the same filling. *Drosophila* homeobox-containing genes that are not considered as genuine *Hox* genes are indicated with unfilled boxes. Genes whose location has not yet been reported are indicated with question marks. The transcriptional directionality of the clusters is indicated below.

For example, the incorrect activity of the *Antennapedia* gene leads to the substitution of antennas by legs (Gehring, 1966; Schneuwly, Klemenz, & Gehring, 1987; Struhl, 1981). Such homeotic transformations have been observed whenever *Drosophila* and vertebrate *Hox* genes were inappropriately activated, thus illustrating their role in organizing segmental identities along the developing AP axis (e.g., Chisaka & Capecchi, 1991; Kessel, Balling, & Gruss, 1990; Morata, Botas, Kerridge, & Struhl, 1983).

*Hox* genes provide these regional identities via their discrete patterns of sustained functional activities (Fig. 4.1). This is achieved either by domain-specific transcriptional controls, like stripes in *Drosophila* (e.g., Akam & Martinez-Arias, 1985; Gaunt et al., 1988; Kosman et al., 2004; Levine, Hafen, Garber, & Gehring, 1983), or by using the functional dominance of some posterior HOX proteins over more anterior ones, a feature that turns largely overlapping expression territories into restricted functional domains ("posterior prevalence"; Duboule & Morata, 1994). The initial transcriptional activation of *Hox* genes requires general cell signaling pathways (e.g., Wnt, Notch, FGF; reviewed in Deschamps & van Nes, 2005), as well as the combined readout of regulatory elements (e.g., in *Drosophila*: Maeda & Karch, 2006, 2010). Segmental identity is subsequently maintained by fixing these expression patterns. This mechanism seems to be particularly stable, as illustrated by adult human cell lines derived from different body levels, which continue to transcribe *Hox* combinations resembling, to some extent, their embryonic body levels of origin (Chang et al., 2002; Rinn, Bondre, Gladstone, Brown, & Chang, 2006).

In bilateria, *Hox* genes are often linked in a genomic cluster and the ever increasing availability of novel animal genomes suggests that early bilaterian animals carried an ancestral *Hox* cluster that contained a rather small number of genes (Chourrout et al., 2006). Horizontal gene duplication and subsequent reorganization gave rise to a more complex cluster (Fig. 4.2), which may have permitted the evolution of the wide diversity in bilaterian body plans. *Drosophila* subspecies contain eight *Hox* genes, localized in two to three split segments of an ancestral cluster (see Akam, 1989; Negre, Ranz, Casals, Caceres, & Ruiz, 2003; Ranz, Segarra, & Ruiz, 1997). On the other hand, the primitive chordate Amphioxus has a much-expanded *Hox* cluster that contains 15 genes, each being somewhat orthologous to either one of the *Drosophila Hox* genes (Fig. 4.2; Holland, Albalat, et al., 2008). In vertebrates, multiple rounds of genome duplications, which occurred at the basis of this taxon (Ohno, 1970), generated four paralogous *Hox* clusters (and more in most fish species; see Kuraku & Meyer, 2009), which are structurally related to that of Amphioxus (Lynch & Wagner, 2009; Ohno, 1970). Within the vertebrate gene clusters, the high functional equivalence of the HOX proteins (e.g., Greer, Puetz, Thomas, & Capecchi, 2000) may have allowed for substantial structural reorganization and fine tuning of the transcriptional regulatory programs.

Notwithstanding the common themes in *Hox* gene organization, the genomic aspects of *Hox* clusters can vary considerably between animals (Fig. 4.3; Duboule, 2007). For example, mammalian *Hox* clusters display the strictest internal organization: (1) *Hox* genes are present in uninterrupted and compact genomic clusters, (2) they are all transcribed in the same orientation, (3) they contain few and short introns, and (4) they are generally depleted from repetitive elements (Fried, Prohaska, & Stadler, 2004). In Amphioxus, as well as in many invertebrates, a single cluster exists that is similarly organized, yet with a less stringent structure. As a result, the gene

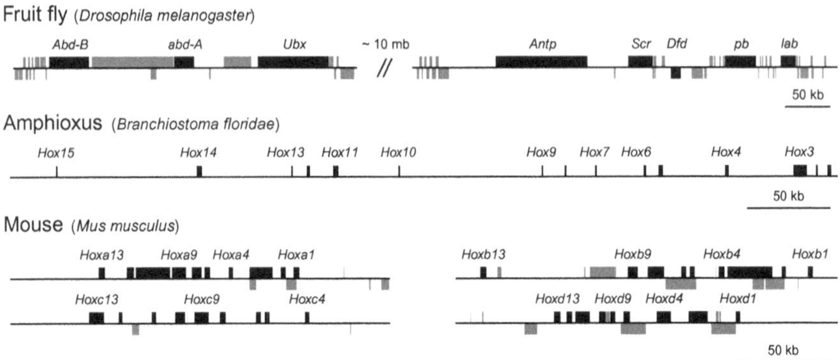


**Figure 4.3** Genomic organization of *Hox* clusters in *Drosophila*, Amphioxus, and mouse. *Hox* genes are depicted as black boxes and other genes (transcripts) are shown as grey boxes. Scale bars for each species are indicated below the clusters. Due to the incomplete annotation of the Amphioxus genome, no information is available for other transcripts in the *Hox* cluster.

cluster is usually larger and may contain some repeats. Finally, in *Drosophila* species, *Hox* genes are even less strictly organized. Depending on the species, the genes are found in two to three sub-clusters, with several possible internal breakpoints (Fig. 4.2; Negre et al., 2003; Ranz et al., 1997). In *D. melanogaster*, the ANT-C cluster contains from the *lab* to the *Antp* genes, whereas the BX-C cluster contains from *Ubx* to *Abd-B*. In other *Drosophila* species, though, *Ubx* can be linked to ANT-C. Besides the split of the ancestral cluster, the organization of both the semi-clusters and the *Hox* genes themselves appear less structured. *Drosophila Hox* genes are spread out over larger genomic regions, they may be transcribed in either orientations and they may contain large introns (Fig 4.3). Yet, *Drosophila* spatial collinearity is maintained, in spite of this highly derived organization. As such, regulation of collinearity in *Drosophila* requires less of a cluster-wide type of regulation, when compared to Amphioxus or mouse. This situation is even more striking in the tunicate Oikopleura, where *Hox* genes are completely isolated from one another, yet they still keep some traces of collinear regulation (Seo et al., 2004; see Section 4.3).

*Drosophila* and mouse *Hox* gene activities differ further in their distribution along the AP axis (Fig. 4.1). *Drosophila Hox* genes are generally expressed in non-overlapping domains, with the exception of the partially overlapping *Ubx* and *abd-A* transcript territories (Kosman et al., 2004). In contrast, the transcription of murine *Hox* genes is usually maintained along the AP axis, such that the spatial distribution of the expression domains includes progressively more genes toward more caudal parts of the embryo, along with the sequential activation process (temporal collinearity, like Russian dolls; see Fig. 4.1). In this chapter, we discuss recent insights concerning both the epigenetic and the 3D chromatin organization of *Hox* clusters, and in particular how these variables may help translating a specific genomic topology into diverse collinear transcriptional programs.

## 3. THE MANY FACES OF COLLINEARITY

During *Drosophila* embryogenesis, the *labial* (*lab*) gene, located at one extremity of ANT-C, is the most anteriorly expressed *Hox* gene, whereas *Abd-B*, a gene located at an extremity of BX-C, is active in the most posterior part of the embryo (Figs. 4.1 and 4.2; Kosman et al., 2004). For their initial activation, *Drosophila Hox* genes require local enhancers (e.g., Maeda & Karch, 2006, 2010). At later stages, active and inactive transcriptional states are maintained by the Trithorax and Polycomb protein complexes (see below).

In vertebrate embryos, collinear mechanisms are more diverse and observed in a wide range of developing structures. Collinearity along the primary AP axis, which is generally considered as the evolutionary most ancient function (Duboule, 2007), can be divided in temporal and spatial modalities (Deschamps et al., 1999; Deschamps & van Nes, 2005; Kmita & Duboule, 2003). These collinear modalities are implemented at all four paralogous gene clusters. The first *Hox* gene activity is detected early on, in mouse embryos at embryonic stage 7.2 (E7.2), when the 3′-located group 1 *Hox* genes are activated in the primitive streak (Fig. 4.1). In chick embryos, early activation is also observed in gastrulating mesoderm cells (Iimura & Pourquie, 2006). In a temporal sequence, *Hox* genes are subsequently activated one after the other (the "*Hox* clock"; Duboule, 1994), up to the most 5′-located group 13 genes at around E8.5.

*Hox* gene activity in vertebrates is maintained up to the stage of somitogenesis and later, which results in domains where the transcription of increasingly more *Hox* genes overlaps (Fig. 4.1). This process coincides with axial elongation and the accompanying segmentation clock (Pourquie, 2003), and hence, both clocks must be tightly coordinated to achieve proper body patterning. In Amphioxus, both temporal and spatial collinearities are observed as well, thus confirming that these mechanisms are already present in primitive vertebrates and predated the genome duplication events (reviewed in Holland, Holland, & Gilland, 2008). Indeed, the presence of an uninterrupted *Hox* gene cluster was previously proposed to be indispensable for a temporal sequence in AP patterning (Duboule, 1992), a relationship that still holds for any animal species displaying such a developmental strategy.

To understand the underlying molecular processes, both temporal and spatial collinearities have been intensively studied at the mouse *HoxD* cluster by using a combination of genetic and molecular tools (Tschopp & Duboule, 2011; Tschopp, Tarchini, Spitz, Zakany, & Duboule, 2009). Interestingly, internal deletions in the gene cluster affect the outcomes of temporal and spatial collinearities in different ways. Regarding temporal collinearity, internal deletions generally accelerated the activation of genes located 5′ to the deletion breakpoint, whereas in certain instances it delayed the transcriptional activation of genes located 3 to the breakpoint (Tschopp et al., 2009). Therefore, the timing of gene activation was proposed to rely on a balance between a 3′-located positive regulation and a 5′-located repressive influence (Fig. 4.4A). In some cases, the separation of the *HoxD* cluster from its 5′-located flanking sequences accelerated gene expression, further supporting the presence of repressive elements on this side (Tschopp & Duboule, 2011). Importantly though, *Hoxd* genes were still activated following the correct order, indicating that additional repressive influences are acting from within *HoxD* cluster to delay the transcription of the most 5′-located members. At later stages of development, the same internal deletions mostly affected genes located near the breakpoints, suggesting that local enhancers and silencers contribute to the maintenance of spatial collinear patterns (Fig. 4.4B; Tschopp et al., 2009).

In vertebrates, *Hox* genes also function in axial structures other than the major body axis, where they implement newly acquired collinear transcriptional programs (Favier & Dolle, 1997; Hombria & Lovegrove, 2003). The most thoroughly studied example is that of limb patterning, where analyses of the mouse *HoxA* and *HoxD* cluster have provided insights into how these co-opted mechanisms may differ from the original collinear program along the primary body axis (Kmita & Duboule, 2003). Proper patterning of the mouse limb involves the localized expression of a pair of *Hoxa* genes, as well as the implementation of two opposite collinear programs at the *HoxD* cluster (Kmita et al., 2005; reviewed in Woltering & Duboule, 2010; Zakany & Duboule, 2007).

At early stages of limb budding, a temporal and spatial collinear program is deployed that resembles in many respects the early collinear mechanisms implemented along the developing trunk axis. This mostly involves genes from the central part of the cluster, from *Hoxd8* to *Hoxd11* (Tarchini &

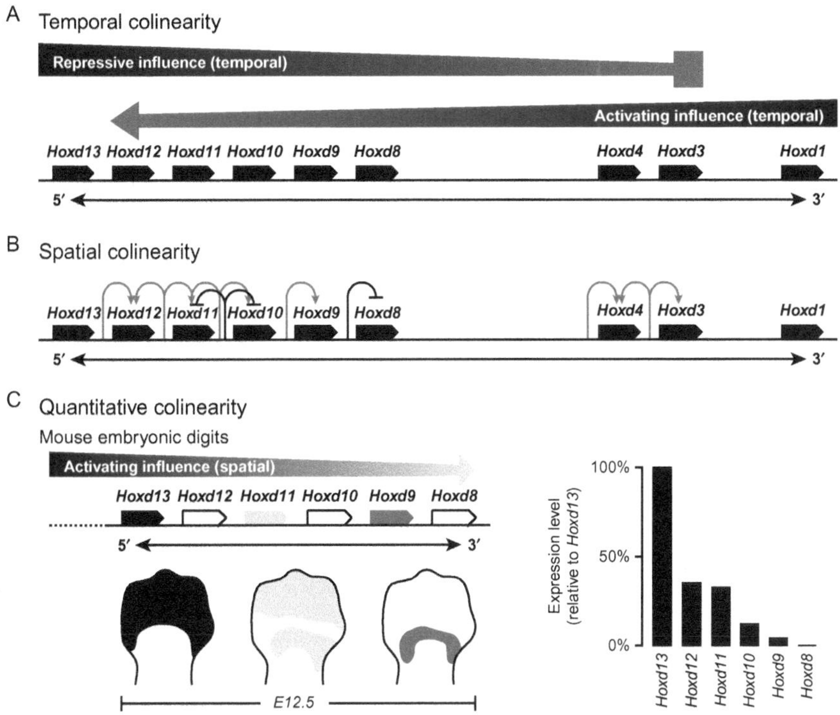


**Figure 4.4** Examples of collinear programs at the mouse *HoxD* cluster. (A) Directional regulatory influences modulating temporal collinearity in the early embryonic trunk. (B) Examples of local regulatory influences maintaining transcriptional states during spatial collinearity in the embryonic trunk, at later stages. (C) Distant 5′-located enhancers, imposing quantitative collinear expression of *Hoxd* genes in embryonic digits. The resulting expression domains are indicated on the left and the quantitation of steady-state mRNA levels in digits on the right. *Panels (A and B) and (C) data are taken from Tschopp et al. (2009) and Montavon, Le Garrec, Kerszberg, and Duboule (2008).*

Duboule, 2006; Zakany, Kmita, & Duboule, 2004). Activation of this phase of transcription requires sequences located in the 3′ neighborhood of the gene cluster (Spitz et al., 2001; Spitz, Herkenne, Morris, & Duboule, 2005). Subsequently, when digits are laid down, a quantitative collinear mechanism is observed, which results in a progressive decrease in transcriptional outcome, relative to genes' position (Fig. 4.4C). In developing digits, *Hoxd13* is expressed at maximum level, whereas *Hoxd9* and *Hoxd8* are transcribed with minimal efficiency (Kmita, Fraudeau, Herault, & Duboule, 2002; Montavon et al., 2008). In contrast to the early phase of activation, this second mechanism requires sequences located in the 5′ neighborhood of the gene cluster (Montavon et al., 2011; Spitz et al., 2005).

This increased diversity and complexity in collinear programs has been hypothesized to require a strict topological organization of *Hox* clusters as observed, for example, in mouse (discussed in detail in Duboule, 2007). In this model, split and fragmented clusters are unable to implement temporal collinearity. In contrast, some basic traits of spatial collinearity can be achieved, which thus appear to be more gene-autonomous. The temporal process necessitates an intact and well-organized gene cluster, as well as some global regulatory influences located outside of it. The duplication of this "consolidated" organization in primitive vertebrates, following the 2R events, may have provided a window of opportunity to evolve new collinear programs by simply positioning strong regulatory elements in the neighborhoods of the gene clusters. This is supported by the genomic organization of both the *HoxD* and the *HoxA* clusters, which are surrounded by large gene deserts containing conserved regulatory elements (Lee, Koh, Tay, Brenner, & Venkatesh, 2006; Lehoczky, Williams, & Innis, 2004). These newly acquired collinear programs in turn may have induced a further structuring phase, leading to the observed difference between these genomic loci in vertebrates and early chordates (Fig. 4.2).

## 4. ARE POLYCOMB AND TRITHORAX MEDIATORS OF COLLINEARITY?

For embryonic structures to be properly patterned, *Hox* gene transcription must be tightly regulated. In *Drosophila*, the spatial succession of expression domains requires a regulatory mechanism that can act as a switch. Yet at the same time, collinear activation of these genes ought to be dynamic and should allow sufficient selectivity among these closely spaced transcription units. Protein complexes belonging to two groups with opposing functions regulate and maintain *Hox* gene activity and may thus fulfill some of these requirements. Polycomb group proteins can maintain the repressed state of these genes, whereas Trithorax group proteins maintain their active state. Both Polycomb and Trithorax group proteins exert their function through modifications of histones, thereby inducing variable states of chromatin compaction. Mutations in genes from either group can result in homeotic transformations, indicating their importance for *Hox* gene regulation (Fig. 4.5A; reviewed in Paro, 1990; Schuettengruber et al., 2007).

Proteins belonging to both groups are found in many multiprotein complexes (Fig. 4.5A). *Drosophila* contains three major classes of Polycomb complexes: PRC2, PRC1, and PhoRC (Bantignies & Cavalli, 2011; Schwartz &

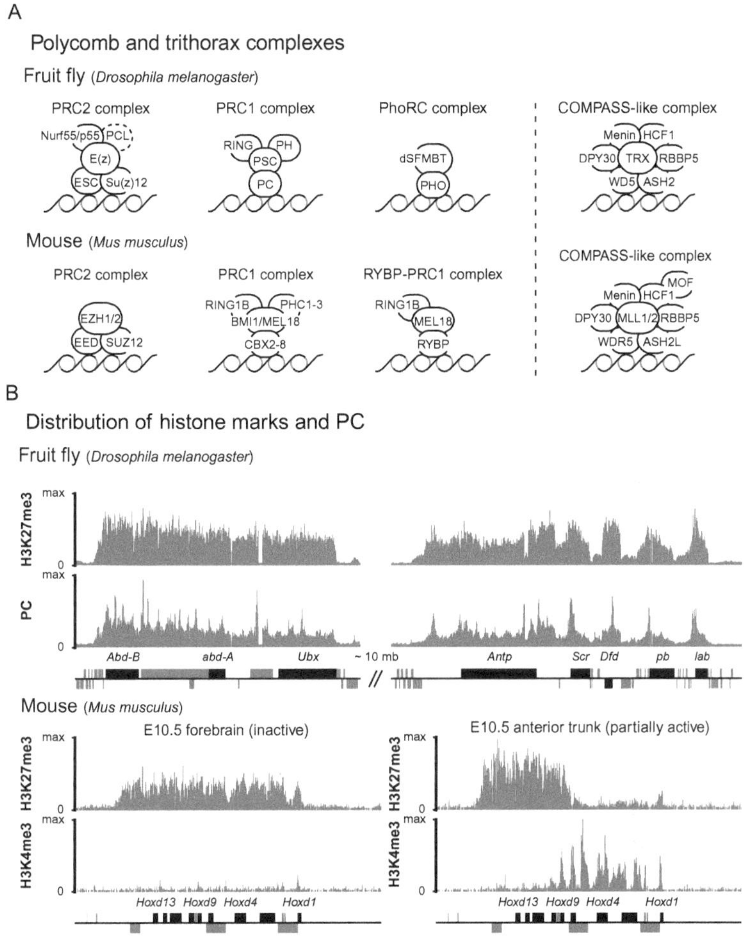


**Figure 4.5** Polycomb and Trithorax complexes and the distribution of modified histones on *Hox* clusters. (A) Polycomb (left) and Trithorax (right) complexes in *Drosophila* and mouse with described associations in *Hox* gene regulation. (B) Top: Distribution of the repressive H3K27me3 histone mark and PC (Polycomb) over BX-C and ANT-C in 4–12 h *Drosophila* embryos. Bottom: Distribution of repressive H3K27me3 and activating H3K4me3 marks on the *HoxD* cluster in the forebrain and the anterior part of the trunk in E10.5 mouse embryos. The anterior trunk sample is a dissection of the most dorsal part of the embryo, where most cells express the *Hoxd1–8* genes, following spatial collinearity. *Panel (A) compiled from Bantignies and Cavalli (2011), Mohan, Lin, Guest, and Shilatifard (2010), Sawarkar and Paro (2010), Schuettengruber, Chourrout, Vervoort, Leblanc, and Cavalli (2007), Schuettengruber, Martinez, Iovino, and Cavalli (2011), Schwartz and Pirrotta (2007, 2008), and Tavares et al. (2012). Panels (B, top) ChIP-on-chip data from Schuettengruber et al. (2009) and (B, bottom) ChIP-seq data from Noordermeer et al. (2011).*

Pirrotta, 2007, 2008). The PRC2 protein complex, besides other functions, trimethylates lysine 27 of histone H3 tail (H3K27me3), a modification associated with transcriptional repression. The H3K27me3 mark can subsequently recruit the Polycomb (PC) subunit of the PRC1 complex, which in turn ubiquitinates lysine 199 of histone H2A (H2AK199ub), a modification that helps to induce a compacted chromatin state (see Section 4.5). The PhoRC complex can bind DNA and may associate with the PRC2 complex, though this seems insufficient to explain targeting to all Polycomb targets (Wang, Brown, et al., 2004). In mammals, similar PRC complexes are present, though with a larger variety in their subunits (Fig. 4.5A).

Several Trithorax complexes are present in *Drosophila* and mammals, which may function as chromatin remodelers, histone methyltransferases, and histone acetyltransferases (Mohan et al., 2010; Schuettengruber et al., 2007, 2011). In *Drosophila,* the Trithorax (TRX) containing COMPASS-like complex is required for maintained *Hox* gene activity (Breen & Harte, 1991) and can deposit the activating H3K4me3 histone mark (Fig. 4.5A; Czermin et al., 2002; Eissenberg & Shilatifard, 2010; Smith et al., 2004). In mammals, four COMPASS-like complexes are found, of which only the MLL1 and MLL2 containing complexes can deposit the H3K4me3 mark at *Hox* loci (Wang et al., 2009).

The distribution of both Polycomb and Trithorax components at *Hox* clusters in *Drosophila*, as well as the presence of H3K27me3 and H3K4me3 marks, has been determined using ChIP-on-chip (Fig. 4.5B; Schuettengruber et al., 2009; Schwartz et al., 2006). Large domains of H3K27me3 coat both the BX-C and the *Antp* and *Scr* genes in embryos and embryonic cell lines. In contrast, the presence of H3K27me3 marks at the more anteriorly expressed *lab*, *pb*, and *Dfd* is more gene specific (Fig. 4.5B; Schuettengruber et al., 2009). Also, the distribution of PRC1 and PhoRC subunits is restricted to somewhat smaller regions within both BX-C and ANT-C (Fig. 4.5B; Kahn, Schwartz, Dellino, & Pirrotta, 2006; Schuettengruber et al., 2009; Schwartz et al., 2006). These regions generally overlap with Polycomb Response Elements (PREs), which are defined as *cis*-regulatory elements required to maintain *Hox* gene silencing (Chan, Rastelli, & Pirrotta, 1994; Muller & Kassis, 2006; Ringrose & Paro, 2007).

The nature and function of PREs have been intensively studied at the *Ubx* locus (Papp & Muller, 2006; Tillib et al., 1999). Components of the PhoRC, PRC2, and PRC1 complexes can bind the *Ubx* PREs, irrespective of its state of transcriptional activity. This behavior may be gene- or cell type-specific though, since the *Abd-B* PREs, in an embryonic cell line, are bound by

Polycomb complexes in the repressed state only (Schwartz et al., 2006). Interestingly, the Trithorax-component TRX is also constitutively bound to these PREs (Papp & Muller, 2006) and similar co-occupation was detected at many Polycomb targets loci in *Drosophila* embryos, using ChIP-on-chip (Schuettengruber et al., 2009). Polycomb and Trithorax group components bind closely neighboring sequences though, as determined in the case of one *Ubx* PRE. Despite their apparent colocalization, these two protein complexes therefore appear to bind independently from one another (Tillib et al., 1999). An exception to the constitutive binding of these protein complexes is found at the *Ubx* locus by the Trithorax protein ASH1, which can deposit the H3K4me3 mark, yet it is not a member of a COMPASS-like complex (Beisel, Imhof, Greene, Kremmer, & Sauer, 2002; Schuettengruber et al., 2011). ASH1 strongly binds the *Ubx* promoter only when active, whereas its binding to PREs is still elusive (Papp & Muller, 2006; Sanchez-Elsner, Gou, Kremmer, & Sauer, 2006).

Interestingly, *Drosophila* PREs are depleted in nucleosomes and consequently, they cannot carry H3K27me3 or H3K4me3 marks by themselves (Papp & Muller, 2006). Therefore, they may act as docking platforms for Polycomb complexes, which would distribute the H3K27me3 mark over distant *Hox* promoters and gene-bodies. Upon activation, binding of ASH1 to the *Ubx* promoter, together with TRX bound to the PREs, may override the H3K27me3-mediated repression, despite the presence of Polycomb components at PREs (Papp & Muller, 2006). However, how PRC2 is recruited to PREs in *Drosophila* is not yet fully understood and likely depends upon an intricate cross talk between protein complexes. While the PhoRC complex indeed displays sequence-specific DNA binding activity and associates with the PRC2 complex (Wang, Brown, et al., 2004), its binding alone is not sufficient to explain the recruitment of PRC2 to all targets. Furthermore, PhoRC associates with other protein complexes, including PRC1 (Mohd-Sarip, Venturini, Chalkley, & Verrijzer, 2002).

In the mouse, the distribution of H3K27me3 and H3K4me3 has been studied at *Hox* loci *in vivo*, during temporal and spatial collinearities (Fig. 4.5C; Noordermeer et al., 2011; Soshnikova & Duboule, 2009). During temporal collinear activation, two dynamic and mutually exclusive domains of histone marks cover the *Hoxd* cluster in the embryonic tailbud. At an early time point, H3K27me3 marks coat the inactive genes, whereas H3K4me3 marks are scored over active genes. At a later time point, the H3K4me3 domain has spread further, coinciding with a shrinking of the H3K27me3 domain and accompanying the transcriptional activation of additional genes (Soshnikova & Duboule, 2009). During spatial collinearity, that is when the regulatory programs are maintained in later stage embryos, similar domains of histone modifications are present along the primary AP axis (Fig. 4.5C, anterior trunk). H3K4me3-marked domains coat active *Hox* genes, with notable enrichments at the promoters. In contrast, contiguous H3K27me3-marked domains remain over *Hox* genes that are still inactive (Noordermeer et al., 2011). Therefore, these mutually exclusive domains demarcate *Hox* genes during both their initial temporal collinear activation and their subsequently maintained states of activities.

Fully inactive *Hox* clusters may carry two different patterns of histone modifications. In differentiated cells where all *Hox* genes are constitutively repressed, such as in the fetal brain, H3K27me3-marked domains coat the entire gene clusters, along with very low amounts (if any) of H3K4me3 (Fig. 4.5C, forebrain; Noordermeer et al., 2011). In pluripotent embryonic stem (ES) cells, where *Hox* genes are also inactive, moderate levels of H3K27me3 modifications coat the full gene clusters too, yet considerable H3K4me3 signals are detected over *Hox* gene promoters (Soshnikova & Duboule, 2009). In ES cells, such "bivalent" domains are a common feature of promoters for genes that encode developmental regulators. They are thought to maintain a repressed—yet transcriptionally competent—state of activity (Azuara et al., 2006; Bernstein et al., 2006). This bivalent chromatin state at *Hox* promoters is not observed in *Drosophila*, though the simultaneous occupancy of PREs by Polycomb and Trithorax components may provide a similar flexibility in responsiveness (Papp & Muller, 2006; Schuettengruber et al., 2009).

The recruitment of Polycomb and Trithorax complexes to mammalian *Hox* gene clusters is not well understood (Beisel & Paro, 2011; Schuettengruber & Cavalli, 2009). While some examples of mammalian PREs have been reported (Mendenhall et al., 2010; Sing et al., 2009; Woo, Kharchenko, Daheron, Park, & Kingston, 2010), a comprehensive understanding of which genomic feature(s) help recruit Polycomb is lacking. Three nonexclusive mechanisms have been proposed so far. First, PRC2 would associate with a variety of sequence-specific DNA binding factors (reviewed in Schuettengruber & Cavalli, 2009). For example, DNA binding of the mammalian PHO ortholog YY1 may target Polycomb to a specific region in the human *HoxD* cluster in mesenchymal stem cells (Woo et al., 2010). Furthermore, the REST and SNAIL transcription factors may promote the deposition of H3K27me3 on selected promoters in a neural precursor cell line (Arnold et al., 2012).

A second mechanism relies on the presence of GC-rich sequences, which are often bound by Polycomb components (Ku et al., 2008; Mendenhall et al., 2010). Such sequences do not necessarily need to be CpG islands, whose methylation is also associated with repressed gene activity. The majority of CpG islands are not bound by PRC2 (Ku et al., 2008) and the binding of PRC2 can precede their methylation (Mohn et al., 2008). Furthermore, CpG islands in the *HOXC* cluster are mostly unmethylated in human brain cells (Illingworth et al., 2008) similar to CpG islands around *HOX* gene promoters in human lung fibroblast cells (Weber et al., 2007). The dynamic recruitment of PRC2 to GC rich sequences may thus complement the repressive effect of DNA methylation.

A third mechanism of targeted Polycomb recruitment involves the function of large noncoding RNAs, as exemplified by HOTAIR (Gupta et al., 2010; Rinn et al., 2007). HOTAIR is transcribed from a region between the *HOXC12* and *HOXC11* genes. It was shown to recruit PRC2 in *trans*, to many target sites in the genome. HOTAIR knockdown in cellular systems mildly upregulated several genes, including the 5′-located *HOXD* genes (Gupta et al., 2010; Rinn et al., 2007). However, the direct impact of HOTAIR over the repression of posterior *HOXD* genes *in vivo*, via the recruitment of PRC2, could not be observed in mice carrying a deletion including this LncRNA (Schorderet & Duboule, 2011). Furthermore, HOTAIR binds the same unique site in the *HOXD* cluster in two different cell types, despite the fact that different *HOXD* genes are repressed in these cell types (Chu, Qu, Zhong, Artandi, & Chang, 2011). Also, this RNA is poorly conserved between human and mouse (Schorderet & Duboule, 2011) and, hence, its function may be in part specific to humans. Altogether, while this RNA may be of importance for recruiting PRC2 at many genomic loci, it is likely of little relevance in the dynamic recruitment of PRC2 at the *HOXD* locus during collinearity.

To our knowledge, a systematic mapping of Trithorax components at mammalian *Hox* clusters during development has not yet been reported. At these genomic loci, the H3K4me3 mark is deposited by the COMPASS-like MLL1 and MLL2 complexes that contain the unique Menin subunit, which distinguishes them from the MLL3 and MLL4 complexes (Wang et al., 2009). Menin has been proposed to target Trithorax complexes to *Hox* clusters via its association either with LEDGF, a factor required for full activity of *HOXA9* in cancer cells (Yokoyama & Cleary, 2008), or with ASH2L, a COMPASS-like subunit that binds the *HOXA* cluster in human foreskin fibroblasts (Chen et al., 2011; Hughes et al., 2004). In addition, the human LncRNA HOTTIP has been proposed to associate with the Trithorax components ASH2L and WDR5 and to activate *HOXA* genes in *cis*. Knockdown of HOTTIP interfered with the MLL1, WDR5, and ASH2L distribution over active *HOXA* genes in human foreskin fibroblast cells (Chen et al., 2011; Wang et al., 2011). Such noncoding RNAs have been reported to act in a similar way at a *Drosophila Ubx* PRE, via the recruitment of ASH1 (Sanchez-Elsner et al., 2006).

In summary, the presence of both H3K27me3 and H3K4me3 marks is tightly associated with collinear transcriptional programs, through their dynamic coating of *Hox* clusters. Polycomb complexes are likely instrumental in these processes, by mediating—or tightening—the repression of *Hox* genes at times and in places where given HOX proteins would be functionally detrimental to the developing embryo. Similarly, the Trithorax-containing COMPASS-like complexes appear instrumental in maintaining active states of *Drosophila Hox* genes and the deposition of the H3K4me3 mark at mouse *Hox* clusters (Breen & Harte, 1991; Wang et al., 2009). Importantly though, the majority of H3K4me3 in *Drosophila* and mammals is deposited at active promoters by COMPASS-like complexes, which may associate with RNA polymerases and contain neither the Trithorax nor the MLL1/MLL2 subunits (Ardehali et al., 2011; Lee & Skalnik, 2005, 2008; Lee, Tate, You, & Skalnik, 2007). Therefore, deposition of the majority of H3K4me3 marks at active *Hox* genes may merely be a readout of transcriptional activity, rather than a driving force.

While the functional organization of *Hox* clusters is generally conserved from *Drosophila* to mammals (Duboule & Dolle, 1989; Graham et al., 1989), the recruitment and distribution of Polycomb and Trithorax complexes at *Hox* clusters thus appears to differ substantially, due to some fundamental differences in the underlying developmental principles (Duboule, 1994).

## 5. DOWNSTREAM OF POLYCOMB AND TRITHORAX: A COMPACTED CHROMATIN ARCHITECTURE

Polycomb and Trithorax complexes and their associated histone modifications may help memorize the various collinear transcriptional states, through differential compaction (or condensation) of chromatin. Chromatin compaction as induced by PRC1 has been studied in some detail both *in vitro* and *in vivo*. *In vitro*, the addition of PRC1 leads to the compaction a human nucleosomal arrays, as determined by electron microscopy (Francis, Kingston, & Woodcock, 2004). This *in vitro* compaction requires the

PSC subunit, yet binding to the nucleosomes is independent of the histone N-terminal tails. In a cell line based *in vivo* system, the PRC1 subunit RING1B was essential for full compaction of mammalian *Hox* clusters, as seen by fluorescent microscopy (Eskeland et al., 2010). Intriguingly, however, this compaction is independent from the H2AK199 ubiquitin ligase activity of RING1B, suggesting that the latter subunit has a dual function. Chromatin compaction directly influences silencing, since the absence of RING1B leads to the derepression of both the *Drosophila Ubx* and mammalian *Hox* genes (Eskeland et al., 2010; Wang, Wang, et al., 2004).

On the other hand, a direct role for Trithorax complexes in chromatin decompaction has not been reported. The collinear activation of *Hox* genes coincides with the transcription of many large noncoding RNAs though, which may indirectly induce chromatin decompaction (reviewed in Hekimoglu & Ringrose, 2009). When *Drosophila Hox* genes are active, the surrounding chromatin, including the PREs, is transcribed following a similar collinear pattern (Bae, Calhoun, Levine, Lewis, & Drewell, 2002; Sanchez-Herrero & Akam, 1989). Also, when enforced, the transcription through PREs can derepress target genes and thus interfere with Polycomb-mediated silencing (Bender & Fitzgerald, 2002; Hogga & Karch, 2002; Rank, Prestel, & Paro, 2002). The production of noncoding RNAs may thus provide *Drosophila* Trithorax complexes an indirect means of antagonizing chromatin compaction. In mammals, many noncoding RNAs are located within *Hox* clusters and their activities coincide with the transcriptional state of the surrounding chromatin (Rinn et al., 2007; Sessa et al., 2007), like any promoter artificially introduced at the vicinity of *Hox* genes (see e.g., Herault, Kmita, Sawaya, & Duboule, 2002). While the transcription of surrounding chromatin in mammalian *Hox* cluster may thus serve a similar decompacting function, functional evidences are still lacking.

The dynamics of chromatin compaction accompanying *Hox* gene collinearity has been extensively studied using fluorescent microscopy (Chambeyron & Bickmore, 2004; Chambeyron, Da Silva, Lawson, & Bickmore, 2005; Eskeland et al., 2010; Morey, Da Silva, Perry, & Bickmore, 2007). Various states of chromatin compaction can be visualized using differentially labeled probes located on the 5′ and 3′ ends of murine *Hox* clusters. In cells where the clusters are inactive, the inter-probes distance within the *HoxB* and *HoxD* clusters is minimal, indicating compacted states. In *in vitro* differentiated ES cells, as well as in E9.5 embryonic cells, where 3′ located *Hox* genes are active, the inter-probes distances are significantly increased. Therefore, partially activated *Hox* clusters display a more decompacted state (Chambeyron & Bickmore, 2004; Chambeyron et al., 2005; Morey et al., 2007). In cells lacking RING1B, the derepressed *Hox* clusters also appear less compacted, despite being coated with H3K27me3 marks (Eskeland et al., 2010), suggesting that chromatin compaction is specifically maintained by the PRC1 complex, rather than by PRC2 and the associated H3K27me3 mark. In distal E10.5 limb bud cells, where the *HoxD* cluster is partially active and decompacted, RING1B occupancy is indeed strongly decreased (Williamson et al., 2012). In such cells, however, this decrease in RING1B occupancy is paralleled by a decrease in H3K27me3 occupancy as well (Montavon et al., 2011).

## 6. 3D CHROMATIN ORGANIZATION AND COLLINEARITY IN *DROSOPHILA*

The development of dedicated techniques like the Chromosome Conformation Capture (3C) approach, its derivatives 4C, 5C, and Hi-C, and the Dam-ID technique, has allowed to assess the 3D organization of genomic loci at high resolution (Simonis, Kooren, & de Laat, 2007; van Steensel & Dekker, 2010). In *Drosophila*, Dam-ID, 3C, and 4C have been used to investigate the chromatin architecture of BX-C in embryos, larvae, and cell lines (Fig. 4.6; Bantignies et al., 2011; Cleard, Moshkin, Karch, & Maeda, 2006; Lanzuolo et al., 2007). The 3D organization of the inactive BX-C was assessed by using the *Fab-7* PRE as a viewpoint in S2 cells. *Fab-7* interacts with other well-known PREs located in the cluster, as well as with the promoters and transcription termination sites of BX-C *Hox* genes. Interactions were also scored with interspersed sequences within the cluster, though at considerably lower levels. Likewise, the *abd-A* gene contacted various PREs and genes from the cluster, both in inactive S2 cells and in a mixed population of embryonic cells (Lanzuolo et al., 2007). A similar interaction between the *Fab-7* PRE and the inactive *Abd-B* promoter was previously identified in the head of adult flies, using a smaller Dam-ID screen (Cleard et al., 2006).

Of note, the interactions within BX-C are significantly higher than with the chromatin located outside, as shown by 3C and 4C (Fig. 4.6B; Bantignies et al., 2011; Lanzuolo et al., 2007). Consequently, when inactive, BX-C adopts a compartmentalized conformation, which is nucleated around the PREs, the promoters and the transcription termination sites (Fig. 4.6C). The extent of this 3D compartment largely matches the domain decorated with H3K27me3 marks. Accordingly, the H3K27me3 marked domains is

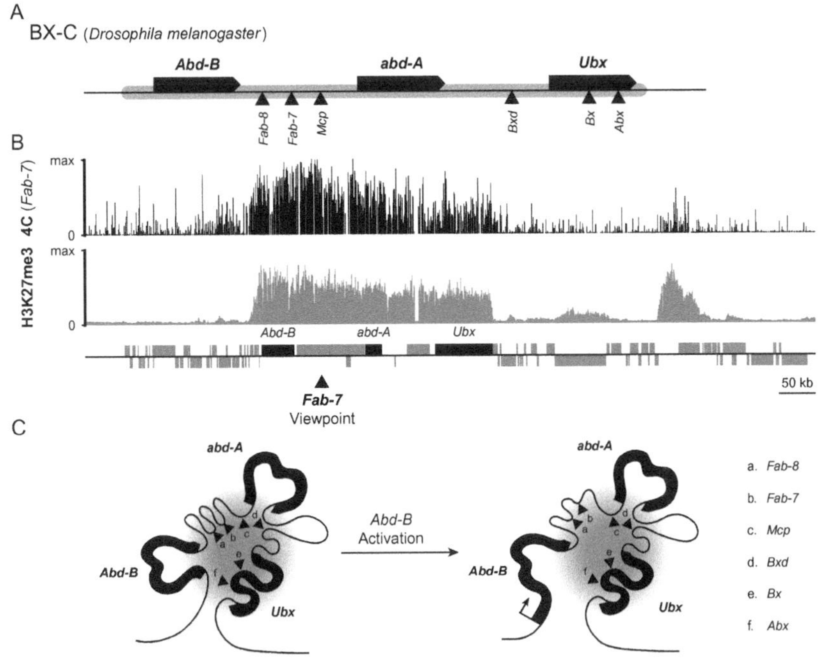


**Figure 4.6** 3D organization of the *Drosophila* BX-C. (A) Schematic representation of BX-C with genes indicated as arrows and well-characterized PREs as arrowheads. The thick grey line indicates the approximate extent of the H3K27me3 coated domain. (B) Unprocessed 4C signal in the inactive part of BX-C, from third-instar larval brain and anterior discs. The *Fab-7* PRE, regulating *Abd-B* and located between *Abd-B* and *abd-A*, was used as viewpoint (indicated below). For comparison, the distribution of the H3K27me3 marks in 4- to 12-h-old embryos is depicted below. (C) Schematic 2D representation of the BX-C 3D organization. On the left is the 3D organization when all genes are inactive and on the right is the organization when the *Abd-B* gene is active. Grey area represents the H3K27me3 marked 3D compartment. *Panel (B) 4C data from Bantignies et al. (2011) and ChIP-on-chip data from Schuettengruber et al. (2009). Panel (C) based on 3C data from Lanzuolo, Roure, Dekker, Bantignies, and Orlando (2007).*

physically separated from the surrounding chromatin. Within this 3D compartment, the repressive Polycomb machinery is present at elevated amount, which may strengthen and/or stabilize repression. Physical interactions within this chromatin compartment are also dependent on the Polycomb complex, as its depletion strongly reduces the frequency of these interactions (Lanzuolo et al., 2007).

When the *Abd-B* gene is active, BX-C adopts a different 3D organization. In S3 cells, the frequency of interactions between the *Fab-7* PRE and the active *Abd-B* promoter is strongly decreased, when compared to the inactive situation in S2 cells (Fig. 4.6C), though the interactions between the *Fab-7* PRE and other PREs (and the inactive genes) were essentially maintained. Likewise, in the same context, the *bxd* PRE kept most of its interactions, yet with a decreased frequency with both *Fab-7* and *Fab-8* PREs and an almost complete loss of interactions with the *Abd-B* promoter (Lanzuolo et al., 2007). A loss of interactions had been previously observed between the active *Abd-B* gene and *Fab-7* in the abdomen of adult flies, when compared to the "inactive" brain tissue (Cleard et al., 2006). The active *Abd-B* promoter thus loops out of the H3K27me3 labeled chromatin domain, whereas it locates inside when inactive (Fig. 4.6C). Whether or not such looping out from the inactive BX-C and ANT-C 3D compartments is a common feature during collinear expression in *Drosophila* remains to be determined.

4C studies with *Drosophila* larvae have identified significant interaction frequencies, though at low level, between Polycomb bound genomic regions, including BX-*C* and ANT-C (Bantignies et al., 2011; Tolhuis et al., 2011). Fluorescent microscopy in embryos confirmed that about 20 percent of those chromosomes carrying both *Abd-B* and *Antp* inactive alleles visually overlapped. This proximity decreased below ten percent when one of the two alleles was active. The deletion of *Fab-7*, located in the BX-C, only moderately decreased the interactions between the two gene clusters, but seemed to derepress *Antp* (Bantignies et al., 2011). In the *Drosophila* nucleus, the genomic regions marked by Polycomb and H3K27me3 are present in foci called "Polycomb bodies" (reviewed in Bantignies & Cavalli, 2011; Pirrotta & Li, 2012). This apparent clustering of repressed genes is mostly stochastic, though the size of the H3K27me3 domain, their proximity and their location on the same chromosome arm impact upon the interaction frequency (Bantignies et al., 2011; Sexton et al., 2012; Tolhuis et al., 2011). Also, such long-range clustering might be further organized by insulators, genomic elements that provide some spatial restriction to PREs (Li et al., 2011). Clustering at Polycomb bodies may provide another level where the repressive machinery can be concentrated, and hence, it may further secure and/or maintain repression.

To summarize, in the nuclear space, inactive *Hox* genes from the same complex cluster together. They may also contact other Polycomb targets, though at a lower frequency and on a more stochastic basis. At the time of their activation, *Hox* genes loop out from these inactive 3D structures. Because *Drosophila Hox* gene are regulated individually, by distinct combinations of factors inherited from the maternally driven segmentation strategy, it is possible that genomic clustering of *Hox* genes is mostly used to efficiently maintain their repressed states, via 3D compartmentalization, rather than to help coordinating their transcriptional activation, as may be the case in vertebrates (see below). This "repressive" function of *Hox* clustering in flies may explain why these genes are still clustered in different subspecies where the homeotic complexes have been reorganized (Fig. 4.2).

## 7. A 3D CHROMATIN TIMER FOR VERTEBRATE COLLINEARITY?

The 3D conformations of mammalian *Hox* clusters have been determined by 4C during the implementation of both spatial and quantitative collinearities. In E10.5 embryonic brain cells, the interaction profiles revealed that inactive *Hox* clusters are organized as distinct local 3D compartments (Fig. 4.7A; Noordermeer et al., 2011). The genomic limits of these compartments strikingly coincide with the domains of H3K27me3 modifications. As in *Drosophila*, the efficient maintenance of *Hox* gene repression appears to be accompanied by a 3D compartmentalization mechanism, which may thus be an ancestral feature of bilateria embryogenesis. Little substructure is observed within these inactive compartments, suggesting that the chromatin displays a random 3D organization. However, the mammalian CTCF insulator protein (Herold, Bartkuhn, & Renkawitz, 2012), may provide some scaffolding for local 3D compartmentalization. Mammalian *HOX* clusters contain series of CTCF binding sites, due to the unusually high concentration of GC islands, with the majority of them being located in between the (posterior) 5′-located genes. 3D modeling, using a dataset from cells where *Hox* clusters are inactive, indicated that CTCF binding sites are found in close spatial proximity. As CTCF sites are numerous at the 5′ extremity of the clusters, 3D compartments may partially nucleate around these sites (Ferraiuolo et al., 2010).

The activation of distinct subsets of *Hox* genes, at different AP domains, is paralleled by a dynamic 3D reorganization of chromatin, different from what is observed in *Drosophila*. In E10.5 cells obtained from the developing primary AP axis, the inactive 5′-located *Hox* genes remain organized in a local 3D compartment of restricted size, expectedly matching the H3K27me3 domain. In addition, active genes located at more 3′ positions and marked by H3K4me3 are also organized in a discrete local 3D compartment, which is distinctly separated from the negative domain (Fig. 4.7B; Noordermeer et al., 2011). At a later developmental stage, the *HoxC* cluster

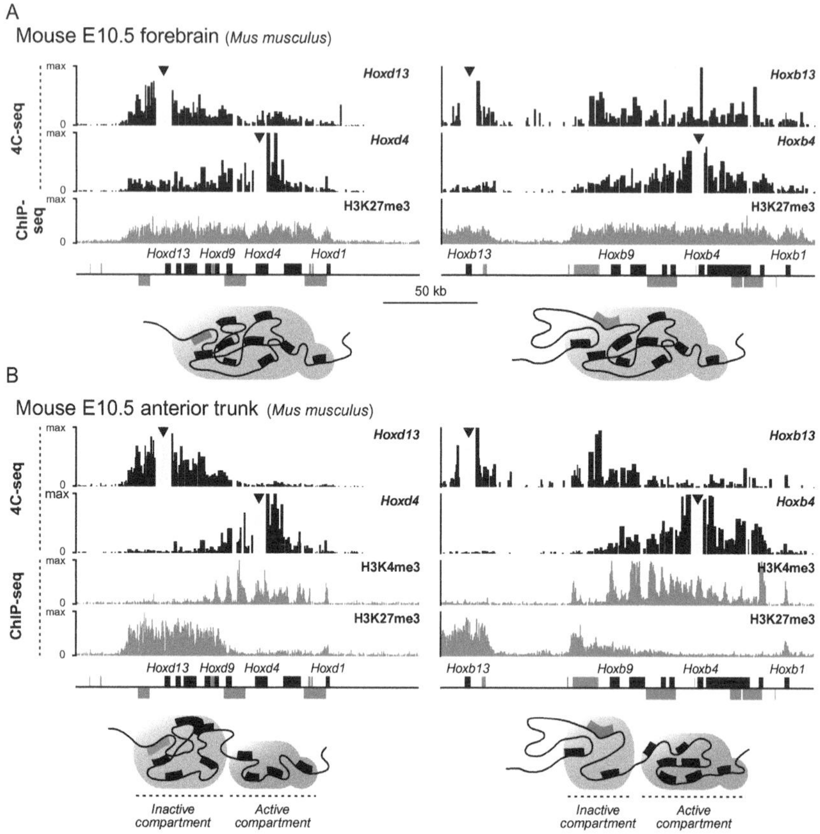


**Figure 4.7** 3D organization of the murine *Hox* clusters during spatial collinearity. (A) Local compartmentalization of inactive *Hox* clusters in the E10.5 mouse brain. Quantitative 4C-seq (Circular Chromosome Conformation Capture) signal and distribution of the H3K27me3 mark are shown. The *HoxD* cluster is on the left and *HoxB* on the right. The viewpoints *Hoxd13, Hoxd4, Hoxb13,* and *Hoxb4* are indicated. Local 3D compartmentalization of the inactive *HoxD* and *HoxB* clusters is schematized below. (B) Bimodal local compartmentalization of *Hox* clusters along the E10.5 mouse embryonic AP axis. Quantitative 4C-seq signals and the distribution of the H3K4me3 and H3K27me3 marks are shown. *Hox* genes from group 1–8 are active at this body level, in agreement with spatial collinearity. The same viewpoints as in (A) are shown. Local 3D compartmentalization of the *HoxD* and *HoxB* clusters along the AP axis is schematized below. *Panels (A) and (B) data from Noordermeer et al. (2011).*

shows a similar organization, suggesting that this bimodal structure is maintained along the AP axis during embryonic development (Min, Lee, & Kim, 2012). The size of these local compartments varies along the AP axis. In the caudal part of the embryo, where gradually more *Hox* genes are transcribed, the active 3D compartment extends further toward the 5′ end of the cluster, whereas the inactive compartment has retracted accordingly (Noordermeer et al., 2011). This bimodal organization leads to a physical separation both between active and inactive *Hox* genes, and between the *Hox* clusters and their chromatin environments. The 3D compartmentalization of active genes may involve activating factors, whose elevated concentration may in turn reinforce the transcriptional outcome and the recycling of the transcription machinery.

So far, the local compartmentalization of active chromatin has been reported only in mouse and human *Hox* gene clusters (Montavon et al., 2011; Noordermeer et al., 2011; Wang et al., 2011). It may be that the MLL1/MLL2 containing COMPASS-like complexes are involved in this spatial organization, as they specifically mediate H3K4me3 deposition at these loci (Wang et al., 2009). Whether or not active *Hox* genes in *Drosophila* are also organized within such 3D domains remains to be determined. However, unlike in vertebrates, the expression domains of *Hox* genes in *Drosophila* do not systematically overlap and, hence, 3D compartments, if any, would need to be more locally restricted.

Unlike other *Hox* loci, *Hoxb13* is separated from the *HoxB* cluster by a relatively large repeat-containing piece of DNA (Fig. 4.3; Zeltser, Desplan, & Heintz, 1996). When inactive, the *HoxB* cluster forms a 3D compartment that includes all genes, yet the intervening DNA loops out (Fig. 4.7A, right). Upon gene activation, this cluster expectedly adopts a bimodal 3D organization, which, however, does not include the intervening DNA in either compartment (Fig. 4.7B, right). Interestingly, the temporal and spatial expression pattern of *Hoxb13* is not affected by the deletion of the rest of the *HoxB* cluster (Medina-Martinez, Bradley, & Ramirez-Solis, 2000), suggesting that it is regulated autonomously from its closely related *Hox* cluster. In this case, as in *Drosophila*, the proximity to the *HoxB* cluster may help strengthening the repressive state.

By extension, vertebrate *Hox* clusters may also be organized as dynamic 3D compartments during the implementation of temporal collinearity. However, a progressive shift of 3D compartments in the same cell lineage along with time has not yet been reported. While spatial collinearity in mammals depends—at least in part—upon local enhancers, comparable to the case of *Drosophila* (Maeda & Karch, 2010; Tschopp et al., 2009), temporal collinearity seems to be guided by a global mechanism, also influenced by the neighborhood of the gene cluster (Tschopp et al., 2009). Noteworthy, should temporal collinearity rely upon a dynamic shift between local 3D compartments, this would give a mechanistic ground to the tight association between temporal collinearity and the presence of an uninterrupted *Hox* gene cluster (Duboule, 1994). In Amphioxus, indeed, the time-sequence in *Hox* genes activation occurs on the top of a relatively well organized *Hox* gene cluster (Fig. 4.3; Holland, Albalat, et al., 2008; Holland, Holland, et al., 2008b). In this view, the sequential transition of every gene from one domain to the other would act as a timer (the *Hox* clock). However, the underlying mechanism(s) that is the driving force responsible for this unidirectional transcriptional activation remains elusive.

Temporal collinearity correlates with the step-wise removal of H3K27me3 marks, from one extremity of the cluster to the other, at the same time H3K4 becomes trimethylated. As a result, local enhancers may be progressively mobilized, making *Hox* genes available for transcription (Noordermeer et al., 2011; Tschopp et al., 2009). The directionality of this mechanism may rely on an intrinsic polarity of *Hox* clusters, either for repressive or for activating (or both) protein complexes. Polarized effects have been reported when the overall amount of PRC1 components was decreased, via the deletion either of CBX2 (one of the four mammalian PC orthologous genes) or of BMI1 (one of two mammalian PSC orthologous genes). In both cases, *Hox* gene activity was anteriorized (Bel-Vialar et al., 2000; van der Lugt, Alkema, Berns, & Deschamps, 1996), whereas the opposite effect was scored in mice overexpressing PRC1 components (Alkema, van der Lugt, Bobeldijk, Berns, & van Lohuizen, 1995; van der Lugt et al., 1996). In both cases though, collinearity was maintained.

## 8. A REGULATORY ARCHIPELAGO AND COLLINEARITY IN DEVELOPING DIGITS

During the development of tetrapod digits, the five genes located at the posterior extremity of the *HoxD* cluster (*Hoxd9* to *Hoxd13*) are transcribed with an intensity decreasing along with their relative genomic position, with a maximal expression for *Hoxd13* (Fig. 4.4C; Dolle, Izpisua-Belmonte, Falkenstein, Renucci, & Duboule, 1989; Montavon et al., 2008). By using a targeted enhancer trap system, two regulatory regions, Prox and GCR, located centromeric to the gene cluster, were identified as digit enhancers (Fig. 4.8A; Gonzalez, Duboule, & Spitz, 2007; Spitz, Gonzalez, & Duboule, 2003). Subsequently, 4C experiments using *Hoxd13* as a viewpoint identified five additional islands of high interaction, all located in the centromeric gene desert (Fig. 4.8A; Montavon et al., 2011). Scanning deletion studies carried out *in embryo*, as well as transgenic reporter assays

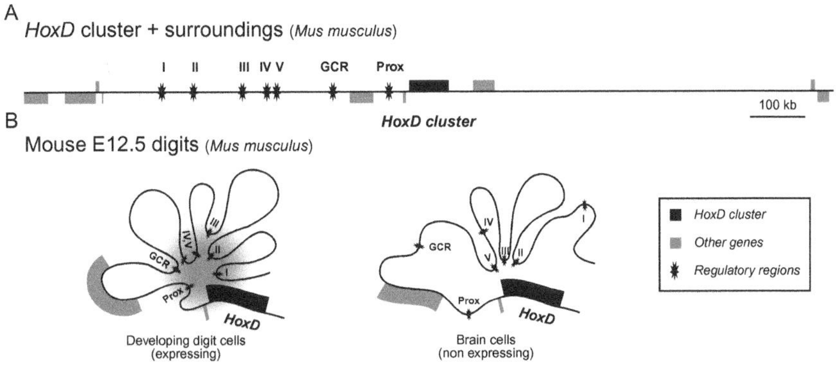


**Figure 4.8** 3D organization of the mouse *HoxD* cluster during quantitative collinearity in embryonic digits. (A) Genomic organization of the mouse *HoxD* cluster and surrounding gene deserts. The *HoxD* cluster is indicated in black and other genes in grey. Regulatory regions involved in developing digits are indicated by black stars. (B) Schematic 2D representation of long-range chromatin interactions at the *HoxD* cluster in developing digits (left) and nonexpressing brain cells (right), as determined by Chromosome Conformation Capture-on-Chip. *Panel (B) based on data from Montavon et al. (2011).*

confirmed the functional importance of at least five of these elements for *Hox* gene in the developing digits.

Targeted deletions of increasing numbers of regulatory elements within this "regulatory archipelago" had additive effects on *Hoxd* gene activity (Fig. 4.8B; Montavon et al., 2011), whereas various deletions within the 5′-located target *Hoxd* genes resulted in regulatory reallocations, leading to the upregulation of the remaining genes (Montavon et al., 2008). Also, the integration of a supernumerary copy of *Hoxd9* into the centromeric gene desert led to a decreased transcription of the native target genes, via a titration effect (Monge, Kondo, & Duboule, 2003).

Whether these complex and multiple enhancer–promoter interactions are somehow determined, for example by forming a large and rather static structure, or are stochastic and dynamic with various interactions occurring at different times and frequencies, is difficult to evaluate with the currently available technology. It is clear, however, that such distal enhancers can specifically activate the 5′-located *Hoxd* genes, while 3′-located genes are concomitantly repressed. As such, these interactions override the regulatory polarity of the *HoxD* cluster as observed along the embryonic AP axis (see above).

This situation illustrates how a novel collinear transcriptional program can be co-opted by using the preexisting functional organization of *Hox* clusters. In this particular case, the evolution of strong enhancer sequences outside the gene cluster induced an "opposite" directionality in the collinear mechanism, thus overriding the polarity as observed along the AP axis. Analogous to the situation in the embryonic AP axis, genes heavily transcribed in developing digits form a discrete 3D compartment, as detected by 4C analyses (Montavon et al., 2011; Noordermeer et al., 2011). This compartmentalization may help active genes to engage for interactions with distant regulatory elements in the nuclear space, in addition to increasing the local concentration of activating factors.

In embryonic brain cells, *Hoxd13* also contacted several regulatory islands even though all *Hox* genes are fully inactive (Fig. 4.8B, right; Montavon et al., 2011). These interactions may be part of a constitutive "prestructure" (groundstate) that provides scaffolding for efficient formation of long-range interactions in developing digits. Conversely, the inactive *Hoxd4* gene, in brain cells, established contacts with the telomeric neighborhood of the *HoxD* cluster. This polarity in the contacts is similar to what was observed in the anterior part of the trunk, where *Hoxd13* is inactive, whereas *Hoxd4* is active (Noordermeer et al., 2011). In the case of the developing limbs, this structural polarity observed at the *HoxD* cluster matches a partitioning of the regulatory landscapes. The early collinear activation of *Hoxd* genes in the future proximal part of the limb relies on regulatory elements located telomeric to the cluster, whereas subsequent transcriptional control, during digit development originates from the centromeric gene desert (Spitz et al., 2005).

Therefore, while inactive *Hox* clusters appear as single, largely homogeneous and local 3D compartments (see above), a directionality exists in long-range contacts, outside the cluster itself, regardless whether active or inactive cells are considered. Such genomic domains of constitutive long-range interactions (topological domains) are a common feature of the mouse and human ES cell genome (Dixon et al., 2012; Nora et al., 2012) and have been proposed to form "enhancer–promoter units," in which genes and their cell type-specific enhancers are contained (Nora et al., 2012; Shen et al., 2012). Besides facilitating the formation of long-range interactions between regulatory elements and target genes, these domains may also reduce interactions with nontarget genes, located outside the domain. A further reinforcement of regulatory maintenance may thus be achieved by long-range compartmentalization. In this context, it is noteworthy that the interaction domain observed between *Hoxd13* and its centromeric regulatory archipelago matches one such topological domain as reported by Ren and colleagues in ES cells (Dixon et al., 2012).

# 9. CLUSTERING, COATING, COMPACTION, COMPARTMENTALIZATION, AND CONTACTS: THE FIVE C's OF COLLINEARITY?

Recent studies have looked at chromatin dynamics at *Hox* clusters during collinear expression and drastic changes were observed in both *Drosophila* and mammals, concerning the profiles of histone posttranslational modifications (and associated proteins) and the higher order chromatin organization. While based on similar grounds, the collinear mechanisms come as different flavors in various animal classes, depending on which early developmental strategies are used. These findings nevertheless start to shape a framework wherein the multiple declinations of collinearity can be understood and which incorporates the parameters clustering, coating, compaction, compartmentalization, and contacts. We argue that these five parameters are important, if not sufficient, to account for the different outcomes of collinear programs in bilateria.

## 9.1. Clustering

Collinearity is the translation of a genomic topology into coordinated transcription programs. In this context, clustering of *Hox* genes may help to secure and enhance the necessary repression of *Hox* genes, at developmental times and in embryonic territories where these genes need to be transcriptionally inactive. Also, an uninterrupted *Hox* gene cluster is necessary for temporal collinearity to be fully implemented.

## 9.2. Coating

The coating, either of active *Hox* genes by H3K4me3 or of inactive genes by H3K27me3, labels the progression of collinear programs. Members of the Trithorax and Polycomb complexes deposit these marks over the *Hox* clusters in both *Drosophila* and mammals, yet their recruitment and distribution differ considerably. Selectivity appears to be achieved by two nonexclusive mechanisms. The first involves the binding of transcriptional regulators to local enhancers, whereas the second has a polar component and may provide both the entry point and the dynamics for temporal collinearity. In the course of vertebrate evolution, the co-optation of collinear programs involved distant regulatory elements, which could override the built-in polarity of this coating.

## 9.3. Compaction

The state of chromatin compaction, via the binding of PRC1, may strongly influence the transcriptional competence of *Hox* loci. The maintenance of the correct compaction state, as well as the decompaction process, in space and time, is thus essential. The active induction of local chromatin decompaction may guide transcriptional activities, yet the opposite process is equally possible, with transcription of the surrounding chromatin inducing the eviction of Polycomb complexes.

## 9.4. Compartmentalization

*Hox* clusters form dynamic local 3D compartments labeled by either H3K27me3 or H3K4me3 marks. These compartments physically separate active from inactive genes and the gene cluster from the surrounding chromatin. Polycomb components are concentrated into the inactive compartments, whereas Trithorax components and the transcription machinery are concentrated into the active compartment, where they likely synergize to promote and maintain transcription itself. The step-wise transition of *Hox* genes from the inactive to the active compartments may act as a timer accompanying temporal collinearity. The pace of transition may reflect the various affinities for the repressive and activating machineries, polarized toward the two extremities of the gene cluster.

## 9.5. Contacts

*Hox* clusters are involved in positive and negative long-range contacts. Collinear programs co-opted along with the radiation of vertebrates generally involve remote regulatory information, which can override the cluster-intrinsic repressive polarity. On the other hand, inactive *Hox* genes coated by H3K27me3 can establish low-frequency contacts with other *Hox* clusters. These arguably stochastic contacts, which occur at Polycomb bodies in *Drosophila*, may serve to further concentrate the repressive machinery.

# ACKNOWLEDGMENTS

We apologize to all authors whose work has been excluded due to space constraints. *Drosophila* ChIP-on-chip data for Figs. 4.5 and 4.6 was downloaded from http://cav-ouranos.igh.cnrs.fr/viewer-0.3_public/index.php. Data for Topological domains as discussed in section 4.8 was obtained from http://chromosome.sdsc.edu/mouse/hi-c/database.php. We thank members of the Duboule labs for useful discussion and acknowledge the financial support from the Ecole Polytechnique Fédérale (Lausanne), the University of Geneva, the Swiss National Research Fund, the National Research Centre "Frontiers in Genetics," and the European Research Council grant "SystemsHox.ch."

## REFERENCES


Akam, M. (1989). Hox and HOM: Homologous gene clusters in insects and vertebrates. *Cell*, *57*, 347–349.

Akam, M. E., & Martinez-Arias, A. (1985). The distribution of Ultrabithorax transcripts in Drosophila embryos. *The EMBO Journal*, *4*, 1689–1700.

Alkema, M. J., van der Lugt, N. M., Bobeldijk, R. C., Berns, A., & van Lohuizen, M. (1995). Transformation of axial skeleton due to overexpression of bmi-1 in transgenic mice. *Nature*, *374*, 724–727.

Ardehali, M. B., Mei, A., Zobeck, K. L., Caron, M., Lis, J. T., & Kusch, T. (2011). Drosophila Set1 is the major histone H3 lysine 4 trimethyltransferase with role in transcription. *The EMBO Journal*, *30*, 2817–2828.

Arnold, P., Scholer, A., Pachkov, M., Balwierz, P., Jorgensen, H., Stadler, M. B., et al. (2012). Modeling of epigenome dynamics identifies transcription factors that mediate Polycomb targeting. *Genome Research*, *23*, 60–73.

Azuara, V., Perry, P., Sauer, S., Spivakov, M., Jorgensen, H. F., John, R. M., et al. (2006). Chromatin signatures of pluripotent cell lines. *Nature Cell Biology*, *8*, 532–538.

Bae, E., Calhoun, V. C., Levine, M., Lewis, E. B., & Drewell, R. A. (2002). Characterization of the intergenic RNA profile at abdominal-A and Abdominal-B in the Drosophila bithorax complex. *Proceedings of the National Academy of Sciences of the United States of America*, *99*, 16847–16852.

Bantignies, F., & Cavalli, G. (2011). Polycomb group proteins: Repression in 3D. *Trends in Genetics*, *27*, 454–464.

Bantignies, F., Roure, V., Comet, I., Leblanc, B., Schuettengruber, B., Bonnet, J., et al. (2011). Polycomb-dependent regulatory contacts between distant Hox loci in Drosophila. *Cell*, *144*, 214–226.

Beisel, C., Imhof, A., Greene, J., Kremmer, E., & Sauer, F. (2002). Histone methylation by the Drosophila epigenetic transcriptional regulator Ash1. *Nature*, *419*, 857–862.

Beisel, C., & Paro, R. (2011). Silencing chromatin: Comparing modes and mechanisms. *Nature Reviews. Genetics*, *12*, 123–135.

Bel-Vialar, S., Core, N., Terranova, R., Goudot, V., Boned, A., & Djabali, M. (2000). Altered retinoic acid sensitivity and temporal expression of Hox genes in polycomb-M33-deficient mice. *Developmental Biology*, *224*, 238–249.

Bender, W., & Fitzgerald, D. P. (2002). Transcription activates repressed domains in the Drosophila bithorax complex. *Development*, *129*, 4923–4930.

Bernstein, B. E., Mikkelsen, T. S., Xie, X., Kamal, M., Huebert, D. J., Cuff, J., et al. (2006). A bivalent chromatin structure marks key developmental genes in embryonic stem cells. *Cell*, *125*, 315–326.

Breen, T. R., & Harte, P. J. (1991). Molecular characterization of the trithorax gene, a positive regulator of homeotic gene expression in Drosophila. *Mechanisms of Development*, *35*, 113–127.

Chambeyron, S., & Bickmore, W. A. (2004). Chromatin decondensation and nuclear reorganization of the HoxB locus upon induction of transcription. *Genes & Development*, *18*, 1119–1130.

Chambeyron, S., Da Silva, N. R., Lawson, K. A., & Bickmore, W. A. (2005). Nuclear reorganisation of the Hoxb complex during mouse embryonic development. *Development*, *132*, 2215–2223.

Chan, C. S., Rastelli, L., & Pirrotta, V. (1994). A Polycomb response element in the Ubx gene that determines an epigenetically inherited state of repression. *The EMBO Journal*, *13*, 2553–2564.

Chang, H. Y., Chi, J. T., Dudoit, S., Bondre, C., van de Rijn, M., Botstein, D., et al. (2002). Diversity, topographic differentiation, and positional memory in human fibroblasts. *Proceedings of the National Academy of Sciences of the United States of America*, *99*, 12877–12882.

Chen, Y., Wan, B., Wang, K. C., Cao, F., Yang, Y., Protacio, A., et al. (2011). Crystal structure of the N-terminal region of human Ash2L shows a winged-helix motif involved in DNA binding. *EMBO Reports*, *12*, 797–803.

Chisaka, O., & Capecchi, M. R. (1991). Regionally restricted developmental defects resulting from targeted disruption of the mouse homeobox gene hox-1.5. *Nature*, *350*, 473–479.

Chourrout, D., Delsuc, F., Chourrout, P., Edvardsen, R. B., Rentzsch, F., Renfer, E., et al. (2006). Minimal ProtoHox cluster inferred from bilaterian and cnidarian Hox complements. *Nature*, *442*, 684–687.

Chu, C., Qu, K., Zhong, F. L., Artandi, S. E., & Chang, H. Y. (2011). Genomic maps of long noncoding RNA occupancy reveal principles of RNA-chromatin interactions. *Molecular Cell*, *44*, 667–678.

Cleard, F., Moshkin, Y., Karch, F., & Maeda, R. K. (2006). Probing long-distance regulatory interactions in the Drosophila melanogaster bithorax complex using Dam identification. *Nature Genetics*, *38*, 931–935.

Czermin, B., Melfi, R., McCabe, D., Seitz, V., Imhof, A., & Pirrotta, V. (2002). Drosophila enhancer of Zeste/ESC complexes have a histone H3 methyltransferase activity that marks chromosomal Polycomb sites. *Cell*, *111*, 185–196.

de Rosa, R., Grenier, J. K., Andreeva, T., Cook, C. E., Adoutte, A., Akam, M., et al. (1999). Hox genes in brachiopods and priapulids and protostome evolution. *Nature*, *399*, 772–776.

Deschamps, J., van den Akker, E., Forlani, S., De Graaff, W., Oosterveen, T., Roelen, B., et al. (1999). Initiation, establishment and maintenance of Hox gene expression patterns in the mouse. *The International Journal of Developmental Biology*, *43*, 635–650.

Deschamps, J., & van Nes, J. (2005). Developmental regulation of the Hox genes during axial morphogenesis in the mouse. *Development*, *132*, 2931–2942.

Dixon, J. R., Selvaraj, S., Yue, F., Kim, A., Li, Y., Shen, Y., et al. (2012). Topological domains in mammalian genomes identified by analysis of chromatin interactions. *Nature*, *485*, 376–380.

Dolle, P., Izpisua-Belmonte, J. C., Falkenstein, H., Renucci, A., & Duboule, D. (1989). Coordinate expression of the murine Hox-5 complex homoeobox-containing genes during limb pattern formation. *Nature*, *342*, 767–772.

Duboule, D. (1992). The vertebrate limb: A model system to study the Hox/HOM gene network during development and evolution. *Bioessays*, *14*, 375–384.

Duboule, D. (1994). Temporal colinearity and the phylotypic progression: A basis for the stability of a vertebrate Bauplan and the evolution of morphologies through heterochrony. *Development. Supplement*, 135–142.

Duboule, D. (2007). The rise and fall of Hox gene clusters. *Development*, *134*, 2549–2560.

Duboule, D., & Dolle, P. (1989). The structural and functional organization of the murine HOX gene family resembles that of Drosophila homeotic genes. *The EMBO Journal*, *8*, 1497–1505.

Duboule, D., & Morata, G. (1994). Colinearity and functional hierarchy among genes of the homeotic complexes. *Trends in Genetics*, *10*, 358–364.

Eissenberg, J. C., & Shilatifard, A. (2010). Histone H3 lysine 4 (H3K4) methylation in development and differentiation. *Developmental Biology*, *339*, 240–249.

Eskeland, R., Leeb, M., Grimes, G. R., Kress, C., Boyle, S., Sproul, D., et al. (2010). Ring1B compacts chromatin structure and represses gene expression independent of histone ubiquitination. *Molecular Cell*, *38*, 452–464.

Favier, B., & Dolle, P. (1997). Developmental functions of mammalian Hox genes. *Molecular Human Reproduction*, *3*, 115–131.

Ferraiuolo, M. A., Rousseau, M., Miyamoto, C., Shenker, S., Wang, X. Q., Nadler, M., et al. (2010). The three-dimensional architecture of Hox cluster silencing. *Nucleic Acids Research*, *38*, 7472–7484.

Francis, N. J., Kingston, R. E., & Woodcock, C. L. (2004). Chromatin compaction by a polycomb group protein complex. *Science, 306*, 1574–1577.
Fried, C., Prohaska, S. J., & Stadler, P. F. (2004). Exclusion of repetitive DNA elements from gnathostome Hox clusters. *Journal of Experimental Zoology. Part B, Molecular and Developmental Evolution, 302*, 165–173.
Garcia-Fernandez, J. (2005). The genesis and evolution of homeobox gene clusters. *Nature Reviews. Genetics, 6*, 881–892.
Gaunt, S. J., Sharpe, P. T., & Duboule, D. (1988). Spatially restricted domains of homeogene transcripts in mouse embryos: Relation to a segmented body plan. *Development. Supplement, 104*, 169–179.
Gehring, W. (1966). Bildung eines vollständigen Mittelbeines mit Sternopleura in der Antennenregion bei der Mutante Nasobemia (Ns) von Drosophila melanogaster. *Arch Julius Klaus Stift Vererbungsforsch Sozialanthropol Rassenhyg, 41*, 44–54.
Gonzalez, F., Duboule, D., & Spitz, F. (2007). Transgenic analysis of Hoxd gene regulation during digit development. *Developmental Biology, 306*, 847–859.
Graham, A., Papalopulu, N., & Krumlauf, R. (1989). The murine and Drosophila homeobox gene complexes have common features of organization and expression. *Cell, 57*, 367–378.
Greer, J. M., Puetz, J., Thomas, K. R., & Capecchi, M. R. (2000). Maintenance of functional equivalence during paralogous Hox gene evolution. *Nature, 403*, 661–665.
Gupta, R. A., Shah, N., Wang, K. C., Kim, J., Horlings, H. M., Wong, D. J., et al. (2010). Long non-coding RNA HOTAIR reprograms chromatin state to promote cancer metastasis. *Nature, 464*, 1071–1076.
Hekimoglu, B., & Ringrose, L. (2009). Non-coding RNAs in polycomb/trithorax regulation. *RNA Biology, 6*, 129–137.
Herault, Y., Kmita, M., Sawaya, C. C., & Duboule, D. (2002). A nested deletion approach to generate Cre deleter mice with progressive Hox profiles. *The International Journal of Developmental Biology, 46*, 185–191.
Herold, M., Bartkuhn, M., & Renkawitz, R. (2012). CTCF: Insights into insulator function during development. *Development, 139*, 1045–1057.
Hogga, I., & Karch, F. (2002). Transcription through the iab-7 cis-regulatory domain of the bithorax complex interferes with maintenance of Polycomb-mediated silencing. *Development, 129*, 4915–4922.
Holland, L. Z., Albalat, R., Azumi, K., Benito-Gutierrez, E., Blow, M. J., Bronner-Fraser, M., et al. (2008). The amphioxus genome illuminates vertebrate origins and cephalochordate biology. *Genome Research, 18*, 1100–1111.
Holland, L. Z., Holland, N. D., & Gilland, E. (2008). Amphioxus and the evolution of head segmentation. *Integrative and Comparative Biology, 48*, 630–646.
Hombria, J. C., & Lovegrove, B. (2003). Beyond homeosis—HOX function in morphogenesis and organogenesis. *Differentiation, 71*, 461–476.
Hughes, C. M., Rozenblatt-Rosen, O., Milne, T. A., Copeland, T. D., Levine, S. S., Lee, J. C., et al. (2004). Menin associates with a trithorax family histone methyltransferase complex and with the hoxc8 locus. *Molecular Cell, 13*, 587–597.
Iimura, T., & Pourquie, O. (2006). Collinear activation of Hoxb genes during gastrulation is linked to mesoderm cell ingression. *Nature, 442*, 568–571.
Illingworth, R., Kerr, A., Desousa, D., Jorgensen, H., Ellis, P., Stalker, J., et al. (2008). A novel CpG island set identifies tissue-specific methylation at developmental gene loci. *PLoS Biology, 6*, e22.
Izpisua-Belmonte, J. C., Falkenstein, H., Dolle, P., Renucci, A., & Duboule, D. (1991). Murine genes related to the Drosophila AbdB homeotic genes are sequentially expressed during development of the posterior part of the body. *The EMBO Journal, 10*, 2279–2289.

Kahn, T. G., Schwartz, Y. B., Dellino, G. I., & Pirrotta, V. (2006). Polycomb complexes and the propagation of the methylation mark at the Drosophila ubx gene. *The Journal of Biological Chemistry, 281*, 29064–29075.

Kessel, M., Balling, R., & Gruss, P. (1990). Variations of cervical vertebrae after expression of a Hox-1.1 transgene in mice. *Cell, 61*, 301–308.

Kmita, M., & Duboule, D. (2003). Organizing axes in time and space; 25 years of colinear tinkering. *Science, 301*, 331–333.

Kmita, M., Fraudeau, N., Herault, Y., & Duboule, D. (2002). Serial deletions and duplications suggest a mechanism for the collinearity of Hoxd genes in limbs. *Nature, 420*, 145–150.

Kmita, M., Tarchini, B., Zakany, J., Logan, M., Tabin, C. J., & Duboule, D. (2005). Early developmental arrest of mammalian limbs lacking HoxA/HoxD gene function. *Nature, 435*, 1113–1116.

Kosman, D., Mizutani, C. M., Lemons, D., Cox, W. G., McGinnis, W., & Bier, E. (2004). Multiplex detection of RNA expression in Drosophila embryos. *Science, 305*, 846.

Krumlauf, R. (1994). Hox genes in vertebrate development. *Cell, 78*, 191–201.

Ku, M., Koche, R. P., Rheinbay, E., Mendenhall, E. M., Endoh, M., Mikkelsen, T. S., et al. (2008). Genomewide analysis of PRC1 and PRC2 occupancy identifies two classes of bivalent domains. *PLoS Genetics, 4*, e1000242.

Kuraku, S., & Meyer, A. (2009). The evolution and maintenance of Hox gene clusters in vertebrates and the teleost-specific genome duplication. *The International Journal of Developmental Biology, 53*, 765–773.

Lanzuolo, C., Roure, V., Dekker, J., Bantignies, F., & Orlando, V. (2007). Polycomb response elements mediate the formation of chromosome higher-order structures in the bithorax complex. *Nature Cell Biology, 9*, 1167–1174.

Lee, A. P., Koh, E. G., Tay, A., Brenner, S., & Venkatesh, B. (2006). Highly conserved syntenic blocks at the vertebrate Hox loci and conserved regulatory elements within and outside Hox gene clusters. *Proceedings of the National Academy of Sciences of the United States of America, 103*, 6994–6999.

Lee, J. H., & Skalnik, D. G. (2005). CpG-binding protein (CXXC finger protein 1) is a component of the mammalian Set1 histone H3-Lys4 methyltransferase complex, the analogue of the yeast Set1/COMPASS complex. *The Journal of Biological Chemistry, 280*, 41725–41731.

Lee, J. H., & Skalnik, D. G. (2008). Wdr82 is a C-terminal domain-binding protein that recruits the Setd1A Histone H3-Lys4 methyltransferase complex to transcription start sites of transcribed human genes. *Molecular and Cellular Biology, 28*, 609–618.

Lee, J. H., Tate, C. M., You, J. S., & Skalnik, D. G. (2007). Identification and characterization of the human Set1B histone H3-Lys4 methyltransferase complex. *The Journal of Biological Chemistry, 282*, 13419–13428.

Lehoczky, J. A., Williams, M. E., & Innis, J. W. (2004). Conserved expression domains for genes upstream and within the HoxA and HoxD clusters suggests a long-range enhancer existed before cluster duplication. *Evolution and Development, 6*, 423–430.

Levine, M., Hafen, E., Garber, R. L., & Gehring, W. J. (1983). Spatial distribution of Antennapedia transcripts during Drosophila development. *The EMBO Journal, 2*, 2037–2046.

Lewis, E. B. (1978). A gene complex controlling segmentation in Drosophila. *Nature, 276*, 565–570.

Li, H. B., Muller, M., Bahechar, I. A., Kyrchanova, O., Ohno, K., Georgiev, P., et al. (2011). Insulators, not Polycomb response elements, are required for long-range interactions between Polycomb targets in Drosophila melanogaster. *Molecular and Cellular Biology, 31*, 616–625.

Lynch, V. J., & Wagner, G. P. (2009). Multiple chromosomal rearrangements structured the ancestral vertebrate Hox-bearing protochromosomes. *PLoS Genetics*, *5*, e1000349.

Maeda, R. K., & Karch, F. (2006). The ABC of the BX-C: The bithorax complex explained. *Development*, *133*, 1413–1422.

Maeda, R. K., & Karch, F. (2010). Cis-regulation in the Drosophila bithorax complex. *Advances in Experimental Medicine and Biology*, *689*, 17–40.

McGinnis, W., & Krumlauf, R. (1992). Homeobox genes and axial patterning. *Cell*, *68*, 283–302.

Medina-Martinez, O., Bradley, A., & Ramirez-Solis, R. (2000). A large targeted deletion of Hoxb1-Hoxb9 produces a series of single-segment anterior homeotic transformations. *Developmental Biology*, *222*, 71–83.

Mendenhall, E. M., Koche, R. P., Truong, T., Zhou, V. W., Issac, B., Chi, A. S., et al. (2010). GC-rich sequence elements recruit PRC2 in mammalian ES cells. *PLoS Genetics*, *6*, e1001244.

Min, H., Lee, J. Y., & Kim, M. H. (2012). Structural dynamics and epigenetic modifications of Hoxc loci along the anteroposterior body axis in developing mouse embryos. *International Journal of Biological Sciences*, *8*, 802–810.

Mohan, M., Lin, C., Guest, E., & Shilatifard, A. (2010). Licensed to elongate: A molecular mechanism for MLL-based leukaemogenesis. *Nature Reviews. Cancer*, *10*, 721–728.

Mohd-Sarip, A., Venturini, F., Chalkley, G. E., & Verrijzer, C. P. (2002). Pleiohomeotic can link polycomb to DNA and mediate transcriptional repression. *Molecular and Cellular Biology*, *22*, 7473–7483.

Mohn, F., Weber, M., Rebhan, M., Roloff, T. C., Richter, J., Stadler, M. B., et al. (2008). Lineage-specific polycomb targets and de novo DNA methylation define restriction and potential of neuronal progenitors. *Molecular Cell*, *30*, 755–766.

Monge, I., Kondo, T., & Duboule, D. (2003). An enhancer-titration effect induces digit-specific regulatory alleles of the HoxD cluster. *Developmental Biology*, *256*, 212–220.

Montavon, T., Le Garrec, J. F., Kerszberg, M., & Duboule, D. (2008). Modeling Hox gene regulation in digits: Reverse collinearity and the molecular origin of thumbness. *Genes & Development*, *22*, 346–359.

Montavon, T., Soshnikova, N., Mascrez, B., Joye, E., Thevenet, L., Splinter, E., et al. (2011). A regulatory archipelago controls Hox genes transcription in digits. *Cell*, *147*, 1132–1145.

Morata, G., Botas, J., Kerridge, S., & Struhl, G. (1983). Homeotic transformations of the abdominal segments of Drosophila caused by breaking or deleting a central portion of the bithorax complex. *Journal of Embryology and Experimental Morphology*, *78*, 319–341.

Morey, C., Da Silva, N. R., Perry, P., & Bickmore, W. A. (2007). Nuclear reorganisation and chromatin decondensation are conserved, but distinct, mechanisms linked to Hox gene activation. *Development*, *134*, 909–919.

Muller, J., & Kassis, J. A. (2006). Polycomb response elements and targeting of Polycomb group proteins in Drosophila. *Current Opinion in Genetics and Development*, *16*, 476–484.

Negre, B., Ranz, J. M., Casals, F., Caceres, M., & Ruiz, A. (2003). A new split of the Hox gene complex in Drosophila: Relocation and evolution of the gene labial. *Molecular Biology and Evolution*, *20*, 2042–2054.

Noordermeer, D., Leleu, M., Splinter, E., Rougemont, J., De Laat, W., & Duboule, D. (2011). The dynamic architecture of Hox gene clusters. *Science*, *334*, 222–225.

Nora, E. P., Lajoie, B. R., Schulz, E. G., Giorgetti, L., Okamoto, I., Servant, N., et al. (2012). Spatial partitioning of the regulatory landscape of the X-inactivation centre. *Nature*, *485*, 381–385.

Ohno, S. (1970). *Evolution by gene duplication*. Heidelberg: Springer-Verlag.

Papp, B., & Muller, J. (2006). Histone trimethylation and the maintenance of transcriptional ON and OFF states by trxG and PcG proteins. *Genes & Development, 20*, 2041–2054.
Paro, R. (1990). Imprinting a determined state into the chromatin of Drosophila. *Trends in Genetics, 6*, 416–421.
Pirrotta, V., & Li, H. B. (2012). A view of nuclear Polycomb bodies. *Current Opinion in Genetics and Development, 22*, 101–109.
Pourquie, O. (2003). The segmentation clock: Converting embryonic time into spatial pattern. *Science, 301*, 328–330.
Rank, G., Prestel, M., & Paro, R. (2002). Transcription through intergenic chromosomal memory elements of the Drosophila bithorax complex correlates with an epigenetic switch. *Molecular and Cellular Biology, 22*, 8026–8034.
Ranz, J. M., Segarra, C., & Ruiz, A. (1997). Chromosomal homology and molecular organization of Muller's elements D and E in the Drosophila repleta species group. *Genetics, 145*, 281–295.
Ringrose, L., & Paro, R. (2007). Polycomb/Trithorax response elements and epigenetic memory of cell identity. *Development, 134*, 223–232.
Rinn, J. L., Bondre, C., Gladstone, H. B., Brown, P. O., & Chang, H. Y. (2006). Anatomic demarcation by positional variation in fibroblast gene expression programs. *PLoS Genetics, 2*, e119.
Rinn, J. L., Kertesz, M., Wang, J. K., Squazzo, S. L., Xu, X., Brugmann, S. A., et al. (2007). Functional demarcation of active and silent chromatin domains in human HOX loci by noncoding RNAs. *Cell, 129*, 1311–1323.
Sanchez-Elsner, T., Gou, D., Kremmer, E., & Sauer, F. (2006). Noncoding RNAs of trithorax response elements recruit Drosophila Ash1 to Ultrabithorax. *Science, 311*, 1118–1123.
Sanchez-Herrero, E., & Akam, M. (1989). Spatially ordered transcription of regulatory DNA in the bithorax complex of Drosophila. *Development, 107*, 321–329.
Sawarkar, R., & Paro, R. (2010). Interpretation of developmental signaling at chromatin: The Polycomb perspective. *Developmental Cell, 19*, 651–661.
Schneuwly, S., Klemenz, R., & Gehring, W. J. (1987). Redesigning the body plan of Drosophila by ectopic expression of the homoeotic gene Antennapedia. *Nature, 325*, 816–818.
Schorderet, P., & Duboule, D. (2011). Structural and functional differences in the long noncoding RNA hotair in mouse and human. *PLoS Genetics, 7*, e1002071.
Schuettengruber, B., & Cavalli, G. (2009). Recruitment of polycomb group complexes and their role in the dynamic regulation of cell fate choice. *Development, 136*, 3531–3542.
Schuettengruber, B., Chourrout, D., Vervoort, M., Leblanc, B., & Cavalli, G. (2007). Genome regulation by polycomb and trithorax proteins. *Cell, 128*, 735–745.
Schuettengruber, B., Ganapathi, M., Leblanc, B., Portoso, M., Jaschek, R., Tolhuis, B., et al. (2009). Functional anatomy of polycomb and trithorax chromatin landscapes in Drosophila embryos. *PLoS Biology, 7*, e13.
Schuettengruber, B., Martinez, A. M., Iovino, N., & Cavalli, G. (2011). Trithorax group proteins: Switching genes on and keeping them active. *Nature Reviews. Molecular Cell Biology, 12*, 799–814.
Schwartz, Y. B., Kahn, T. G., Nix, D. A., Li, X. Y., Bourgon, R., Biggin, M., et al. (2006). Genome-wide analysis of Polycomb targets in Drosophila melanogaster. *Nature Genetics, 38*, 700–705.
Schwartz, Y. B., & Pirrotta, V. (2007). Polycomb silencing mechanisms and the management of genomic programmes. *Nature Reviews. Genetics, 8*, 9–22.
Schwartz, Y. B., & Pirrotta, V. (2008). Polycomb complexes and epigenetic states. *Current Opinion in Cell Biology, 20*, 266–273.

Seo, H. C., Edvardsen, R. B., Maeland, A. D., Bjordal, M., Jensen, M. F., Hansen, A., et al. (2004). Hox cluster disintegration with persistent anteroposterior order of expression in Oikopleura dioica. *Nature*, *431*, 67–71.

Sessa, L., Breiling, A., Lavorgna, G., Silvestri, L., Casari, G., & Orlando, V. (2007). Non-coding RNA synthesis and loss of Polycomb group repression accompanies the colinear activation of the human HOXA cluster. *RNA*, *13*, 223–239.

Sexton, T., Yaffe, E., Kenigsberg, E., Bantignies, F., Leblanc, B., Hoichman, M., et al. (2012). Three-dimensional folding and functional organization principles of the Drosophila genome. *Cell*, *148*, 458–472.

Shen, Y., Yue, F., McCleary, D. F., Ye, Z., Edsall, L., Kuan, S., et al. (2012). A map of the cis-regulatory sequences in the mouse genome. *Nature*, *488*, 116–120.

Simonis, M., Kooren, J., & de Laat, W. (2007). An evaluation of 3C-based methods to capture DNA interactions. *Nature Methods*, *4*, 895–901.

Sing, A., Pannell, D., Karaiskakis, A., Sturgeon, K., Djabali, M., Ellis, J., et al. (2009). A vertebrate Polycomb response element governs segmentation of the posterior hindbrain. *Cell*, *138*, 885–897.

Smith, S. T., Petruk, S., Sedkov, Y., Cho, E., Tillib, S., Canaani, E., et al. (2004). Modulation of heat shock gene expression by the TAC1 chromatin-modifying complex. *Nature Cell Biology*, *6*, 162–167.

Soshnikova, N., & Duboule, D. (2009). Epigenetic temporal control of mouse Hox genes in vivo. *Science*, *324*, 1320–1323.

Spitz, F., Gonzalez, F., & Duboule, D. (2003). A global control region defines a chromosomal regulatory landscape containing the HoxD cluster. *Cell*, *113*, 405–417.

Spitz, F., Gonzalez, F., Peichel, C., Vogt, T. F., Duboule, D., & Zakany, J. (2001). Large scale transgenic and cluster deletion analysis of the HoxD complex separate an ancestral regulatory module from evolutionary innovations. *Genes & Development*, *15*, 2209–2214.

Spitz, F., Herkenne, C., Morris, M. A., & Duboule, D. (2005). Inversion-induced disruption of the Hoxd cluster leads to the partition of regulatory landscapes. *Nature Genetics*, *37*, 889–893.

Struhl, G. (1981). A homoeotic mutation transforming leg to antenna in Drosophila. *Nature*, *292*, 635–638.

Tarchini, B., & Duboule, D. (2006). Control of Hoxd genes' collinearity during early limb development. *Developmental Cell*, *10*, 93–103.

Tavares, L., Dimitrova, E., Oxley, D., Webster, J., Poot, R., Demmers, J., et al. (2012). RYBP-PRC1 complexes mediate H2A ubiquitylation at polycomb target sites independently of PRC2 and H3K27me3. *Cell*, *148*, 664–678.

Tillib, S., Petruk, S., Sedkov, Y., Kuzin, A., Fujioka, M., Goto, T., et al. (1999). Trithorax- and Polycomb-group response elements within an Ultrabithorax transcription maintenance unit consist of closely situated but separable sequences. *Molecular and Cellular Biology*, *19*, 5189–5202.

Tolhuis, B., Blom, M., Kerkhoven, R. M., Pagie, L., Teunissen, H., Nieuwland, M., et al. (2011). Interactions among Polycomb domains are guided by chromosome architecture. *PLoS Genetics*, 7, e1001343.

Tschopp, P., & Duboule, D. (2011). A regulatory 'landscape effect' over the HoxD cluster. *Developmental Biology*, *351*, 288–296.

Tschopp, P., Tarchini, B., Spitz, F., Zakany, J., & Duboule, D. (2009). Uncoupling time and space in the collinear regulation of Hox genes. *PLoS Genetics*, *5*, e1000398.

van der Lugt, N. M., Alkema, M., Berns, A., & Deschamps, J. (1996). The Polycomb-group homolog Bmi-1 is a regulator of murine Hox gene expression. *Mechanisms of Development*, *58*, 153–164.

van Steensel, B., & Dekker, J. (2010). Genomics tools for unraveling chromosome architecture. *Nature Biotechnology*, *28*, 1089–1095.

Wang, L., Brown, J. L., Cao, R., Zhang, Y., Kassis, J. A., & Jones, R. S. (2004). Hierarchical recruitment of polycomb group silencing complexes. *Molecular Cell, 14*, 637–646.
Wang, P., Lin, C., Smith, E. R., Guo, H., Sanderson, B. W., Wu, M., et al. (2009). Global analysis of H3K4 methylation defines MLL family member targets and points to a role for MLL1-mediated H3K4 methylation in the regulation of transcriptional initiation by RNA polymerase II. *Molecular and Cellular Biology, 29*, 6074–6085.
Wang, H., Wang, L., Erdjument-Bromage, H., Vidal, M., Tempst, P., Jones, R. S., et al. (2004). Role of histone H2A ubiquitination in Polycomb silencing. *Nature, 431*, 873–878.
Wang, K. C., Yang, Y. W., Liu, B., Sanyal, A., Corces-Zimmerman, R., Chen, Y., et al. (2011). A long noncoding RNA maintains active chromatin to coordinate homeotic gene expression. *Nature, 472*, 120–124.
Weber, M., Hellmann, I., Stadler, M. B., Ramos, L., Paabo, S., Rebhan, M., et al. (2007). Distribution, silencing potential and evolutionary impact of promoter DNA methylation in the human genome. *Nature Genetics, 39*, 457–466.
Williamson, I., Eskeland, R., Lettice, L. A., Hill, A. E., Boyle, S., Grimes, G. R., et al. (2012). Anterior-posterior differences in HoxD chromatin topology in limb development. *Development, 139*, 3157–3167.
Woltering, J. M., & Duboule, D. (2010). The origin of digits: Expression patterns versus regulatory mechanisms. *Developmental Cell, 18*, 526–532.
Woo, C. J., Kharchenko, P. V., Daheron, L., Park, P. J., & Kingston, R. E. (2010). A region of the human HOXD cluster that confers polycomb-group responsiveness. *Cell, 140*, 99–110.
Yokoyama, A., & Cleary, M. L. (2008). Menin critically links MLL proteins with LEDGF on cancer-associated target genes. *Cancer Cell, 14*, 36–46.
Zakany, J., & Duboule, D. (2007). The role of Hox genes during vertebrate limb development. *Current Opinion in Genetics and Development, 17*, 359–366.
Zakany, J., Kmita, M., & Duboule, D. (2004). A dual role for Hox genes in limb anterior-posterior asymmetry. *Science, 304*, 1669–1672.
Zeltser, L., Desplan, C., & Heintz, N. (1996). Hoxb-13: A new Hox gene in a distant region of the HOXB cluster maintains colinearity. *Development, 122*, 2475–2484.

CHAPTER FIVE

# Primordial Germ-Cell Development and Epigenetic Reprogramming in Mammals

**Harry G. Leitch**[*,†,‡], **Walfred W.C. Tang**[*,†], **M. Azim Surani**[*,†,‡,1]
[*]Wellcome Trust/Cancer Research UK Gurdon Institute of Cancer and Developmental Biology, University of Cambridge, Cambridge, United Kingdom
[†]Department of Physiology, Development, and Neuroscience, University of Cambridge, Cambridge, United Kingdom
[‡]Wellcome Trust-Medical Research Council Cambridge Stem Cell Institute, University of Cambridge, Cambridge, United Kingdom
[1]Corresponding author: e-mail address: a.surani@gurdon.cam.ac.uk

## Contents



## Abstract

Primordial germ cells (PGCs) are the embryonic precursors of the gametes and represent the founder cells of the germline. Specification of PGCs is a critical divergent point during embryogenesis. Whereas the somatic lineages will ultimately perish, cells of the germline have the potential to form a new individual and hence progress to the next generation. It is therefore critical that the genome emerges intact and carrying the appropriate epigenetic information during its passage through the germline. To ensure

*Current Topics in Developmental Biology*, Volume 104
ISSN 0070-2153
http://dx.doi.org/10.1016/B978-0-12-416027-9.00005-X

this fidelity of transmission, PGC development encompasses extensive epigenetic reprogramming. The low cell numbers and relative inaccessibility of PGCs present a challenge to those seeking mechanistic understanding of the crucial developmental and epigenetic processes in this most fascinating of lineages. Here, we present an overview of PGC development in the mouse and compare this with the limited information available for other mammalian species. We believe that a comparative approach will be increasingly important to uncover the extent to which mechanisms are conserved and reveal the critical steps during PGC development in humans.

## 1. EMBRYOLOGY

In the early to mid-twentieth century, embryologists recognized PGCs based on morphological characteristics such as their large size and prominent nucleolus. The meticulous observations by Witschi (1948) meant that human PGC development was among the best characterized of the mammals. Witschi first discriminated PGCs in Carnegie stage (CS) 11 embryos (≈embryonic day (E)24) in the extraembryonic yolk sac, close to the junction with the allantois. In more advanced embryos of the same CS, PGCs are found in the yolk sac as well as the developing hindgut and by CS12 (≈E26) all the PGCs reside in the hindgut. In CS13 embryos (≈E28), some PGCs are observed leaving the hindgut and entering the region of the developing genital ridges. By CS16 (≈E37), all of the PGCs have left the gut and its mesentery and can be found in the genital ridges. Witschi noted that the PGCs in the hindgut form pseudopodia and "show signs of locomotor activity." He suggested that migration at this stage may be due to "active movements" of individual cells, observations confirmed in the mouse over 50 years later (Molyneaux, Stallock, Schaible, & Wylie, 2001). Witschi noted that throughout the migratory period, from yolk sac to genital ridge, PGCs increase in number. He also concluded that "No facts and no convincing theoretical arguments in favor of secondary germ-cell formation during the period covered by this investigation were found. The germ cells are strictly specific and do not change into somatic elements." Sexual differentiation of the gonadal somatic cells begins at the beginning of the 7th week followed by the expression of sex determination genes *Sry* and *Sox9* (Hanley et al., 2000). Thereafter, germ cells are referred to as gonocytes. In female, gonocytes enter meiosis at the 10th week (Gondos, Westergaard, & Byskov, 1986), whereas male germ cells do not undergo meiosis until after puberty.

In the mouse, Chiquoine used alkaline phosphatase (AP) staining to identify PGCs, a significant advancement over the previously used morphological criteria (Chiquoine, 1954). He was able to identify PGCs as early as E8 residing in the allantoic and yolk sac mesoderm and at the caudal end of the primitive streak (Chiquoine, 1954), the same location as the earliest PGCs in human embryos (Fig. 5.1A). Indeed, mouse PGCs were found to follow a similar course along the "germ tract" from yolk sac to genital ridges via the hindgut and its mesentery. In mouse, PGCs begin to enter the developing hindgut at around E8.5 and reside there by E9, while entry to the genital ridges begins at E10 and is largely complete by E11. Sex-specific differences in the gonads are evident by E12.5 and this represents a convenient endpoint of PGC development (Sasaki & Matsui, 2008).

Modifications to the AP staining protocol enabled mouse PGCs to be identified a day earlier at E7 and allowed a detailed description of their location during gastrulation (Ginsburg, Snow, & McLaren, 1990).The earliest PGCs that can be discerned are in the posterior primitive streak in the

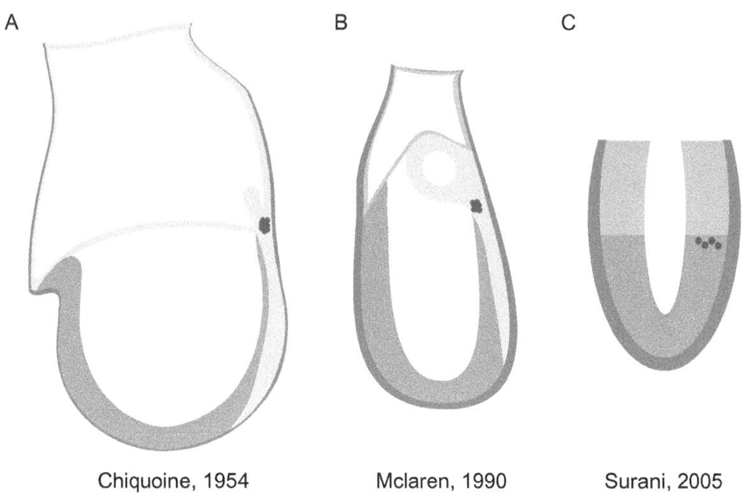


**Figure 5.1** History of mouse PGC specification. Schematic diagrams of embryos depicting the earliest developmental stage at which PGCs were observed in landmark studies. (A) Chiquoine observed PGCs in early E8 embryos in the allantoic and yolk sac mesoderm (Chiquoine, 1954). (B) Modifications to the AP staining protocol allowed PGCs to be detected a day earlier emerging from the posterior primitive streak, in extra-embryonic mesoderm (Ginsburg, Snow, & McLaren, 1990). (C) Using Blimp1-transgenic reporter mice, four to eight PGCs were observed in the E6.25 prestreak epiblast (Ohinata et al., 2005). (See Color Insert.)

extraembryonic mesoderm of mid-primitive streak stage embryos (Fig. 5.1B). Thereafter, they are found in extraembryonic mesoderm at the base of the developing allantois and then subsequently in the developing hindgut endoderm (Ginsburg et al., 1990). With the advent of transgenic technologies, PGCs have been traced back to the epiblast at E6.25 when a few cells positive for Blimp1 (see later) can be identified (Fig. 5.1C). Recent lineage tracing experiments suggest that the whole germ-cell lineage may be descended from as few as three initial PGCs (Ueno, Turnbull, & Weissman, 2009).

Although the relevance of the high AP activity in PGCs remains unclear, it appears to be well conserved, marking PGCs not only in mice but also rats (Kemper & Peters, 1987), humans (Fujimoto, Miyayama, & Fuyuta, 1977), cows (Cherny et al., 1994), sheep (Ledda, Bogliolo, Bebbere, Ariu, & Pirino, 2010), and pigs (Takagi, Talbot, Rexroad, & Pursel, 1997), although the enzymatic activity has been reported to be lower in goats (Kühholzer, Baguisi, & Overström, 2000).

Rat PGC development seems to follow a similar gross embryological time course to that of the mouse apart from differences in timing, due to their longer gestation. At E10, rat PGCs are found at the base of the allantoic mesoderm, entering the developing hindgut (equivalent to E8.5 in the mouse). By E13.5, colonization of the genital ridges is essentially completed (equivalent to E11.5 in the mouse) (Kemper & Peters, 1987).

More recently, PGC development in the pig has been studied using specific marker genes, such as Oct4 (see later). PGCs can first be detected in early primitive streak stage embryos (E12) at the posterior pole of the epiblast (Wolf, Serup, & Hyttel, 2010). In late primitive streak stage embryos (E13–14), PGCs are found in the yolk sac/allantoic mesoderm and endoderm, presumably after passing through the streak (Vejlsted, Offenberg, Thorup, & Maddox-Hyttel, 2006; Wolf et al., 2010). The early stages of porcine PGC development therefore appear to be largely similar to mouse and human. The E15 pig embryo appears comparable to the E8.5 mouse, with germ cells commencing migration into the developing gut. At around E18, colonization of the genital ridge commences (Hyldig, Ostrup, Vejlsted, & Thomsen, 2011). There appears to be disparity between timing of development of the genital ridge/gonad and PGCs in the pig (in comparison with the mouse), with PGCs having left the hindgut by E21–22 but morphological genital ridges not being evident until E25 (Hyldig, Ostrup, et al., 2011). Overt sexual differentiation of the gonad occurs by E28 (Pilon, Behdjani, Daneau, Lussier, & Silversides, 1998).

# 2. PGC SPECIFICATION

## 2.1. Induction of PGCs from epiblast

Gene knockout studies in mice have revealed the critical role of bone morphogenetic protein (BMP) signaling in PGC induction from the epiblast. *Bmp4* and *Bmp8b* are expressed in the extraembryonic ectoderm and both knockouts exhibit loss of PGCs (Lawson et al., 1999; Ying, Liu, Marble, Lawson, & Zhao, 2000). Mice heterozygous for either *Bmp* also have reduced numbers of PGCs, as do *Bmp2* knockout mice (Ying & Zhao, 2001), emphasizing the importance of BMP dosage for PGC induction. Similarly, the BMP type 1 receptor *Alk2* (de Sousa Lopes, 2004) and the intracellular BMP-signaling transducers *Smad5* (Chang & Matzuk, 2001) and *Smad1* (Tremblay, Dunn, & Robertson, 2001) all exhibit reduced numbers of PGCs when heterozygous and loss of PGCs in homozygous mutants. Induction of PGCs from cultured E5.5 epiblasts requires the presence of the extraembryonic ectoderm, but not of the visceral endoderm (VE; Yoshimizu, Obinata, & Matsui, 2001). However, PGCs do emerge from the isolated E6.5 epiblast in culture (Yoshimizu et al., 2001). Exposure to BMP4 or 8b increases the induction of PGCs in cultured E6 epiblasts, with or without an intact VE, demonstrating the direct effect of these molecules on the epiblast (Ying, Qi, & Zhao, 2001). These combined results suggest that BMP molecules released from the extraembryonic ectoderm are required to induce PGCs but not for subsequent specification events.

A recent comprehensive study of the signaling pathways required to induce PGCs in cultured epiblasts has confirmed and expanded upon these previous findings (Ohinata et al., 2009). The authors report that BMP4 is sufficient to induce PGCs in cultured epiblast in the absence of the VE. BMP2 can also induce PGCs although less efficiently than BMP4, while BMP8b does not act directly to induce PGCs. However, when epiblasts are cultured in the presence of the VE, addition of BMP8b is necessary for BMP4-mediated induction. The proposed mechanism is that BMP8b restricts inhibitory signals emanating from the anterior VE. In keeping with this, *Bmp8b* null embryos have an expanded *Cer1* expression domain, and their defect in PGC specification can be rescued by culturing *Bmp8b* null epiblasts in the absence of VE. Interestingly, *FoxH1* null embryos, which have a defect in anterior VE formation, appear to have increased numbers of PGCs. Critically, PGCs induced in this manner exhibit an appropriate gene expression pattern and, after further culture in a cocktail of supportive growth factors, can repopulate the testis of recipient mice and subsequently produce healthy offspring following intracytoplasmic sperm injection (Ohinata et al., 2009). More recently, a similar strategy has been utilized to induce PGCs from embryonic stem (ES) cells (Hayashi, Ohta, Kurimoto, Aramaki, & Saitou, 2011).

As of yet the role of BMPs in PGC specification has not been confirmed in any other mammalian species although BMP4 may play a conserved role in axolotl PGC induction (Johnson et al., 2003).

## 2.2. Specification genes

Initial attempts to uncover genes required for mouse PGC specification utilized a single-cell differential expression screen between PGCs and somatic neighboring cells isolated at E7.0–E7.5 (Saitou, Barton, & Surani, 2002). Two differentially expressed genes *Fragilis* (*Ifitm3*) and *Stella* (*Pgc7*, *Dppa3*) were characterized further and both exhibited expression patterns consistent with a role in germ-cell specification. However, knockout mice for *Fragilis* are fertile, as are mice with a deletion of the entire *Ifitm* locus encompassing *Fragilis* 1–5 (Lange et al., 2008). Similarly, *Stella* knockout mice exhibit no gross PGC defects and are fertile (Payer et al., 2003), although *Stella* does seem to play a role in the protection of maternal imprints in the zygote (Nakamura et al., 2007, 2012). Nevertheless, *Stella* remains a useful marker gene and transgenic reporter mice have proved valuable tools for isolating and tracking PGCs (Ohinata, Sano, Shigeta, Yamanaka, & Saitou, 2008; Payer et al., 2006; West et al., 2009). Further screening of differentially expressed genes led to the discovery of a new candidate—*Blimp1* (*Prdm1*). *Blimp1* knockout leads to an early and profound loss of PGCs (Ohinata et al., 2005; Vincent et al., 2005). A small cluster of less than 20 weakly staining Stella- and AP-positive cells can be found at the base of the allantois in *Blimp1* null embryos at the 2–10 somite stage. However, these cells do not appear to migrate appropriately from this position, and after this stage, no more than one or two PGC-like cells can be detected (Ohinata et al., 2005; Vincent et al., 2005). In contrast to wild-type PGCs, these nascent PGC-like cells fail to repress expression of *Hoxa1* or *Hoxb1* by the mid-bud stage (Kurimoto et al., 2008; Ohinata et al., 2005). This has led to the hypothesis that a principal role of Blimp1 during PGC specification is to repress the somatic program. This suggestion is consistent with Blimp1's known function as a transcriptional repressor (Keller & Maniatis, 1991) and subsequent global gene expression profiling of *Blimp1* null PGCs (Kurimoto et al., 2008). *Blimp1* expression is first detected in approximately six cells in the prestreak E6.25 epiblast (Ohinata et al., 2005). Lineage tracing using a *Blimp1*-cre transgenic mouse indicates that Blimp1-positive cells in the epiblast are germline lineage-restricted. This apparently contradicts previous lineage tracing experiments that found no lineage-restricted PGCs at E6.5 (Lawson & Hage, 1994). However, the small number of Blimp1-positive cells at E6.5 makes it possible that these escaped labeling in the earlier experiments, explaining why Lawson failed to find any lineage-restricted cells at this stage. If recruitment to the germ-cell lineage is still occurring at E6.5, then labeled Blimp1-negative cells may still divide to give a Blimp1-positive, lineage-restricted PGC and a somatic sister cell (McLaren & Lawson, 2005). PGC induction at E6.5 is still possible in the context of heterotrophic transplants (Tam & Zhou, 1996) suggesting that ongoing recruitment to the lineage is indeed feasible.

A global gene expression profiling study during the process of PGC specification, from E6.25 until E8.25, has provided an invaluable resource (Kurimoto et al., 2008). The authors also included *Blimp1* null PGC-like cells from two different developmental stages, allowing them to expand on the previous characterization of these cells. In total, 134 single cells were analyzed by microarray. This constitutes a huge data set but some important patterns are particularly noteworthy. A significant developmental switch appears to occur at the late-streak/no-bud stage (E6.75–E7.0) when Blimp1-positive, HoxB1-positive cells become HoxB1-negative. Following principal component analysis, Blimp1-positive, HoxB1-positive cells segregate with somatic mesodermal cells across the first principal component (PC1), whereas Blimp1-positive, HoxB1 negative cells cluster closely with all later stages of PGC development. As PC1 accounts for the greatest degree of variability in the transcriptome data, it suggests that this switch correlates with a global gene expression change that represents the definitive specification of PGCs.

Genes upregulated during the transition from a Blimp1-positive, HoxB1-positive to Blimp1-positive, HoxB1-negative state include pluripotency-associated genes *Nanog*, *Sox2*, *Klf2*, and *Stella*. Indeed, the expression of pluripotency genes in PGCs has been noted before. For instance, Oct4 is expressed throughout PGC development (Schöler, Dressler, Balling, Rohdewohld, & Gruss, 1990; Yeom et al., 1996), while Nanog protein can be detected from E7.75 (Yamaguchi, Kimura, Tada, Nakatsuji, & Tada, 2005) and Sox2 and Stella protein at E7.5 (Kurimoto et al., 2008; Saitou et al., 2002). Thus, with the exception of Klf4 (Durcova-Hills,

Tang, Doody, Tooze, & Surani, 2008), all the major nodes in the pluripotency network become reexpressed during PGC specification (Chen et al., 2008; Niwa, 2007).

The transcriptome of *Blimp1* null PGCs indicates that a minority of specification-associated genes are still induced in the absence of *Blimp1*, suggesting that there are Blimp1-independent mechanisms during specification (Kurimoto et al., 2008). However, in the absence of *Blimp1*, there is almost complete failure to repress somatic differentiation genes, of which over 70% remain expressed at levels comparable to somatic neighbors. Thus, the initial speculation that Blimp1's primary role is the suppression of the somatic mesodermal program is supported by transcriptome analysis (Kurimoto et al., 2008; Ohinata et al., 2005).

Lin28 has been reported to function upstream of Blimp1 during PGC specification (West et al., 2009). *Lin28* knockdown in ES cells reduces their contribution to PGCs in chimeric mice, while overexpression increases contribution (West et al., 2009). Blimp1 has been reported to be post-transcriptionally regulated by let-7 miRNA targeting of its 3′-untranslated region (UTR; Nie et al., 2008). The authors propose that Lin28 acts to inhibit let-7-mediated suppression of Blimp1. Consistent with this, overexpression of Blimp1 lacking its 3′-UTR can rescue the contribution of *Lin28* knockdown ES cells to PGCs (West et al., 2009). However, recently *Lin28* has been knocked out and the phenotype is more consistent with a role in regulating germ cell proliferation rather than in PGC specification (Shinoda et al., 2013)

Only two further genes with an early PGC phenotype have been reported. Prdm14, like Blimp1, is a PR/SET domain containing protein which had previously been reported to play a role in suppressing differentiation markers in human ES cells (Tsuneyoshi et al., 2008). *Prdm14* knockout mice are sterile and show a progressive loss of PGCs commencing around E7.5 and continuing until E12.5, although a few AP-positive cells do remain even at this latter time point (Yamaji et al., 2008). The authors examined single-cell cDNA libraries from E7.5 PGCs and found that *Prdm14* null PGCs had specifically failed to upregulate Sox2 and had reduced levels of Stella. Differentiation markers were appropriately repressed leading the authors to conclude that Prdm14, in contrast to Blimp1, does not function to repress the somatic program but is required for reacquisition of potential pluripotency. However, notably the pluripotency genes *Nanog* and *Oct4*, and the PGC genes *Dnd1* and *Nanos3* were not affected by loss of *Prmd14* at this stage. Yamaji and colleagues also reported that *Prdm14* null PGCs do not undergo some of the characteristic epigenetic changes that occur in PGCs soon after specification (see Section 4). Tcfap2c/Ap-2γ is expressed in PGCs from E7.25 and based on expression profiling is a putative Blimp1 target (Kurimoto et al., 2008; Weber et al., 2010). *Tcfap2c* knockout results in a loss of PGCs around E8.0 (Weber et al., 2010). The authors found that PGCs generated from *Tcfap2c* null ES cells by an embryoid body differentiation protocol had increased levels of some differentiation markers, including *Hoxa1*, *Hoxb1*, and *T*, and that *HOXA1*, *HOXB1*, *MYOD1*, and *HAND1* were upregulated upon *TCFAP2C* knockdown in a human seminoma cell line. Weber and colleagues conclude that Tcfap2c may function downstream of Blimp1 to suppress mesodermal differentiation.

There is scant literature detailing gene expression changes during specification in other mammals. Blimp1 is expressed in rat PGCs by E8.5 (Leitch et al., 2010), but its expression at earlier stages is not known. Due to limited access to early stage human embryos, the gene expression pattern during human PGCs specification, which occurs before the 4th week of development, are largely unknown. However, LIN28 (Childs, Kinnell, He, & Anderson, 2012; Gillis et al., 2011), BLIMP1 (Eckert et al., 2008) and TCFAP2C (Pauls et al., 2005) are reported to be expressed in human PGCs/gonocytes at around week 6 to 10 postconception (CS17 onward), respectively. Given the importance of these factors in early mouse PGC development, the expression of LIN28, BLIMP1, and TCFAP2C in human PGCs may indicate conserved roles in PGC specification and maintenance.

## 2.3. Pluripotency-associated genes and specification

As previously noted, the entire network of pluripotency genes, with the exception of *Klf4*, becomes specifically upregulated in nascent PGCs. Their expression is relatively specific to the PGC period of germline development, in general lasting until around the time of sex-specific differentiation. Furthermore, *Oct4* and *Nanog* have been reported to be necessary for PGC development (Chambers et al., 2007; Kehler et al., 2004; Okamura, Tokitake, Niwa, & Matsui, 2008; Yamaguchi et al., 2009). However, due to their requirement for preimplantation development (Mitsui et al., 2003; Nichols et al., 1998), conditional approaches have been necessary to assess their function in PGCs. Deletion of *Oct4* using *tissue non-specific alkaline phosphatase* (*TNAP*)-cre leads to apoptotic death of most PGCs as they colonize the gonads (Kehler et al., 2004). The few surviving germ cells are found to be Oct4-positive consistent with the relatively low efficiency of this Cre transgenic line (Lomelí, Ramos-Mejía, Gertsenstein, Lobe, & Nagy, 2000). Furthermore, *TNAP*-cre only becomes active around E9.5–E10.5, making this an unsuitable model for investigating Oct4 function in PGCs at earlier stages. An earlier requirement for Oct4 during PGCs specification has been reported (Okamura et al., 2008). Generation of chimeric mice from ES cells maintained by an *Oct4* transgene resulted in constitutive repression of the transgene and therefore Oct4 downregulation. This generated *Oct4* null epiblast cells, and these appeared to be capable of expressing *Blimp1* at E7.0. However, no descendants of these cells were found to be expressing Stella at E7.25 or contribute to the germline thereafter (Okamura et al., 2008). These results provide some indication that Oct4 may be required to complete PGC specification. In contrast, *Nanog* null ES cells appear to be able to colonize the germline and survive until E11.5 in chimeric animals (Chambers et al., 2007). *Nanog* null PGCs expressing Oct4 and mouse Vasa homologue (Mvh) can be detected in the gonads of chimeric animals at E11.5. However, these are lost by E12.5. This suggests that Nanog is not required for specification or indeed, the vast majority of PGC development, but may be required to complete the epigenetic reprogramming occurring at E11 (see later). However, a subsequent report that utilized tamoxifen-induced, short-hairpin (sh) RNA knockdown to deplete Nanog in PGCs reported an earlier phenotype (Yamaguchi et al., 2009). Tamoxifen administration at E7.5 resulted in Nanog depletion by E8.5 and extensive apoptosis at E9.5, both *in vivo* and *in vitro*. However, the authors report that tamoxifen administration at E9.5 did not result in significant germ-cell depletion (Yamaguchi et al., 2009). The discrepancy between the two papers is most likely due to the differing experimental approaches. It is possible that *Nanog* null ES cells undergo some compensatory changes to enable their survival in culture and that this results in resistance to apoptosis in the context of PGCs. Alternatively, the knockdown approach may result in off-target effects or a more general toxicity in PGCs.

In mouse PGCs, Oct4 appears to be the only pluripotency factor which is expressed continuously during the transition of epiblast cells to specified PGCs. However, in pig embryos both *Oct4* and *Nanog* exhibit a similar pattern of expression, with the earliest PGCs, located at the posterior epiblast, expressing both these factors (Wolf et al., 2010). *Oct4* expression is maintained until at least E42 in the pig, similar data is not yet available for *Nanog* (Hyldig, Croxall, Contreras, Thomsen, & Alberio, 2011).

This marked difference in pig PGC specification is of great interest and merits further investigation in other organisms. It will also be interesting to measure the expression of other pluripotency genes in the porcine germline to ascertain whether this early expression pattern is unique to *Nanog*. *Oct4* expression has been demonstrated in rat PGCs at E10 (Leitch et al., 2010) and in rat genital ridges (Northrup et al., 2011). Notably, rat genital ridges also appear to express Klf4 (Northrup et al., 2011). Although specific expression in PGCs has not been demonstrated, this may well represent a significant difference with mouse PGCs in which Klf4 is not expressed (Durcova-Hills et al., 2008). The pluripotency-associated cell-surface marker SSEA-1 seems to be a well-conserved marker of PGCs in a range of species including mouse (Fox, Damjanov, Martinez-Hernandez, Knowles, & Solter, 1981), pig (Takagi et al., 1997), sheep (Ledda et al., 2010), goats (Kühholzer et al., 2000), and humans (see below), but not in the rat (Encinas, Zogbi, & Stumpp, 2012).

Human PGCs/gonocytes express pluripotency markers, including OCT4, NANOG, c-KIT, SSEA-1, and SSEA-4 (Gaskell, Esnal, Robinson, Anderson, & Saunders, 2004; Kerr, Hill, Blumenthal, & Gearhart, 2008a, 2008b). However, the pluripotency-associated transcription factor SOX2 is absent in human PGCs (Perrett et al., 2008). SOX2 has been shown to cooperate with OCT4 and NANOG to regulate pluripotency in both human and mouse ES cells (Boyer et al., 2005; Fong, Hohenstein, & Donovan, 2008; Wang et al., 2006), and is expressed in mouse inner cell mass (ICM) and PGCs (Avilion et al., 2003), as well as in human ICM (Cauffman, De Rycke, Sermon, Liebaers, & Van de Velde, 2009). Interestingly, *Sox2* expression appears to be downregulated during sheep PGC development, despite the maintenance of *Oct4* and *Nanog* (Ledda et al., 2010). Studies in a range of other mammals will help to ascertain whether low or absent Sox2 expression in PGCs is a more general feature.

In mouse PGCs, most of the specification- and pluripotency-associated genes are downregulated before the onset of sex-specific differentiation. However, in human, these genes, including BLIMP1, PRMT5, TCFAP2C, LIN28, OCT4, and NANOG, are still expressed in a portion of germ cells at the second or even the third trimester (Childs et al., 2012; Eckert et al., 2008; Gillis et al., 2011; Kerr et al., 2008a, 2008b; Pauls et al., 2005). In particular, a PGC-like population of OCT4+/NANOG+/c-KIT+ cells can still be observed in the human fetal ovary at week 13 postconception, 3 weeks after the onset of meiosis in some female germ cells (Kerr et al., 2008a). However, these OCT+/LIN28+ germ cells do not coexpress the meiosis marker SYCP3 (Childs et al., 2012). This suggests that a subpopulation of germ cells in the fetal ovary has yet to enter meiosis, but has retained expression of pluripotency factors and remained proliferative (Stoop et al., 2005). Indeed, meiosis onset in human fetal ovary is considered to be highly asynchronous (Skrzypczak, Pisarski, Biczysko, & Kedzia, 1981), compared to mouse in which female germ cells enter meiosis in a synchronized manner at E13.5. It is possible that some cells in this PGC-like subpopulation may escape meiosis and apoptosis (Fulton, Martins da Silva, Bayne, & Anderson, 2005) and are a possible source of the recently reported mitotically active germ cells purified from adult human ovaries (White et al., 2012).

Despite these collective findings across a range of different species, the exact function of pluripotency genes in the context of PGCs remains unclear. Nanog expression in the preimplantation epiblast correlates with X-chromosome reactivation (Silva et al., 2009) and a direct role for pluripotency genes in regulating X-chromosome inactivation status has been reported (Donohoe, Silva, Pinter, Xu, & Lee, 2009; Navarro et al., 2008, 2010). It is therefore possible that reexpression of pluripotency genes is involved in the early reprogramming events, such as X-chromosome reactivation, in PGCs. They may also act, in parallel with Blimp1, to prevent somatic differentiation of early PGCs. Further mechanistic studies in the mouse combined with careful observations in other mammalian species will be particularly informative to begin to uncover their precise role.

## 2.4. Other early PGC factors

Nanos proteins are evolutionarily conserved RNA-binding proteins that regulate germline development (Tsuda et al., 2003). In *Drosophila*, nanos is involved in appropriate migration of germ cells to the gonad (Kobayashi, Yamada, Asaoka, & Kitamura, 1996). In mice, Nanos2 is required for male gametogenesis (Suzuki & Saga, 2008), whereas Nanos3 plays an earlier role in PGCs. Although first expressed as early as E7.5, the requirement Nanos3 becomes evident during migration at around E9.5 when PGC numbers are decreased in *Nanos3* null embryos and by E12.5 few germ cells can be found in mutants (Suzuki, Tsuda, Kiso, & Saga, 2008). This loss of germ cells is due to apoptosis and appears to be partially dependent on Bax. Double knockout *Nanos3−/−*; *Bax−/−* mice are fertile but they have vastly reduced germ-cell numbers, suggesting multiple pathways regulate PGC survival (Suzuki et al., 2008).

The *Ter* mutation in *Dnd1* leads to germ-cell loss (Youngren et al., 2005). Although *Dnd1* expression is not specific to PGCs, the *Ter* mutation leads to a reduction in PGC numbers when homozygous, commencing as early as E8.0 (Sakurai, Iguchi, Moriwaki, & Noguchi, 1995) and few germ cells remain by E13.5 (Cook, Munger, Nadeau, & Capel, 2011). However, in strain 129 mice, the remaining PGCs frequently give rise to teratocarcinomas (Cook et al., 2011; Stevens & Little, 1954). *Dnd1* is also required in zebrafish PGCs where it has a vital function in controlling PGC migration (Weidinger et al., 2003). Dnd1 is also an RNA-binding protein that counteracts the function of miRNA-mediated translational repression in germ cells (Kedde et al., 2007). Thus, posttranscriptional regulation of RNA function by specific binding proteins is a feature of early germ cells, as well as later stages of germline development.

## 3. REGULATION OF GERM-CELL SURVIVAL AND MIGRATION

Starting from approximately E8.0, PGCs migrate through the hindgut endoderm toward their final position in the genital ridges. To make it to the genital ridges PGCs must survive, be actively motile, have access to an appropriate pathway to migrate along, and have directional cues to guide them. Steel factor (SF) has been demonstrated to play a critical role in both survival and migration. Steel-J, Grizzle Belly, Steel-$8^{H}$, Steel-$10^{H}$, and Steel-$18^{H}$ are null alleles at the Steel locus, that contain the *Kitlg* gene encoding SF (Copeland et al., 1990; Huang et al., 1990; Zsebo et al., 1990). Mice homozygous for these alleles exhibit loss of germ cells starting at around E9 and few if any PGCs make it to the genital ridge (Bennett, 1956; McCoshen & McCallion, 1975; Silver, 1979a). In contrast, the Steel-Dickie allele has a deletion in the transmembrane and cytoplasmic domains of SF leading to a specific loss of the membrane-bound form while maintaining a biologically active secreted protein (Flanagan, Chan, & Leder, 1991; Molyneaux et al., 2001). Homozygous Steel-Dickie mice are viable but profoundly anemic and infertile. The germ-cell phenotype essentially phenocopies the null alleles, emphasizing the importance of membrane-bound SF for survival and proliferation *in vivo* (Chiquoine, 1954; Mahakali Zama, Hudson, & Bedell, 2005). A similar PGC phenotype occurs in mice with homozygous deletions alleles at the *W* locus (*dominant spotting*) (Buehr, McLaren, Bartley, & Darling, 1993; Mintz & Russell, 1957; Silver, 1979b) that encodes the SF receptor c-Kit (Chabot, Stephenson, Chapman,

Besmer, & Bernstein, 1988; Geissler & Ryan, 1988; Huang et al., 1990; Williams et al., 1990). c-Kit is expressed in PGCs throughout their development (Manova & Bachvarova, 1991; Orr-Urtreger et al., 1990), while SF is expressed along the migratory pathway and in the genital ridges (Matsui, Zsebo, & Hogan, 1990). SF is specifically downregulated in the midline between E9.5 and E10.5, resulting in apoptosis of PGCs that have failed to move laterally toward the genital ridges (Runyan et al., 2006). Thus, the expression pattern of SF appears to define the boundaries for PGC migration, although cell adhesion to laminins may also play a role (García-Castro, Anderson, Heasman, & Wylie, 1997). Blocking PGC apoptosis in *Steel* mutant embryos, by simultaneously knocking out the proapoptotic gene *Bax* (Knudson, Tung, Tourtellotte, Brown, & Korsmeyer, 1995; Stallock & Wylie, 2003), confirms that SF is required for appropriate migration and reveals that SF also regulates PGC proliferation *in vivo* (Runyan et al., 2006). The migration of PGCs also requires directionality. The initial movement away from the base of the allantois may in part be due to morphogenetic tissue movement (Mclaren, 2003). However, in the hindgut, PGCs are highly motile (Molyneaux et al., 2001). They move first ventrally into the dorsal midline between E9.0–9.5, before moving laterally to the genital ridges during the E10–10.5 period (Molyneaux et al., 2001). The possibility that the genital ridges release a chemoattractant molecule is supported by the observation that PGCs actively migrate toward dissected genital ridges (Godin, Wylie, & Heasman, 1990). In keeping with this, the ligand/receptor pair SDF1, expressed by the genital ridge, and CXCR4, expressed by PGCs, are critical players for colonization of the developing gonad (Molyneaux et al., 2003). Very few of the PGCs found in the gut mesentery at E10.5 migrate into the genital ridge (Molyneaux et al., 2001) and they most likely undergo apoptosis (Runyan et al., 2006).

Upon reaching the genital ridges, PGCs commence expression of *Mvh*, a specific germline marker until the postmeiotic stage in males and females (Toyooka et al., 2000). *Dazl* and *Sycp3* exhibit a similar expression pattern (Cooke, Lee, Kerr, & Ruggiu, 1996; Di Carlo, Travia, & de Felici, 2000; Osterlund, Töhönen, Forslund, & Nordqvist, 2000). All three genes are regulated by promoter DNA methylation (Maatouk & Resnick, 2006) with their reexpression possibly reflecting global changes in DNA methylation occurring at this time (see later). Indeed, recent evidence suggests that many genes implicated in later stages of germ-cell development, such as gametogenesis and meiosis, are regulated by promoter DNA methylation

(Borgel et al., 2010; Guibert, Forne, & Weber, 2012). Although *Dazl*, *Mvh*, and *Sycp3* appear not to be required during PGC development in the mouse, they play an essential role in later germ cells. *Dazl* is required for gametogenesis in both males and females (Ruggiu et al., 1997), whereas *Mvh* and *Sycp3* play a critical role in spermatogenesis (Tanaka et al., 2000; Yuan et al., 2000) In the rat, Mvh expression is detected in migrating PGCs prior to their colonization of the genital ridges, while GCNA and SSEA-1, PGC markers in the mouse, are not detected in rat PGCs (Encinas et al., 2012). Thus, even between rodents there appear to be some differences in the expression profile of classic late PGC marker genes. As previously discussed, SSEA-1 and c-Kit are detected in human PGCs although the onset and duration of their expression has not been rigorously determined. DAZL, VASA, and SYCP3 are expressed in human gonadal germ cells (Anderson, Fulton, Cowan, Coutts, & Saunders, 2007; Castrillon, Quade, Wang, Quigley, & Crum, 2000; Liu et al., 2007). Similar to rats, VASA is detectable in human migratory PGCs at the gut and mesenteric root (Castrillon et al., 2000). Interestingly, the CpG islands associated with *Dazl* and *Vasa* in the mouse appear to be conserved in humans, and they remain hypomethylated in sperm but not other somatic tissues (Chai, Phillips, Fernandez, & Yen, 1997; Sugimoto et al., 2009). Whether regulation of germline-specific genes by promoter methylation is a general feature in humans remains to be elucidated.

A summary of the expression of a range of PGC marker genes is presented in Table 5.1. We hope that in the coming years this will be completed and expanded upon. A more detailed analysis of the expression of germ-cell marker genes in a range of mammals, including human, would greatly enhance our understanding of the most conserved, and therefore perhaps most critical, progressions in gene expression during mammalian PGC development.

## 4. EPIGENETIC REPROGRAMMING

Epigenetic reprogramming is a key feature of PGC development. Perhaps most notably, this allows erasure of the DNA methylation marks associated with imprinted genes, allowing establishment of sex-specific imprints during gametogenesis. There is also extensive reprogramming of histone modifications both prior to and during imprint erasure. These chromatin changes may be necessary to facilitate DNA demethylation but may have other functions such as the erasure of "somatic" epigenetic marks established during postimplantation development, to allow X-chromosome

**Table 5.1** Comparison of PGC gene expression in different mammals

| | | Mouse | Human | Rat | Others |
|---|---|---|---|---|---|
| Specification genes | Blimp1 | + | + | + | |
| | Prdm14 | + | + | ND | |
| | Tcfap2c | + | + | ND | |
| | Lin28 | + | + | ND | |
| | Stella | + | ND | + | |
| Pluripotency genes | Oct4 | + | + | + | Sheep: +, Pig: + |
| | Nanog | + | + | + | Sheep: +, Pig: + |
| | Sox2 | + | – | + | Sheep: low |
| | Klf4 | – | ND | + | |
| | SSEA1 | + | + | – | Sheep: +, Pig: +, Goat: + |
| | c-Kit | + | + | ND | |
| Germ-cell-specific genes and others | Prmt5 | + | + | ND | |
| | Nanos3 | + | ND | + | |
| | Dnd1 | + | ND | ND | |
| | Vasa/ Mvh | + | + | + | |
| | Dazl | + | + | + | |
| | AP | + | + | + | Sheep: +, Pig: +, Cow: +, Goat: low |

ND, not determined; +, expressed; –, not expressed.

reactivation in females or to prevent inheritance of epimutations. Epigenetic reprogramming is usually considered to occur in two distinct waves and we will stick to this convention. However, whether the early and late stages of epigenetic reprogramming fulfill discrete biological function(s) is not clear.

## 4.1. Early

A number of epigenetic changes have been documented to occur immediately following specification, at around E8.0–8.5. There is a marked reduction in histone H3 lysine 9 dimethylation (H3K9me2), starting at around

E7.5 and continuing until E8.75, followed by an increase in histone H3 lysine 27 trimethylation (H3K27me3) between E8.25 and E9.5 (Seki et al., 2007). The decrease in H3K9me2 is most likely due to the reduction in GLP, a SET domain protein that forms a heteromeric complex with G9a (Tachibana et al., 2005), as both G9a and GLP are required for the H3K9 methyltransferase (HMT) activity of the complex (Tachibana et al., 2005). *Glp* transcripts are reduced in PGCs from E7.25 (Yabuta, Kurimoto, Ohinata, Seki, & Saitou, 2006), with a reduction in immunofluorescence (IF) staining intensity already visible at E7.75 (Seki et al., 2007). The reduction in H3K9me2 correlates with a period of mitotic arrest in G2 phase of the cell-cycle, shown by a decrease in bromodeoxyuridine (BrdU) incorporation and the expression of high level of Cyclin B1, and a period of transcriptional quiescence, as adjudged by a decrease BrUTP incorporation and a reduction in phosphorylation of Ser 2 and Ser 5 of the C-terminal domain of RNA polymerase II in comparison to neighboring cells (Seki et al., 2007). There is also a decrease in Dnmt3a and 3b transcript (Yabuta et al., 2006) and protein (Seki et al., 2005) at E7.25 and E8, respectively. This is accompanied by a reported decrease in 5-methylcytosine (5mC) staining in PGCs at E8, in comparison with somatic neighbors (Seki et al., 2005).

Prmt5, an arginine-specific HMT that mediates symmetrical dimethylation of arginine 3 on histone H2A and/or H4 tails (H2A/H4R3me2s) is enriched in PGCs from E8.5 onward, and the H2A/H4R3me2s shows higher accumulation in PGCs than soma at E10.5 (Ancelin et al., 2006). Blimp1 and Prmt5 can be coimmunoprecipitated in 293T cells suggesting that they may form a complex in PGCs.

Finally, reactivation of the inactive X chromosome in females also commences at an early stage of PGC development. PGCs at E6.5 exhibit nuclear H3K27me3 foci and random inactivation of an X-GFP transgene, consistent with an inactivated X chromosome (Chuva De Sousa Lopes et al., 2008). However, starting from as early as E7 there is a progressive loss of the *Xist* signal and at E10.5 PGCs are *Xist*-negative (Sugimoto & Abe, 2007). During migration to the genital ridges, the H3K27me3 foci are also gradually lost, and X-GFP is expressed in all PGCs by E13.5 (Chuva De Sousa Lopes et al., 2008). Thus, there appears to be a global epigenetic transition that begins in PGCs soon after specification. However, the function of, or necessity for, these global epigenetic changes has not been assessed and remains unclear.

A recent study in pigs has reported that H3K27me3 is high in migratory PGCs (E15–E21), whereas H3K9me2 levels are low at E15 (the onset of migration) and undetectable by IF staining at E21 (Hyldig, Croxall, et al., 2011). Furthermore, in PGCs at E17 an increased proportion of cells were found in G2 of the cell cycle, possibly paralleling the G2 arrest observed in mice. The early stage of reprogramming has not been studied in other mammals and is particularly difficult to study in humans due to the lack of human material available at this early stage of development. However, PRMT5 is coexpressed with BLIMP1 in human gonocytes (Eckert et al., 2008) suggesting some similarities with the mouse, at least in the expression of this epigenetic regulator.

## 4.2. Late

Once PGCs have arrived in the genital ridges, a major epigenetic reprogramming event ensues at E11. There is genome-wide DNA demethylation, changes in chromatin structure, and loss of numerous histone modifications brought about by genome-wide histone replacement (Hajkova et al., 2002, 2008). DNA demethylation in PGCs appears to occur rapidly, suggesting an active process, and may entail the DNA repair pathway (Hajkova et al., 2010). This reprogramming results in demethylation of many repetitive elements but critically includes the erasure of imprints (Hajkova et al., 2002), allowing the establishment of sex-specific imprints during gametogenesis (Reik, 2001). The mechanism of DNA demethylation remains controversial (Ooi & Bestor, 2008) and has been extensively reviewed elsewhere (Branco, Ficz, & Reik, 2012; Hackett, Zylicz, & Surani, 2012; Hajkova, 2011; Saitou, Kagiwada, & Kurimoto, 2012; Wu & Zhang, 2010).

Very little is known about the process of imprint erasure in other mammals. The most detailed analysis to date has been in pig embryos (Hyldig et al., 2011a,b). As observed for mouse embryos, there is a global decrease in methylation at repetitive sequences occurring on arrival at the genital ridge. This seems to coincide with demethylation of the *Igf2/H19* and *Igf2r* imprinted genes. However, in male embryos, *Igf2r* demethylation appears to be occurring during migration, an interesting difference that merits further investigation. There is an intimate relationship between H3K9me2 and DNA methylation, and it is possible that low H3K9me2 facilitates DNA demethylation in PGCs. Mouse PGC development lasts less than a week, but in the pig it progresses over 2–3 weeks and is likely even more prolonged in humans. Thus, in species with longer gestation, it is possible that the unique histone modification profile established during early reprogramming may play a more prominent role in facilitating DNA demethylation. Nevertheless, arrival at the genital ridge is accompanied by a large global decrease in DNA methylation in pig PGCs, a pattern similar to the mouse.

In humans, very few observations have been reported with regard to chromatin changes and DNA demethylation in PGCs. Indeed, a definitive timeline of when these events occur in humans has not been established. As genome-wide epigenetic reprogramming is completed prior to the onset of meiosis in mouse female germ cells, it is speculated that reprogramming of human germ cells, including DNA demethylation and chromatin remodeling, should indeed be completed before week 10 (commencement of meiosis in some female germ cells). However, published observations of epigenetic changes in human PGCs have almost exclusively examined later stage gonocytes, rather than PGCs. An immunohistochemical study of global 5mC level shows that both female and male human fetal germ cells are hypomethylated at week 13 postconception, approximately 3 weeks after the commencement of meiosis in female (Wermann et al., 2010). As transposon-related elements occupy about 40% of the genome while genes only occupy 2–3% (Lander & International Human Genome Sequencing Consortium, 2001; Waterston & Mouse Genome Sequencing Consortium, 2002), the low 5mC level detected by immunohistochemical method suggests that germ cells have already undergone genome-wide demethylation of retrotransposons by this stage. Indeed, the number of hypomethylated male fetal germ cells decline from week 13 to the time of birth, indicating gradual remethylation of the genome after global DNA demethylation (Wermann et al., 2010). Regarding chromatin states, human male germ cells at around 16–20 weeks postconception demonstrate absence/low levels of repressive H3K9me2, H3K27me3, and H3K9me3 modifications but high levels of permissive H3K9ac and H2A.Z marks (Almstrup et al., 2010; Bartkova et al., 2011). In contrast, H3K27me3 and H3K9me3 are present in E13.5 mouse PGCs (after reprogramming), while H3K9ac and H2A.Z are low/absent. The functional significance of these intriguing discrepancies in chromatin status between mouse and human germ cells remain to be explored. It will also be critical to study earlier human PGCs to establish a more accurate timeline of reprogramming (Fig. 5.2), for instance, when genome-wide demethylation is initiated and for how long it lasts. As human PGC development is more protracted than the mouse, it is likely that this will take place over a longer time frame which may facilitate dissection of the molecular events during reprogramming. Due to the differences in gestation between mice and humans, further

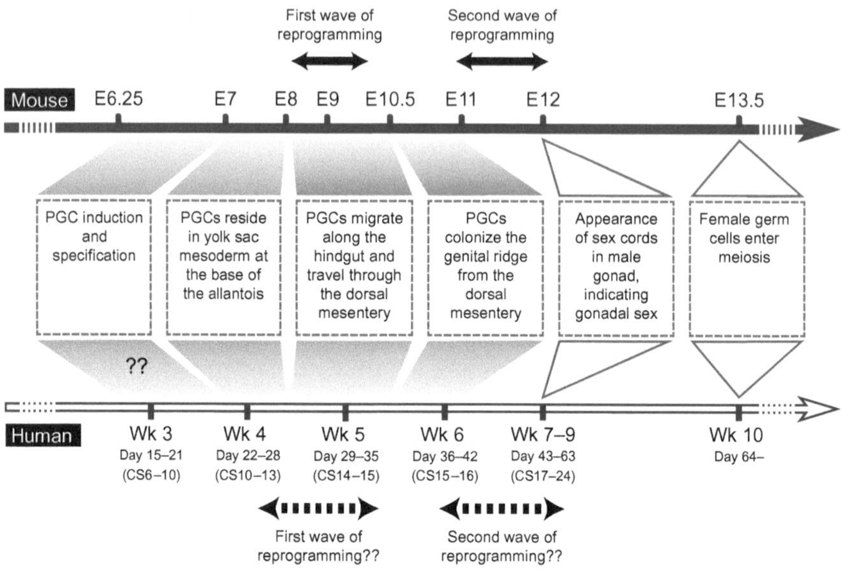


**Figure 5.2** Timeline of mouse and human PGC development. Established embryological landmarks are compared between mouse and human. The exact timing of human PGC specification is unknown (indicated by "??"). The timing and molecular features of epigenetic reprogramming in humans has not been established. Note that the timelines are not drawn to scale, and the mouse timeline is aligned to the human timeline according to the timing of various embryological events. CS, Carnegie stage; E, embryonic day; Wk, week.

comparison with large domestic animals and, if information emerges, nonhuman primates will be particularly informative in combination with direct studies in the human. We hope that further studies will allow a more rigorous comparison of PGC development and epigenetic reprogramming between mouse and man in the future (Fig. 5.2).

## 5. PGC CULTURE AND EMBRYONIC GERM-CELL DERIVATION

Efficient *in vitro* culture of PGCs has long been a goal of researchers as a tool to study germ-cell development in greater molecular detail. Investigations aimed at improving *in vitro* culture of PGCs also led to the serendipitous discovery of embryonic germ (EG) cells (Matsui, Zsebo, & Hogan, 1992; Resnick, Bixler, Cheng, & Donovan, 1992). Although derived from PGCs, EG cells exhibit growth requirements and properties indistinguishable from ES cells and are thus classified as pluripotent stem cells. As a result, *in vitro* investigations have focused largely on two different goals; continued attempts to culture PGCs themselves (and/or potentially progress them toward gametogenesis) or to establish pluripotent EG cell lines from them. Any modification that improves the survival or proliferation of PGCs in culture would be of potential benefit to both these ends. However, PGCs have hitherto not been amenable to culture, and the limited success has largely relied on the use of undefined components such as feeders and serum, as well as the addition of further growth factors. (de Felici & McLaren, 1983; Donovan, 1994; Donovan & de Miguel, 2003).

## 5.1. Requirement for feeders and serum

Although PGCs can survive and proliferate in serum-free medium (Ffrench-Constant, Hollingsworth, Heasman, & Wylie, 1991; Godin & Wylie, 1991), the use of serum is still routine. Irrespective of culture condition, PGCs do not adhere readily to tissue culture plastic or gelatin (de Felici & McLaren, 1983; Heath, 1978), and although mouse and rat PGCs can survive for up to a week in suspension, they exhibit only limited proliferation (de Felici & McLaren, 1983; Heath, 1978). Robust proliferation of PGCs requires adhesion to a feeder cell layer (Dolci et al., 1991; Donovan, Stott, Cairns, Heasman, & Wylie, 1986). Various feeder cell lines have been shown to support PGC proliferation including STO (Donovan et al., 1986), Sl4-m220 (Matsui et al., 1991), and the Sertoli cell line $TM_4$ (de Felici & Dolci, 1991). PGCs do adhere to a number of extracellular matrices including fibronectin, laminin, collagen IV, and Matrigel (de Felici & Dolci, 1989; de Felici, Pesce, Giustiniani, & Di Carlo, 1998; Ffrench-Constant et al., 1991; García-Castro et al., 1997). However, cell fragmentation, consistent with the occurrence of apoptosis, occurs within hours of culture on fibronectin, laminin, and Matrigel (de Felici et al., 1998). Koshimizu et al. reported limited cell survival of PGCs after 48 h of culture on Matrigel (Koshimizu et al., 1996). As Matrigel is an undefined basement membrane preparation extracted from Engelbreth-Holm-Swarm mouse sarcoma cells, its use has little benefit over that of a feeder layer. In the presence of serum and a cocktail of growth factors, PGCs do adhere to and can survive for up to 5 days on a transwell membrane insert (Farini, Scaldaferri, Iona, La Sala, & de Felici, 2005). However, there is extensive growth of somatic cells in this system that may play a supportive role during this period (Farini et al., 2005). Thus, the necessity of a feeder layer remains a major obstacle in the development of fully defined culture system for PGCs.

Serum and feeder-based strategies for culturing PGCs of other mammals have been attempted. Limited survival of pig PGCs on a STO feeder layer, in serum-based medium, has been observed (Takagi et al., 1997), and similar results have been reported in rabbits (Kakegawa et al., 2008), cow (Cherny et al., 1994), goats (Kühholzer et al., 2000), and sheep (Ledda et al., 2010).

A number of growth factors have been reported to increase the survival and/or proliferation of PGCs or the derivation efficiency of EG cells in feeder cultures.

## 5.2. Growth factors

In 1991, three papers reported a positive effect of SF on PGC cultures (Dolci et al., 1991; Godin et al., 1991; Matsui et al., 1991). Dolci et al. and Godin et al. concluded that SF improves PGC survival but not proliferation. While they provide good evidence for the former claim, the latter is complicated by the fact that both studies performed BrdU incorporation assays on a STO feeder layer which itself expresses high levels of SF (Dolci et al., 1991). Matsui et al. performed the same assay on a SF null feeder line. They show that addition of SF increases BrdU incorporation to a level comparable to that measured on STO cells, providing a clear demonstration that SF can act both as a survival factor and a mitogen (Matsui et al., 1991). Furthermore, they demonstrate that membrane-bound SF increases PGC number further, even when saturating concentrations of soluble SF are added. These findings are consistent with the phenotypes of the various Steel alleles described in mice (see earlier).

Leukemia inhibitory factor (LIF) can also act as both a survival factor and a mitogen for PGCs and is capable of acting synergistically with SF (Matsui et al., 1991). A subsequent report concluded that LIF acts as a "putative survival factor for proliferating PGCs" (de Felici & Dolci, 1991). Indeed, both SF and LIF can act on PGCs to block apoptosis (Pesce, Farrace, Piacentini, Dolci, & de Felici, 1993), although this effect does not preclude other roles for either factor. When basic fibroblast growth factor (bFGF) is combined with LIF and soluble SF, there is a dramatic increase in the number of PGCs (Matsui et al., 1992) but reprogramming to EG cells only occurs in the presence of membrane-bound SF (Matsui et al., 1992; Resnick et al., 1992). bFGF can bind to, and act directly on, germ cells to increase their numbers (Resnick, Ortiz, Keller, & Donovan, 1998) and is widely believed to act as a mitogen (de Felici, Dolci, & Pesce, 1993; Matsui et al., 1992; Resnick et al., 1992; Takeuchi, Molyneaux, Runyan, Schaible, & Wylie, 2005) despite the lack of direct evidence that it increases PGC proliferation rather than survival.

The plant extract Forskolin (FK) acts as a potent mitogen for PGCs (de Felici et al., 1993; Koshimizu, Watanabe, & Nakatsuji, 1995). FK has been shown to activate protein kinase A (PKA) but has pleiotropic effects on cells (Laurenza, Sutkowski, & Seamon, 1989; Lippincott-Schwartz et al., 1991). However, FK treatment increases cAMP in PGCs (Dolci, Pesce, & de Felici, 1993), and blockade of PKA eliminates the effect of FK on mitotic activity (de Felici et al., 1993). Furthermore, other factors that raise cAMP also increase PGC proliferation (Pesce, Canipari, Ferri, Siracusa, & de Felici, 1996). Thus, there is strong evidence that FK acts through PKA/cAMP and that this pathway is potentially an important regulator of PGC proliferation. However, the discovery of growth factors that act through PKA/cAMP *in vivo* have not been forthcoming (de Felici & Pesce, 1994). Retinoic acid (RA) is an equally potent PGC mitogen and can act synergistically with LIF to further increase proliferation (Koshimizu et al., 1995). Both FK and RA can replace bFGF in EG cell derivation protocols and, in fact, lead to an increase in derivation efficiency (Koshimizu et al., 1996). This provides indirect evidence that, like FK and RA, bFGF acts as a mitogen in the context of EG cell derivation. Furthermore, although LIF and SF can increase the mitotic activity of PGCs, proliferation may be a limiting factor during conversion of PGCs to EG cells.

Other factors have also been reported to influence PGC culture or EG cell derivation. LIF can be substituted by Oncostatin-M and IL-6 (in combination with soluble IL-6 receptor), both of which activate the gp130 receptor (Koshimizu et al., 1996). Similarly, other FGFs can substitute for bFGF to stimulate EG cell derivation. However, while FGF5 seems to have similar activity to bFGF on E8.5 PGCs, FGF9 and FGF10 are less efficient at generating EG cell colonies (Durcova-Hills, Adams, Barton, Surani, & Mclaren, 2006). These findings are not adequately accounted for by the different receptor affinities of the FGF molecules. However, it is noteworthy that mutants in FGFR2-IIIB have reduced numbers of PGCs at E11.5 and that addition of FGF7, a ligand for this receptor, causes a marginal increase in PGC numbers in embryo slice cultures (Takeuchi et al., 2005).

TGFβ (Godin & Wylie, 1991) and Activin (Richards, Enders, & Resnick, 1999) have been shown to have a negative effect on PGC proliferation, while TNFα (Kawase, Yamamoto, Hashimoto, & Nakatsuji, 1994) and IL-4 (Cooke, Heasman, & Wylie, 1996) have a positive effect. However, no subsequent work has built on these observations, and the role of these pathways during reprogramming to EG cells has not been investigated. Finally, Trichostatin (TSA), a histone deacetylase inhibitor, can replace bFGF in EG cell derivation (Durcova-Hills et al., 2008). However, given the diverse cellular targets of HDACs including nonhistones (Bolden, Peart, & Johnstone, 2006; Drummond et al., 2005), the exact manner in which TSA acts during reprogramming is unclear. A major caveat to these studies is the inability to draw firm mechanistic conclusions in a heterogeneous system. The presence of feeders and contaminating somatic cells mean that any factor added to the culture could be acting directly on PGCs, indirectly on feeders or somatic cells, or any combination of these options (Fig. 5.3). For instance, bFGF has been shown to increase the expression of both membrane-bound and soluble LIF by feeder cells (Rathjen et al., 1990). The possibility remains that this may create a high local concentration or that the signal from membrane-bound LIF may be quite different to the soluble form, as has been demonstrated for SF (Matsui et al., 1991, 1992). Furthermore, RA, FK, and TSA could all be exerting their effects indirectly through bFGF, by stimulating its release from feeders, somatic cells, or even PGCs themselves (Fig. 5.3). Some factors have been shown to influence survival or proliferation of PGCs in the absence of feeders including RA (Koshimizu et al., 1995), FK (de Felici et al., 1993), LIF (Koshimizu et al., 1996), and bFGF (Resnick et al., 1998). However, in these reports, PGCs are not fully purified and so somatic cells could mediate indirect effects (Fig. 5.3). More accurate assessment of the growth factors required by PGCs, and for EG cell derivation, requires a system to culture purified PGCs without a feeder layer.

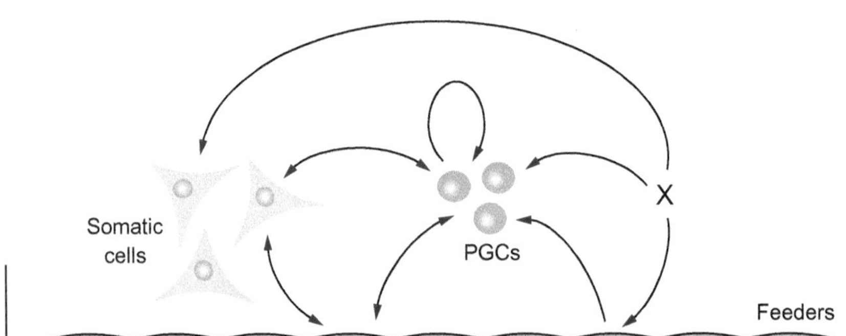


**Figure 5.3** Schematic of possible interactions in a heterogeneous culture system. (For color version of this figure, the reader is referred to the online version of this chapter.)

## 5.3. Signaling pathways

The signal transduction pathways downstream of growth factors have been less thoroughly investigated in PGCs. The Nakano laboratory has demonstrated that conditional deletion of *Pten* in PGCs leads to testicular teratomas in mice and also increases the efficiency of EG cell derivation (Kimura et al., 2003). Deletion of *Pten* in ES cells leads to an increase in phosphatidylinositol (3,4,5)-triphosphate (PIP3) and activation of the downstream Akt signaling pathway (Sun et al., 1999). In keeping with this, Akt is hyperphosphorylated in *Pten*-deleted PGCs and teratomas (Kimura et al., 2003). PGCs in *Pten*-deleted mice display hyperproliferation, increased survival, and tumor formation (Kimura et al., 2003). Overexpression of Akt has been demonstrated to improve the survival of PGCs in culture (de Miguel, Cheng, Holland, Federspiel, & Donovan, 2002), and forced Akt activation in cultured PGCs leads to an increased EG cell derivation efficiency, phenocopying *Pten*-deletion (Kimura et al., 2008). This constitutively active Akt construct also obviates the need for bFGF during derivation. The authors also demonstrated that *p53* deletion similarly improves EG cell derivation efficiency in the presence or absence of bFGF (Kimura et al., 2008). Taken together, these findings indicate that the PI3K/Akt pathway is a critical regulator of PGC proliferation and viability, and provide compelling evidence that activation of this axis is an important step to enable establishment of EG cells. This suggests a model of the signal transduction pathway downstream of bFGF during EG cell derivation (Fig. 5.4). In addition, it has been reported that bFGF is required for only the first 24 h during derivation (Durcova-Hills et al., 2006), as is the forced activation of Akt (Kimura et al., 2008). If bFGF is omitted during this time, EG cells cannot be derived even if bFGF is subsequently added back (Durcova-Hills et al., 2006). This may indicate that Akt activation by bFGF provides an early stimulus to PGCs and acts during a critical time window. However, whether the bFGF/PI3K/Akt axis acts purely to increase proliferation of PGCs or in some way directly triggers reprogramming is unclear.

## 5.4. EG cells

Upon blastocyst injection, EG cells contribute to chimeric mice and exhibit germline transmission, confirming their pluripotency (Labosky, Barlow, & Hogan, 1994; Stewart, Gadi, & Bhatt, 1994). In contrast, PGCs do not contribute to chimeras following blastocyst injection (Durcova-Hills et al., 2006; Papaioannou & Gardner, personal communication; Rossant, 1993)

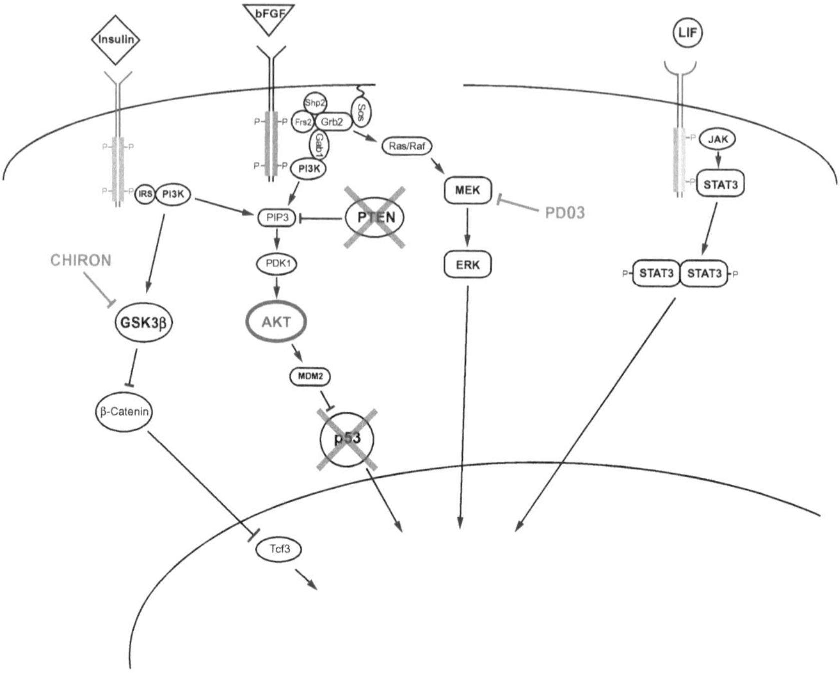


**Figure 5.4** PGC signaling pathways. (See Color Insert.)

and so are not considered pluripotent. Using the protocol established in mice, EG cells have been reported in range of other species, including pigs, goats, and buffalo (Durcova-Hills & Surani, 2008). Although some contribution to chimeric animals has been achieved, EG cells derived by traditional means from mammals other than the mouse have not exhibited germline transmission. Notably, a number of groups have reported derivation of EG cell-like colonies from pigs, suggesting that they may be a permissive species for future studies. It may be that using more modern culture conditions (see below) it will be possible to derive EG cells capable of producing high contribution chimeras, including access to the porcine germline (Alberio & Perez, 2012). Human EG cells have been also been reported (Shamblott et al., 1998; Turnpenny et al., 2003). However, although these cultures express many pluripotency markers, they exhibit only limited proliferative potential and therefore cannot be considered bona fide stem cells (Turnpenny et al., 2006).

Recently, we have reported the derivation of germline-competent rat EG cells (Blair et al., 2012; Leitch et al., 2010) utilizing the 2i culture system

(Fig. 5.4). 2i consists of basal medium supplemental with small molecule inhibitors of MAPK signaling and of glycogen synthase kinase-3b and has been demonstrated to maintain ES cells in a self-renewing ground state (Ying et al., 2008). The 2i culture system, supplemented with LIF, has enabled efficient culture and derivation of ES cells from all strains of mice tested, including 129, C57BL/6, CBA, and NOD (Batlle-Morera, Smith, & Nichols, 2008; Kiyonari, Kaneko, Abe, & Aizawa, 2010; Nichols et al., 2009; Ying et al., 2008). Furthermore, reprogramming of mouse somatic cells to induced pluripotency is also enhanced using 2i/LIF (Buehr et al., 2008; Li et al., 2008). 2i/LIF has also enabled derivation, for the first time, of germline-competent ES cells from rats (Buehr et al., 2008; Li et al., 2008). We found that addition of 2i/LIF on the third day of EG cell derivation led to a fivefold increase in EG cell derivation in the mouse (Leitch et al., 2010). This same protocol allowed the derivation of rat EG cells which self-renew in 2i/LIF conditions and exhibit properties indistinguishable from rat ES cells, including extensive contribution to chimeras (Leitch et al., 2010). Subsequently, rat EG cells have also been established from later stage gonadal PGCs (Northrup et al., 2011). Intriguingly, these authors also reported that efficiency of EG cell derivation is increased in rats carrying a mutation in *Dnd1* (Northrup et al., 2011). Rat EG cells have also been reported to express Vasa (Northrup et al., 2011), a notable difference to mouse EG cells which do not (Toyooka et al., 2000). Crucially, rat EG cells derived from both early and gonadal PGCs have been demonstrated to undergo germline transmission (Blair et al., 2012; Northrup et al., 2011). This work establishes that the capacity of PGCs to convert to pluripotent stem cells is not a mouse-specific phenomenon and demonstrates that EG cells represent a feasible route to establishing transgenic technologies in other mammals. We propose that application of 2i may prove a useful tool to extend these findings to non-rodents, including humans. Although transgenic technologies could not be applied to humans, establishment of naïve pluripotent human EG cells would be useful in attempts to derive human PGC-like cells in culture (Hayashi et al., 2011).

## 6. CONCLUSION

Our understanding of mammalian germ-cell development has advanced significantly since the early observations of embryologists. Transgenic technologies have facilitated studies in the mouse, with relatively few studies in other organisms. Due to the challenging nature of PGC research, there remains much to learn about murine germline development and epigenetic reprogramming, in particular. As transgenic tools become available in a broader spectrum of mammals, it is possible that mechanistic studies will extend beyond the mouse. The recent report of the first knockout rat generated by homologous recombination in ES cells is particularly noteworthy in this regard (Tong, Li, Wu, Yan, & Ying, 2010). However, it is likely that the mouse will remain the organism of choice for functional studies in the years to come. Nevertheless, it is crucial that further efforts are made to broaden and extend our knowledge of PGC development in a wide range of different mammals. This will be of particular utility in delineating conserved developmental and epigenetic mechanisms from species-specific quirks. This in turn will serve as a useful guide to investigators with access to precious human samples. The development of *in vitro* systems for PGC culture and EG cell derivation will prove an important tool especially if recent advances can be extended to domestic animal and human PGCs.

## ACKNOWLEDGMENTS

We thank all of the members of Surani lab past and present, and the Wellcome Trust for funding. W. W. C. T. is supported by a PhD scholarship jointly funded by the Croucher Foundation (Hong Kong) and the Cambridge Overseas Trust (United Kingdom).

## REFERENCES

Alberio, R., & Perez, A. R. (2012). Recent advances in stem and germ cell research: Implications for the derivation of pig pluripotent cells. *Reproduction in Domestic Animals*, *47*, 98–106.

Almstrup, K., Nielsen, J. E., Mlynarska, O., Jansen, M. T., Jørgensen, A., Skakkebæk, N. E., et al. (2010). Carcinoma in situ testis displays permissive chromatin modifications similar to immature foetal germ cells. *British Journal of Cancer*, *103*, 1269–1276.

Ancelin, K., Lange, U. C., Hajkova, P., Schneider, R., Bannister, A. J., Kouzarides, T., et al. (2006). Blimp1 associates with Prmt5 and directs histone arginine methylation in mouse germ cells. *Nature Cell Biology*, *8*, 623–630.

Anderson, R. A., Fulton, N., Cowan, G., Coutts, S., & Saunders, P. T. (2007). Conserved and divergent patterns of expression of DAZL, VASA and OCT4 in the germ cells of the human fetal ovary and testis. *BMC Developmental Biology*, *7*, 136.

Avilion, A. A., Nicolis, S. K., Pevny, L. H., Perez, L., Vivian, N., & Lovell-Badge, R. (2003). Multipotent cell lineages in early mouse development depend on SOX2 function. *Genes & Development*, *17*, 126–140.

Bartkova, J., Moudry, P., Hodny, Z., Lukas, J., Rajpert-De Meyts, E., & Bartek, J. (2011). Heterochromatin marks HP1γ, HP1α and H3K9me3, and DNA damage response activation in human testis development and germ cell tumours. *International Journal of Andrology*, *34*, e103–e113.

Batlle-Morera, L., Smith, A., & Nichols, J. (2008). Parameters influencing derivation of embryonic stem cells from murine embryos. *Genesis*, *46*, 758–767.

Bennett, D. (1956). Developmental analysis of a mutation with pleiotropic effects in the mouse. *Journal of Morphology*, *98*, 199–233.

Blair, K., Leitch, H. G., Mansfield, W., Dumeau, C. E., Humphreys, P., & Smith, A. G. (2012). Culture parameters for stable expansion, genetic modification and germline transmission of rat pluripotent stem cells. *Biology Open*, *1*, 58–65.

Bolden, J. E., Peart, M. J., & Johnstone, R. W. (2006). Anticancer activities of histone deacetylase inhibitors. *Nature Reviews. Drug Discovery*, *5*, 769–784.

Borgel, J., Guibert, S., Li, Y., Chiba, H., Schübeler, D., Sasaki, H., et al. (2010). Targets and dynamics of promoter DNA methylation during early mouse development. *Nature Genetics*, *42*, 1093–1100.

Branco, M. R., Ficz, G., & Reik, W. (2012). Uncovering the role of 5-hydroxymethylcytosine in the epigenome. *Nature Reviews Genetics*, *13*, 7–13.

Boyer, L. A., Lee, T. I., Cole, M. F., Johnstone, S. E., Levine, S. S., Zucker, J. P., et al. (2005). Core transcriptional regulatory circuitry in human embryonic stem cells. *Cell*, *122*, 947–956.

Buehr, M., McLaren, A., Bartley, A., & Darling, S. (1993). Proliferation and migration of primordial germ cells in We/We mouse embryos. *Developmental Dynamics*, *198*, 182–189.

Buehr, M., Meek, S., Blair, K., Yang, J., Ure, J., Silva, J., et al. (2008). Capture of authentic embryonic stem cells from rat blastocysts. *Cell*, *135*, 1287–1298.

Castrillon, D. H., Quade, B. J., Wang, T. Y., Quigley, C., & Crum, C. P. (2000). The human VASA gene is specifically expressed in the germ cell lineage. *Proceedings of the National Academy of Sciences of the United States of America*, *97*, 9585–9590.

Cauffman, G., De Rycke, M., Sermon, K., Liebaers, I., & Van de Velde, H. (2009). Markers that define stemness in ESC are unable to identify the totipotent cells in human preimplantation embryos. *Human Reproduction (Oxford, England)*, *24*, 63–70.

Chabot, B., Stephenson, D. A., Chapman, V. M., Besmer, P., & Bernstein, A. (1988). The proto-oncogene c-kit encoding a transmembrane tyrosine kinase receptor maps to the mouse W locus. *Nature*, *335*, 88–89.

Chai, N. N., Phillips, A., Fernandez, A., & Yen, P. H. (1997). A putative human male infertility gene DAZLA: Genomic structure and methylation status. *Molecular Human Reproduction*, *3*, 705–708.

Chambers, I., Silva, J., Colby, D., Nichols, J., Nijmeijer, B., Robertson, M., et al. (2007). Nanog safeguards pluripotency and mediates germline development. *Nature*, *450*, 1230–1234.

Chang, H., & Matzuk, M. M. (2001). Smad5 is required for mouse primordial germ cell development. *Mechanisms of Development*, *104*, 61–67.

Chen, X., Xu, H., Yuan, P., Fang, F., Huss, M., Vega, V. B., et al. (2008). Integration of external signaling pathways with the core transcriptional network in embryonic stem cells. *Cell*, *133*, 1106–1117.

Cherny, R. A., Stokes, T. M., Merei, J., Lom, L., Brandon, M. R., & Williams, R. L. (1994). Strategies for the isolation and characterization of bovine embryonic stem cells. *Reproduction, Fertility, and Development*, *6*, 569–575.

Childs, A. J., Kinnell, H. L., He, J., & Anderson, R. A. (2012). LIN28 is selectively expressed by primordial and pre-meiotic germ cells in the human fetal ovary. *Stem Cells and Development*, *21*, 2343–2349.

Chiquoine, A. D. (1954). The identification, origin, and migration of the primordial germ cells in the mouse embryo. *The Anatomical Record*, *118*, 135–146.

Chuva De Sousa Lopes, S. M., Hayashi, K., Shovlin, T. C., Mifsud, W., Surani, M. A., & Mclaren, A. (2008). X chromosome activity in mouse XX primordial germ cells. *PLoS Genetics*, *4*, e30.

Cook, M. S., Munger, S. C., Nadeau, J. H., & Capel, B. (2011). Regulation of male germ cell cycle arrest and differentiation by DND1 is modulated by genetic background. *Development, 138*, 23–32.
Cooke, J. E., Heasman, J., & Wylie, C. C. (1996). The role of interleukin-4 in the regulation of mouse primordial germ cell numbers. *Developmental Biology, 174*, 14–21.
Cooke, H. J., Lee, M., Kerr, S., & Ruggiu, M. (1996). A murine homologue of the human DAZ gene is autosomal and expressed only in male and female gonads. *Human Molecular Genetics, 5*, 513–516.
Copeland, N. G., Gilbert, D. J., Cho, B. C., Donovan, P. J., Jenkins, N. A., Cosman, D., et al. (1990). Mast cell growth factor maps near the steel locus on mouse chromosome 10 and is deleted in a number of steel alleles. *Cell, 63*, 175–183.
de Felici, M., & Dolci, S. (1989). In vitro adhesion of mouse fetal germ cells to extracellular matrix components. *Cell Differentiation and Development, 26*, 87–96.
de Felici, M., & Dolci, S. (1991). Leukemia inhibitory factor sustains the survival of mouse primordial germ cells cultured on TM4 feeder layers. *Developmental Biology, 147*, 281–284.
de Felici, M., Dolci, S., & Pesce, M. (1993). Proliferation of mouse primordial germ cells in vitro: A key role for cAMP. *Developmental Biology, 157*, 277–280.
de Felici, M., & McLaren, A. (1983). In vitro culture of mouse primordial germ cells. *Experimental Cell Research, 144*, 417–427.
de Felici, M., & Pesce, M. (1994). Growth factors in mouse primordial germ cell migration and proliferation. *Progress in Growth Factor Research, 5*, 135–143.
de Felici, M., Pesce, M., Giustiniani, Q., & Di Carlo, A. (1998). In vitro adhesiveness of mouse primordial germ cells to cellular and extracellular matrix component substrata. *Microscopy Research and Technique, 43*, 258–264.
de Miguel, M. P., Cheng, L., Holland, E. C., Federspiel, M. J., & Donovan, P. J. (2002). Dissection of the c-Kit signaling pathway in mouse primordial germ cells by retroviral-mediated gene transfer. *Proceedings of the National Academy of Sciences of the United States of America, 99*, 10458–10463.
de Sousa Lopes, S. M. C. (2004). BMP signaling mediated by ALK2 in the visceral endoderm is necessary for the generation of primordial germ cells in the mouse embryo. *Genes and Development, 18*, 1838–1849.
Di Carlo, A. D., Travia, G., & de Felici, M. (2000). The meiotic specific synaptonemal complex protein SCP3 is expressed by female and male primordial germ cells of the mouse embryo. *The International Journal of Developmental Biology, 44*, 241–244.
Dolci, S., Pesce, M., & de Felici, M. (1993). Combined action of stem cell factor, leukemia inhibitory factor, and cAMP on in vitro proliferation of mouse primordial germ cells. *Molecular Reproduction and Development, 35*, 134–139.
Dolci, S., Williams, D. E., Ernst, M. K., Resnick, J. L., Brannan, C. I., Lock, L. F., et al. (1991). Requirement for mast cell growth factor for primordial germ cell survival in culture. *Nature, 352*, 809–811.
Donohoe, M. E., Silva, S. S., Pinter, S. F., Xu, N., & Lee, J. T. (2009). The pluripotency factor Oct4 interacts with Ctcf and also controls X-chromosome pairing and counting. *Nature, 460*, 128–132.
Donovan, P. J. (1994). Growth factor regulation of mouse primordial germ cell development. *Current Topics in Developmental Biology, 29*, 189–225.
Donovan, P. J., & de Miguel, M. P. (2003). Turning germ cells into stem cells. *Current Opinion in Genetics and Development, 13*, 463–471.
Donovan, P. J., Stott, D., Cairns, L. A., Heasman, J., & Wylie, C. C. (1986). Migratory and postmigratory mouse primordial germ cells behave differently in culture. *Cell, 44*, 831–838.
Drummond, D. C., Noble, C. O., Kirpotin, D. B., Guo, Z., Scott, G. K., & Benz, C. C. (2005). Clinical development of histone deacetylase inhibitors as anticancer agents*. *Annual Review of Pharmacology and Toxicology, 45*, 495–528.

Durcova-Hills, G., Adams, I. R., Barton, S. C., Surani, M. A., & Mclaren, A. (2006). The role of exogenous fibroblast growth factor-2 on the reprogramming of primordial germ cells into pluripotent stem cells. *Stem Cells*, *24*, 1441–1449.
Durcova-Hills, G., & Surani, A. (2008). *Current protocols in stem cell biology*. pp. 1–20.
Durcova-Hills, G., Tang, F., Doody, G., Tooze, R., & Surani, M. A. (2008). Reprogramming primordial germ cells into pluripotent stem cells. *PLoS One*, *3*, e3531.
Eckert, D., Biermann, K., Nettersheim, D., Gillis, A. J. M., Steger, K., Jäck, H.-M., et al. (2008). Expression of BLIMP1/PRMT5 and concurrent histone H2A/H4 arginine 3 dimethylation in fetal germ cells, CIS/IGCNU and germ cell tumors. *BMC Developmental Biology*, *8*, 106.
Encinas, G., Zogbi, C., & Stumpp, T. (2012). Detection of four germ cell markers in rats during testis morphogenesis: Differences and similarities with mice. *Cells, Tissues, Organs*, *195*, 443–455.
Farini, D., Scaldaferri, M. L., Iona, S., La Sala, G., & de Felici, M. (2005). Growth factors sustain primordial germ cell survival, proliferation and entering into meiosis in the absence of somatic cells. *Developmental Biology*, *285*, 49–56.
Ffrench-Constant, C., Hollingsworth, A., Heasman, J., & Wylie, C. C. (1991). Response to fibronectin of mouse primordial germ cells before, during and after migration. *Development*, *113*, 1365–1373.
Flanagan, J. G., Chan, D. C., & Leder, P. (1991). Transmembrane form of the kit ligand growth factor is determined by alternative splicing and is missing in the Sld mutant. *Cell*, *64*, 1025–1035.
Fox, N., Damjanov, I., Martinez-Hernandez, A., Knowles, B. B., & Solter, D. (1981). Immunohistochemical localization of the early embryonic antigen (SSEA-1) in postimplantation mouse embryos and fetal and adult tissues. *Developmental Biology*, *83*, 391–398.
Fong, H., Hohenstein, K. A., & Donovan, P. J. (2008). Regulation of self-renewal and pluripotency by Sox2 in human embryonic stem cells. *Stem Cells*, *26*, 1931–1938.
Fujimoto, T., Miyayama, Y., & Fuyuta, M. (1977). The origin, migration and fine morphology of human primordial germ cells. *The Anatomical Record*, *188*, 315–330.
Fulton, N., Martins da Silva, S. J., Bayne, R. A. L., & Anderson, R. A. (2005). Germ cell proliferation and apoptosis in the developing human ovary. *The Journal of Clinical Endocrinology and Metabolism*, *90*, 4664–4670.
García-Castro, M. I., Anderson, R., Heasman, J., & Wylie, C. (1997). Interactions between germ cells and extracellular matrix glycoproteins during migration and gonad assembly in the mouse embryo. *The Journal of Cell Biology*, *138*, 471–480.
Gaskell, T. L., Esnal, A., Robinson, L. L. L., Anderson, R. A., & Saunders, P. T. K. (2004). Immunohistochemical profiling of germ cells within the human fetal testis: Identification of three subpopulations. *Biology of Reproduction*, *71*, 2012–2021.
Geissler, E., & Ryan, M. (1988). The dominant-white spotting (W) locus of the mouse encodes the c-kit proto-oncogene. *Cell*, *55*, 185–192.
Gillis, A. J. M., Stoop, H., Biermann, K., van Gurp, R. J. H. L. M., Swartzman, E., Cribbes, S., et al. (2011). Expression and interdependencies of pluripotency factors LIN28, OCT3/4, NANOG and SOX2 in human testicular germ cells and tumours of the testis. *International Journal of Andrology*, *34*, e160–e174.
Ginsburg, M., Snow, M., & McLaren, A. (1990). Primordial germ cells in the mouse embryo during gastrulation. *Development*, *110*, 521–528.
Godin, I., Deed, R., Cooke, J., Zsebo, K., Dexter, M., & Wylie, C. C. (1991). Effects of the steel gene product on mouse primordial germ cells in culture. *Nature*, *352*, 807–809.
Godin, I., & Wylie, C. C. (1991). TGF beta 1 inhibits proliferation and has a chemotropic effect on mouse primordial germ cells in culture. *Development*, *113*, 1451–1457.

Godin, I., Wylie, C., & Heasman, J. (1990). Genital ridges exert long-range effects on mouse primordial germ cell numbers and direction of migration in culture. *Development, 108*, 357–363.

Gondos, B., Westergaard, L., & Byskov, A. G. (1986). Initiation of oogenesis in the human fetal ovary: ultrastructural and squash preparation study. *American Journal of Obstetrics and Gynecology, 155*, 189–195.

Guibert, S., Forne, T., & Weber, M. (2012). Global profiling of DNA methylation erasure in mouse primordial germ cells. *Genome Research, 22*, 633–641.

Hackett, J. A., Zylicz, J. J., & Surani, M. A. (2012). Parallel mechanisms of epigenetic reprogramming in the germline. *Trends in Genetics, 28*, 164–174.

Hajkova, P. (2011). Epigenetic reprogramming in the germline: Towards the ground state of the epigenome. *Philosophical Transactions of the Royal Society B Biological Sciences, 366*, 2266–2273.

Hajkova, P., Ancelin, K., Waldmann, T., Lacoste, N., Lange, U. C., Cesari, F., et al. (2008). Chromatin dynamics during epigenetic reprogramming in the mouse germ line. *Nature, 452*, 877–881.

Hajkova, P., Erhardt, S., Lane, N., Haaf, T., El-Maarri, O., Reik, W., et al. (2002). Epigenetic reprogramming in mouse primordial germ cells. *Mechanisms of Development, 117*, 15–23.

Hajkova, P., Jeffries, S. J., Lee, C., Miller, N., Jackson, S. P., & Surani, M. A. (2010). Genome-wide reprogramming in the mouse germ line entails the base excision repair pathway. *Science, 329*, 78–82.

Hanley, N. A., Hagan, D. M., Clement-Jones, M., Ball, S. G., Strachan, T., Salas-Cortes, L., et al. (2000). SRY, SOX9, and DAX1 expression patterns during human sex determination and gonadal development. *Mechanisms of Development, 91*, 403–407.

Hayashi, K., Ohta, H., Kurimoto, K., Aramaki, S., & Saitou, M. (2011). Reconstitution of the mouse germ cell specification pathway in culture by pluripotent stem cells. *Cell, 146*, 519–532.

Heath, J. K. (1978). Mammalian primordial germ cells. In M. Johnson (Ed.), *Development in mammals* (pp. 267–298). New York: North Holland Biomedical Press.

Huang, E., Nocka, K., Beier, D. R., Chu, T. Y., Buck, J., Lahm, H. W., et al. (1990). The hematopoietic growth factor KL is encoded by the Sl locus and is the ligand of the c-kit receptor, the gene product of the W locus. *Cell, 63*, 225–233.

Hyldig, S. M. W., Croxall, N., Contreras, D. A., Thomsen, P. D., & Alberio, R. (2011). Epigenetic reprogramming in the porcine germ line. *BMC Developmental Biology, 11*, 11.

Hyldig, S. M. W., Ostrup, O., Vejlsted, M., & Thomsen, P. D. (2011). Changes of DNA methylation level and spatial arrangement of primordial germ cells in embryonic day 15 to embryonic day 28 pig embryos. *Biology of Reproduction, 84*, 1087–1093.

Johnson, A. D., Crother, B., White, M. E., Patient, R., Bachvarova, R. F., Drum, M., et al. (2003). Regulative germ cell specification in axolotl embryos: A primitive trait conserved in the mammalian lineage. *Philosophical Transactions of the Royal Society B Biological Sciences, 358*, 1371–1379.

Kakegawa, R., Teramura, T., Takehara, T., Anzai, M., Mitani, T., Matsumoto, K., et al. (2008). Isolation and culture of rabbit primordial germ cells. *The Journal of Reproduction and Development, 54*, 352–357.

Kawase, E., Yamamoto, H., Hashimoto, K., & Nakatsuji, N. (1994). Tumor necrosis factor-alpha (TNF-alpha) stimulates proliferation of mouse primordial germ cells in culture. *Developmental Biology, 161*, 91–95.

Kedde, M., Strasser, M. J., Boldajipour, B., Vrielink, J. A. F. O., Slanchev, K., le Sage, C., et al. (2007). RNA-binding protein Dnd1 inhibits microRNA access to target mRNA. *Cell, 131*, 1273–1286.

Kehler, J., Tolkunova, E., Koschorz, B., Pesce, M., Gentile, L., Boiani, M., et al. (2004). Oct4 is required for primordial germ cell survival. *EMBO Reports*, *5*, 1078–1083.

Keller, A. D., & Maniatis, T. (1991). Identification and characterization of a novel repressor of beta-interferon gene expression. *Genes and Development*, *5*, 868–879.

Kemper, C. H., & Peters, P. W. (1987). Migration and proliferation of primordial germ cells in the rat. *Teratology*, *36*, 117–124.

Kerr, C. L., Hill, C. M., Blumenthal, P. D., & Gearhart, J. D. (2008a). Expression of pluripotent stem cell markers in the human fetal ovary. *Human Reproduction*, *23*, 589–599.

Kerr, C. L., Hill, C. M., Blumenthal, P. D., & Gearhart, J. D. (2008b). Expression of pluripotent stem cell markers in the human fetal testis. *Stem Cells*, *26*, 412–421.

Kimura, T., Suzuki, A., Fujita, Y., Yomogida, K., Lomelí, H., Asada, N., et al. (2003). Conditional loss of PTEN leads to testicular teratoma and enhances embryonic germ cell production. *Development*, *130*, 1691–1700.

Kimura, T., Tomooka, M., Yamano, N., Murayama, K., Matoba, S., Umehara, H., et al. (2008). AKT signaling promotes derivation of embryonic germ cells from primordial germ cells. *Development*, *135*, 869–879.

Kiyonari, H., Kaneko, M., Abe, S.-I., & Aizawa, S. (2010). Three inhibitors of FGF receptor, ERK, and GSK3 establishes germline-competent embryonic stem cells of C57BL/6N mouse strain with high efficiency and stability. *Genesis*, *48*, 317–327.

Knudson, C. M., Tung, K. S., Tourtellotte, W. G., Brown, G. A., & Korsmeyer, S. J. (1995). Bax-deficient mice with lymphoid hyperplasia and male germ cell death. *Science*, *270*, 96–99.

Kobayashi, S., Yamada, M., Asaoka, M., & Kitamura, T. (1996). Essential role of the posterior morphogen nanos for germline development in Drosophila. *Nature*, *380*, 708–711.

Koshimizu, U., Taga, T., Watanabe, M., Saito, M., Shirayoshi, Y., Kishimoto, T., et al. (1996). Functional requirement of gp130-mediated signaling for growth and survival of mouse primordial germ cells in vitro and derivation of embryonic germ (EG) cells. *Development*, *122*, 1235–1242.

Koshimizu, U., Watanabe, M., & Nakatsuji, N. (1995). Retinoic acid is a potent growth activator of mouse primordial germ cells in vitro. *Developmental Biology*, *168*, 683–685.

Kühholzer, B., Baguisi, A., & Overström, E. W. (2000). Long-term culture and characterization of goat primordial germ cells. *Theriogenology*, *53*, 1071–1079.

Kurimoto, K., Yabuta, Y., Ohinata, Y., Shigeta, M., Yamanaka, K., & Saitou, M. (2008). Complex genome-wide transcription dynamics orchestrated by Blimp1 for the specification of the germ cell lineage in mice. *Genes and Development*, *22*, 1617–1635.

Labosky, P. A., Barlow, D. P., & Hogan, B. L. (1994). Mouse embryonic germ (EG) cell lines: Transmission through the germline and differences in the methylation imprint of insulin-like growth factor 2 receptor (Igf2r) gene compared with embryonic stem (ES) cell lines. *Development*, *120*, 3197–3204.

Lander, E. S., & International Human Genome Sequencing Consortium, (2001). Initial sequencing and analysis of the human genome. *Nature*, *409*, 860–921.

Lange, U. C., Adams, D. J., Lee, C., Barton, S., Schneider, R., Bradley, A., et al. (2008). Normal germ line establishment in mice carrying a deletion of the Ifitm/Fragilis gene family cluster. *Molecular and Cellular Biology*, *28*, 4688–4696.

Laurenza, A., Sutkowski, E. M., & Seamon, K. B. (1989). Forskolin: A specific stimulator of adenylyl cyclase or a diterpene with multiple sites of action? *Trends in Pharmacological Sciences*, *10*, 442–447.

Lawson, K. A., Dunn, N. R., Roelen, B. A., Zeinstra, L. M., Davis, A. M., Wright, C. V., et al. (1999). Bmp4 is required for the generation of primordial germ cells in the mouse embryo. *Genes and Development*, *13*, 424–436.

Lawson, K., & Hage, W. (1994). Clonal analysis of the origin of primordial germ cells in the mouse. *CIBA Foundation Symposium, 182*, 68–91.

Ledda, S., Bogliolo, L., Bebbere, D., Ariu, F., & Pirino, S. (2010). Characterization, isolation and culture of primordial germ cells in domestic animals: Recent progress and insights from the ovine species. *Theriogenology, 74*, 534–543.

Leitch, H. G., Blair, K., Mansfield, W., Ayetey, H., Humphreys, P., Nichols, J., et al. (2010). Embryonic germ cells from mice and rats exhibit properties consistent with a generic pluripotent ground state. *Development, 137*, 2279–2287.

Li, P., Tong, C., Mehrian-Shai, R., Jia, L., Wu, N., Yan, Y., et al. (2008). Germline competent embryonic stem cells derived from rat blastocysts. *Cell, 135*, 1299–1310.

Lippincott-Schwartz, J., Glickman, J., Donaldson, J. G., Robbins, J., Kreis, T. E., Seamon, K. B., et al. (1991). Forskolin inhibits and reverses the effects of brefeldin A on Golgi morphology by a cAMP-independent mechanism. *The Journal of Cell Biology, 112*, 567–577.

Liu, Y., Wu, C., Lyu, Q., Yang, D., Albertini, D. F., Keefe, D. L., et al. (2007). Germline stem cells and neo-oogenesis in the adult human ovary. *Developmental Biology, 306*, 112–120.

Lomelí, H., Ramos-Mejía, V., Gertsenstein, M., Lobe, C. G., & Nagy, A. (2000). Targeted insertion of Cre recombinase into the TNAP gene: Excision in primordial germ cells. *Genesis, 26*, 116–117.

Maatouk, D. M., & Resnick, J. L. (2006). DNA methylation is a primary mechanism for silencing postmigratory primordial germ cell genes in both germ cell and somatic cell lineages. *Development, 133*, 3411–3418.

Mahakali Zama, A., Hudson, F. P., & Bedell, M. A. (2005). Analysis of hypomorphic KitlSl mutants suggests different requirements for KITL in proliferation and migration of mouse primordial germ cells. *Biology of Reproduction, 73*, 639–647.

Manova, K., & Bachvarova, R. F. (1991). Expression of c-kit encoded at the W locus of mice in developing embryonic germ cells and presumptive melanoblasts. *Developmental Biology, 146*, 312–324.

Matsui, Y., Toksoz, D., Nishikawa, S., Nishikawa, S., Williams, D., Zsebo, K., et al. (1991). Effect of Steel factor and leukaemia inhibitory factor on murine primordial germ cells in culture. *Nature, 353*, 750–752.

Matsui, Y., Zsebo, K. M., & Hogan, B. L. (1990). Embryonic expression of a haematopoietic growth factor encoded by the Sl locus and the ligand for c-kit. *Nature, 347*, 667–669.

Matsui, Y., Zsebo, K., & Hogan, B. L. (1992). Derivation of pluripotential embryonic stem cells from murine primordial germ cells in culture. *Cell, 70*, 841–847.

McCoshen, J. A., & McCallion, D. J. (1975). A study of the primordial germ cells during their migratory phase in Steel mutant mice. *Experientia, 31*, 589–590.

Mclaren, A. (2003). Primordial germ cells in the mouse. *Developmental Biology, 262*, 1–15.

McLaren, A., & Lawson, K. (2005). How is the mouse germ-cell lineage established? *Differentiation, 73*, 435–437.

Mintz, B., & Russell, E. (1957). Gene-induced embryological modifications of primordial germ cells in the mouse. *The Journal of Experimental Zoology, 134*, 207–237.

Mitsui, K., Tokuzawa, Y., Itoh, H., Segawa, K., Murakami, M., Takahashi, K., et al. (2003). The homeoprotein Nanog is required for maintenance of pluripotency in mouse epiblast and ES cells. *Cell, 113*, 631–642.

Molyneaux, K. A., Stallock, J., Schaible, K., & Wylie, C. (2001). Time-lapse analysis of living mouse germ cell migration. *Developmental Biology, 240*, 488–498.

Molyneaux, K. A., Zinszner, H., Kunwar, P. S., Schaible, K., Stebler, J., Sunshine, M. J., et al. (2003). The chemokine SDF1/CXCL12 and its receptor CXCR4 regulate mouse germ cell migration and survival. *Development, 130*, 4279–4286.

Nakamura, T., Arai, Y., Umehara, H., Masuhara, M., Kimura, T., Taniguchi, H., et al. (2007). PGC7/Stella protects against DNA demethylation in early embryogenesis. *Nature Cell Biology, 9*, 64–71.

Nakamura, T., Liu, Y.-J., Nakashima, H., Umehara, H., Inoue, K., Matoba, S., et al. (2012). PGC7 binds histone H3K9me2 to protect against conversion of 5mC to 5hmC in early embryos. *Nature, 486*, 415–419.

Navarro, P., Chambers, I., Karwacki-Neisius, V., Chureau, C., Morey, C., Rougeulle, C., et al. (2008). Molecular coupling of Xist regulation and pluripotency. *Science, 321*, 1693–1695.

Navarro, P., Oldfield, A., Legoupi, J., Festuccia, N., Dubois, A., Attia, M., et al. (2010). Molecular coupling of Tsix regulation and pluripotency. *Nature, 468*, 457–460.

Nichols, J., Jones, K., Phillips, J. M., Newland, S. A., Roode, M., Mansfield, W., et al. (2009). Validated germline-competent embryonic stem cell lines from nonobese diabetic mice. *Nature Medicine, 15*, 814–818.

Nichols, J., Zevnik, B., Anastassiadis, K., Niwa, H., Klewe-Nebenius, D., Chambers, I., et al. (1998). Formation of pluripotent stem cells in the mammalian embryo depends on the POU transcription factor Oct4. *Cell, 95*, 379–391.

Nie, K., Gomez, M., Landgraf, P., Garcia, J.-F., Liu, Y., Tan, L. H. C., et al. (2008). Micro-RNA-mediated down-regulation of PRDM1/Blimp-1 in Hodgkin/Reed-Sternberg cells: A potential pathogenetic lesion in Hodgkin lymphomas. *The American Journal of Pathology, 173*, 242–252.

Niwa, H. (2007). How is pluripotency determined and maintained? *Development, 134*, 635–646.

Northrup, E., Eisenblätter, R., Glage, S., Rudolph, C., Dorsch, M., Schlegelberger, B., et al. (2011). Loss of Dnd1 facilitates the cultivation of genital ridge-derived rat embryonic germ cells. *Experimental Cell Research, 317*, 1885–1894.

Ohinata, Y., Ohta, H., Shigeta, M., Yamanaka, K., Wakayama, T., & Saitou, M. (2009). A signaling principle for the specification of the germ cell lineage in mice. *Cell, 137*, 571–584.

Ohinata, Y., Payer, B., O'carroll, D., Ancelin, K., Ono, Y., Sano, M., et al. (2005). Blimp1 is a critical determinant of the germ cell lineage in mice. *Nature, 436*, 207–213.

Ohinata, Y., Sano, M., Shigeta, M., Yamanaka, K., & Saitou, M. (2008). A comprehensive, non-invasive visualization of primordial germ cell development in mice by the Prdm1-mVenus and Dpp a3-ECFP double transgenic reporter. *Reproduction, 136*, 503–514.

Okamura, D., Tokitake, Y., Niwa, H., & Matsui, Y. (2008). Requirement of Oct3/4 function for germ cell specification. *Developmental Biology, 317*, 576–584.

Ooi, S. K. T., & Bestor, T. H. (2008). The colorful history of active DNA demethylation. *Cell, 133*, 1145–1148.

Orr-Urtreger, A., Avivi, A., Zimmer, Y., Givol, D., Yarden, Y., & Lonai, P. (1990). Developmental expression of c-kit, a proto-oncogene encoded by the W locus. *Development, 109*, 911–923.

Osterlund, C., Töhönen, V., Forslund, K. O., & Nordqvist, K. (2000). Mage-b4, a novel melanoma antigen (MAGE) gene specifically expressed during germ cell differentiation. *Cancer Research, 60*, 1054–1061.

Pauls, K., Jäger, R., Weber, S., Wardelmann, E., Koch, A., Büttner, R., et al. (2005). Transcription factor AP-2gamma, a novel marker of gonocytes and seminomatous germ cell tumors. *International Journal of Cancer, 115*, 470–477.

Payer, B., Chuva De Sousa Lopes, S. M., Barton, S. C., Lee, C., Saitou, M., & Surani, M. A. (2006). Generation of stella-GFP transgenic mice: A novel tool to study germ cell development. *Genesis, 44*, 75–83.

Payer, B., Saitou, M., Barton, S. C., Thresher, R., Dixon, J. P. C., Zahn, D., et al. (2003). Stella is a maternal effect gene required for normal early development in mice. *Current Biology*, *13*, 2110–2117.

Perrett, R. M., Turnpenny, L., Eckert, J. J., O'Shea, M., Sonne, S. B., Cameron, I. T., et al. (2008). The early human germ cell lineage does not express SOX2 during in vivo development or upon in vitro culture. *Biology of Reproduction*, *78*, 852–858.

Pesce, M., Canipari, R., Ferri, G. L., Siracusa, G., & de Felici, M. (1996). Pituitary adenylate cyclase-activating polypeptide (PACAP) stimulates adenylate cyclase and promotes proliferation of mouse primordial germ cells. *Development*, *122*, 215–221.

Pesce, M., Farrace, M. G., Piacentini, M., Dolci, S., & de Felici, M. (1993). Stem cell factor and leukemia inhibitory factor promote primordial germ cell survival by suppressing programmed cell death (apoptosis). *Development*, *118*, 1089–1094.

Pilon, N., Behdjani, R., Daneau, I., Lussier, J. G., & Silversides, D. W. (1998). Porcine steroidogenic factor-1 gene (pSF-1) expression and analysis of embryonic pig gonads during sexual differentiation. *Endocrinology*, *139*, 3803–3812.

Rathjen, P. D., Nichols, J., Toth, S., Edwards, D. R., Heath, J. K., & Smith, A. G. (1990). Developmentally programmed induction of differentiation inhibiting activity and the control of stem cell populations. *Genes and Development*, *4*, 2308–2318.

Reik, W. (2001). Epigenetic reprogramming in mammalian development. *Science*, *293*, 1089–1093.

Resnick, J. L., Bixler, L. S., Cheng, L., & Donovan, P. J. (1992). Long-term proliferation of mouse primordial germ cells in culture. *Nature*, *359*, 550–551.

Resnick, J. L., Ortiz, M., Keller, J. R., & Donovan, P. J. (1998). Role of fibroblast growth factors and their receptors in mouse primordial germ cell growth. *Biology of Reproduction*, *59*, 1224–1229.

Richards, A. J., Enders, G. C., & Resnick, J. L. (1999). Activin and TGFbeta limit murine primordial germ cell proliferation. *Developmental Biology*, *207*, 470–475.

Rossant, J. (1993). Immortal germ cells? *Current Biology*, *3*, 47–49.

Ruggiu, M., Speed, R., Taggart, M., McKay, S. J., Kilanowski, F., Saunders, P., et al. (1997). The mouse Dazla gene encodes a cytoplasmic protein essential for gametogenesis. *Nature*, *389*, 73–77.

Runyan, C., Schaible, K., Molyneaux, K., Wang, Z., Levin, L., & Wylie, C. (2006). Steel factor controls midline cell death of primordial germ cells and is essential for their normal proliferation and migration. *Development*, *133*, 4861–4869.

Saitou, M., Barton, S. C., & Surani, M. A. (2002). A molecular programme for the specification of germ cell fate in mice. *Nature*, *418*, 293–300.

Saitou, M., Kagiwada, S., & Kurimoto, K. (2012). Epigenetic reprogramming in mouse preimplantation development and primordial germ cells. *Development*, *139*, 15–31.

Sakurai, T., Iguchi, T., Moriwaki, K., & Noguchi, M. (1995). The ter mutation first causes primordial germ cell deficiency in ter/ter mouse embryos at 8 days of gestation. *Development, Growth and Differentiation*, *37*, 293–302.

Sasaki, H., & Matsui, Y. (2008). Epigenetic events in mammalian germ-cell development: Reprogramming and beyond. *Nature Reviews Genetics*, *9*, 129–140.

Schöler, H. R., Dressler, G. R., Balling, R., Rohdewohld, H., & Gruss, P. (1990). Oct-4: A germline-specific transcription factor mapping to the mouse t-complex. *The EMBO Journal*, *9*, 2185–2195.

Seki, Y., Hayashi, K., Itoh, K., Mizugaki, M., Saitou, M., & Matsui, Y. (2005). Extensive and orderly reprogramming of genome-wide chromatin modifications associated with specification and early development of germ cells in mice. *Developmental Biology*, *278*, 440–458.

Seki, Y., Yamaji, M., Yabuta, Y., Sano, M., Shigeta, M., Matsui, Y., et al. (2007). Cellular dynamics associated with the genome-wide epigenetic reprogramming in migrating primordial germ cells in mice. *Development*, *134*, 2627–2638.

Shamblott, M. J., Axelman, J., Wang, S., Bugg, E. M., Littlefield, J. W., Donovan, P. J., et al. (1998). Derivation of pluripotent stem cells from cultured human primordial germ cells. *Proceedings of the National Academy of Sciences of the United States of America, 95*, 13726–13731.

Shinoda, G., de Soysa, T. Y., Seligson, M. T., Yabuuchi, A., Fujiwara, Y., Yi Huang, P., et al. (2013). Lin28a Regulates Germ Cell Pool Size and Fertility. *Stem Cells*. http://dx.doi.org/10.1002/stem.1343. [Epub ahead of print].

Silva, J., Nichols, J., Theunissen, T. W., Guo, G., van Oosten, A. L., Barrandon, O., et al. (2009). Nanog is the gateway to the pluripotent ground state. *Cell, 138*, 722–737.

Silver, W. K. (1979a). Steel, flexed-tailed, splotch, and varitint-waddler. In *The coat colors of mice: A model for mammalian gene action and interaction* (pp. 243–267). New York: Springer-Verlag.

Silver, W. K. (1979b). Dominant spotting, patch, and rump-white. In *The coat colors of mice: A model for mammalian gene action and interaction*. New York: Springer-Verlag.

Skrzypczak, J., Pisarski, T., Biczysko, W., & Kedzia, H. (1981). Evaluation of germ cells development in gonads of human fetuses and newborns. *Folia Histochemica Et Cytochemica (Krakow), 19*, 17–24.

Stallock, J., & Wylie, C. (2003). The pro-apoptotic gene Bax is required for the death of ectopic primordial germ cells during their migration in the mouse embryo. *Development, 130*, 6589–6597.

Stevens, L. C., & Little, C. C. (1954). Spontaneous testicular teratomas in an inbred strain of mice. *Proceedings of the National Academy of Sciences of the United States of America, 40*, 1080–1087.

Stewart, C. L., Gadi, I., & Bhatt, H. (1994). Stem cells from primordial germ cells can reenter the germ line. *Developmental Biology, 161*, 626–628.

Stoop, H., Honecker, F., Cools, M., de Krijger, R., Bokemeyer, C., & Looijenga, L. H. J. (2005). Differentiation and development of human female germ cells during prenatal gonadogenesis: An immunohistochemical study. *Human Reproduction, 20*, 1466–1476.

Sugimoto, M., & Abe, K. (2007). X chromosome reactivation initiates in nascent primordial germ cells in mice. *PLoS Genetics, 3*, e116.

Sugimoto, K., Koh, E., Sin, H.-S., Maeda, Y., Narimoto, K., Izumi, K., et al. (2009). Tissue-specific differentially methylated regions of the human VASA gene are potentially associated with maturation arrest phenotype in the testis. *Journal of Human Genetics, 54*, 450–456.

Sun, H., Lesche, R., Li, D. M., Liliental, J., Zhang, H., Gao, J., et al. (1999). PTEN modulates cell cycle progression and cell survival by regulating phosphatidylinositol 3,4,5,-trisphosphate and Akt/protein kinase B signaling pathway. *Proceedings of the National Academy of Sciences of the United States of America, 96*, 6199–6204.

Suzuki, A., & Saga, Y. (2008). Nanos2 suppresses meiosis and promotes male germ cell differentiation. *Genes and Development, 22*, 430–435.

Suzuki, H., Tsuda, M., Kiso, M., & Saga, Y. (2008). Nanos3 maintains the germ cell lineage in the mouse by suppressing both Bax-dependent and -independent apoptotic pathways. *Developmental Biology, 318*, 133–142.

Tachibana, M., Ueda, J., Fukuda, M., Takeda, N., Ohta, T., Iwanari, H., et al. (2005). Histone methyltransferases G9a and GLP form heteromeric complexes and are both crucial for methylation of euchromatin at H3-K9. *Genes and Development, 19*, 815–826.

Takagi, Y., Talbot, N. C., Rexroad, C. E., & Pursel, V. G. (1997). Identification of pig primordial germ cells by immunocytochemistry and lectin binding. *Molecular Reproduction and Development, 46*, 567–580.

Takeuchi, Y., Molyneaux, K., Runyan, C., Schaible, K., & Wylie, C. (2005). The roles of FGF signaling in germ cell migration in the mouse. *Development, 132*, 5399–5409.

Tam, P. P., & Zhou, S. X. (1996). The allocation of epiblast cells to ectodermal and germline lineages is influenced by the position of the cells in the gastrulating mouse embryo. *Developmental Biology, 178*, 124–132.

Tanaka, S. S., Toyooka, Y., Akasu, R., Katoh-Fukui, Y., Nakahara, Y., Suzuki, R., et al. (2000). The mouse homolog of Drosophila Vasa is required for the development of male germ cells. *Genes and Development*, *14*, 841–853.

Tong, C., Li, P., Wu, N. L., Yan, Y., & Ying, Q.-L. (2010). Production of p53 gene knockout rats by homologous recombination in embryonic stem cells. *Nature*, *467*, 211–213.

Toyooka, Y., Tsunekawa, N., Takahashi, Y., Matsui, Y., Satoh, M., & Noce, T. (2000). Expression and intracellular localization of mouse Vasa-homologue protein during germ cell development. *Mechanisms of Development*, *93*, 139–149.

Tremblay, K. D., Dunn, N. R., & Robertson, E. J. (2001). Mouse embryos lacking Smad1 signals display defects in extra-embryonic tissues and germ cell formation. *Development*, *128*, 3609–3621.

Tsuda, M., Sasaoka, Y., Kiso, M., Abe, K., Haraguchi, S., Kobayashi, S., et al. (2003). Conserved role of nanos proteins in germ cell development. *Science*, *301*, 1239–1241.

Tsuneyoshi, N., Sumi, T., Onda, H., Nojima, H., Nakatsuji, N., & Suemori, H. (2008). PRDM14 suppresses expression of differentiation marker genes in human embryonic stem cells. *Biochemical and Biophysical Research Communications*, *367*, 899–905.

Turnpenny, L., Brickwood, S., Spalluto, C. M., Piper, K., Cameron, I. T., Wilson, D. I., et al. (2003). Derivation of human embryonic germ cells: An alternative source of pluripotent stem cells. *Stem Cells*, *21*, 598–609.

Turnpenny, L., Spalluto, C. M., Perrett, R. M., O'shea, M., Hanley, K. P., Cameron, I. T., et al. (2006). Evaluating human embryonic germ cells: Concord and conflict as pluripotent stem cells. *Stem Cells*, *24*, 212–220.

Ueno, H., Turnbull, B. B., & Weissman, I. L. (2009). Two-step oligoclonal development of male germ cells. *Proceedings of the National Academy of Sciences of the United States of America*, *106*, 175–180.

Vejlsted, M., Offenberg, H., Thorup, F., & Maddox-Hyttel, P. (2006). Confinement and clearance of OCT4 in the porcine embryo at stereomicroscopically defined stages around gastrulation. *Molecular Reproduction and Development*, *73*, 709–718.

Vincent, S. D., Dunn, N. R., Sciammas, R., Shapiro-Shalef, M., Davis, M. M., Calame, K., et al. (2005). The zinc finger transcriptional repressor Blimp1/Prdm1 is dispensable for early axis formation but is required for specification of primordial germ cells in the mouse. *Development*, *132*, 1315–1325.

Wang, J., Rao, S., Chu, J., Shen, X., Levasseur, D. N., Theunissen, T. W., et al. (2006). A protein interaction network for pluripotency of embryonic stem cells. *Nature*, *444*, 364–368.

Waterston, R. H., & Mouse Genome Sequencing Consortium, (2002). Initial sequencing and comparative analysis of the mouse genome. *Nature*, *420*, 520–562.

Weber, S., Eckert, D., Nettersheim, D., Gillis, A. J. M., Schäfer, S., Kuckenberg, P., et al. (2010). Critical function of AP-2 gamma/TCFAP2C in mouse embryonic germ cell maintenance. *Biology of Reproduction*, *82*, 214–223.

Weidinger, G., Stebler, J., Slanchev, K., Dumstrei, K., Wise, C., Lovell-Badge, R., et al. (2003). Dead end, a novel vertebrate germ plasm component, is required for zebrafish primordial germ cell migration and survival. *Current Biology*, *13*, 1429–1434.

Wermann, H., Stoop, H., Gillis, A. J. M., Honecker, F., van Gurp, R. J. H. L. M., Ammerpohl, O., et al. (2010). Global DNA methylation in fetal human germ cells and germ cell tumours: Association with differentiation and cisplatin resistance. *The Journal of Pathology*, *221*, 433–442.

West, J. A., Viswanathan, S. R., Yabuuchi, A., Cunniff, K., Takeuchi, A., Park, I.-H., et al. (2009). A role for Lin28 in primordial germ-cell development and germ-cell malignancy. *Nature*, *460*, 909–913.

White, Y. A. R., Woods, D. C., Takai, Y., Ishihara, O., Seki, H., & Tilly, J. L. (2012). Oocyte formation by mitotically active germ cells purified from ovaries of reproductive-age women. *Nature Medicine*, *18*, 413–421.

Williams, D. E., Eisenman, J., Baird, A., Rauch, C., Van Ness, K., March, C. J., et al. (1990). Identification of a ligand for the c-kit proto-oncogene. *Cell*, *63*, 167–174.

Witschi, E. (1948). Migration of the germ cells of human embryos from the yolk sac to the primitive gonadal folds. *Contributions to Embryology*, *32*, 67–80.

Wolf, X. A., Serup, P., & Hyttel, P. (2010). Three-dimensional localisation of NANOG, OCT4, and E-cadherin in porcine pre- and peri-implantation embryos. *Developmental Dynamics*, *240*, 204–210.

Wu, S. C., & Zhang, Y. (2010). Active DNA demethylation: Many roads lead to Rome. *Nature Reviews Molecular Cell Biology*, *11*, 607–620.

Yabuta, Y., Kurimoto, K., Ohinata, Y., Seki, Y., & Saitou, M. (2006). Gene expression dynamics during germline specification in mice identified by quantitative single-cell gene expression profiling. *Biology of Reproduction*, *75*, 705–716.

Yamaguchi, S., Kimura, H., Tada, M., Nakatsuji, N., & Tada, T. (2005). Nanog expression in mouse germ cell development. *Gene Expression Patterns*, *5*, 639–646.

Yamaguchi, S., Kurimoto, K., Yabuta, Y., Sasaki, H., Nakatsuji, N., Saitou, M., et al. (2009). Conditional knockdown of Nanog induces apoptotic cell death in mouse migrating primordial germ cells. *Development*, *136*, 4011–4020.

Yamaji, M., Seki, Y., Kurimoto, K., Yabuta, Y., Yuasa, M., Shigeta, M., et al. (2008). Critical function of Prdm14 for the establishment of the germ cell lineage in mice. *Nature Genetics*, *40*, 1016–1022.

Yeom, Y. I., Fuhrmann, G., Ovitt, C. E., Brehm, A., Ohbo, K., Gross, M., et al. (1996). Germline regulatory element of Oct-4 specific for the totipotent cycle of embryonal cells. *Development*, *122*, 881–894.

Ying, Y., Liu, X. M., Marble, A., Lawson, K. A., & Zhao, G. Q. (2000). Requirement of Bmp8b for the generation of primordial germ cells in the mouse. *Molecular Endocrinology*, *14*, 1053–1063.

Ying, Y., Qi, X., & Zhao, G. Q. (2001). Induction of primordial germ cells from murine epiblasts by synergistic action of BMP4 and BMP8B signaling pathways. *Proceedings of the National Academy of Sciences of the United States of America*, *98*, 7858–7862.

Ying, Q.-L., Wray, J., Nichols, J., Batlle-Morera, L., Doble, B., Woodgett, J., et al. (2008). The ground state of embryonic stem cell self-renewal. *Nature*, *453*, 519–523.

Ying, Y., & Zhao, G. Q. (2001). Cooperation of endoderm-derived BMP2 and extraembryonic ectoderm-derived BMP4 in primordial germ cell generation in the mouse. *Developmental Biology*, *232*, 484–492.

Yoshimizu, T., Obinata, M., & Matsui, Y. (2001). Stage-specific tissue and cell interactions play key roles in mouse germ cell specification. *Development*, *128*, 481–490.

Youngren, K. K., Coveney, D., Peng, X., Bhattacharya, C., Schmidt, L. S., Nickerson, M. L., et al. (2005). The Ter mutation in the dead end gene causes germ cell loss and testicular germ cell tumours. *Nature*, *435*, 360–364.

Yuan, L., Liu, J. G., Zhao, J., Brundell, E., Daneholt, B., & Höög, C. (2000). The murine SCP3 gene is required for synaptonemal complex assembly, chromosome synapsis, and male fertility. *Molecular Cell*, *5*, 73–83.

Zsebo, K. M., Wypych, J., McNiece, I. K., Lu, H. S., Smith, K. A., Karkare, S. B., et al. (1990). Identification, purification, and biological characterization of hematopoietic stem cell factor from buffalo rat liver—Conditioned medium. *Cell*, *63*, 195–201.

CHAPTER SIX

# Epigenetics and Development in Plants: Green Light to Convergent Innovations

**Daniel Grimanelli*[,1], François Roudier†[,1]**
*Institut de Recherche pour le Développement, UMR 232, Université de Montpellier II, Montpellier, France
†Institut de Biologie de l'Ecole Normale Supérieure, CNRS-UMR 8197/INSERM U1024, Paris, France
[1]Corresponding authors: e-mail address: daniel.grimanelli@ird.fr; roudier@biologie.ens.fr

## Contents



## Abstract

Plants are sessile organisms that must constantly adjust to their environment. In contrast to animals, plant development mainly occurs postembryonically and is characterized by continuous growth and extensive phenotypic plasticity. Chromatin-level regulation of transcriptional patterns plays a central role in the ability of plants to adapt to internal and external cues. Here, we review selected examples of chromatin-based mechanisms involved in the regulation of key aspects of plant development. These illustrate that, in addition to mechanisms conserved between plants and animals, plant-specific innovations lead to particular chromatin dynamics related to their developmental and life strategies.

*Current Topics in Developmental Biology*, Volume 104
ISSN 0070-2153
http://dx.doi.org/10.1016/B978-0-12-416027-9.00006-1

# 1. INTRODUCTION

Spatial and temporal patterns of gene expression are central to developmental programs in multicellular organisms. In plants, development and growth are continuous processes that occur mostly postembryonically. Following seed germination, two primary stem cell populations that are established during embryogenesis, the root and shoot meristems, generate the different organs throughout plant life. In the absence of cellular migration, due to the presence of a rigid surrounding wall, organ patterning mainly results from the precise coordination of cell division, expansion, and differentiation.

Plants are characterized by a remarkable phenotypic plasticity that meets the constraints of a sessile lifestyle. They have developed several strategies to perceive environmental changes, integrate them with endogenous cues, and adjust developmental pathways accordingly. Indeed, during their life cycle, plants undergo several major developmental transitions, many of which are determined by the environment. Chromatin-based regulation of transcriptional patterns, in conjunction with a large repertoire of transcription factors, likely underpins most of this developmental flexibility. Here, we review our current knowledge on chromatin organization and dynamics in plants, with a special emphasis on the role of chromatin-based mechanisms in regulating key aspects of *Arabidopsis* development.

# 2. PLANT EPIGENOMES: COMPONENTS AND ORGANIZATION

Packaging of DNA into chromatin is pivotal for the regulation of genome activity in eukaryotes, and plants are no exception. The basic unit of chromatin is the nucleosome, which is composed of 147 bp of DNA wrapped around a protein octamer composed of two molecules each of the core histones H2A, H2B, H3, and H4. Histone H1 is further associated with linker DNA, which contributes to higher levels of chromatin organization. The functional properties, positioning, and occupancy of nucleosomes can be modulated in several ways. Thus, covalent modifications of core histones, incorporation of histone variants, and other factors, such as chromatin-modifying and remodeling enzymes, DNA methylation, and small RNAs, all contribute to defining dynamic chromatin states that modulate access to DNA and confer distinct transcriptional outcomes (Reviewed in Berger, 2007; Jenuwein & Allis, 2001; Strahl & Allis, 2000).

In plants, chromatin has been studied most extensively in *Arabidopsis*, owing to its small and near-complete genome sequence, and a comprehensive set of genomic and genetic tools. The emerging view is a subtle mix of features conserved between plants and metazoans together with plant-specific innovations (Reviewed in Feng & Jacobsen, 2011). In particular, plants are characterized by one of the largest arrays of proteins involved in chromatin metabolism, by DNA methylation in all sequence contexts (CG, CHG, and CHH, where H corresponds to A, T, or C), and by an extensive diversification of small RNA-based regulatory pathways (Pontier et al., 2012; reviewed in Berr, Shafiq, & Shen, 2011; He, Elling, & Deng, 2011; Pontier et al., 2012; Simon & Meyers, 2011), based in part on the invention of plant-specific RNA polymerases (Reviewed in Haag & Pikaard, 2011).

## 2.1. Chromatin components

As in other eukaryotes, canonical plant histones are typically expressed during the S phase of the cell cycle and their incorporation is dependent on the replication process (Reviewed in Costas, Desvoyes, & Gutierrez, 2011). By contrast, expression and incorporation of histone variants occur independently of replication. Assigning a function to the different histones is complicated by the fact that plants have experienced a remarkable amplification of the repertoire of both canonical histones and their variant isoforms, which are often encoded by multigene families (Reviewed in Talbert et al., 2012). Plant genomes contain clear homologues of the key histone chaperone complexes found in other eukaryotes, including the H3–H4 chaperones CHROMATIN ASSEMBLY FACTOR-1, HISTONE REGULATORY HOMOLOG A, and ANTISILENCING FUNCTION1 (Exner, Taranto, Schonrock, Gruissem, & Hennig, 2006; Phelps-Durr, Thomas, Vahab, & Timmermans, 2005; Zhu et al., 2011), as well as the H2A–H2B chaperones NUCLEOSOME ASSEMBLY PROTEIN1 (Zhu et al., 2006). However, their developmental role remains, in general, poorly understood.

A striking feature of plant nucleosomes is their responsiveness to environmental conditions, which can affect histone occupancy, histone variant incorporation, and histone modifications (Reviewed in Zhu, Dong, & Shen, 2012). This is best exemplified by the identification of the H2A.Z variant as an important component of thermal sensing in *Arabidopsis* (Kumar & Wigge, 2010). Thus, increase in temperature triggers the eviction of H2A.Z in a locus-specific manner, which results in either activation or repression of the affected loci, suggesting that H2A.Z occupancy negatively regulates the accessibility of chromatin to both activators and repressors (Kumar & Wigge, 2010).

Responsiveness is likely important for short-term adaptation as it provides flexible means for large-scale reprogramming of gene expression. Yet, this creates challenges with respect to the preservation of long-term epigenetic memory. While stresses can modify chromatin organization and regulation, the epigenome returns rapidly to a basal state once the stress is relieved (Pecinka et al., 2010; Tittel-Elmer et al., 2010). Interestingly, stress-induced relaxation of chromatin states affects nucleosome structure but leaves DNA methylation patterns largely unchanged. This suggests that DNA methylation might provide a source of stable information for reestablishing basal chromatin states during recovery.

In plants, cytosine methylation is not restricted to CG sites but also occurs at CHG and CHH sites. The molecular mechanisms that shape DNA methylation patterns are well understood, thanks to the ability of plants to withstand dramatically altered methylomes. At least three families of DNA methyltransferases (DNMTs) are responsible for establishing or maintaining cytosine methylation (Reviewed in Law & Jacobsen, 2010; Teixeira & Colot, 2010). DOMAINS REARRANGED METHYL TRANSFERASE2 (DRM2), a homologue of mammalian DNMT3, is the main *de novo* DNMT and affects all sequence contexts (Cao & Jacobsen, 2002). DRM2 also plays a role, together with the plant-specific CHROMOMETHYLTRANSFERASE3 (CMT3), in the maintenance of methylation at non-CG sites (Cao et al., 2003). Along with DRM2, small interfering RNAs (siRNAs) and proteins of the RNAi pathway are involved in maintaining methylation at asymmetrical CHH sites via the so-called RNA-directed DNA methylation (RdDM) pathway (Huettel et al., 2007). CMT3 interacts with H3K9me2-containing nucleosomes and methylates CHG sites preferentially. METHYLTRANSFERASE1, the homologue of mammalian DNMT1, is involved in maintaining CG methylation. Several additional DNMTs are present in the *Arabidopsis* genome, but their functions are unclear (Cao et al., 2000). DNA methylation is enzymatically removed by the activity of methylcytosine DNA glycosylases, four of which are present in *Arabidopsis* and named REPRESSOR OF SILENCING1, DEMETER (DME), DEMETER-LIKE2 and 3 (Reviewed in Law & Jacobsen, 2010; Teixeira & Colot, 2010).

*Arabidopsis* is the first organism for which a whole-genome single-nucleotide resolution methylome was extensively characterized (Cokus et al., 2008; Lister et al., 2008). Patterns of DNA methylation differ substantially between genes and transposable elements (TEs) or other repeated sequences (Reviewed in Saze & Kakutani, 2011). Within repeats, which are usually stably silenced, individual cytosines are methylated in all contexts with a typically higher level of methylation at CG than at CHG and CHH sites (90%, 30–80%, and 30%, respectively) in somatic cells. By contrast, methylation over protein-coding genes is almost exclusively found at CG sites in the transcribed region and is associated with moderate expression rather than silencing. Although conserved between plant orthologs (Takuno & Gaut, 2013), functional role of gene-body methylation remains elusive as its loss has only minor effects on expression (Zilberman, Gehring, Tran, Ballinger & Henikoff, 2007).

Histones can be subject to a vast range of posttranslational covalent modifications including acetylation, methylation, ubiquitinylation, or sumoylation of specific residues, most of which are present in all eukaryotes (Reviewed in Kouzarides, 2007). Epigenomic profiling in *Arabidopsis* has notably established that, as in other eukaryotes, trimethylation of histone H3 at lysine 27 (H3K27me3) in plants is associated with gene repression and depends on the evolutionary conserved Polycomb Repressive Complex 2 (PRC2) (Turck et al., 2007; Zhang, Clarenz et al., 2007; Bouyer et al., 2011; Lafos et al., 2011; Schubert et al., 2006). Four Polycomb-group proteins (PcG), originally identified in Drosophila, form the core of PRC2: the histone methyltransferase Enhancer of Zeste (E(Z)), Extra sex combs (ESC), Suppressor of Zeste-12 (Su(Z)12), and the nucleosome remodeling factor 55 (NURF-55) (Margueron & Reinberg, 2011). In *Arabidopsis*, with the exception of the ESC homologue, FERTILIZATION-INDEPENDENT ENDOSPERM (FIE), all other PcG components belong to protein families, suggesting the existence of multiple PRC2 complexes. Thus, the *Arabidopsis* genome encodes three homologues of E(Z), MEDEA, CURLY LEAF (CLF) and SWINGER, five NURF-55-like proteins, named MULTICOPY SUPPRESSOR OF IRA1-5, and three Su(Z)12 homologues: FERTILIZATION-INDEPENDENT SEED2 (FIS2), EMBRYONIC FLOWER2 (EMF2), and VERNALIZATION 2 (VRN2) (Reviewed in Hennig & Derkacheva, 2009). Furthermore, at least three PRC2 complexes have been distinguished in *Arabidopsis* based on the fact that FIS2 has a role during reproductive development, whereas EMF2 and VRN2 control vegetative development and are partially nonredundant (Reviewed in Bemer & Grossniklaus, 2012; Hennig & Derkacheva, 2009).

In *Drosophila* and mammals, additional PcG proteins form a second complex, called PRC1, which participates in stabilizing PcG-mediated repression by binding to H3K27me3 and catalyzing the monoubiquitylation of histone H2A (Reviewed in Schuettengruber, Chourrout, Vervoort,

Leblanc, & Cavalli, 2007; Simon & Kingston, 2009). In addition to this sequential interaction, PRC2 and PRC1 can also act redundantly and independently of each other (Leeb et al., 2010; Tavares et al., 2012). Although PRC1 components are less well conserved over evolution (Schuettengruber et al., 2007), several lines of evidence indicate that plants also have PRC1-like complexes. Thus, three homologues of BMI1 (AtBMI1A, B, and C) and two homologues of RING1 (AtRING1A and B) have been identified in *Arabidopsis* and are involved in gene repression. In addition, EMBRYONIC FLOWER 1 and LIKE HETEROCHROMATIN PROTEIN1 (LHP1) interact with RING1 and BMI1 homologues, and LHP1 has been proposed to fulfill a function analogous to Polycomb (Pc), for which there is no *bona fide* homologue in plants (Bratzel, Lopez-Torrejon, Koch, Del Pozo, & Calonje, 2010; Chen, Molitor, Liu, & Shen, 2010; Kim, Lee, Eshed-Williams, Zilberman, & Sung, 2012; Li et al., 2011; Mylne et al., 2006; Turck et al., 2007; Xu & Shen, 2008; Zhang, Germann et al., 2007).

In *Arabidopsis*, between 15% and 30% of all genes are targeted by PRC2, depending on the tissue or organ analyzed (Lafos et al., 2011; Turck et al., 2007; Zhang, Clarenz et al., 2007). H3K27me3 marking differs considerably between meristematic and differentiated cells, suggesting dynamic recruitment of PcG proteins (Lafos et al., 2011). Only a limited subset of PRC2 targets appears to depend on PRC1 for stable transcriptional repression (Bratzel et al., 2010). This indicates that, like in mammals, Pc-based regulation in *Arabidopsis* does not necessarily imply cooperation between the two complexes. As in other eukaryotes, PcG-mediated repression is antagonized by the activity of JumonjiC domain-containing proteins that demethylates H3K27me3 (Lu, Cui, Zhang, Jenuwein, & Cao, 2011) and by proteins of the Trithorax group (TrxG). TrxG proteins include the histone methyltransferase *ARABIDOPSIS* TRITHORAX1 (ATX1) (Alvarez-Venegas et al., 2003), which catalyzes trimethylation of H3K4me3 and the SAND domain protein ULTRAPETALA1 (ULT1) (Carles & Fletcher, 2009). Two SWI2/SNF2 chromatin remodeling ATPases, SPLAYED (SYD) and BRAHMA (BRM), also contribute to Trx function (Smaczniak et al., 2012; Wu et al., 2012).

## 2.2. Epigenome organization in plants

Despite the bewildering variety of chromatin marks and chromatin-associated proteins, large-scale epigenomic studies in several eukaryotes including *Caenorhabditis elegans*, *Drosophila*, *Arabidopsis*, and human cells identified a limited combinatorial repertoire of chromatin modifications corresponding to a relatively small number of chromatin states (Reviewed in Baker, 2011; van Steensel, 2011). A first integrative view of the *Arabidopsis* epigenome, based on the combinatorial analysis of 12 major chromatin modifications, revealed a short-range organization around four predominant chromatin states (CS1–4) with distinct biochemical, transcriptional, and sequence properties (Roudier et al., 2011). CS1 corresponds to transcriptionally active genes and is typically enriched in the trimethylated forms of H3K4, H3K9, and H3K36 as well as H2Bub. Local distribution of these different marks over genes suggests that CS1 could be subdivided into additional chromatin signatures. Thus, H3K4me3 and H3K56Ac peak around the transcription start site of active genes and are possibly linked with the initiation of transcription. On the other hand, H2Bub, H3K9me3, and H3K36me3 are found over the transcribed region and are likely associated with transcriptional elongation. Three other states, CS2–4, correspond to distinct types of chromatin with very low transcriptional activity (Roudier et al., 2011). H3K27me3-marked chromatin (CS2) is mainly associated with genes under PRC2-mediated repression. H3K9me2- and H4K20me1-marked repressive chromatin (CS3) corresponds to classical heterochromatin and is almost exclusively located over silent repeated sequences. The fourth chromatin state (CS4) is characterized by the absence of any prevalent mark among the 12 studied, is preferentially associated with intergenic regions, and is also found over ~10% of genes showing weak expression levels.

The seemingly simple organization of *Arabidopsis* chromatin into four main states shows similarities with that obtained in *Drosophila*, where five principal chromatin types have been defined (Filion et al., 2010; van Steensel, 2011). *Arabidopsis* CS2 and CS3 resemble *Drosophila* "BLUE" and "GREEN" repressive chromatin types, which also have counterparts in other metazoans (Reviewed in Baker, 2011). Conservation of the Pc pathway and classical heterochromatin between plants and animals is remarkable, given the mechanistic divergences associated with the absence of Pc and of the canonical HP1a- and H3K9me3-based heterochromatin pathway in plants. CS4 is reminiscent of *Drosophila* "BLACK" chromatin, which constitutes a repressive environment distinct from heterochromatin. In contrast to the single transcriptionally active chromatin state described in *Arabidopsis* (CS1), two distinct types exist in *Drosophila* that differ in several ways, including the enrichment of H3K36me3 in "YELLOW" but not in "RED" chromatin.

The organization of plant chromosomes into small and well-demarcated chromatin domains that often coincide with single genes points to a predominant role for transcription in shaping the chromatin landscape (Li et al., 2008; Wang et al., 2009). Noticeably, the short-range distribution of CS2, H3K27me3-enriched, domains that rarely span more than one or two genes (Roudier et al., 2011; Turck et al., 2007; Zhang, Clarenz, et al., 2007) is in sharp contrast to the large patches of H3K27me3, which encompass several genes reported in *Drosophila* and mammals (Bernstein et al., 2006; Schwartz et al., 2006; Tolhuis et al., 2006). This suggests organism-specific variations in the use of otherwise conserved mechanisms.

Further refinement in the classification of *Arabidopsis* chromatin states can be expected, possibly as a result of the analysis of additional chromatin marks and proteins but also, and more importantly, through an increase in the cellular resolution of epigenomic maps. This is clearly a necessary step for determining the dynamics of chromatin states in a developmental context, which is of key importance to reveal the functional impact of chromatin-level regulation during organogenesis and development.

## 3. CHROMATIN-BASED PROCESSES DURING VEGETATIVE DEVELOPMENT

While many regulators of chromatin-associated pathways have been characterized in *Arabidopsis*, their specific impact during development remains poorly understood in many cases. Extensive analysis of the role of PcG proteins has shown that PRC2-mediated gene repression affects most developmental transitions in *Arabidopsis* (Reviewed in Bemer & Grossniklaus, 2012; Holec & Berger, 2012; Zheng & Chen, 2011). Moreover, loss of PRC2 or PRC1 activities leads to dedifferentiation and active proliferation as well as to an unbalance between division and differentiation in meristems (Aichinger, Villar, Di Mambro, Sabatini, & Kohler, 2011; Bouyer et al., 2011; Bratzel et al., 2010; Schubert et al., 2006), which suggests that, like in metazoans, PcG proteins are central regulators of cell fate acquisition and maintenance in plants (Reviewed in Hennig & Derkacheva, 2009; Köhler & Hennig, 2010). Among all developmental transitions, flowering has crystallized many research efforts that are starting to provide a refined understanding of how the environment affects gene expression, either transiently or over longer periods.

### 3.1. Aligning developmental transitions with seasonal changes of the environment

The transition to flowering is a major developmental switch in the plant life cycle. Regulation of flowering time in response to seasonal changes is critical for reproductive success, as it ensures the completion of seed production in favorable environmental conditions. Multiple pathways control flowering time in *Arabidopsis*, including the autonomous pathway, the photoperiod pathway, and the vernalization pathway, which form a regulatory network that converges on two main regulators, FLOWERING LOCUS C (FLC) and FLOWERING LOCUS T (FT) (Reviewed in Andres & Coupland, 2012). The regulation of *FLC* has become a paradigm to understand the general roles of chromatin in the transcription process in plants and has illuminated the paramount contribution of chromatin-based mechanisms in sensing and monitoring the environment. In addition, it is one of the very few examples in which initiation, maintenance, and resetting of epigenetic control can be dissected in a developmental context.

## 3.2. License to flower

*FLC* encodes a MADS-box transcription factor that acts, in a dosage-dependent manner, as a potent repressor of the floral transition (Michaels & Amasino, 1999). In winter-annual *Arabidopsis*, *FLC* expression is gradually turned off and epigenetically silenced by vernalization, which refers to the acquisition of the ability to flower (in spring) after prolonged exposure to cold temperatures in winter (Chouard, 1960). The vernalization process can be subdivided into three main phases that correspond to three distinct chromatin landscapes at the *FLC* locus: (1) high expression level before exposure to cold, (2) cold-induced silencing, and (3) epigenetic maintenance of the repressed state after return to warm conditions (Fig. 6.1).

The winter-annual growth habit results from high expression level of *FLC* in young seedlings that depends on the FRIGIDA (FRI) coiled-coiled protein and multiple components of conserved chromatin-modifying complexes (Reviewed in Crevillen & Dean, 2011). Recruitment of the ATPase chromatin remodeling SWR1 complex by the FRI complex leads to the incorporation of nucleosomes containing H2A.Z around the transcription start site, thus rendering *FLC* chromatin permissive for transcription. The activity of different WDR5a-containing COMPASS complexes then triggers the deposition of H3K4me3/2 and H3K36me3/2 (Jiang, Gu, & He, 2009). In addition, high *FLC* expression relies on the cyclic monoubiquitylation of histone H2B by the ubiquitin ligases UBC1-3 and HUB1-2 as well as the ubiquitin protease UBP26 (Cao, Dai, Cui, & Ma, 2008; Gu, Jiang, Wang, Bachmair, & He, 2009; Schmitz, Tamada, Doyle, Zhang, & Amasino, 2009). This leads to the establishment of an active chromatin state that is tightly associated with high transcription of *FLC*.

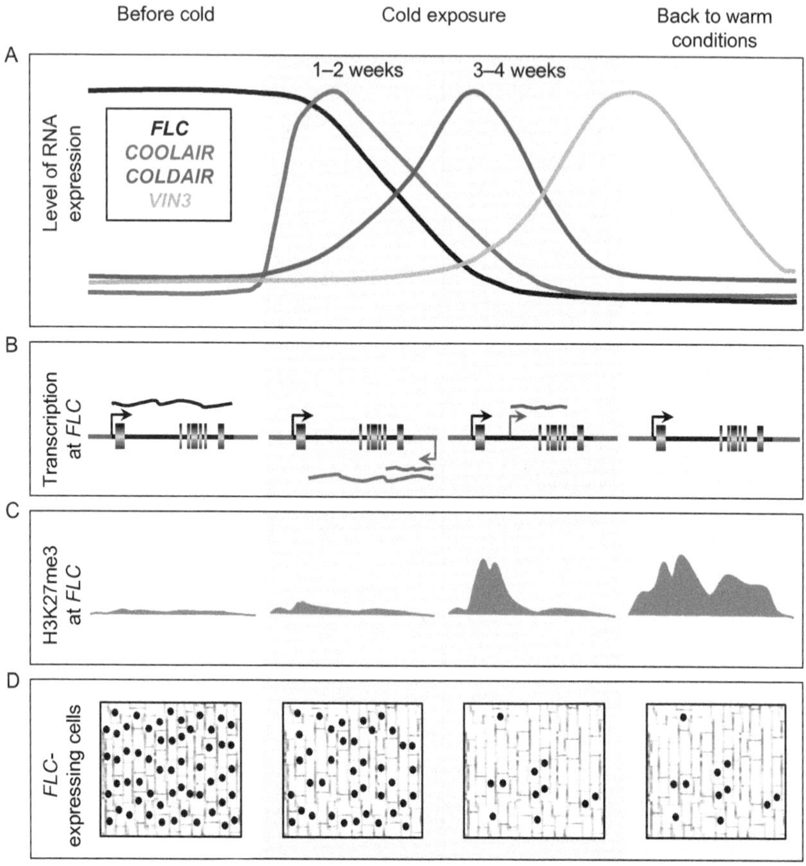


**Figure 6.1** Chromatin-based repression of *FLC* during vernalization. The following features are depicted at different stages before, during and after prolonged exposure to cold. (A) Expression levels of *FLC* mRNA, *COOLAIR* noncoding antisense RNAs, *COLDAIR* noncoding sense RNA, and *VIN3* mRNA. (B) Location of the different transcripts over the *FLC* locus. The transcription start site of the three types of transcripts is indicated by colored arrows. (C) H3K27me3 enrichment over the *FLC* locus. (D) Bistable expression of *FLC* in individual cells within a tissue. Black dots represent nuclei of cells expressing *FLC*. Before cold exposure, *FLC* sense mRNA levels are high in all cells. Soon after cold exposure, *FLC* expression is downregulated in some cells concomitantly with the transcription of COOLAIR. Longer exposure to cold triggers the expression of COLDAIR that recruits a catalytically active PRC2 complex at the nucleation region around the TSS of *FLC*, which results in the local enrichment of H3K27me3. This leads to an increasing number of cells in which *FLC* switches to a repressed state. Return to warm temperature after at least 4 weeks of cold exposure results in stable *FLC* silencing associated with the spreading of VIN3-PRC2 and H3K27me3 over the whole locus. (See Color Insert.)

During vernalization, cold temperatures (~4 °C) gradually induce the downregulation of *FLC* in a quantitative manner (Fig. 6.1A). Two long noncoding RNAs (lncRNAs) and the activities of distinct PRC2 contribute to the radical change in chromatin state at *FLC*. About 2 weeks after cold exposure, *FLC* mRNA levels drop concomitantly with the transient upregulation of a set of antisense noncoding transcripts, referred to as COOLAIR, which initiate from the 3′ end of the *FLC* gene (Fig. 6.1A and B) (Swiezewski, Liu, Magusin, & Dean, 2009). COOLAIR transcripts participate in the recruitment of the H3K4me2/3 demethylase FLD (Liu et al., 2007), which acts in concert with the histone deacetylase HDA6, the MSI4 and MSI5 proteins (Pazhouhandeh, Molinier, Berr, & Genschik, 2011), and PRC2-mediated H3K27me3 deposition to repress *FLC* through the vernalization-independent, autonomous pathway. Indeed, although COOLAIR accumulation in response to cold leads to downregulation of *FLC*, it is, however, neither sufficient nor required for vernalization (Helliwell, Robertson, Finnegan, Buzas, & Dennis, 2011). Nevertheless, COOLAIR-triggered transcriptional repression of *FLC* might provide a bridge between the autonomous and vernalization pathways by promoting the transient expression of a second lncRNA named COLDAIR (Heo & Sung, 2011). Expression of this sense RNA encoded in the first intron of *FLC* peaks about 3 weeks after the start of vernalization (Fig. 6.1A and B). Contrary to COOLAIR, COLDAIR expression is required for the stable silencing of *FLC* after cold exposure, and this has been proposed to be associated with an ability to recruit PRC2 (Heo & Sung, 2011). The gradual decrease of *FLC* mRNA during the first 4 weeks of cold exposure is indeed paralleled by the progressive accumulation of H3K27me3 catalyzed by a specific PRC2 that comprises CLF and plant homeodomain (PHD) proteins (De Lucia, Crevillen, Jones, Greb, & Dean, 2008). H3K27me3 deposition by PHD–PRC2 first occurs at a short nucleation region within the first intron of *FLC* (Fig. 6.1C), which overlaps with a region necessary for stable silencing (Sheldon, Conn, Dennis, & Peacock, 2002; Sung et al., 2006).

Remarkably, polymorphism in sequences spanning the nucleation region was shown to account for the different vernalization requirements that characterize *Arabidopsis* strains growing under different latitudes, such as Col (a laboratory strain) and Lov-1 (an accession from northern Sweden) (Coustham et al., 2012). Indeed, fine-mapping of a quantitative trait locus for vernalization in a segregating population between these two accessions identified a combination of four single-nucleotide polymorphisms that quantitatively affects H3K27me3 deposition. Thus, in comparison to Col, the Lov-1 *FLC* allele requires a much longer period of cold to accumulate high levels of H3K27me3 and achieve full *FLC* silencing. This extended delay in flowering matches the longer winters encountered in northern Europe. This work illustrates how chromatin-based mechanisms can be modulated by a few changes in the regulatory sequences of a gene with strong phenotypic impact and how epigenetic regulation can be linked to adaptation.

Providing that plants have experienced cold for long enough, typically several weeks, stable silencing of *FLC* is maintained and passed on through mitosis during the rest of the life cycle, even when temperatures become warmer. This epigenetic "memory of winter" requires the activity of the PRC2–PHD complex that spreads H3K27me3 across the entire *FLC* locus (De Lucia et al., 2008) (Fig. 6.1C). Association of PHD–PRC2 with an additional PHD protein named VERNALIZATION INSENSITIVE3 (VIN3), the expression of which peaks at the end of the vernalization period, appears essential to achieve stable silencing of *FLC* (Heo & Sung, 2011; Sung & Amasino, 2004). Maintenance of *FLC* repression also relies on LHP1 and AtBMI1C, two members of a PRC1-like complex (Li et al., 2011; Mylne et al., 2006; Sung et al., 2006).

Mathematical modeling of the vernalization process has highlighted general principles that might pertain to other chromatin-based "memory" systems (Angel, Song, Dean, & Howard, 2011; Satake & Iwasa, 2012; Song, Angel, Howard, & Dean, 2012). A key feature is that cold-induced epigenetic silencing of *FLC* is cell-autonomous and bistable, that is, in each cell, the locus can exist in one of the two stable states, active or silent. The switch is probabilistic and relies on a positive feedback, whereby a local change (the nucleation of H3K27me3) recruits catalytic activities (PRC2 and PRC1) that reinforce this initial event and stabilize the silent state through cell division. Another key feature is that the probability of this cell-autonomous switch increases with the duration of the triggering signal (cold), thus providing a biological measure of time based on the fraction of cells that have switched (Fig. 6.1D). Experimental testing of this hypothesis indicates indeed that the quantitative relationship between *FLC* repression and H3K27me3 accumulation appears to act at the cell population level, with an increasing number of cells in which *FLC* is silent resulting from longer cold exposure (Angel et al., 2011). This population averaging strategy, which provides a means to buffer noisy changes in the environment, likely contributes to the robustness of the response whereby a license to flower is only granted by *FLC* silencing after winter has passed (Satake & Iwasa, 2012; Song et al., 2012).

Intriguingly, mitotic activity is necessary to achieve spreading of H3K27me3 over the *FLC* locus (Finnegan & Dennis, 2007). A possible explanation for this requirement is that the epigenetic switch relies on the replication-dependent deposition of a H3.1 variant, usually found over transcriptionally silent chromatin (Stroud et al., 2012; Wollmann et al., 2012) and preferentially enriched in H3K27me3 in comparison with H3.3 (Johnson et al., 2004). The mechanism by which epigenetic silencing is maintained at *FLC* through DNA replication is unknown. The observation that some PcG proteins such as CLF or LHP1 appear to be bound to chromatin during most of the cell cycle, though not on metaphasic chromosomes (Libault et al., 2005; Schubert et al., 2006) and interact with replication-associated proteins (Barrero, Gonzalez-Bayon, del Pozo, Ponce, & Micol, 2007) might indicate that it is the propagation of PRC2 complexes, rather than the H3K27me3 mark itself, that maintains the repressive chromatin state, which is in keeping with recent models proposed in metazoans (Hansen et al., 2008; Petruk et al., 2012).

Importantly, the vernalized state is not transmitted to the progeny, which ensures that each generation requires proper cold exposure to flower. *FLC* silencing is relieved during gametogenesis upon the removal of H3K27me3, and *FLC* expression is reactivated during early embryo development (Choi et al., 2009; Sheldon et al., 2008). This resetting could result either from a genome-wide reprogramming associated with histone turnover as shown in the zygote (see Section 4) or from the local recruitment of H3K27me3 demethylases or Trithorax-associated ATP-dependent chromatin remodelers.

## 3.3. Time to bloom

The vernalization process can be viewed as the conversion of digital epigenetic information stored at the cellular level into a quantitative response of the whole organism. Plants mainly sense environmental conditions through organs such as leaves, but the developmental decision to flower occurs in the stem cell niche at the apex. This implies that the epigenetic silencing of *FLC* is relayed by a signal promoting flowering at the level of the organism and over a distance. *FT* is an immediate downstream target of FLC repression and encodes a small globular protein that translocates from the leaves to the shoot apex, thus representing a key component of this mobile flowering signal (Corbesier et al., 2007; reviewed in Wigge, 2011). The decision of when to flower largely depends on the level of *FT* expression, which is strongly influenced by seasonal cues such as the length of the photoperiod and warm temperatures (Reviewed in Andres & Coupland, 2012).

Modulation of the chromatin state at the *FT* locus is another illustration of how chromatin contributes to precise transcriptional regulation by sensing environmental signals. In unfavorable conditions, *FT* expression is kept to a basal level via the H3K4me2/3 demethylase activity of the plant-specific, Jumonji C-domain JMJ14, which might act in conjunction with PcG-mediated repression (Adrian et al., 2010; Jeong et al., 2009; Jiang, Wang, & He, 2008; Lu, Cui, Zhang, Liu, & Cao, 2010; Yang, Jiang, Jiang, & He, 2010). This likely prevents precocious flowering and sets up the ground for a gradual response to warm temperatures that relies in part on H2A.Z dynamics, as described above (Kumar & Wigge, 2010). Thus, a rise from 12 to 27 °C triggers the gradual eviction of H2A.Z-containing nucleosomes at the regulatory regions of *FT*, which facilitates the binding of the PIF4 transcriptional activator (Kumar et al., 2012). H2A.Z occupancy appears therefore to provide a mechanism for sensing a graded increase in ambient temperature by modulating chromatin accessibility to PIF4 at the *FT* locus, thereby inducing *FT* expression and flowering initiation.

## 3.4. Emerging regulatory principles coming in from the cold and more

How chromatin-modifying complexes are recruited to their specific targets remains a key question. Polycomb Response Elements (PRE) have been described originally in *Drosophila* as DNA sequences bound by PcG proteins that create a domain for H3K27 trimethylation and convey transcriptional repression (Reviewed in Schuettengruber et al., 2007; Simon & Kingston, 2009). Despite the thousands of PRC2 targets detected in *Arabidopsis*, only two genomic regions have been identified that meet most of the criteria expected for PRE-like sequences. One coincides with the nucleation region described above for *FLC* (Buzas, Tamada, & Kurata, 2012; Sheldon et al., 2002; Sung et al., 2006) and the other one corresponds to a 50-bp sequence that is required for PRC2-mediated repression of *LEC2* during late embryogenesis and vegetative development (Berger, Dubreucq, Roudier, Dubos, & Lepiniec, 2011). Identification of additional PRE-like sequences and of the proteins that specifically recognize them is obviously of considerable interest.

Analysis of the vernalization pathway also led to the emergence of new regulatory principles in plants, such as the complex interplay between lncRNA transcription and the recruitment of specific chromatin-modifying complexes. The COOLAIR antisense lncRNA was named with reference to the HOTAIR lncRNA that is required for the silencing of *HOX* loci through the targeted enrichment of PRC2 in mammals (Rinn et al., 2007).

In this respect, COLDAIR, rather than COOLAIR, appears to be functionally equivalent to HOTAIR in the regulation of *FLC* during vernalization. However, HOTAIR also recruits the H3K4me2 demethylase LSD1, which contributes to gene silencing (Tsai et al., 2010). Similarly, COOLAIR participates in the recruitment of the H3K4me2/3 demethylase FLD, one of the *Arabidopsis* LSD1 homologue, which is also involved in *FLC* repression via the autonomous pathway. Taken together, these observations therefore suggest an intriguing evolutionary twist whereby two lncRNA instead of one would coordinate two independent pathways at *FLC* through the recruitment of synergistic chromatin-modifying activities.

Targeted recruitment of PRC2 via a lncRNA, either directly or indirectly, has also been proposed to be involved in X-chromosome inactivation in female mammals (Reviewed in Augui, Nora, & Heard, 2011) and thus may be a conserved mechanism in metazoans. It remains to be determined whether such a mechanism is widespread in plants.

## 4. THE CENTRAL ROLE OF EPIGENETICS IN PLANT REPRODUCTIVE DEVELOPMENT

Chromatin-based regulations are an essential determinant of reproductive development in plants. Unlike animals, plants do not set aside germ cells early in development. Instead, plant germ cells differentiate from somatic cells following a prolonged vegetative life and produce multicellular gametophytes that contain the gametes (Fig. 6.2). In the ovules, usually a single archesporial (germ) cell differentiates from maternal somatic tissues. It undergoes meiosis to produce four spores, only one of which survives. In the majority of flowering plants, the surviving spore goes through three rounds of nuclear division without cytokinesis. Subsequent cellularization and differentiation result in a seven-celled embryo sac (female gametophyte) comprising two gametic cells, the haploid egg and the homodiploid central cells (generated from two haploid nuclei), as well as five accessory cells, two synergids, involved in interaction with the male gametophyte and three antipodal cells of unknown functions. Similarly, the male gametophyte (the pollen grain) is produced in the anthers following the late definition of the male germ cells and comprises one accessory (vegetative) cell and two sperm cells derived from a single haploid spore. Fertilization of the egg cell by one of the sperm cells gives rise to the zygote, while fertilization of the central cell by the second sperm cell produces the endosperm, an embryo-supporting tissue. Embryo and endosperm develop in a coordinated manner together with surrounding maternal integuments, forming the seed.

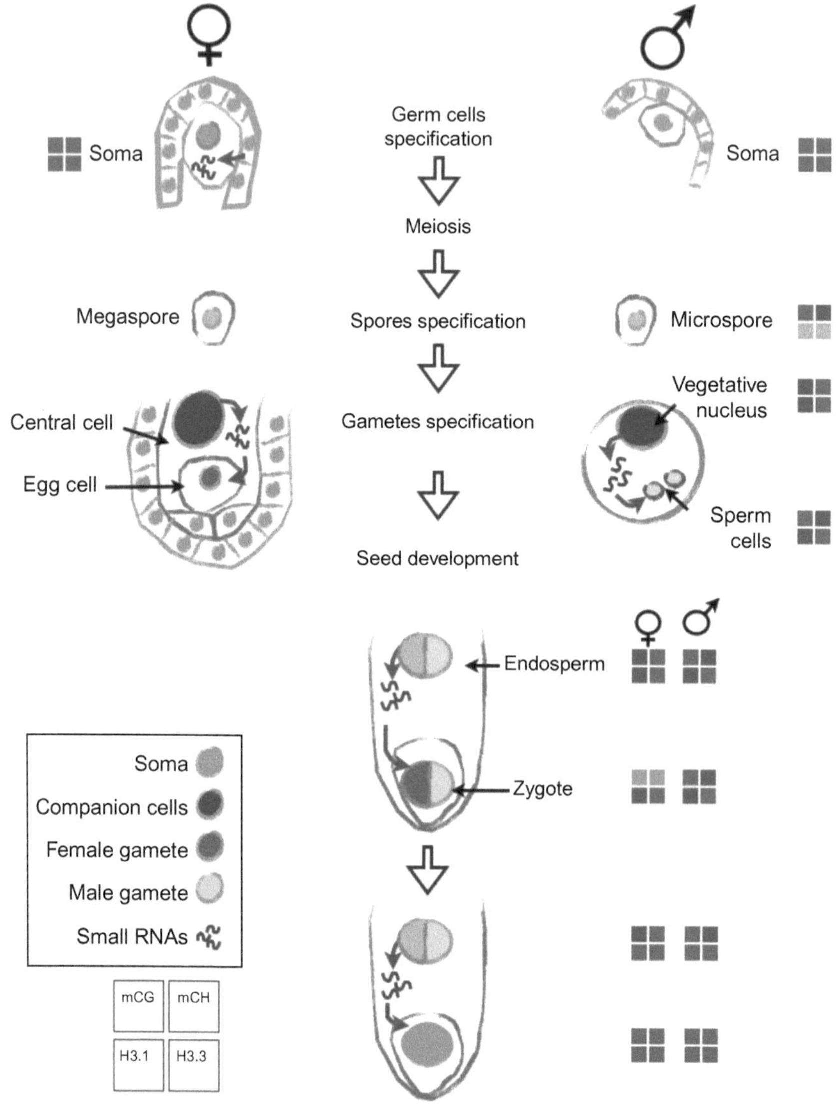


**Figure 6.2** Chromatin dynamics during germ cells development and reprogramming. While much remains to be determined, existing data suggest highly dynamic chromatin changes starting with the specification of germ cells. Known patterns of CG methylation (mCG), non-CG methylation (mCH), canonical H3 (H3.1), and variant H3 (H3.3) deposition are indicated in the squares for the different cell types. Red indicates enrichment, while blue indicates depletion. Gray boxes indicate missing data. Virtually nothing is known with respect to male or female sporogenesis. Small RNA molecules have been identified at different time points, originating potentially from both somatic and companion cells, but their biological role remains unclear. A non cell-autonomous mode of action,

## 4.1. Chromatin changes at the core of reproductive development transitions

Plant reproductive development thus follows a series of developmental transitions: (1) from the soma to the precursor cells of the male and female gametes, equivalent to the animals' primordial germ cells; (2) from the haploid spores to the gametes; and (3) from the gametes to the embryo upon fertilization (Fig. 6.2). These transitions entail global epigenomic changes that affect chromatin in a cell-specific manner and result in the establishment of germ cell and zygote-specific chromatin states (Autran et al., 2011; Calarco et al., 2012; Ibarra et al., 2012; Ingouff et al., 2010; Pillot et al., 2010). Recent progress is starting to shed light on the importance of these chromatin states for gamete and seed development, and also for the reprogramming of genome activity.

The first developmental transition during plant reproduction is the specification of germ cells from the somatic tissues. In the ovules, this transition is governed at least partially by small RNA-dependent pathways acting on DNA methylation in the precursors of the germ cells, the archespore (Garcia-Aguilar, Michaud, Leblanc, & Grimanelli, 2010; Olmedo-Monfil et al., 2010; Singh et al., 2011). Defective alleles of *ARGONAUTE9* (*AGO9*) in *Arabidopsis* produce ovules with multiple germ cells indicating that this RNAi-associated function likely contributes to repress germ cell specification in the ovule to a single archespore. AGO104 deficiency in maize turns meiosis into a mitosis-like division at a high frequency, indicating a role in repressing somatic (mitotic) fate in the germ cells. AGO104 and AGO9 protein accumulation is restricted to sporophytic cells that surround the gametic cells, which might indicate non-cell-autonomous spreading of a mobile silencing signal from the somatic cells. AGO9 and AGO104 bind heterochromatic 24-nt siRNA and are involved in the establishment or maintenance of

involving movement of the small RNAs from the soma, has been proposed during germ cell specification but may also act at later stages of gamete development (not indicated). In the mature gametophytes, small RNAs originating from companion cells target gametic chromatin in the egg and the sperm cells. The gametes are characterized by distinct epigenomes indicated by different colors. The epigenetic states of the paternally (♂) and maternally (♀) inherited genomes following fertilization are indicated separately. The extent of epigenome reprogramming in the embryo remains a matter of debate, but the picture assumes a phase of reprogramming before the first division of the zygote, resulting in the reestablishment of a mostly "somatic" genome where parental origins have been largely erased. (See Color Insert.)

non-CG methylation (Duran-Figueroa & Vielle-Calzada, 2010; Havecker et al., 2010; Singh et al., 2011). This suggests that non-CG methylation in plants plays a role in defining germ cell identity in the ovule (Garcia-Aguilar et al., 2010).

The second transition, from spores to gametophytes, is also marked by dramatic chromatin changes (Fig. 6.2). In both male and female reproductive organs, the gametes are enclosed in a multicellular gametophyte together with "accessory companion cells" that participate in reproduction but are not transmitted to the next generation. Analyses of tagged versions of the Histone Three Related (HTR) gene family indicate that female gamete specification is accompanied by the eviction of most somatic H3 isoforms (Ingouff et al., 2010). At maturity, a limited repertoire of HTR proteins is retained: the H3.3 variant HTR5 is the only one detected in the egg, and only one H3.1 protein (HTR3) and two H3.3 variants (HTR8 and HTR14) are found in the central cell (Fig. 6.2). Thus, the two female gametes show distinct histone-based signatures. Similarly, they are dimorphic for two histone marks associated with silent chromatin, H3K9me2 and H3K27me3 (Pillot et al., 2010). In the sperm cells, only HTR5 and HTR10 (a sperm-cell specific H3.3 variant) are found, whereas HTR8 and HTR14 are detected in the vegetative nuclei (Ingouff et al., 2010). Genome-wide analyses indicate that the formation of the sperm and vegetative cells also involves dynamic changes and distinct DNA methylation patterns. Interestingly, plant sperm cells show high levels of both CG and non-CG methylation (Calarco et al., 2012; Ibarra et al., 2012), suggesting that contrary to mammals, DNA methylation is mostly retained during reproduction. However, whereas CG methylation remains relatively constant during sperm development, CHH methylation is transiently reduced at TE loci in the spores and reestablished later. By contrast, the vegetative nuclei showed loss of CG methylation but maintenance of high levels of CHH methylation. Thus, similar to the female gametophyte, the male gametophyte produces epigenetically divergent cell types. Patterns of DNA methylation in female gametophytic cell types are less well defined. Active and passive DNA demethylation takes place in the central cell, through the action of DNA glycosylases (Choi et al., 2002; Gehring, Bubb, & Henikoff, 2009; Hsieh et al., 2009) and the repression of maintenance DNMTs in young ovules (Jullien, Susaki, Yelagandula, Higashiyama, & Berger, 2012; Jullien et al., 2008). The central cell is therefore expected to have lower levels of DNA methylation.

Finally, the transition from the gametes to the embryo and the endosperm also implies abrupt chromatin changes (Fig. 6.2). Gametic H3 proteins are rapidly eliminated following fertilization and replaced by newly synthesized histones (Ingouff et al., 2010). DNA methylation and histone PTMs also likely undergo changes in very early embryos (Autran et al., 2011). Expression patterns of known DNMTs indicate that both the key maintenance enzymes and *de novo* DNMTs are expressed in the embryo (Jullien et al., 2012), which results in the rapid acquisition of DNA methylation patterns that are similar to those of the soma for both parental genomes (Calarco et al., 2012; Ibarra et al., 2012). However, strong methylome differences are observed in the endosperm between the maternally and paternally inherited genomes, the maternal genome being significantly less methylated than the paternal genome in the CG context, with "hot spots" of extensive demethylation. Most of the differences in CG methylation between the two genomes derive from the activity of the DNA glycosylase DME, in the central cell but not in the egg. In contrast to CG, CHG methylation patterns are similar for both parental genomes, except for targets of DME that has context-independent activity.

Thus, a key feature of plant gametophytic development is the establishment in the mature gametophytes of epigenetically dimorphic cell types that distinguishes germ cells from cells that have a transitory role in reproduction. The resulting dimorphism emerges as a key innovation of plant reproductive biology, as exemplified by its importance for the regulation of genomic imprinting and genome reprogramming.

## 4.2. DNA methylation and small RNAs determine parental imprinting in the seed

Imprinting refers to the differential expression of the maternal and paternal alleles of a gene. In plants, imprinting mostly occurs in the endosperm (which from that perspective is analogous to the extra-embryonic tissues of mammals), affecting an estimated 50 loci (Gehring, Missirian, & Henikoff, 2011; Gehring et al., 2009; Hsieh et al., 2011). A key actor in this process is DME, the activity of which results in differentially methylated regions (DMR) for most imprinted genes in the endosperm (Calarco et al., 2012; Choi et al., 2002; Gehring et al., 2009, 2006; Ibarra et al., 2012). Importantly, DME-dependent demethylation mostly targets small, AT-rich repeats. In both rice and *Arabidopsis*, such sequence targets are mostly found near genes (Ibarra et al., 2012). This suggests that imprinting is often a direct or indirect consequence of mechanisms involved in transposon silencing. This idea is further supported by genome-wide comparisons of DNA methylation patterns between sperm cells and the endosperm, which indicate that maternally and paternally imprinted genes are preferentially associated with DMRs. This model applies correctly to most of the maternally imprinted genes but is not universal (Wohrmann et al., 2012). Interestingly, maternally imprinted genes are also associated with strong accumulation of 24-nt siRNAs that target neighboring TE sequences in sperm cells, which likely contributes to silence the paternal allele prior to fertilization via RdDM. Most TEs targeted by 24-nt siRNAs in the sperm cells also correspond to targets of DME in the vegetative nucleus. This raises the possibility that loss of silencing in the vegetative nucleus might generate small RNAs that are translocated to the sperm cells, where they act to reestablish silencing via RdDM (Calarco et al., 2012; Slotkin et al., 2009).

In addition to DNA methylation, imprinting also relies on the activity of PRC2 and the deposition of H3K27me3 (Reviewed in Köhler & Grossniklaus, 2005; Köhler et al., 2012). This is the case at the paternally imprinted *PHERES* locus, where imprinting requires both the demethylation of a repeated sequence and PRC2 targeting to the promoter region of the maternal *PHERES* allele. The current model suggests that DNA methylation protects against PRC2 targeting, whereas hypomethylation of repeats and TEs favors PRC2-mediated repression, which in turn promotes non-CG methylation in the endosperm, possibly via RdDM. PRC2 is further involved in the regulation of maternal imprinting by repressing paternal alleles at several maternally imprinted genes. Whether specific patterns of DNA methylation or siRNA targeting are similarly involved in the recruitment of PRC2 proteins to their targets is unknown.

### 4.3. Epigenetic reprogramming in germ cells

The establishment of specific chromatin states in the different gametophytic cell types might also play an important role in genome reprogramming (Calarco et al., 2012; Ibarra et al., 2012; Jullien et al., 2012; Slotkin et al., 2009), that is, the resetting of the differentiated state of the gametes, the selective erasure of chromatin marks accumulated during vegetative growth, and the establishment of a fully totipotent state in the newly formed organism.

In animals, it is generally accepted that DNA methylation patterns are reprogrammed through waves of active and passive demethylation and *de novo* methylation during both gamete formation and early stages of embryogenesis. Current knowledge of reprogramming in plants remains limited (Reviewed in Bourc'his & Voinnet, 2010; Wollmann & Berger, 2012).

As mentioned above, the repertoire of parental H3 proteins is largely reset following fertilization. Yet, it is generally assumed that at least part of the chromatin-based information present in the parental genomes can be inherited in plants. This is the case, for example, for DNA methylation states at transgenes and repeats (Matzke, Kanno, Daxinger, Huettel, & Matzke, 2009). This has long been viewed as an argument against a global resetting of epigenetic state through generations. However, this does not imply that DNA methylation is not reprogrammed during reproduction, and the stable inheritance of epigenetic states might also imply successive steps of erasure and selective reestablishment of marks. Analyzing the dynamics of chromatin changes during plant reproductive development is technically challenging, and the current data lack sufficient temporal resolution to discriminate between the two possibilities. However, it has been shown that some reprogramming (a decrease followed by reestablishment) of DNA methylation at non-CG sites occurs during the late stages of sperm development and is dependent on RNAi (Calarco et al., 2012; Ibarra et al., 2012). Similarly, genetically hypomethylated repeats can be gradually remethylated through generations by an RNAi-dependent process that likely occurs during reproduction, possibly in the early embryo (Teixeira & Colot, 2010; Teixeira et al., 2009). This indicates at least that DNA methylation states are not passively transmitted through the germ cells.

In addition, genome-wide DNA demethylation of companion cells as a consequence of both active (DNA glycosylate dependent) and passive (absence of maintenance methyltransferase) mechanisms may serve to stabilize the germline epigenome. DNA demethylation of the vegetative cell in the pollen grain and the central cell in the female gametophyte are proposed to trigger massive production of TE-associated small RNAs. These small RNA molecules possibly move into the germ cells and the embryo, with a role not only in protecting the genome against transposable elements but also in reconfiguring the embryo epigenome via RdDM activity (Calarco et al., 2012; Gehring et al., 2009; Hsieh et al., 2009; Ibarra et al., 2012; Slotkin et al., 2009). As most siRNAs present in the seed are of maternal origin (Mosher et al., 2009), these are possibly involved in restorating CHH methylation patterns in the embryo, where maternal RdDM components regulate genome activity according to parental origin

(Autran et al., 2011). In addition, it has been shown that loss of AGO9 function, which is expressed exclusively in the maternal tissues surrounding the germ cells, results in the reactivation of transposable elements in the egg (Olmedo-Monfil et al., 2010), suggesting that somatic tissues play an additional role in protecting the gametic genome.

## 4.4. Reproduction without sex

Sexual reproduction is crucial for the definition of embryonic fate. It is thus all the more surprising that plants have also developed numerous alternatives to biparental sexual reproduction, most of which allow the formation of totipotent embryos with neither meiosis nor fertilization.

A particularly intriguing form of asexual reproduction is apomixis, which refers to diverse mechanisms of reproduction resulting in clonal reproduction through seeds and that can be found in over 400 angiosperm species (Reviewed in Grimanelli, 2012; Rodriguez-Leal & Vielle-Calzada, 2012). Apomictic plants bypass both meiotic reduction and egg fertilization, developing parthenogenetic embryos that are genetic clones of their mother. Interestingly, apomixis and sex are not mutually exclusive and can coexist in the same species and even the same individual, which illustrates a high degree of developmental plasticity. The mechanisms driving reproduction toward either sexual or asexual reproduction remain unclear, but recent results suggest a key role for chromatin in the process. In particular, loss-of-function mutants of *AGO9* and of several RdDM components in *Arabidopsis* phenocopy essential aspects of apomictic reproduction, producing spores that skip meiosis altogether (Olmedo-Monfil et al., 2010). Similarly, mutations in *AGO104* and *DMT103*, the maize homologue of *Arabidopsis* AGO9 and DRM2, respectively, produce ameiotic spores and gametophytes, a key step toward apomictic development (Singh et al., 2011; Garcia-Aguilar et al., 2010). Interestingly, these mutations do not abolish sexual reproduction, which coexists with alternative development in the mutants. Rather, they seemingly endow the genome with added developmental flexibility. Thus, establishment or maintenance of specific patterns of non-CG methylation in the ovule is likely essential in defining reproductive strategies in angiosperms. Because most of these genes are also involved in reprogramming, these observations raise important questions regarding the stability of the genome during asexual reproduction. Incidentally, an unstable epigenome during asexual reproduction might provide an interesting explanation to the surprising ecological success of clonal reproduction in plant, as an unexpected source of adaptive variation.

## 5. TRANSGENERATIONAL INHERITANCE OF EPIGENETIC VARIATION

Epigenetics historically refers to the stable propagation of chromatin states during DNA replication. However, alternative epigenetic states at genetically identical alleles (epialleles) can be stably transmitted through meiosis in rare instances. No more than a dozen natural epialleles with clear phenotypic impact have been reported in plants (Weigel & Colot, 2012). This suggests that transgenerational epigenetic inheritance (TEI) may in fact be a rather uncommon source of phenotypic diversity in natural populations. On the other hand, natural strains of *Arabidopsis* differ for numerous epialleles (Vaughn et al., 2007). In addition, spontaneous (Reviewed in Becker et al., 2011; Schmitz et al., 2011) or experimentally induced epialleles (Johannes et al., 2009; Reinders et al., 2009; Teixeira et al., 2009) are easily identified in laboratory conditions and can be meiotically propagated across many generations. This suggests that epialleles are widespread but rarely identified, that they are rapidly eliminated upon occurrence, or, alternatively, that the conditions that favor the emergence of epialleles are severely constrained during plant development in natural populations.

Changes in DNA methylation patterns have been associated to all molecularly characterized epialleles (Reviewed in Becker & Weigel, 2012; Weigel & Colot, 2012). Moreover, differential methylation in or around TEs or repeat sequences is often associated with the occurrence of epialleles. Somewhat conflicting with this observation, long-term analyses of single cytosine methylation stability in *Arabidopsis* lines indicate that, while spontaneous single cytosine methylation variants arise at a relatively high frequency (several orders of magnitude higher than nucleotidic mutations), DNA methylation changes in TEs and repeats (typically large segments of contiguous methylation) are rare, with a spontaneous epimutation rate in the range of nucleotidic mutation rate (Becker et al., 2011; Schmitz et al., 2011). As expected, siRNAs are associated with most differentially methylated TEs, but not with single cytosine methylation polymorphisms. Thus, siRNA-dependent maintenance of DNA methylation may account for the stability of repeat-associated epialleles. This, however, generates an interesting paradox: while repeats are protected against epigenetic alterations through RNAi, they apparently generate most evolutionary relevant instances of TEI.

One possibility to reconcile these observations is that repeat-associated siRNA themselves might contribute to TEI, as suggested by the phenomenon of paramutation. First described in maize, paramutation refers to the epigenetic transfer of information through *trans*-interactions between homologous sequences resulting in distinct states of gene expression that are heritable through generations (Reviewed in Arteaga-Vazquez & Chandler, 2010; Hollick, 2012). Paramutations are unique among epialleles for their "infectious" behavior: when an allele capable of inducing paramutation (paramutagenic) is crossed to a paramutable allele, the latter is changed to the paramutagenic state. Using the *B* locus in maize as a model, it has been shown that sequences required for paramutation are direct tandem repeats ~100-kb upstream of the transcription start site (Stam, Belele, Dorweiler, & Chandler, 2002). In addition, clonal analyses indicate that paramutated states are established during reproduction (Coe, 1966). Paramutation in plants involves proteins implicated in RdDM (Reviewed in Arteaga-Vazquez & Chandler, 2010; Hollick, 2012). Indeed, siRNAs generated from tandem repeats are necessary for paramutation, suggesting that it might be an "extreme" illustration of RdDM (Teixeira & Colot, 2010). There are, however, unique and distinct properties of paramutation that cannot be accounted for by RdDM. Importantly, loci silenced by RdDM are generally not paramutagenic. Also, whereas siRNAs are necessary for paramutation, they are not sufficient, as all alleles at *b* (paramutagenic, paramutable, or neutral) produce the same amount of siRNA (Arteaga-Vazquez et al., 2010). Yet, paramutations illustrate the possibility that RNA-induced chromatin changes at repeats, when established during reproductive development, might contribute to TEI. This possibility is strongly supported by the recent observation that maternal Piwi-interacting RNAs are necessary and sufficient to induce paramutation-like behavior in *Drosophila* (de Vanssay et al., 2012). By extension, it is tempting to speculate that paramutation in plants results from an RNAi-induced, indiscriminate reprogramming of both the paramutagenic and paramutable alleles during reproduction. This might imply a phased reprogramming step during reproduction when both alleles acquire a "neutral" epigenetic state that is then modified by homology-dependent RNAi toward the paramutagenic state. This would fit with a model in which, at least for some loci, reprogramming involves successive phases of erasure and reestablishment of information based on extra chromosomal, siRNA-based inheritance.

## 6. CONCLUDING REMARKS AND FUTURE OUTLOOK

Genetic, genomic, and biochemical studies in a variety of organisms have begun to reveal that similar chromatin-based strategies are employed in plants and metazoans to control in a timely and accurate manner the expression of a large number of genes during development. Whereas some pathways are based on seemingly conserved mechanisms, others rely on more divergent regulators that might have evolved to accompany distinct interpretations of multicellularity. Significant advances in our understanding of epigenome organization and chromatin-associated pathways indicate that divergent innovations might be exacerbated in plants, perhaps due to their extensive developmental plasticity and the necessity to adjust to environmental changes. Although some principles are beginning to emerge, thanks in part to large-scale epigenomic mapping and also to the extensive analysis of isolated loci, such as *FLC*, several key questions remain. At the molecular level, the mechanisms underlying the recruitment or targeting of most chromatin modifiers are largely unknown. Understanding these mechanisms and the activities responsible for removing chromatin marks or erasing chromatin states will be pivotal in order to interpret the dynamics of epigenomes during developmental stages. Determining the extent of this dynamics requires substantial improvement of the temporal and spatial resolution of epigenomic data, which can be achieved by using cell type-specific methodologies and by developing refined model systems. In addition, the interplay between epigenome dynamics and upstream signal transduction pathways as well as parallel gene regulatory networks is poorly understood. Finally, analyzing the epigenomes of *Arabidopsis* relatives and more distant species is necessary in order to reveal conservation or diversity in the relationship between genotype, epigenotype and phenotype.

Recent efforts in *Arabidopsis* have suggested that epigenetic variation on its own can be a substantial source of heritable phenotypic variation (Becker et al., 2011; Johannes et al., 2009; Richards, 2006, 2011; Schmitz et al., 2011; Weigel & Colot, 2012). The ability of the plant epigenome to respond to the environment and the fact that plants do not segregate their germline early in development open up the exciting possibility that epigenetic variation provides a flexible means of contributing significantly to plant adaptation and evolution.

## ACKNOWLEDGMENTS

We apologize to those researchers whose work could not be covered due to space limitations. We thank V. Colot and M. Mirouze for critical reading of this chapter. Work in the D.G. lab is supported by ANR (Agence Nationale de la Recherche) and the Institut de Recherche pour le Développement (IRD). Work in the F.R. lab is supported by ANR, the Centre National de la Recherche Scientifique (CNRS) and the European Union Network of Excellence "EpigeneSys".

## REFERENCES


Adrian, J., Farrona, S., Reimer, J. J., Albani, M. C., Coupland, G., & Turck, F. (2010). cis-Regulatory elements and chromatin state coordinately control temporal and spatial expression of FLOWERING LOCUS T in *Arabidopsis*. *The Plant Cell, 22*, 1425–1440.

Aichinger, E., Villar, C. B., Di Mambro, R., Sabatini, S., & Kohler, C. (2011). The CHD3 chromatin remodeler PICKLE and polycomb group proteins antagonistically regulate meristem activity in the *Arabidopsis* root. *The Plant Cell, 23*, 1047–1060.

Alvarez-Venegas, R., Pien, S., Sadder, M., Witmer, X., Grossniklaus, U., & Avramova, Z. (2003). ATX-1, an *Arabidopsis* homolog of trithorax, activates flower homeotic genes. *Current Biology, 13*, 627–637.

Andres, F., & Coupland, G. (2012). The genetic basis of flowering responses to seasonal cues. *Nature Reviews. Genetics, 13*, 627–639.

Angel, A., Song, J., Dean, C., & Howard, M. (2011). A polycomb-based switch underlying quantitative epigenetic memory. *Nature, 476*, 105–108.

Arteaga-Vazquez, M. A., & Chandler, V. L. (2010). Paramutation in maize: RNA mediated trans-generational gene silencing. *Current Opinion in Genetics & Development, 20*, 156–163.

Arteaga-Vazquez, M., Sidorenko, L., Rabanal, F. A., Shrivistava, R., Nobuta, K., Green, P. J., et al. (2010). RNA-mediated trans-communication can establish paramutation at the b1 locus in maize. *Proceedings of the National Academy of Sciences of the United States of America, 107*, 12986–12991.

Augui, S., Nora, E. P., & Heard, E. (2011). Regulation of X-chromosome inactivation by the X-inactivation centre. *Nature Reviews. Genetics, 12*, 429–442.

Autran, D., Baroux, C., Raissig, M. T., Lenormand, T., Wittig, M., Grob, S., et al. (2011). Maternal epigenetic pathways control parental contributions to *Arabidopsis* early embryogenesis. *Cell, 145*, 707–719.

Baker, M. (2011). Making sense of chromatin states. *Nature Methods, 8*, 717–722.

Barrero, J. M., Gonzalez-Bayon, R., del Pozo, J. C., Ponce, M. R., & Micol, J. L. (2007). INCURVATA2 encodes the catalytic subunit of DNA Polymerase alpha and interacts with genes involved in chromatin-mediated cellular memory in *Arabidopsis thaliana*. *The Plant Cell, 19*, 2822–2838.

Becker, C., Hagmann, J., Muller, J., Koenig, D., Stegle, O., Borgwardt, K., et al. (2011). Spontaneous epigenetic variation in the *Arabidopsis thaliana* methylome. *Nature, 480*, 245–249.

Becker, C., & Weigel, D. (2012). Epigenetic variation: Origin and transgenerational inheritance. *Current Opinion in Plant Biology, 15*, 562–567.

Bemer, M., & Grossniklaus, U. (2012). Dynamic regulation of Polycomb group activity during plant development. *Current Opinion in Plant Biology, 15*, 523–529.

Berger, S. L. (2007). The complex language of chromatin regulationduring transcription. *Nature, 447*, 407–412.

Berger, N., Dubreucq, B., Roudier, F., Dubos, C., & Lepiniec, L. (2011). Transcriptional regulation of *Arabidopsis* LEAFY COTYLEDON2 involves RLE, a cis-element that regulates trimethylation of histone H3 at lysine-27. *The Plant Cell, 23*, 4065–4078.

Bernstein, E., Duncan, E. M., Masui, O., Gil, J., Heard, E., & Allis, C. D. (2006). Mouse polycomb proteins bind differentially to methylated histone H3 and RNA and are enriched in facultative heterochromatin. *Molecular and Cellular Biology, 26*, 2560–2569.

Berr, A., Shafiq, S., & Shen, W. H. (2011). Histone modifications in transcriptional activation during plant development. *Biochimica et Biophysica Acta, 1809*, 567–576.

Bourc'his, D., & Voinnet, O. (2010). A small-RNA perspective on gametogenesis, fertilization, and early zygotic development. *Science, 330*, 617–622.

Bouyer, D., Roudier, F., Heese, M., Andersen, E. D., Gey, D., Nowack, M. K., et al. (2011). Polycomb repressive complex 2 controls the embryo-to-seedling phase transition. *PLoS Genetics*, *7*, e1002014.

Bratzel, F., Lopez-Torrejon, G., Koch, M., Del Pozo, J. C., & Calonje, M. (2010). Keeping cell identity in *Arabidopsis* requires PRC1 RING-finger homologs that catalyze H2A monoubiquitination. *Current Biology*, *20*, 1853–1859.

Buzas, D. M., Tamada, Y., & Kurata, T. (2012). FLC: A hidden polycomb response element shows up in silence. *Plant & Cell Physiology*, *53*, 785–793.

Calarco, J. P., Borges, F., Donoghue, M. T., Van Ex, F., Jullien, P. E., Lopes, T., et al. (2012). Reprogramming of DNA methylation in pollen guides epigenetic inheritance via small RNA. *Cell*, *151*, 194–205.

Cao, X., Aufsatz, W., Zilberman, D., Mette, M. F., Huang, M. S., Matzke, M., et al. (2003). Role of the DRM and CMT3 methyltransferases in RNA-directed DNA methylation. *Current Biology*, *13*, 2212–2217.

Cao, Y., Dai, Y., Cui, S., & Ma, L. (2008). Histone H2B monoubiquitination in the chromatin of FLOWERING LOCUS C regulates flowering time in *Arabidopsis*. *The Plant Cell*, *20*, 2586–2602.

Cao, X., & Jacobsen, S. E. (2002). Role of the *Arabidopsis* DRM methyltransferases in de novo DNA methylation and gene silencing. *Current Biology*, *12*, 1138–1144.

Cao, X., Springer, N. M., Muszynski, M. G., Phillips, R. L., Kaeppler, S., & Jacobsen, S. E. (2000). Conserved plant genes with similarity to mammalian de novo DNA methyltransferases. *Proceedings of the National Academy of Sciences of the United States of America*, *97*, 4979–4984.

Carles, C. C., & Fletcher, J. C. (2009). The SAND domain protein ULTRAPETALA1 acts as a trithorax group factor to regulate cell fate in plants. *Genes & Development*, *23*, 2723–2728.

Chen, D., Molitor, A., Liu, C., & Shen, W. H. (2010). The *Arabidopsis* PRC1-like ring-finger proteins are necessary for repression of embryonic traits during vegetative growth. *Cell Research*, *20*, 1332–1344.

Choi, Y., Gehring, M., Johnson, L., Hannon, M., Harada, J. J., Goldberg, R. B., et al. (2002). DEMETER, a DNA glycosylase domain protein, is required for endosperm gene imprinting and seed viability in *Arabidopsis*. *Cell*, *110*, 33–42.

Choi, J., Hyun, Y., Kang, M. J., In Yun, H., Yun, J. Y., Lister, C., et al. (2009). Resetting and regulation of Flowering Locus C expression during *Arabidopsis* reproductive development. *The Plant Journal*, *57*, 918–931.

Chouard, P. (1960). Vernalization and its relations to dormancy. *Annual Review of Plant Physiology*, *11*, 191–238.

Coe, E. H. (1966). The properties, origin, and mechanism of conversion-type inheritance at the B locus in maize. *Genetics*, *53*, 1035–1063.

Cokus, S. J., Feng, S., Zhang, X., Chen, Z., Merriman, B., Haudenschild, C. D., et al. (2008). Shotgun bisulphite sequencing of the *Arabidopsis* genome reveals DNA methylation patterning. *Nature*, *452*, 215–219.

Corbesier, L., Vincent, C., Jang, S., Fornara, F., Fan, Q., Searle, I., et al. (2007). FT protein movement contributes to long-distance signaling in floral induction of *Arabidopsis*. *Science*, *316*, 1030–1033.

Costas, C., Desvoyes, B., & Gutierrez, C. (2011). A chromatin perspective of plant cell cycle progression. *Biochimica et Biophysica Acta*, *1809*, 379–387.

Coustham, V., Li, P., Strange, A., Lister, C., Song, J., & Dean, C. (2012). Quantitative modulation of polycomb silencing underlies natural variation in vernalization. *Science*, *337*, 584–587.

Crevillen, P., & Dean, C. (2011). Regulation of the floral repressor gene FLC: The complexity of transcription in a chromatin context. *Current Opinion in Plant Biology, 14*, 38–44.

De Lucia, F., Crevillen, P., Jones, A. M., Greb, T., & Dean, C. (2008). A PHD-polycomb repressive complex 2 triggers the epigenetic silencing of FLC during vernalization. *Proceedings of the National Academy of Sciences of the United States of America, 105*, 16831–16836.

de Vanssay, A., Bouge, A. L., Boivin, A., Hermant, C., Teysset, L., Delmarre, V., et al. (2012). Paramutation in Drosophila linked to emergence of a piRNA-producing locus. *Nature, 490*, 112–115.

Duran-Figueroa, N., & Vielle-Calzada, J. P. (2010). ARGONAUTE9-dependent silencing of transposable elements in pericentromeric regions of *Arabidopsis*. *Plant Signaling & Behavior, 5*, 1476–1479.

Exner, V., Taranto, P., Schonrock, N., Gruissem, W., & Hennig, L. (2006). Chromatin assembly factor CAF-1 is required for cellular differentiation during plant development. *Development, 133*, 4163–4172.

Feng, S., & Jacobsen, S. E. (2011). Epigenetic modifications in plants: An evolutionary perspective. *Current Opinion in Plant Biology, 14*, 179–186.

Filion, G. J., van Bemmel, J. G., Braunschweig, U., Talhout, W., Kind, J., Ward, L. D., et al. (2010). Systematic protein location mapping reveals five principal chromatin types in Drosophila cells. *Cell, 143*, 212–224.

Finnegan, E. J., & Dennis, E. S. (2007). Vernalization-induced trimethylation of histone H3 lysine 27 at FLC is not maintained in mitotically quiescent cells. *Current Biology, 17*, 1978–1983.

Garcia-Aguilar, M., Michaud, C., Leblanc, O., & Grimanelli, D. (2010). Inactivation of a DNA methylation pathway in maize reproductive organs results in apomixis-like phenotypes. *The Plant Cell, 22*, 3249–3267.

Gehring, M., Bubb, K. L., & Henikoff, S. (2009). Extensive demethylation of repetitive elements during seed development underlies gene imprinting. *Science, 324*, 1447–1451.

Gehring, M., Huh, J. H., Hsieh, T. F., Penterman, J., Choi, Y., Harada, J. J., et al. (2006). DEMETER DNA glycosylase establishes MEDEA polycomb gene self-imprinting by allele-specific demethylation. *Cell, 124*, 495–506.

Gehring, M., Missirian, V., & Henikoff, S. (2011). Genomic analysis of parent-of-origin allelic expression in *Arabidopsis thaliana* seeds. *PLoS One, 6*, e23687.

Grimanelli, D. (2012). Epigenetic regulation of reproductive development and the emergence of apomixis in angiosperms. *Current Opinion in Plant Biology, 15*, 57–62.

Gu, X., Jiang, D., Wang, Y., Bachmair, A., & He, Y. (2009). Repression of the floral transition via histone H2B monoubiquitination. *The Plant Journal, 57*, 522–533.

Haag, J. R., & Pikaard, C. S. (2011). Multisubunit RNA polymerases IV and V: Purveyors of non-coding RNA for plant gene silencing. *Nature Reviews. Molecular Cell Biology, 12*, 483–492.

Hansen, K. H., Bracken, A. P., Pasini, D., Dietrich, N., Gehani, S. S., Monrad, A., et al. (2008). A model for transmission of the H3K27me3 epigenetic mark. *Nature Cell Biology, 10*, 1291–1300.

Havecker, E. R., Wallbridge, L. M., Hardcastle, T. J., Bush, M. S., Kelly, K. A., Dunn, R. M., et al. (2010). The *Arabidopsis* RNA-directed DNA methylation argonautes functionally diverge based on their expression and interaction with target loci. *The Plant Cell, 22*, 321–334.

He, G., Elling, A. A., & Deng, X. W. (2011). The epigenome and plant development. *Annual Review of Plant Biology, 62*, 411–435.

Helliwell, C. A., Robertson, M., Finnegan, E. J., Buzas, D. M., & Dennis, E. S. (2011). Vernalization-repression of *Arabidopsis* FLC requires promoter sequences but not antisense transcripts. *PLoS One, 6*, e21513.

Hennig, L., & Derkacheva, M. (2009). Diversity of Polycomb group complexes in plants: Same rules, different players? *Trends in Genetics, 25*, 414–423.
Heo, J. B., & Sung, S. (2011). Vernalization-mediated epigenetic silencing by a long intronic noncoding RNA. *Science, 331*, 76–79.
Holec, S., & Berger, F. (2012). Polycomb group complexes mediate developmental transitions in plants. *Plant Physiology, 158*, 35–43.
Hollick, J. B. (2012). Paramutation: A trans-homolog interaction affecting heritable gene regulation. *Current Opinion in Plant Biology, 15*, 536–543.
Hsieh, T. F., Ibarra, C. A., Silva, P., Zemach, A., Eshed-Williams, L., Fischer, R. L., et al. (2009). Genome-wide demethylation of *Arabidopsis* endosperm. *Science, 324*, 1451–1454.
Hsieh, T. F., Shin, J., Uzawa, R., Silva, P., Cohen, S., Bauer, M. J., et al. (2011). Regulation of imprinted gene expression in *Arabidopsis* endosperm. *Proceedings of the National Academy of Sciences of the United States of America, 108*, 1755–1762.
Huettel, B., Kanno, T., Daxinger, L., Bucher, E., van der Winden, J., Matzke, A. J., et al. (2007). RNA-directed DNA methylation mediated by DRD1 and Pol IVb: A versatile pathway for transcriptional gene silencing in plants. *Biochimica et Biophysica Acta, 1769*, 358–374.
Ibarra, C. A., Feng, X., Schoft, V. K., Hsieh, T. F., Uzawa, R., Rodrigues, J. A., et al. (2012). Active DNA demethylation in plant companion cells reinforces transposon methylation in gametes. *Science, 337*, 1360–1364.
Ingouff, M., Rademacher, S., Holec, S., Soljic, L., Xin, N., Readshaw, A., et al. (2010). Zygotic resetting of the HISTONE 3 variant repertoire participates in epigenetic reprogramming in *Arabidopsis*. *Current Biology, 20*, 2137–2143.
Jenuwein, T., & Allis, C. D. (2001). Translating the histone code. *Science, 293*, 1074–1080.
Jeong, J. H., Song, H. R., Ko, J. H., Jeong, Y. M., Kwon, Y. E., Seol, J. H., et al. (2009). Repression of FLOWERING LOCUS T chromatin by functionally redundant histone H3 lysine 4 demethylases in *Arabidopsis*. *PLoS One, 4*, e8033.
Jiang, D., Gu, X., & He, Y. (2009). Establishment of the winter-annual growth habit via FRIGIDA-mediated histone methylation at FLOWERING LOCUS C in *Arabidopsis*. *The Plant Cell, 21*, 1733–1746.
Jiang, D., Wang, Y., & He, Y. (2008). Repression of FLOWERING LOCUS C and FLOWERING LOCUS T by the *Arabidopsis* Polycomb repressive complex 2 components. *PLoS One, 3*, e3404.
Johannes, F., Porcher, E., Teixeira, F. K., Saliba-Colombani, V., Simon, M., Agier, N., et al. (2009). Assessing the impact of transgenerational epigenetic variation on complex traits. *PLoS Genetics, 5*, e1000530.
Johnson, L., Mollah, S., Garcia, B. A., Muratore, T. L., Shabanowitz, J., Hunt, D. F., et al. (2004). Mass spectrometry analysis of *Arabidopsis* histone H3 reveals distinct combinations of post-translational modifications. *Nucleic Acids Research, 32*, 6511–6518.
Jullien, P. E., Mosquna, A., Ingouff, M., Sakata, T., Ohad, N., & Berger, F. (2008). Retinoblastoma and its binding partner MSI1 control imprinting in *Arabidopsis*. *PLoS Biology, 6*, e194.
Jullien, P. E., Susaki, D., Yelagandula, R., Higashiyama, T., & Berger, F. (2012). DNA methylation dynamics during sexual reproduction in *Arabidopsis thaliana*. *Current Biology, 22*, 1825–1830.
Kim, S. Y., Lee, J., Eshed-Williams, L., Zilberman, D., & Sung, Z. R. (2012). EMF1 and PRC2 cooperate to repress key regulators of *Arabidopsis* development. *PLoS Genetics, 8*, e1002512.
Köhler, C., & Grossniklaus, U. (2005). Seed development and genomic imprinting in plants. *Progress in Molecular and Subcellular Biology, 38*, 237–262.
Köhler, C., & Hennig, L. (2010). Regulation of cell identity by plant Polycomb and trithorax group proteins. *Current Opinion in Genetics & Development, 20*, 541–547.

Köhler, C., Wolff, P., & Spillane, C. (2012). Epigenetic mechanisms underlying genomic imprinting in plants. *Annual Review of Plant Biology, 63*, 331–352.
Kouzarides, T. (2007). Chromatin modifications and their function. *Cell, 128*, 693–705.
Kumar, S. V., Lucyshyn, D., Jaeger, K. E., Alos, E., Alvey, E., Harberd, N. P., et al. (2012). Transcription factor PIF4 controls the thermosensory activation of flowering. *Nature, 484*, 242–245.
Kumar, S. V., & Wigge, P. A. (2010). H2A.Z-containing nucleosomes mediate the thermosensory response in *Arabidopsis*. *Cell, 140*, 136–147.
Lafos, M., Kroll, P., Hohenstatt, M. L., Thorpe, F. L., Clarenz, O., & Schubert, D. (2011). Dynamic regulation of H3K27 trimethylation during *Arabidopsis* differentiation. *PLoS Genetics*, 7, e1002040.
Law, J. A., & Jacobsen, S. E. (2010). Establishing, maintaining and modifying DNA methylation patterns in plants and animals. *Nature Reviews. Genetics, 11*, 204–220.
Leeb, M., Pasini, D., Novatchkova, M., Jaritz, M., Helin, K., & Wutz, A. (2010). Polycomb complexes act redundantly to repress genomic repeats and genes. *Genes & Development, 24*, 265–276.
Li, X., Wang, X., He, K., Ma, Y., Su, N., He, H., et al. (2008). High-resolution mapping of epigenetic modifications of the rice genome uncovers interplay between DNA methylation, histone methylation, and gene expression. *The Plant Cell, 20*, 259–276.
Li, W., Wang, Z., Li, J., Yang, H., Cui, S., Wang, X., et al. (2011). Overexpression of AtBMI1C, a polycomb group protein gene, accelerates flowering in *Arabidopsis*. *PLoS One, 6*, e21364.
Libault, M., Tessadori, F., Germann, S., Snijder, B., Fransz, P., & Gaudin, V. (2005). The *Arabidopsis* LHP1 protein is a component of euchromatin. *Planta, 222*, 910–925.
Lister, R., O'Malley, R. C., Tonti-Filippini, J., Gregory, B. D., Berry, C. C., Millar, A. H., et al. (2008). Highly integrated single-base resolution maps of the epigenome in *Arabidopsis*. *Cell, 133*, 523–536.
Liu, F., Quesada, V., Crevillen, P., Baurle, I., Swiezewski, S., & Dean, C. (2007). The *Arabidopsis* RNA-binding protein FCA requires a lysine-specific demethylase 1 homolog to downregulate FLC. *Molecular Cell, 28*, 398–407.
Lu, F., Cui, X., Zhang, S., Jenuwein, T., & Cao, X. (2011). *Arabidopsis* REF6 is a histone H3 lysine 27 demethylase. *Nature Genetics, 43*, 715–719.
Lu, F., Cui, X., Zhang, S., Liu, C., & Cao, X. (2010). JMJ14 is an H3K4 demethylase regulating flowering time in *Arabidopsis*. *Cell Research, 20*, 387–390.
Margueron, R., & Reinberg, D. (2011). The Polycomb complex PRC2 and its mark in life. *Nature, 469*, 343–349.
Matzke, M., Kanno, T., Daxinger, L., Huettel, B., & Matzke, A. J. (2009). RNA-mediated chromatin-based silencing in plants. *Current Opinion in Cell Biology, 21*, 367–376.
Michaels, S. D., & Amasino, R. M. (1999). FLOWERING LOCUS C encodes a novel MADS domain protein that acts as a repressor of flowering. *The Plant Cell, 11*, 949–956.
Mosher, R. A., Melnyk, C. W., Kelly, K. A., Dunn, R. M., Studholme, D. J., & Baulcombe, D. C. (2009). Uniparental expression of PolIV-dependent siRNAs in developing endosperm of *Arabidopsis*. *Nature, 460*, 283–286.
Mylne, J. S., Barrett, L., Tessadori, F., Mesnage, S., Johnson, L., Bernatavichute, Y. V., et al. (2006). LHP1, the *Arabidopsis* homologue of HETEROCHROMATIN PROTEIN1, is required for epigenetic silencing of FLC. *Proceedings of the National Academy of Sciences of the United States of America, 103*, 5012–5017.
Olmedo-Monfil, V., Duran-Figueroa, N., Arteaga-Vazquez, M., Demesa-Arevalo, E., Autran, D., Grimanelli, D., et al. (2010). Control of female gamete formation by a small RNA pathway in *Arabidopsis*. *Nature, 464*, 628–632.
Pazhouhandeh, M., Molinier, J., Berr, A., & Genschik, P. (2011). MSI4/FVE interacts with CUL4-DDB1 and a PRC2-like complex to control epigenetic regulation of flowering

time in *Arabidopsis*. *Proceedings of the National Academy of Sciences of the United States of America*, *108*, 3430–3435.

Pecinka, A., Dinh, H. Q., Baubec, T., Rosa, M., Lettner, N., & Mittelsten Scheid, O. (2010). Epigenetic regulation of repetitive elements is attenuated by prolonged heat stress in *Arabidopsis*. *The Plant Cell*, *22*, 3118–3129.

Petruk, S., Sedkov, Y., Johnston, D. M., Hodgson, J. W., Black, K. L., Kovermann, S. K., et al. (2012). TrxG and PcG proteins but not methylated histones remain associated with DNA through replication. *Cell*, *150*, 922–933.

Phelps-Durr, T. L., Thomas, J., Vahab, P., & Timmermans, M. C. (2005). Maize rough sheath2 and its *Arabidopsis* orthologue ASYMMETRIC LEAVES1 interact with HIRA, a predicted histone chaperone, to maintain knox gene silencing and determinacy during organogenesis. *The Plant Cell*, *17*, 2886–2898.

Pillot, M., Baroux, C., Vazquez, M. A., Autran, D., Leblanc, O., Vielle-Calzada, J. P., et al. (2010). Embryo and endosperm inherit distinct chromatin and transcriptional states from the female gametes in *Arabidopsis*. *The Plant Cell*, *22*, 307–320.

Pontier, D., Picart, C., Roudier, F., Garcia, D., Lahmy, S., Azevedo, J., et al. (2012). NERD, a plant-specific GW protein, defines an additional RNAi-dependent chromatin-based pathway in *Arabidopsis*. *Molecular Cell*, *48*, 121–132.

Reinders, J., Wulff, B. B., Mirouze, M., Mari-Ordonez, A., Dapp, M., Rozhon, W., et al. (2009). Compromised stability of DNA methylation and transposon immobilization in mosaic *Arabidopsis* epigenomes. *Genes & Development*, *23*, 939–950.

Richards, E. J. (2006). Inherited epigenetic variation—Revisiting soft inheritance. *Nature Reviews. Genetics*, *7*, 395–401.

Richards, E. J. (2011). Natural epigenetic variation in plant species: A view from the field. *Current Opinion in Plant Biology*, *14*, 204–209.

Rinn, J. L., Kertesz, M., Wang, J. K., Squazzo, S. L., Xu, X., Brugmann, S. A., et al. (2007). Functional demarcation of active and silent chromatin domains in human HOX loci by noncoding RNAs. *Cell*, *129*, 1311–1323.

Rodriguez-Leal, D., & Vielle-Calzada, J. P. (2012). Regulation of apomixis: Learning from sexual experience. *Current Opinion in Plant Biology*, *15*, 549–555.

Roudier, F., Ahmed, I., Berard, C., Sarazin, A., Mary-Huard, T., Cortijo, S., et al. (2011). Integrative epigenomic mapping defines four main chromatin states in *Arabidopsis*. *The EMBO Journal*, *30*, 1928–1938.

Satake, A., & Iwasa, Y. (2012). A stochastic model of chromatin modification: Cell population coding of winter memory in plants. *Journal of Theoretical Biology*, *302*, 6–17.

Saze, H., & Kakutani, T. (2011). Differentiation of epigenetic modifications between transposons and genes. *Current Opinion in Plant Biology*, *14*, 81–87.

Schmitz, R. J., Schultz, M. D., Lewsey, M. G., O'Malley, R. C., Urich, M. A., Libiger, O., et al. (2011). Transgenerational epigenetic instability is a source of novel methylation variants. *Science*, *334*, 369–373.

Schmitz, R. J., Tamada, Y., Doyle, M. R., Zhang, X., & Amasino, R. M. (2009). Histone H2B deubiquitination is required for transcriptional activation of FLOWERING LOCUS C and for proper control of flowering in *Arabidopsis*. *Plant Physiology*, *149*, 1196–1204.

Schubert, D., Primavesi, L., Bishopp, A., Roberts, G., Doonan, J., Jenuwein, T., et al. (2006). Silencing by plant Polycomb-group genes requires dispersed trimethylation of histone H3 at lysine 27. *The EMBO Journal*, *25*, 4638–4649.

Schuettengruber, B., Chourrout, D., Vervoort, M., Leblanc, B., & Cavalli, G. (2007). Genome regulation by polycomb and trithorax proteins. *Cell*, *128*, 735–745.

Schwartz, Y. B., Kahn, T. G., Nix, D. A., Li, X. Y., Bourgon, R., Biggin, M., et al. (2006). Genome-wide analysis of Polycomb targets in Drosophila melanogaster. *Nature Genetics*, *38*, 700–705.

Sheldon, C. C., Conn, A. B., Dennis, E. S., & Peacock, W. J. (2002). Different regulatory regions are required for the vernalization-induced repression of FLOWERING LOCUS C and for the epigenetic maintenance of repression. *The Plant Cell, 14*, 2527–2537.

Sheldon, C. C., Hills, M. J., Lister, C., Dean, C., Dennis, E. S., & Peacock, W. J. (2008). Resetting of FLOWERING LOCUS C expression after epigenetic repression by vernalization. *Proceedings of the National Academy of Sciences of the United States of America, 105*, 2214–2219.

Simon, J. A., & Kingston, R. E. (2009). Mechanisms of polycomb gene silencing: Knowns and unknowns. *Nature Reviews. Molecular Cell Biology, 10*, 697–708.

Simon, S. A., & Meyers, B. C. (2011). Small RNA-mediated epigenetic modifications in plants. *Current Opinion in Plant Biology, 14*, 148–155.

Singh, M., Goel, S., Meeley, R. B., Dantec, C., Parrinello, H., Michaud, C., et al. (2011). Production of viable gametes without meiosis in maize deficient for an ARGONAUTE protein. *The Plant Cell, 23*, 443–458.

Slotkin, R. K., Vaughn, M., Borges, F., Tanurdzic, M., Becker, J. D., Feijo, J. A., et al. (2009). Epigenetic reprogramming and small RNA silencing of transposable elements in pollen. *Cell, 136*, 461–472.

Smaczniak, C., Immink, R. G., Muino, J. M., Blanvillain, R., Busscher, M., Busscher-Lange, J., et al. (2012). Characterization of MADS-domain transcription factor complexes in *Arabidopsis* flower development. *Proceedings of the National Academy of Sciences of the United States of America, 109*, 1560–1565.

Song, J., Angel, A., Howard, M., & Dean, C. (2012). Vernalization—A cold-induced epigenetic switch. *Journal of Cell Science, 125*, 3723–3731.

Stam, M., Belele, C., Dorweiler, J. E., & Chandler, V. L. (2002). Differential chromatin structure within a tandem array 100 kb upstream of the maize b1 locus is associated with paramutation. *Genes & Development, 16*, 1906–1918.

Strahl, B. D., & Allis, C. D. (2000). The language of covalent histone modifications. *Nature, 403*, 41–45.

Stroud, H., Otero, S., Desvoyes, B., Ramirez-Parra, E., Jacobsen, S. E., & Gutierrez, C. (2012). Genome-wide analysis of histone H3.1 and H3.3 variants in *Arabidopsis thaliana. Proceedings of the National Academy of Sciences of the United States of America, 109*, 5370–5375.

Sung, S., & Amasino, R. M. (2004). Vernalization in *Arabidopsis thaliana* is mediated by the PHD finger protein VIN3. *Nature, 427*, 159–164.

Sung, S., He, Y., Eshoo, T. W., Tamada, Y., Johnson, L., Nakahigashi, K., et al. (2006). Epigenetic maintenance of the vernalized state in *Arabidopsis thaliana* requires LIKE HETEROCHROMATIN PROTEIN 1. *Nature Genetics, 38*, 706–710.

Swiezewski, S., Liu, F., Magusin, A., & Dean, C. (2009). Cold-induced silencing by long antisense transcripts of an *Arabidopsis* Polycomb target. *Nature, 462*, 799–802.

Takuno, S., & Gaut, B. S. (2013). Gene body methylation is conserved between plant orthologs and is of evolutionary consequence. *Proceedings of the National Academy of Sciences of the United States of America, 110*, 1797–1802.

Talbert, P. B., Ahmad, K., Almouzni, G., Ausio, J., Berger, F., Bhalla, P. L., et al. (2012). A unified phylogeny-based nomenclature for histone variants. *Epigenetics Chromatin, 5*, 7.

Tavares, L., Dimitrova, E., Oxley, D., Webster, J., Poot, R., Demmers, J., et al. (2012). RYBP-PRC1 complexes mediate H2A ubiquitylation at polycomb target sites independently of PRC2 and H3K27me3. *Cell, 148*, 664–678.

Teixeira, F. K., & Colot, V. (2010). Repeat elements and the *Arabidopsis* DNA methylation landscape. *Heredity (Edinb), 105*, 14–23.

Teixeira, F. K., Heredia, F., Sarazin, A., Roudier, F., Boccara, M., Ciaudo, C., et al. (2009). A role for RNAi in the selective correction of DNA methylation defects. *Science, 323*, 1600–1604.

Tittel-Elmer, M., Bucher, E., Broger, L., Mathieu, O., Paszkowski, J., & Vaillant, I. (2010). Stress-induced activation of heterochromatic transcription. *PLoS Genetics*, *6*, e1001175.
Tolhuis, B., Muijrers, I., de Wit, E., Teunissen, H., Talhout, W., van Steensel, B., et al. (2006). Genome-wide profiling of PRC1 and PRC2 Polycomb chromatin binding in Drosophila melanogaster. *Nature Genetics*, *38*, 694–699.
Tsai, M. C., Manor, O., Wan, Y., Mosammaparast, N., Wang, J. K., Lan, F., et al. (2010). Long noncoding RNA as modular scaffold of histone modification complexes. *Science*, *329*, 689–693.
Turck, F., Roudier, F., Farrona, S., Martin-Magniette, M. L., Guillaume, E., Buisine, N., et al. (2007). *Arabidopsis* TFL2/LHP1 specifically associates with genes marked by trimethylation of histone H3 lysine 27. *PLoS Genetics*, *3*, e86.
van Steensel, B. (2011). Chromatin: Constructing the big picture. *The EMBO Journal*, *30*, 1885–1895.
Vaughn, M. W., Tanurd Ic, M., Lippman, Z., Jiang, H., Carrasquillo, R., Rabinowicz, P. D., et al. (2007). Epigenetic natural variation in *Arabidopsis thaliana*. *PLoS Biology*, *5*, e174.
Wang, X., Elling, A. A., Li, X., Li, N., Peng, Z., He, G., et al. (2009). Genome-wide and organ-specific landscapes of epigenetic modifications and their relationships to mRNA and small RNA transcriptomes in maize. *The Plant Cell*, *21*, 1053–1069.
Weigel, D., & Colot, V. (2012). Epialleles in plant evolution. *Genome Biology*, *13*, 249.
Wigge, P. A. (2011). FT, a mobile developmental signal in plants. *Current Biology*, *21*, R374–R378.
Wohrmann, H. J., Gagliardini, V., Raissig, M. T., Wehrle, W., Arand, J., Schmidt, A., et al. (2012). Identification of a DNA methylation-independent imprinting control region at the *Arabidopsis* MEDEA locus. *Genes & Development*, *26*, 1837–1850.
Wollmann, H., & Berger, F. (2012). Epigenetic reprogramming during plant reproduction and seed development. *Current Opinion in Plant Biology*, *15*, 63–69.
Wollmann, H., Holec, S., Alden, K., Clarke, N. D., Jacques, P. E., & Berger, F. (2012). Dynamic deposition of histone variant H3.3 accompanies developmental remodeling of the *Arabidopsis* transcriptome. *PLoS Genetics*, *8*, e1002658.
Wu, M. F., Sang, Y., Bezhani, S., Yamaguchi, N., Han, S. K., Li, Z., et al. (2012). SWI2/SNF2 chromatin remodeling ATPases overcome polycomb repression and control floral organ identity with the LEAFY and SEPALLATA3 transcription factors. *Proceedings of the National Academy of Sciences of the United States of America*, *109*, 3576–3581.
Xu, L., & Shen, W. H. (2008). Polycomb silencing of KNOX genes confines shoot stem cell niches in *Arabidopsis*. *Current Biology*, *18*, 1966–1971.
Yang, W., Jiang, D., Jiang, J., & He, Y. (2010). A plant-specific histone H3 lysine 4 demethylase represses the floral transition in *Arabidopsis*. *The Plant Journal*, *62*, 663–673.
Zhang, X., Clarenz, O., Cokus, S., Bernatavichute, Y. V., Pellegrini, M., Goodrich, J., et al. (2007). Whole-genome analysis of histone H3 lysine 27 trimethylation in *Arabidopsis*. *PLoS Biology*, *5*, e129.
Zhang, X., Germann, S., Blus, B. J., Khorasanizadeh, S., Gaudin, V., & Jacobsen, S. E. (2007). The *Arabidopsis* LHP1 protein colocalizes with histone H3 Lys27 trimethylation. *Nature Structural & Molecular Biology*, *14*, 869–871.
Zheng, B., & Chen, X. (2011). Dynamics of histone H3 lysine 27 trimethylation in plant development. *Current Opinion in Plant Biology*, *14*, 123–129.
Zhu, Y., Dong, A., Meyer, D., Pichon, O., Renou, J. P., Cao, K., et al. (2006). *Arabidopsis* NRP1 and NRP2 encode histone chaperones and are required for maintaining post-embryonic root growth. *The Plant Cell*, *18*, 2879–2892.
Zhu, Y., Dong, A., & Shen, W. H. (2012). Histone variants and chromatin assembly in plant abiotic stress responses. *Biochimica et Biophysica Acta*, *1819*, 343–348.

Zhu, Y., Weng, M., Yang, Y., Zhang, C., Li, Z., Shen, W. H., et al. (2011). *Arabidopsis* homologues of the histone chaperone ASF1 are crucial for chromatin replication and cell proliferation in plant development. *The Plant Journal*, *66*, 443–455.

Zilberman, D., Gehring, M., Tran, R. K., Ballinger, T., & Henikoff, S. (2007). Genome-wide analysis of *Arabidopsis thaliana* DNA methylation uncovers an interdependence between methylation and transcription. *Nature Genetics*, *39*, 61–69.

CHAPTER SEVEN

# Reprogramming and the Pluripotent Stem Cell Cycle

**Tomomi Tsubouchi*, Amanda G. Fisher†,[1]**
*MRC Genome Damage and Stability Centre, University of Sussex, Falmer, United Kingdom
†MRC Clinical Sciences Centre, Imperial College London, London, United Kingdom
[1]Corresponding author: e-mail address: amanda.fisher@csc.mrc.ac.uk

## Contents



## Abstract

Embryonic stem cells (ESCs) can self renew and retain the potential to differentiate into each of the cell types within the body. During experimental reprogramming, many of the features of ESCs can be acquired by differentiated target cells. One of these is the unusual cell division cycle that characterizes ESCs in which the Gap (G) phases are short and DNA Synthesis (S) phase predominates. Growing evidence has suggested that this atypical cell-cycle structure may be important for maintaining pluripotency and for enhancing pluripotent conversion. Here, we review current knowledge of cell-cycle regulation in ESCs and outline how this unique cell-cycle structure might contribute to successful reprogramming.

## 1. INTRODUCTION

The isolation of embryonic stem cells (ESCs) from the inner cell mass of mouse (Evans & Kaufman, 1981; Martin, 1981) and human blastocysts (Thomson et al., 1998) has revolutionized modern experimental genetics and our understanding of pluripotency. ESCs can differentiate into all the cell types within the body (i.e. they are pluripotent) and can divide

*Current Topics in Developmental Biology*, Volume 104
ISSN 0070-2153
http://dx.doi.org/10.1016/B978-0-12-416027-9.00007-3

indefinitely in a suitable environment enabling pluripotency to be maintained. The pluripotent state of ESCs is sustained by a core network of transcription factors and by chromatin remodeling factors that perpetuate an environment that is permissive for transcription (reviewed in Boyer, Mathur, & Jaenisch, 2006; Cole & Young, 2008; Niwa, 2007). Although we do not yet know how these core transcription factors and chromatin remodelers act in preserving pluripotency during stem cell self-renewal, the presence of "bivalent" chromatin at the promoters of lineage-specific genes has been implicated in establishing a chromatin context in which multiple lineage options are primed in readiness for subsequent developmental cues (reviewed in Spivakov & Fisher, 2007). This bivalent chromatin is characterized by being marked with opposing histone modifications that correlate with both "repressive" and "active" gene expression. ESCs grown in the presence of differentiation inhibitors (so-called 2i; Ying et al., 2008) show a reduction in bivalent marking (Marks et al., 2012) and are pluripotent but are less prone to differentiation.

Under specific conditions, differentiated cells can reset their lineage affiliation and revert to a multi- or pluripotent state. This was first shown in amphibians by transferring nuclei of late stage embryos into enucleated oocytes (Briggs & King, 1952; Gurdon, 1962; Gurdon, Laskey, & Reeves, 1975; King & Briggs, 1955). More recently, the isolation of so-called iPSCs (induced Pluripotent Stem Cells) by the forced expression of four pluripotency-associated factors has independently confirmed that terminally differentiated cells can reassume pluripotency (Takahashi & Yamanaka, 2006; reviewed in Stadtfeld & Hochedlinger, 2010). This has raised expectations that patient-specific pluripotent stem cells could be "tailor made" for cell replacement therapy and catalyzed scientific efforts to understand the molecular mechanisms of reprogramming as well as those that underwrite ESC proliferation.

One of the most remarkable features of ESCs is their rapid cell division. Fast cell division is accomplished by a selective reduction in the Gap (G) phases of the cell cycle (G1 and G2 phase) rather than by shortening DNA Synthesis (S) phase or mitosis (Becker et al., 2006; Fluckiger et al., 2006; Neganova, Zhang, Atkinson, & Lako, 2009; White & Dalton, 2005). Here, we review some recent data describing the unusual features of ES cell cycle and outline the proposal that an altered cell-cycle structure may be critical in determining successful pluripotent reprogramming.

## 2. THE EMBRYONIC STEM CELL CYCLE

The cell-cycle structure of undifferentiated ESCs is characterized as a large proportion of cells in S phase and shortened G1 and G2 phases (Fig. 7.1; Becker et al., 2006; Fluckiger et al., 2006; Neganova et al., 2009; White & Dalton, 2005). While other cell types tend to arrest in G1 or G0 when signals that drive proliferation is lacking, or undergo apoptosis, ESCs do not reenter quiescence or remain in G1 upon growth factor withdrawal (reviewed in White & Dalton, 2005). It has been hypothesized that having a shortened G1 phase somehow allows ESCs to enter cycle without having to wait for cues and that this property is coupled to pluripotent self-renewal (reviewed in Orford & Scadden, 2008). Consistent with this, G1 phase lengthens as mouse ESCs differentiate, a correlation that is also seen upon differentiation of ESCs derived from primates and humans (Fig. 7.1; Fluckiger et al., 2006; Stead et al., 2002; White et al., 2005). Enforced cell-cycle progression in neural stem cell lines has also been shown to interfere with cell differentiation (Lange, Huttner, & Calegari, 2009), suggesting that rapidly cycling cells may be less sensitive to differentiation inducers. The effect of lengthening the overall time of the cell cycle and the effect of altering the structure of the ES cell cycle (i.e., the relative length of each stage) are different; whereas a delay in G1 is detrimental for pluripotency, lengthening the overall cell-cycle time does not appear to adversely impact on self-renewal or differentiation potential (Stead et al., 2002). In line with this, some human ESC and iPSC lines can spend more than twice as long completing the cell cycle compared to other ESC lines (Fig. 7.1, lower panels; Ohtsuka & Dalton, 2008).

Blocking cell-cycle progression in ESCs appears to promote cell differentiation. Overexpression of a cyclin-dependent kinase (CDK) inhibitor p21, for example, causes arrest in G1 and arrest for 3 days is sufficient to induce human ESC differentiation (Ruiz et al., 2011). Upon release from the arrest, cells resume the cell cycle but pluripotency, as judged by a range of markers, is substantially diminished. This suggests that delay of cell-cycle progression is associated with irreversible ESC differentiation (Ruiz et al., 2011). Consistent with this is the demonstration that inhibition of G1 progression in human ESCs compromises pluripotency (Filipczyk, Laslett, Mummery, & Pera, 2007) and treatment of human ESCs with nocodazole (that arrests cells at the G2/M interface) resulted in reduced expression of NANOG and OCT4 (Kallas, Pook, Maimets, Zimmermann, & Maimets, 2011).

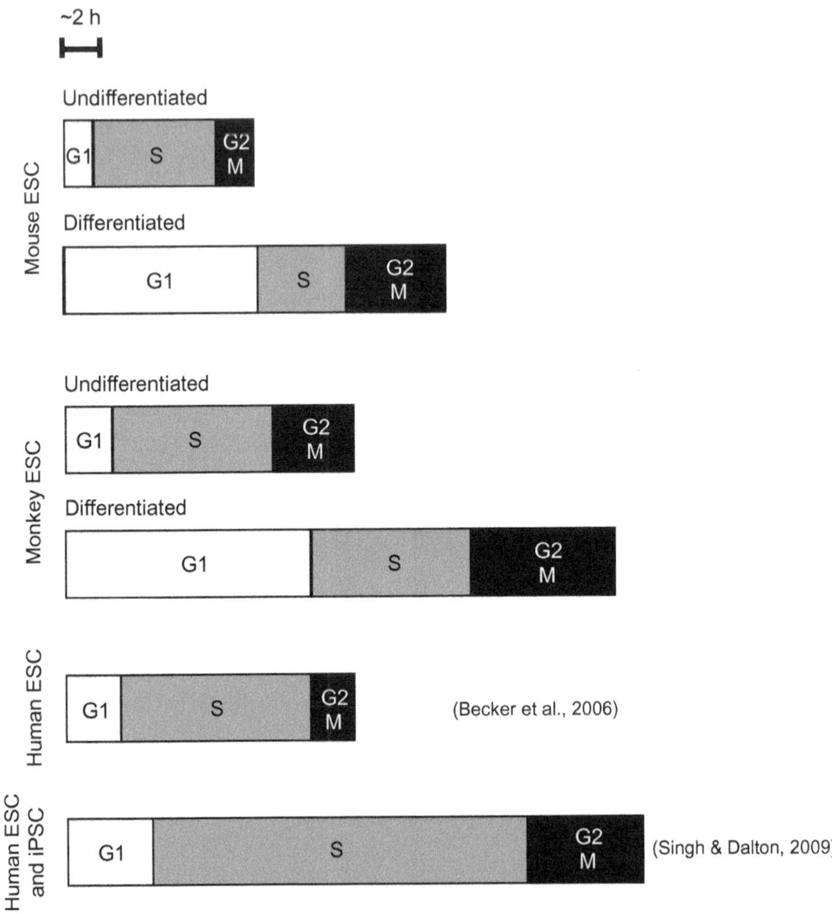


**Figure 7.1** Cell-cycle structure of undifferentiated and differentiated ESCs and iPSCs. Cell-cycle structure of undifferentiated and differentiated ESCs from different species is shown, with boxes representing the length of different cell cycle stages. S phase length of differentiated and undifferentiated ESCs are estimated to be similar. Undifferentiated ESCs have short gap phases with majority of the cell cycle spent in S phase. (Becker et al., 2006; Fluckiger et al., 2006; Ohtsuka & Dalton, 2008; Stead et al., 2002; White & Dalton, 2005).

The mechanisms that normally facilitate somatic cells to pause the cell cycle to deal with damage (i.e., checkpoint mechanisms) also seem to act negatively in maintaining ESCs. For example, somatic cells use a cell-cycle checkpoint to induce apoptosis or pause cell cycle to repair DNA damage when necessary (Aladjem et al., 1998; Zhao & Xu, 2010), and this is mediated by p53. However, murine ESCs lack a G1/S cell-cycle checkpoint;

here, p53 is localized mainly in the cytoplasm (Aladjem et al., 1998), and upon induction of DNA damage, p53 is translocated to the nucleus where it can suppress Nanog expression and thereby interfere with ESC self-renewal (Aladjem et al., 1998; Lin et al., 2005; Solozobova, Rolletschek, & Blattner, 2009). Therefore, rather than promoting genome maintenance by instructing cells to slow down cell cycle in order to repair damage, p53 can instruct ESCs with DNA damage to differentiate. Recent studies using other primate and nonprimate ESCs caution that regulation of the G1- to S-phase transition may vary in different species (Neganova et al., 2009) and that human ESCs may have a partly functioning G1/S checkpoint (Bárta et al., 2010).

At the molecular level, many cell-cycle regulators that, in differentiated cells, show a characteristic oscillation in activity, either do not oscillate in ESCs or do so in a much more muted way (Ballabeni et al., 2011). In somatic cells, different CDKs show sequential peaks of activity restricted to specific stages, allowing cell cycle to progress in an ordered manner (Fig. 7.2, top; reviewed in Bloom & Cross, 2007). CDK activities are closely linked to anaphase promoting complex/cyclosome (APC/C) activity, which is responsible for degradation of key cell-cycle regulators such as cyclins and geminin (reviewed in Bloom & Cross, 2007; Peters, 2006). Geminin inhibits CDT1, a factor that induce DNA synthesis, between S-phase and metaphase–anaphase transition (Wohlschlegel et al., 2000). Such inhibition prevents excess rounds of DNA synthesis. In mouse ESCs, geminin is present throughout most of the cell cycle (Yang et al., 2011). The boundaries restricting CDK activity in ESCs are therefore not distinct (Fig. 7.2, bottom). Other key regulators are also expressed at very high levels compared to differentiated cells, which may also contribute to the unique cell-cycle structure of ESCs (Fig. 7.2, Ballabeni et al., 2011; Fujii-Yamamoto, Kim, Arai, & Masai, 2005; White et al., 2005; Yang et al., 2011).

## 3. METHODS FOR RESTORING PLURIPOTENCY

The identity and transcriptional properties of differentiated cells (or their immediate precursors) are normally preserved through cell division. This mitotic inheritance can, however, be subverted under certain conditions. For example, during *in vivo* regeneration, transdifferentiation and dedifferentiation, cell lineage stage and affiliation are reset. In addition, experimental reprogramming can be used to encourage cells to revisit or adopt new epigenetic and transcriptional programs. Currently, three

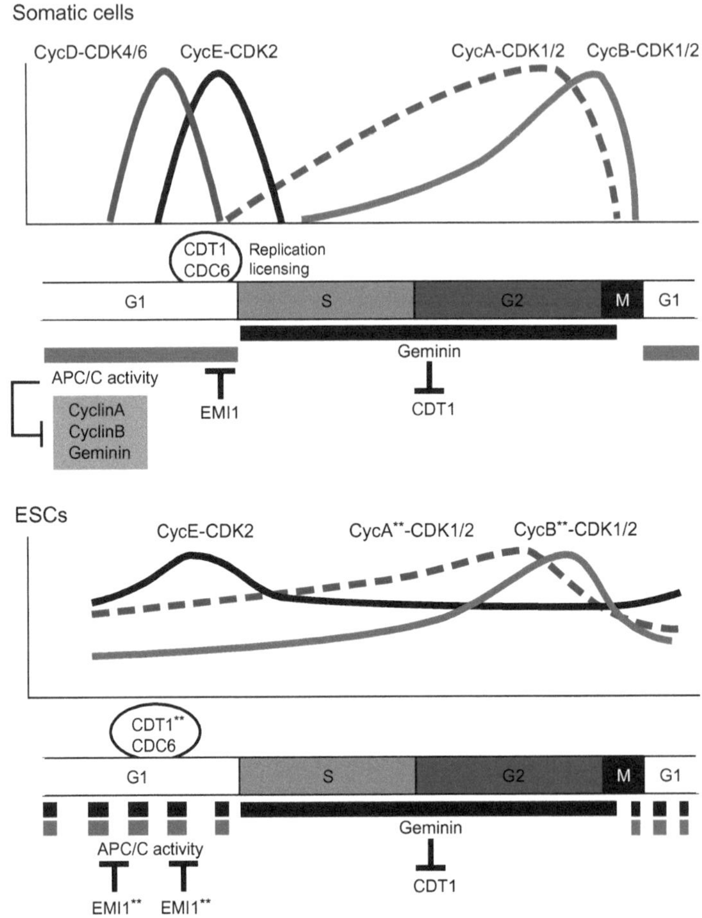


**Figure 7.2** Cell-cycle regulators in somatic cells and ESCs. Differences in the activity of key cell-cycle regulators in somatic and ESCs are shown. In ESCs, CDK activity is high throughout cell cycle and only minimal oscillation is detected (Ballabeni et al., 2011). CycD-associated CDK activity is not detected in ESCs. APC/C activity is attenuated presumably due to abundant EMI1, an inhibitor of APC/C. Geminin prevents extra rounds of DNA synthesis by binding and inhibiting the replication licensing factor CDT1. This interaction is promoted by CDK activity (Ballabeni et al., 2004), which helps protect CDT1 from degrading by APC/C. When geminin is degraded by APC/C during G1, CDT1 is released from geminin to promote replication. Abundant functional CDT1 during G1 may allow ESCs to transition from G1 to S, despite attenuated APC/C function. **Indicates abundant proteins in ESCs due to attenuated APC/C activity.

alternative methods have been devised to "reprogram" differentiated cell toward pluripotency; nuclear transfer (NT), fusion with a pluripotent stem cell partner, and epigenetic reprogramming using defined transcription factors or microRNAs (reviewed in Yamanaka & Blau, 2010).

In NT experiments, differentiated nuclei are reprogrammed by injecting them into the specialized environment of an enucleated oocyte or a fertilized egg. This method was initially developed in amphibians and was used to demonstrate nuclear equivalence; showing that the nucleus of a differentiated cell could generate an entire organism after sequential transfer and conditioning in oocytes (Briggs & King, 1952; Gurdon, 1962; Gurdon et al., 1975; King & Briggs, 1955). Three decades later, the first successfully cloned mammal, Dolly the sheep, was obtained by a similar approach (Wilmut, Schnieke, McWhir, Kind, & Campbell, 1997). Since then, cloning in a variety of species has been reported, although the method remains labor intensive and intrinsically inefficient as most cloned embryos fail to complete gestation (reviewed in Yamanaka, 2008).

A second approach for reprogramming is achieved by fusing a differentiated cell with a pluripotent cell line. Various stem cell partners have been used including ESCs, embryonic germ cells, and embryonic carcinoma cells (Miller & Ruddle, 1976; Tada, Tada, Lefebvre, Barton, & Surani, 1997; Tada, Takahama, Abe, Nakatsuji, & Tada, 2001). In the resulting hybrid, the somatic genome expresses pluripotency-associated markers (such as Oct3/4), acquires an ESC-like epigenetic state, and adopts a DNA methylation and histone modification pattern reminiscent of ESCs (Kimura, Tada, Nakatsuji, & Tada, 2004). Transplantation of these tetraploid cells into nude mice results in the formation of teratomas comprising tissues from all three germ layers (Tada et al., 2001). These data are consistent with the original differentiated cell being reprogrammed following fusion to become pluripotent. Reprogramming using human ESCs as the fusion partner has also been demonstrated (Cowan, Atienza, Melton, & Eggan, 2005; Yu, Vodyanik, He, Slukvin, & Thomson, 2006). Although the pluripotent tetraploids generated by this approach are not suitable for clinical use, the methodology does provide some unique perspectives for those interested in investigating the molecular mechanisms that underlie reprogramming.

Following cell-to-cell fusion, the partner nuclei remain separate for the first 3 days during which reprogramming is initiated. Subsequently, the participating nuclei fuse to generate hybrid cells in which the chromosome content is doubled (Pereira et al., 2008). As the expression of pluripotency-associated genes begins before the nuclei fuse, this gives experimentalists the chance to identify and directly examine what happens within the differentiated nucleus as reprogramming begins (Pereira et al., 2008). As fusion is possible between cells of different species (Harris, 1965; Harris & Watkins, 1965; Harris, Watkins, Campbell, Evans, & Ford, 1965), it is possible to monitor changes in gene expression using species-specific primers so as to discriminate the events that occur in the reprogrammer (ESC) and the reprogrammed (differentiated) targets (Pereira & Fisher, 2009). Such tracking is not possible with cells that belong to the same species, so a range of genetically engineered target cells have become available that express markers (such as OCT4-GFP) in response to pluripotent conversion (e.g., Han et al., 2008). Using such tools to generate heterokaryons with ESCs, it has been possible to chart the earliest events that occur *en route* to successful reprogramming (e.g., Bhutani et al., 2010; Do & Schöler, 2004; Han et al., 2008; Tada et al., 2001) and to show that reprogramming is initiated at a stage immediately prior to nuclear fusion and cell division.

Reprogramming can also be induced by the forced expression of a combination of pluripotency-associated factors. This was first shown by Takahashi and Yamanaka (2006) who reported that the expression of four transcription factors were sufficient to generate ESC-like pluripotent cells from differentiated targets. Since this time, Yamanaka's strategy has been modified and extended so that factor-based induction can be used to reprogram targets to become specific somatic stem cell types (reviewed in Pereira, Lemischka, & Moore, 2012). Despite improvements in generating iPSCs, the efficiency of successful conversion using this approach remains generally low (0.0001–2%). Reprogramming using iPS also takes several days to several weeks to accomplish. As discussed in Section 4, conversion in the iPS system is thought to depend at least partially on yet-to-be-uncovered stochastic events that may facilitate epigenetic reprogramming. These events might be present in only a subset of targets or could occur in most cells provided the inducers are allowed sufficient time and opportunity for cell division (Hanna et al., 2009). As an excess of reprogramming factors is known to be able to trigger cell-cycle arrest, death or inappropriate differentiation in transfected cells, achieving the right balance of reprogramming factors appears to be critical (Banito et al., 2009; Kawamura et al., 2009; Li et al., 2009). Different forms of genomic impairment also seem to block successful reprogramming, as somatic cells with different levels of DNA damage are eliminated by p53-dependent pathway (Marión et al., 2009). Elimination of p53 allows more efficient reprogramming; however, the resulting iPSCs show greater levels of chromosomal instability. As inactivation of the p53/p21 pathway and the *Ink4/Arf* locus is an early step in reprogramming (Li et al., 2009), this inactivation may explain the genome instability reported in some iPSCs (Gore et al., 2011; Mayshar et al., 2010; Mikkelsen et al., 2008).

## 4. CRITICAL STAGES AND EVENTS IN EPIGENETIC REPROGRAMMING

Regardless of the strategy used, reprogramming is widely acknowledged to be a multifactorial process that relies to some level on stochastic events. When reprogramming is attempted using a reduced number of iPS factors, conversion takes weeks to months. This suggests that successful reprogramming may rely on other factors that are either preexisting in a subset of somatic cells or that occur stochastically during the reprogramming process. To examine this question, Hanna et al. characterized the reprogramming efficiency and kinetics of over 1000 somatic-cell-derived monoclonal populations expressing reprogramming factors (Hanna et al., 2009). They found that most pro-B lymphocytes and most monocytes had the potential to generate iPSCs when cultured for extended period of time. This favors the idea that reprogramming depends on stochastic events rather than the preselection of a subset of target cells. Importantly, all iPSCs derived from these cells at different times of culture had normal karyotypes and each was capable of generating teratomas and chimeras.

Once triggered, reprogramming appears to proceed in an ordered manner starting with a downregulation of somatic markers followed by activation of less-stringent pluripotency markers (such as alkaline phosphatase, SSEA-1, and Fbxo1) and then more stringent markers (such as Nanog and Oct4) (Brambrink et al., 2008; Stadtfeld, Maherali, Breault, & Hochedlinger, 2008). Downregulation of somatic markers can occur without successful induction of pluripotency as ESCs lacking PRC2 (that cannot induce full reprogramming) can still extinguish the expression of somatic markers upon fusion with B lymphocytes (Pereira et al., 2010).

A comparison of genome-wide profiles of gene expression and chromatin status between ESCs, iPSCs, somatic cells, and partially reprogrammed cells (pre-iPSCs; Mikkelsen et al., 2008; Sridharan et al., 2009) has revealed that independently isolated iPSCs are similar at molecular level, even when different protocols were applied and cells of different origins (mouse fibroblasts and B lymphocytes) were used. Interestingly, pre-iPSCs also appeared similar, suggesting a common intermediate stage may exist where transiting cells accumulate before conversion is complete. On the other hand, several studies have shown that iPSCs derived from different cell types (mouse fibroblast, hematopoietic, and myogenic cells) have different gene expression patterns, especially at early passages after iPSCs derivation (Polo et al., 2010). Although all early passage iPSCs were shown to have the potential to differentiate into three different germ layers and produce chimeric mice, differences in their differentiation potential were noted that became less evident with successive passage in culture (Polo et al., 2010).

A major roadblock to reprogramming is thought to be the need to reverse DNA methylation at the promoter regions of pluripotency-associated genes such as *OCT4* and *NANOG* (Deb-Rinker, Ly, Jezierski, Sikorska, & Walker, 2005; Gidekel & Bergman, 2002). Demethylation is presumably needed to allow the binding of transcriptional activators that are required for gene reexpression. This step may require multiple rounds of cell division to establish in the iPS system where reprogramming is accomplished with exogenously supplied Oct4, but in the absence of Nanog (Theunissen et al., 2011). Nanog is known to enhance the reprogramming efficiency that can be achieved by Oct4, Sox2, Klf4, and c-myc alone (OSKM) and has been shown to help overcome barriers blocking complete conversion of pre-iPSCs (Theunissen et al., 2011). How DNA demethylation is accomplished during reprogramming remains a hotly debated issue (Hochedlinger & Plath, 2009). One model proposes that DNA methylation is gradually lost by a "passive" dilution where CpG methylation fails to be copied onto newly synthesized DNA during S phase. An opposing model invokes the active removal (by eviction or conversion) of 5-*m*ethyl*cy*tosine (5mC) from the genome. This can be achieved by one of a number of candidate enzymes. A critical difference between the two models is that while the first model depends on DNA replication and cell division for DNA demethylation, the latter model does not. One of the candidate proteins implicated in active DNA demethylation is the 5mC deaminase AID. This enzyme is expressed in activated B cells and in mouse primordial germ cells (Popp et al., 2010) although its expression in ESCs is unclear (Bhutani et al., 2010; Foshay et al., 2012). It has been claimed that AID is essential for the successful reprogramming of human fibroblasts in heterokaryons formed with mouse ESCs (Bhutani et al., 2010). However, others have found no evidence for contribution of AID to reprogramming (Foshay et al., 2012). Recent studies have also shown that Tet (*t*en-*e*leven-*t*ranslocation) family members can oxidize 5mC to 5-hydroxymethylcytosine and potentially to other oxidized forms (Tahiliani et al., 2009; reviewed in

Bhutani, Burns, & Blau, 2011; Niehrs & Schäfer, 2012) and that this may be particularly relevant for reprogramming in early development (Doege et al., 2012; Gu et al., 2011; Inoue & Zhang, 2011; Piccolo et al., 2013). How these modified bases are resolved is still unresolved, although the involvement of DNA repair mechanisms such as base excision repair or nucleotide excision repair has been proposed (Gehring, Reik, & Henikoff, 2009; Niehrs, 2009). Gadd45a (*g*rowth *a*rrest and *D*NA *d*amage 45) has been implicated in active DNA demethylation (Barreto et al., 2007) and have been proposed to provide a link between cellular stress, DNA repair, and epigenetic changes (reviewed in Niehrs & Schäfer, 2012).

## 5. DNA SYNTHESIS AND CHROMATIN REMODELING IN REPROGRAMMING

ESCs have an atypical cell-cycle structure and this unusual profile is adopted by somatic cells upon successful pluripotent reprogramming. To test whether the acquisition of stem cell properties and altered cell-cycle regulation are linked, Ruiz et al. (2011) examined the kinetics of iPSC derivation using cells with modified cell cycle. They found that cycling human fibroblasts and keratinocytes were more easily reprogrammed than counterparts that had been arrested in G1. This was not due to an acceleration in the reprogramming process, since OCT4-GFP positive colonies appeared at similar times using either targets, but rather, it reflected an increase in the number of cells that were amenable to being reprogrammed. Acquisition of a high proliferative rate is not, in itself, a reliable marker of successful reprogramming, since pre-iPSCs showed similarly high proliferation but fail to express the full quota of pluripotency factors (Mikkelsen et al., 2008; Silva et al., 2008; Sridharan et al., 2009). In a similar vein, some human ESCs are reported to have longer cell-cycle times, yet share a truncated G1 phase and high proportion of S-phase cells with other ESCs (Fig. 7.1; Ohtsuka & Dalton, 2008).

The mechanism by which cell division may influence reprogramming is currently unknown. However, there is accumulating evidence that S phase of the cell cycle provides a window of opportunity to change as well as to maintain existing epigenetic states (reviewed in McNairn & Gilbert, 2003). In order to maintain gene expression patterns, a cell must replicate its DNA and copy epigenetic information such as CpG methylation and modified histones. The molecular mechanisms that allow DNA methylation to be templated are, for the most part, understood. This is not the case for nucleosome templating, where the contribution of chromatin writers and chromatin readers has been postulated to be critical in conveying mitotic memory (reviewed in Fisher & Brockdorff, 2012). Many proteins involved in DNA synthesis, such as DNA polymerases, PCNA, and RPA, and in epigenetic inheritance, such as DNA methyltransferase 1 and CAF-1, are known to colocalize at "replication foci" formed during S phase (Rountree, Bachman, & Baylin, 2000; Shibahara & Stillman, 1999; Waga & Stillman, 1998). In addition, a recent report has shown that some enzymes that modify histones (such as Trithorax and Polycomb) continue to associate with their response elements during S phase (Petruk et al., 2012). The authors argue that as the modified histones are lost during S phase, these enzymes may be responsible for reestablishing this epigenetic information on the newly assembled nucleosomes (Petruk et al., 2012). Taken together, these studies suggest that S phase offers an unrivaled opportunity to reset or reprogram gene expression profiles.

Consistent with this idea, Mechali and colleagues have reported that the ability of differentiated nuclei to replicate in *Xenopus* egg extracts is enhanced when a single prior mitosis is permitted (Lemaitre, Danis, Pasero, Vassetzky, & Méchali, 2005). Moreover, preincubation of nuclei with mitotic-phase *Xenopus* egg extracts increased the number of iPSCs obtained using a conventional reprogramming factor cocktail (Ganier et al., 2011; Lemaitre et al., 2005). The authors propose that the chromosome structure of an adult differentiated nucleus is not well adapted for DNA replication and hence preconditioning is necessary for DNA replication to be elicited (Lemaitre et al., 2005). Mitotic conditioning may also allow increased recruitment of replication initiation factors onto chromatin and a shortening of topoisomerase II-dependent chromatin loops. Both these events might lead to the reduced inter-origin spacing that characterize early developmental stages (Walter & Newoirt, 2000; Wu, Yu, & Gilbert, 1997).

We have also seen that cell fusion-based pluripotent reprogramming is more efficient using ESCs that are in S and G2 phases of the cell cycle (Tsubouchi et al., 2013). These ESCs are capable of driving somatic nuclei into precocious DNA synthesis in heterokaryons. Our studies revisit the earlier work of Rao and Johnson (Johnson & Rao, 1970; Rao & Johnson, 1970) in which Hela cells in S and G2/M phases of the cell cycle were shown to induce premature DNA synthesis and chromosome condensation upon fusion with G1 phase targets. The possibility that inducing DNA synthesis facilitates successful reprogramming is also consistent with a report by Sullivan and colleagues showing that ESCs enriched for G2 (by growing to confluence) have an enhanced capacity to generate pluripotent hybrid cells when fused to differentiated fibroblasts or primary thymocytes (Sullivan, Pells, Hooper, Gallagher, & McWhir, 2006). Collectively, these reports provide evidence that the cell-cycle stage of the "reprogrammers" and their targets is likely to be important for reprogramming success and that transition through S phase may facilitate the epigenetic remodeling of differentiated cells.

## 6. CONCLUDING REMARKS

In this chapter, we have outlined data suggesting that altered cell-cycle structure may be important for maintaining pluripotency and for successful reprogramming. How cell cycle is controlled in ESCs is still unclear and it seems likely that this will be the subject of intensive research in the future. It is worth noting that the truncation of cell-cycle gap phases observed in ESCs is reminiscent of what is seen in the early stages of developmental of many other organisms. For example, in frogs and flies, embryogenesis begins with multiple rounds of rapid cell cleavage that occur ahead of zygotic transcription and cell specialization (Etkin, 1988). At this early stage in these organisms, embryonic development is dictated largely by maternally derived factors. In mouse and human, however, transcription from the zygote begins earlier (at the 2- to 8-cell stage) (reviewed in Tadros & Lipshitz, 2009) and cell specification begins as the inner cell mass is formed. It is intriguing to speculate that the atypical cell-cycle structure of ESCs may represent a developmental compromise or adaptation that enables the extensive cell proliferation needed to generate the mammalian embryo to occur, while simultaneously protecting pluripotent function. Somatic cell reprogramming is achieved by a range of approaches that appear to lock into (or imitate) the circuitry of pluripotent self-renewal used by ESCs. Our future challenge will be to better understand these circuits and how they are dismantled as ESCs differentiate. This knowledge will become critical in determining whether human ESCs and iPSCs have the potential for safe use in future cell replacement therapies.

## ACKNOWLEDGMENTS

We would like to thank our colleagues for discussions and the Medical Research Council, UK, and Human Frontier Science Program for support.

## REFERENCES


Aladjem, M. I., Spike, B. T., Rodewald, L. W., Hope, T. J., Klemm, M., Jaenisch, R., et al. (1998). ES cells do not activate p53-dependent stress responses and undergo p53-independent apoptosis in response to DNA damage. *Current Biology, 8*, 145–155.

Ballabeni, A., Melixetian, M., Zamponi, R., Masiero, L., Marinoni, F., & Helin, K. (2004). Human geminin promotes pre-RC formation and DNA replication by stabilizing CDT1 in mitosis. *The EMBO Journal, 23*, 3122–3132.

Ballabeni, A., Park, I. H., Zhao, R., Wang, W., Lerou, P. H., Daley, G. Q., et al. (2011). Cell cycle adaptations of embryonic stem cells. *Proceedings of the National Academy of Sciences of the United States of America, 108*, 19252–19257.

Banito, A., Rashid, S. T., Acosta, J. C., Li, S., Pereira, C. F., Geti, I., et al. (2009). Senescence impairs successful reprogramming to pluripotent stem cells. *Genes and Development, 23*, 2134–2139.

Barreto, G., Schäfer, A., Marhold, J., Stach, D., Swaminathan, S. K., Handa, V., et al. (2007). Gadd45a promotes epigenetic gene activation by repair-mediated DNA demethylation. *Nature, 445*, 671–675.

Bárta, T., Vinarský, V., Holubcová, Z., Dolezalová, D., Verner, J., Pospísilová, S., et al. (2010). Human embryonic stem cells are capable of executing G1/S checkpoint activation. *Stem Cells, 28*, 1143–1152.

Becker, K. A., Ghule, P. N., Therrien, J. A., Lian, J. B., Stein, J. L., van Wijnen, A. J., et al. (2006). Self-renewal of human embryonic stem cells is supported by a shortened G1 cell cycle phase. *Journal of Cellular Physiology, 209*, 883–893.

Bhutani, N., Brady, J. J., Damian, M., Sacco, A., Corbel, S. Y., & Blau, H. M. (2010). Reprogramming towards pluripotency requires AID-dependent DNA demethylation. *Nature, 463*, 1042–1047.

Bhutani, N., Burns, D. M., & Blau, H. M. (2011). DNA demethylation dynamics. *Cell, 146*, 866–872.

Bloom, J., & Cross, F. R. (2007). Multiple levels of cyclin specificity in cell-cycle control. *Nature Reviews Molecular Cell Biology, 8*, 149–160.

Boyer, L. A., Mathur, D., & Jaenisch, R. (2006). Molecular control of pluripotency. *Current Opinion in Genetics and Development, 16*, 455–462.

Brambrink, T., Foreman, R., Welstead, G. G., Lengner, C. J., Wernig, M., Suh, H., et al. (2008). Sequential expression of pluripotency markers during direct reprogramming of mouse somatic cells. *Cell Stem Cell, 2*, 151–159.

Briggs, R., & King, T. J. (1952). Transplantation of living nuclei from blastula cells into enucleated frogs' eggs. *Proceedings of the National Academy of Sciences of the United States of America, 38*, 455–463.

Cole, M. F., & Young, R. A. (2008). Mapping key features of transcriptional regulatory circuitry in embryonic stem cells. *Cold Spring Harbor Symposia on Quantitative Biology, 73*, 183–193.

Cowan, C. A., Atienza, J., Melton, D. A., & Eggan, K. (2005). Nuclear reprogramming of somatic cells after fusion with human embryonic stem cells. *Science, 309*, 1369–1373.

Deb-Rinker, P., Ly, D., Jezierski, A., Sikorska, M., & Walker, P. R. (2005). Sequential DNA methylation of the Nanog and Oct-4 upstream regions in human NT2 cells during neuronal differentiation. *The Journal of Biological Chemistry, 280*, 6257–6260.

Do, J. T., & Schöler, H. R. (2004). Nuclei of embryonic stem cells reprogram somatic cells. *Stem Cells, 22*, 941–949.

Doege, C. A., Inoue, K., Yamashita, T., Rhee, D. B., Travis, S., Fujita, R., et al. (2012). Early-stage epigenetic modification during somatic cell reprogramming by Parp1 and Tet2. *Nature, 488*, 652–655.

Etkin, L. D. (1988). Regulation of the mid-blastula transition in amphibians. *Developmental Biology, 5*, 209–225.

Evans, M. J., & Kaufman, M. H. (1981). Establishment in culture of pluripotential cells from mouse embryos. *Nature*, *292*, 154–156.

Filipczyk, A. A., Laslett, A. L., Mummery, C., & Pera, M. F. (2007). Differentiation is coupled to changes in the cell cycle regulatory apparatus of human embryonic stem cells. *Stem Cell Research*, *1*, 45–60.

Fisher, A. G., & Brockdorff, N. (2012). Epigenetic memory and parliamentary privilege combine to evoke discussions on inheritance. *Development*, *139*, 3891–3896.

Fluckiger, A.-C., Marcy, G., Marchand, M., Negre, D., Cosset, F.-L., Mitalipov, S., et al. (2006). Cell cycle features of primate embryonic stem cells. *Stem Cells*, *24*, 547–556.

Foshay, K. M., Looney, T. J., Chari, S., Mao, F. F., Lee, J. H., Zhang, L., et al. (2012). Embryonic stem cells induce pluripotency in somatic cell fusion through biphasic reprogramming. *Molecular Cell*, *46*, 159–170.

Fujii-Yamamoto, H., Kim, J. M., Arai, K., & Masai, H. (2005). Cell cycle and developmental regulations of replication factors in mouse embryonic stem cells. *The Journal of Biological Chemistry*, *280*, 12976–12987.

Ganier, O., Bocquet, S., Peiffer, I., Brochard, V., Arnaud, P., Puy, A., et al. (2011). Synergic reprogramming of mammalian cells by combined exposure to mitotic Xenopus egg extracts and transcription factors. *Proceedings of the National Academy of Sciences of the United States of America*, *108*, 17331–17336.

Gehring, M., Reik, W., & Henikoff, S. (2009). DNA demethylation by DNA repair. *Trends in Genetics*, *25*, 82–90.

Gidekel, S., & Bergman, Y. (2002). A unique developmental pattern of Oct-3/4 DNA methylation is controlled by a *cis*-demodification element. *The Journal of Biological Chemistry*, *277*, 34521–34530.

Gore, A., Li, Z., Fung, H. L., Young, J. E., Agarwal, S., Antosiewicz-Bourget, J., et al. (2011). Somatic coding mutations in human induced pluripotent stem cells. *Nature*, *471*, 63–67.

Gu, T. P., Guo, F., Yang, H., Wu, H. P., Xu, G. F., Liu, W., et al. (2011). The role of Tet3 DNA dioxygenase in epigenetic reprogramming by oocytes. *Nature*, *477*, 606–610.

Gurdon, J. B. (1962). The developmental capacity of nuclei taken from intestinal epithelium cells of feeding tadpoles. *Journal of Embryology and Experimental Morphology*, *10*, 622–640.

Gurdon, J. B., Laskey, R. A., & Reeves, O. R. (1975). The developmental capacity of nuclei transplanted from keratinized skin cells of adult frogs. *Journal of Embryology and Experimental Morphology*, *34*, 93 112.

Han, D. W., Do, J. T., Gentile, L., Stehling, M., Lee, H. T., & Schöler, H. R. (2008). Pluripotential reprogramming of the somatic genome in hybrid cells occurs with the first cell cycle. *Stem Cells*, *26*, 445–454.

Hanna, J., Saha, K., Pando, B., van Zon, J., Lengner, C. J., Creyghton, M. P., et al. (2009). Direct cell reprogramming is a stochastic process amenable to acceleration. *Nature*, *462*, 595–601.

Harris, H. (1965). Behaviour of differentiated nuclei in heterokaryons of animal cells from different species. *Nature*, *206*, 583–588.

Harris, H., & Watkins, J. F. (1965). Hybrid cells derived from mouse and man: Artificial heterokaryons of mammalian cells from different species. *Nature*, *205*, 640–646.

Harris, H., Watkins, J. F., Campbell, G. L., Evans, E. P., & Ford, C. E. (1965). Mitosis in hybrid cells derived from mouse and man. *Nature*, *207*, 606–608.

Hochedlinger, K., & Plath, K. (2009). Epigenetic reprogramming and induced pluripotency. *Development*, *136*, 509–523.

Inoue, A., & Zhang, Y. (2011). Replication-dependent loss of 5-hydroxymethylcytosine in mouse preimplantation embryos. *Science*, *334*, 194.

Johnson, R. T., & Rao, P. N. (1970). Mammalian cell fusion: Induction of premature chromosome condensation in interphase nuclei. *Nature*, *226*, 717–722.

Kallas, A., Pook, M., Maimets, M., Zimmermann, K., & Maimets, T. (2011). Nocodazole treatment decreases expression of pluripotency markers Nanog and Oct4 in human embryonic stem cells. *PLoS One*, *6*, e19114.

Kawamura, T., Suzuki, J., Wang, Y. V., Menendez, S., Morera, L. B., Raya, A., et al. (2009). Linking the p53 tumour suppressor pathway to somatic cell reprogramming. *Nature*, *460*, 1140–1144.

Kimura, H., Tada, M., Nakatsuji, N., & Tada, T. (2004). Histone code modifications on pluripotential nuclei of reprogrammed somatic cells. *Molecular and Cellular Biology*, *24*, 5710–5720.

King, T. J., & Briggs, R. (1955). Changes in the nuclei of differentiating gastrula cells, as demonstrated by nuclear transplantation. *Proceedings of the National Academy of Sciences of the United States of America*, *41*, 321–325.

Lange, C., Huttner, W. B., & Calegari, F. (2009). Cdk4/cyclinD1 overexpression in neural stem cells shortens G1, delays neurogenesis, and promotes the generation and expansion of basal progenitors. *Cell Stem Cell*, *5*, 320–331.

Lemaitre, J. M., Danis, E., Pasero, P., Vassetzky, Y., & Méchali, M. (2005). Mitotic remodeling of the replicon and chromosome structure. *Cell*, *123*, 787–801.

Li, H., Collado, M., Villasante, A., Strati, K., Ortega, S., Cañamero, M., et al. (2009). The Ink4/Arf locus is a barrier for iPS cell reprogramming. *Nature*, *460*, 1136–1139.

Lin, T., Chao, C., Saito, S., Mazur, S. J., Murphy, M. E., Appella, E., et al. (2005). p53 induces differentiation of mouse embryonic stem cells by suppressing Nanog expression. *Nature Cell Biology*, *7*, 165–171.

Marión, R. M., Strati, K., Li, H., Murga, M., Blanco, R., Ortega, S., et al. (2009). A p53-mediated DNA damage response limits reprogramming to ensure iPS cell genomic integrity. *Nature*, *460*, 1149–1153.

Marks, H., Kalkan, T., Menafra, R., Denissov, S., Jones, K., Hofemeister, H., et al. (2012). The transcriptional and epigenomic foundations of ground state pluripotency. *Cell*, *149*, 590–604.

Martin, G. R. (1981). Isolation of a pluripotent cell line from early mouse embryos cultured in medium conditioned by teratocarcinoma stem cells. *Proceedings of the National Academy of Sciences of the United States of America*, *78*, 7634–7638.

Mayshar, Y., Ben-David, U., Lavon, N., Biancotti, J. C., Yakir, B., Clark, A. T., et al. (2010). Identification and classification of chromosomal aberrations in human induced pluripotent stem cells. *Cell Stem Cell*, *7*, 521–531.

McNairn, A. J., & Gilbert, D. M. (2003). Epigenomic replication: Linking epigenetics to DNA replication. *Bioessays*, *25*, 647–656.

Mikkelsen, T. S., Hanna, J., Zhang, X., Ku, M., Wernig, M., Schorderet, P., et al. (2008). Dissecting direct reprogramming through integrative genomic analysis. *Nature*, *454*, 49–55.

Miller, R. A., & Ruddle, F. H. (1976). Pluripotent teratocarcinoma-thymus somatic cell hybrids. *Cell*, *9*, 45–55.

Neganova, I., Zhang, X., Atkinson, S., & Lako, M. (2009). Expression and functional analysis of G1 to S regulatory components reveals an important role for CDK2 in cell cycle regulation in human embryonic stem cells. *Oncogene*, *28*, 20–30.

Niehrs, C. (2009). Active DNA demethylation and DNA repair. *Differentiation*, *77*, 1–11.

Niehrs, C., & Schäfer, A. (2012). Active DNA demethylation by Gadd45 and DNA repair. *Trends in Cell Biology*, *22*, 220–227.

Niwa, H. (2007). How is pluripotency determined and maintained? *Development*, *134*, 635–646.

Ohtsuka, S., & Dalton, S. (2008). Molecular and biological properties of pluripotent embryonic stem cells. *Gene Therapy*, *15*, 74–81.

Orford, K. W., & Scadden, D. T. (2008). Deconstructing stem cell self-renewal: Genetic insights into cell-cycle regulation. *Nature Reviews Genetics*, *9*, 115–128.

Pereira, C. F., & Fisher, A. G. (2009). Heterokaryon-based reprogramming for pluripotency. *Current Protocols in Stem Cell Biology*, 9:4B.1.1–4B.1.14.

Pereira, C. F., Lemischka, I. R., & Moore, K. (2012). Reprogramming cell fates: Insights from combinatorial approaches. *Annals of the New York Academy of Sciences*, *1266*, 7–17.

Pereira, C. F., Piccolo, F. M., Tsubouchi, T., Sauer, S., Ryan, N. K., Bruno, L., et al. (2010). ESCs require PRC2 to direct the successful reprogramming of differentiated cells toward pluripotency. *Cell Stem Cell*, *6*, 547–556.

Pereira, C. F., Terranova, R., Ryan, N. K., Santos, J., Morris, K. J., Cui, W., et al. (2008). Heterokaryon-based reprogramming of human B lymphocytes for pluripotency requires Oct4 but not Sox2. *PLoS Genetics*, *4*, e1000170.

Peters, J. M. (2006). The anaphase promoting complex/cyclosome: A machine designed to destroy. *Nature Reviews Molecular Cell Biology*, 7, 644–656.

Petruk, S., Sedkov, Y., Johnston, D. M., Hodgson, J. W., Black, K. L., Kovermann, S. K., et al. (2012). TrxG and PcG proteins but not methylated histones remain associated with DNA through replication. *Cell*, *150*, 922–933.

Piccolo, F. M., Bagci, H., Brown, K. E.. Landeira, D., Soza-Ried, J., Feytout, A., et al. (2013). Different Roles for Tet1 and Tet2 Proteins in Reprogramming-Mediated Erasure of Imprints Induced by EGC Fusion. Mol Cell, http://dx.doi.org/10.1016/j.molcel.2013.01.032.

Polo, J. M., Liu, S., Figueroa, M. E., Kulalert, W., Eminli, S., Tan, K. Y., et al. (2010). Cell type of origin influences the molecular and functional properties of mouse induced pluripotent stem cells. *Nature Biotechnology*, *28*, 848–855.

Popp, C., Dean, W., Feng, S., Cokus, S. J., Andrews, S., Pellegrini, M., et al. (2010). Genome-wide erasure of DNA methylation in mouse primordial germ cells is affected by AID deficiency. *Nature*, *463*, 1101–1105.

Rao, P. N., & Johnson, R. T. (1970). Mammalian cell fusion: Studies on the regulation of DNA synthesis and mitosis. *Nature*, *225*, 159–164.

Rountree, M. R., Bachman, K. E., & Baylin, S. B. (2000). DNMT1 binds HDAC2 and a new co-repressor, DMAP1, to form a complex at replication foci. *Nature Genetics*, *25*, 269–277.

Ruiz, S., Panopoulos, A. D., Herrerías, A., Bissig, K. D., Lutz, M., Berggren, W. T., et al. (2011). A high proliferation rate is required for cell reprogramming and maintenance of human embryonic stem cell identity. *Current Biology*, *21*, 45–52.

Shibahara, K., & Stillman, B. (1999). Replication-dependent marking of DNA by PCNA facilitates CAF-1-coupled inheritance of chromatin. *Cell*, *96*, 575–585.

Silva, J., Barrandon, O., Nichols, J., Kawaguchi, J., Theunissen, T. W., & Smith, A. (2008). Promotion of reprogramming to ground state pluripotency by signal inhibition. *PLoS Biology*, *6*, e253.

Solozobova, V., Rolletschek, A., & Blattner, C. (2009). Nuclear accumulation and activation of p53 in embryonic stem cells after DNA damage. *BMC Cell Biology*, *10*, 46.

Spivakov, M., & Fisher, A. G. (2007). Epigenetic signatures of stem-cell identity. *Nature Reviews Genetics*, *8*, 263–271.

Sridharan, R., Tchieu, J., Mason, M. J., Yachechko, R., Kuoy, E., Horvath, S., et al. (2009). Role of the murine reprogramming factors in the induction of pluripotency. *Cell*, *136*, 364–377.

Stadtfeld, M., & Hochedlinger, K. (2010). Induced pluripotency: History, mechanisms, and applications. *Genes and Development*, *24*, 2239–2263.

Stadtfeld, M., Maherali, N., Breault, D. T., & Hochedlinger, K. (2008). Defining molecular cornerstones during fibroblast to iPS cell reprogramming in mouse. *Cell Stem Cell*, *2*, 230–240.

Stead, E., White, J., Faast, R., Conn, S., Goldstone, S., Rathjen, J., et al. (2002). Pluripotent cell division cycles are driven by ectopic Cdk2, cyclin A/E and E2F activities. *Oncogene*, *21*, 8320–8333.

Sullivan, S., Pells, S., Hooper, M., Gallagher, E., & McWhir, J. (2006). Nuclear reprogramming of somatic cells by embryonic stem cells is affected by cell cycle stage. *Cloning and Stem Cells*, *8*, 174–188.

Tada, M., Tada, T., Lefebvre, L., Barton, S. C., & Surani, M. A. (1997). Embryonic germ cells induce epigenetic reprogramming of somatic nucleus in hybrid cells. *The EMBO Journal*, *16*, 6510–6520.

Tada, M., Takahama, Y., Abe, K., Nakatsuji, N., & Tada, T. (2001). Nuclear reprogramming of somatic cells by in vitro hybridization with ES cells. *Current Biology*, *11*, 1553–1558.

Tadros, W., & Lipshitz, H. D. (2009). The maternal-to-zygotic transition: A play in two acts. *Development*, *136*, 3033–3042.

Tahiliani, M., Koh, K. P., Shen, Y., Pastor, W. A., Bandukwala, H., Brudno, Y., et al. (2009). Conversion of 5-methylcytosine to 5-hydroxymethylcytosine in mammalian DNA by MLL partner TET1. *Science*, *324*, 930–935.

Takahashi, K., & Yamanaka, S. (2006). Induction of pluripotent stem cells from mouse embryonic and adult fibroblast cultures by defined factors. *Cell*, *126*, 663–676.

Theunissen, T. W., van Oosten, A. L., Castelo-Branco, G., Hall, J., Smith, A., & Silva, J. C. (2011). Nanog overcomes reprogramming barriers and induces pluripotency in minimal conditions. *Current Biology*, *21*, 65–71.

Thomson, J. A., Itskovitz-Eldor, J., Shapiro, S. S., Waknitz, M. A., Swiergiel, J. J., Marshall, V. S., et al. (1998). Embryonic stem cell lines derived from human blastocysts. *Science*, *282*, 1145–1147.

Tsubouchi, T., Soza-Ried, J., Brown, K., Piccolo, F. M., Cantone, I., Landeira, D., et al. (2013). DNA synthesis is required for reprogramming mediated by stem cell fusion. *Cell*, *152*, 873–883.

Waga, S., & Stillman, B. (1998). The DNA replication fork in eukaryotic cells. *Annual Review of Biochemistry*, *67*, 721–751.

Walter, J., & Newoirt, J. (2000). Initiation of eukaryotic DNA replication: Origin unwinding and sequential chromatin association of Cdc45, RPA, and DNA polymerase alpha. *Molecular Cell*, *5*, 617–627.

White, J., & Dalton, S. (2005). Cell cycle control of embryonic stem cells. *Stem Cell Reviews*, *1*, 131–138.

White, J., Stead, E., Faast, R., Conn, S., Cartwright, P., & Dalton, S. (2005). Developmental activation of the Rb-E2F pathway and establishment of cell cycle-regulated cyclin-dependent kinase activity during embryonic stem cell differentiation. *Molecular Biology of the Cell*, *16*, 2018–2027.

Wilmut, I., Schnieke, A. E., McWhir, J., Kind, A. J., & Campbell, K. H. (1997). Viable offspring derived from fetal and adult mammalian cells. *Nature*, *385*, 810–813.

Wohlschlegel, J. A., Dwyer, B. T., Dhar, S. K., Cvetic, C., Walter, J. C., & Dutta, A. (2000). Inhibition of eukaryotic DNA replication by geminin binding to Cdt1. *Science*, *290*, 2309–2312.

Wu, J. R., Yu, G., & Gilbert, D. M. (1997). Origin-specific initiation of mammalian DNA replication in a Xenopus cell-free system. *Methods*, *13*, 313–324.

Yamanaka, S. (2008). Pluripotency and nuclear reprogramming. *Philosophical Transactions of the Royal Society of London. Series B, Biological Sciences*, *363*, 2079–2087.

Yamanaka, S., & Blau, H. M. (2010). Nuclear reprogramming to a pluripotent state by three approaches. *Nature*, *465*, 704–712.

Yang, V. S., Carter, S. A., Hyland, S. J., Tachibana-Konwalski, K., Laskey, R. A., & Gonzalez, M. A. (2011). Geminin escapes degradation in G1 of mouse pluripotent cells and mediates the expression of Oct4, Sox2, and Nanog. *Current Biology*, *21*, 692–699.

Ying, Q. L., Wray, J., Nichols, J., Batlle-Morera, L., Doble, B., Woodgett, J., et al. (2008). The ground state of embryonic stem cell self-renewal. *Nature*, *453*, 519–523.

Yu, J., Vodyanik, M. A., He, P., Slukvin, I. I., & Thomson, J. A. (2006). Human embryonic stem cells reprogram myeloid precursors following cell–cell fusion. *Stem Cells*, *24*, 168–176.

Zhao, T., & Xu, Y. (2010). p53 and stem cells: New developments and new concerns. *Trends in Cell Biology*, *20*, 170–175.

CHAPTER EIGHT

# H3K9/HP1 and Polycomb: Two Key Epigenetic Silencing Pathways for Gene Regulation and Embryo Development

**Peter Nestorov**[*,†,1], **Mathieu Tardat**[*,1], **Antoine H.F.M. Peters**[*,†,2]
[*]Friedrich Miescher Institute for Biomedical Research, Basel, Switzerland
[†]Faculty of Sciences, University of Basel, Basel, Switzerland
[1]Equal contribution.
[2]Corresponding author: e-mail address: antoine.peters@fmi.ch

## Contents



## Abstract

Proper development of an embryo requires tightly controlled expression of specific sets of genes. In order to generate all the lineages of the adult, populations of pluripotent embryonic stem cells differentiate and activate specific transcriptional programs whereas others are shutdown. The role of transcription factors is obvious in promoting expression of such developmental genes; however maintenance of specific states throughout cell division needs additional mechanisms. Indeed, the nucleoprotein complex of DNA and histones, the chromatin, can act as a facilitator or barrier to transcription depending on its configuration. Chromatin-modifying enzymes regulate accessibility of DNA by establishing specific sets of chromatin, which will be either permissive or repressive to transcription. In this review, we will describe the H3K9/HP1 and Polycomb pathways, which mediate transcriptional repression by modifying chromatin. We discuss how these two major epigenetic silencing modes are dynamically regulated and how they contribute to the early steps of embryo development.

*Current Topics in Developmental Biology*, Volume 104
ISSN 0070-2153
http://dx.doi.org/10.1016/B978-0-12-416027-9.00008-5

# 1. INTRODUCTION AND EVOLUTIONARY PERSPECTIVE

The eukaryotic genome is organized in the nucleus of a cell as chromatin—a dynamic and highly organized DNA–histone complex. In the nucleus of an interphase cell, chromatin appears as two distinct subtypes referred to as euchromatin and heterochromatin. Euchromatin is decondensed, gene rich, replicates early, and is enriched for histone posttranslational modifications (PTMs) associated with active transcription (i.e.. H3 lysine 4 di-/tri-methylation (H3K4me2/3), H3K36me2/3). In contrast, heterochromatin remains compacted throughout the cell cycle, contains repetitive sequences, replicates late in S phase, and is enriched for DNA methylation and histone PTMs related to transcriptional silencing (i.e., H3K9me3, H4K20me3) (Probst, Dunleavy, & Almouzni, 2009). These different chromatin marks have been intensively mapped to the genome of various model organisms and correlated to key biological processes (Campos & Reinberg, 2009; Consortium T.E.P., 2012; Gerstein et al., 2010; Kouzarides, 2007; Roy et al., 2010). Multicellular organisms originate from a single totipotent cell, the zygote that gives rise to a variety of cell types, which share the same genome but differ greatly in their morphology, function, and chromatin structure. This developmental diversity is achieved by complex genome regulation, involving transcription factors and chromatin modifiers.

In this review, we will specifically focus on two major chromatin repressive pathways. The first one is the H3K9/HP1 pathway, which involves lysine-specific methyltransferases (KMTs) that methylate H3K9, a mark recognized by the chromodomain (CD) containing family of HP1 proteins. The second one is the Polycomb repressive pathway, which in mammals is classically related to gene silencing by two distinct complexes, Polycomb repressive complexes 1 and 2 (PRC1 and PRC2) and the associated PTMs H2AK119ub1 and H3K27me3, respectively.

Both pathways are conserved throughout eukaryotic evolution, since homologues for the core components are found in animals, plants, and fungi (Garcia et al., 2007; Krauss, 2008; Schuettengruber, Chourrout, Vervoort, Leblanc, & Cavalli, 2007; Shaver, Casas-Mollano, Cerny, & Cerutti, 2010). In fact, both H3K9 and H3K27 methylation have been found in unicellular algae and protozoa, which suggest that this type of chromatin regulation has a very early origin (Krauss, 2008; Liu et al., 2007; Shaver et al., 2010). Furthermore, it seems that the two pathways not only have a common origin but possibly shared the same biological function. In the protozoa *Tetrahymena thermophila*, Ezl1 (the homologue of the PRC2 KMT Ezh2) catalyzes both H3K9 and H3K27 methylation and is required for the formation of constitutive heterochromatin (Liu et al., 2007). Another example of PRC2-dependent heterochromatin formation comes from the green alga *Chlamydomonas reinhardtii*, where the H3K27 methylation activity is needed for the silencing of repetitive sequences (Shaver et al., 2010).

During evolution and as a result of multiple genome duplication events, the two pathways diverged and were attributed with specific functions in multicellular organisms. Below, we will describe the most important developmental roles of these two pathways and how they contribute to gene regulation.

# 2. THE H3K9/HP1 PATHWAY AND ITS ROLE IN DEVELOPMENT

The seminal discovery of enzymes mediating H3K9 methylation on pericentric heterochromatin (PCH; Rea et al., 2000) and thereby generating a binding site for HP1 (Lachner, O'Carroll, Rea, Mechtler, & Jenuwein, 2001) emphasizes the role of this pathway in regulation of heterochromatin structure (Cheutin et al., 2003). The discovery of the Suv39h enzymes was soon followed by the identification of other H3K9-methylating enzymes (H3K9 KMT) (see Table 8.1) catalyzing different degrees of methylation and regulating repression of different classes of sequences (Krauss, 2008). The concept of division of labor applies also to HP1, as most eukaryotes express different isoforms of HP1 that differ in their nuclear localization (Zeng, Ball, & Yokomori, 2010) and serve partially redundant as well as isoform-specific functions (Rosnoblet, Vandamme, Völkel, & Angrand, 2011). For example, these proteins have been implicated in maintenance of heterochromatin, chromosome segregation, transcriptional silencing/activation (Kwon & Workman, 2011; Schotta, Ebert, & Reuter, 2003), DNA replication (Hayashi, Takahashi, Nakagawa, Nakayama, & Masukata, 2009; Schwaiger, Kohler, Oakeley, Stadler, & Schübeler, 2010), and the DNA damage response (Dinant & Luijsterburg, 2009).

## 2.1. Developmental role through regulation of gene expression

### *2.1.1 Su(var)*

With their pleiotropic roles in diverse biological pathways, H3K9-modifying enzymes and HP1 proteins are of particular interest for embryonic development. Functional studies in several organisms revealed that with the exception of the *Suv39h1* and *Suv39h2* (*Suv39h1/2*) enzymes, removal

**Table 8.1** H3K9/HP1 components

| *M. musculus* | | | *D. melanogaster* | | *A. thaliana* | |
|---|---|---|---|---|---|---|
| *Suv39h1/ KMT1A, Suv39h2/ KMT1B* | DKO animals: viable, reduced size, impaired gametogenesis, increased incidence of lymphoma (Braig et al., 2005; Peters et al., 2001). | DKO ESCs: Normal proliferation, loss of H3K9me3 and HP1 binding on chromocenter, replaced by H3K27me3 (Lehnertz et al., 2003; Peters et al., 2001). | *Su(var) 3–9* | Flies are viable, phenotype similar to mammals, HP1 binding is lost on chromocenters (Schotta et al., 2002; Tschiersch et al., 1994). | *SUVH4/ KYP, SUVH2* | Associate with chromocenters. Mutants show loss of heterochromatic marks, loss of gene silencing. Heterochromatinization if overexpressed (Jackson, Lindroth, Cao, & Jacobsen, 2002; Naumann et al., 2005). |
| *G9a/Ehmt2/ KMT1C, Glp/Ehmt1/ KMT1D* | Embryo: Both *G9a* and *Glp* single KO are lethal around E9.5 due to severe growth defects. Conditional germ line KO impairs development of germ cells and leads to defects in meiosis (Tachibana, Matsumura, Fukuda, Kimura, & Shinkai, 2008; Tachibana, Nozaki, Takeda, & Shinkai, 2007; Tachibana et al., 2002). | ESC: Regulation of euchromatic genes, imprinted genes, rDNA repeats, and transposons. Reduction of DNA methylation in *G9a* KO cells can be rescued by catalytically dead G9a (Dong et al., 2008; Estève et al., 2006; Tachibana et al., 2002; Wagschal et al., 2008). | *dG9a* | Suppressor of PEV, also regulates euchromatic genes. Mutants show developmental defects but are viable, more severe effects when combined with *Su(var) 3-9* mutation (Kato, Kato, Tachibana, Shinkai, & Yamaguchi, 2008; Mis, Ner, & Grigliatti, 2006; Seum, Reo, et al., 2007). | *SUVR4*[a] | Contributes to the silencing of transposons and rDNA repeats (Thorstensen et al., 2006; Veiseth et al., 2011). |

| | | | | | | |
|---|---|---|---|---|---|---|
| *Setdb1*/*ESET*/*KMT1E*, *Setdb2*/*KMT1F* | Embryo: *Setdb1* KO is lethal around E3.5–E5.5 and shows ICM growth defects. No KO mice reported for *Setdb2* (Dodge, Kang, Beppu, Lei, & Li, 2004). | ESC: Setdb1 is involved in transcriptional repression of euchromatic genes and repeats; required for maintenance of pluripotency (Bilodeau, Kagey, Frampton, Rahl, & Young, 2009; Dodge et al., 2004; Lohmann et al., 2010; Matsui et al., 2010). | *dSetdb1* | Homozygous lethal. Localized at euchromatin and on chromosome 4 (Seum, Bontron, Reo, Delattre, & Spierer, 2007). | | |
| *HP1α*/*Cbx5*, *HP1β*/*Cbx1*, *HP1γ*/*Cbx3* | Embryo: *HP1β* KO is perinatal lethal and leads to genomic instability. *HP1γ* KO mice are infertile and show impaired spermatogenesis, cell cycle and meiosis defects. No *HP1α* KO mice have been reported (Abe et al., 2011; Aucott et al., 2008; Brown et al., 2010; Naruse, Fukusumi, Kakiuchi, & Asano, 2007; Takada et al., 2011). | ESC: Modest upregulation of ERV (Maksakova et al., 2011). | *Su(var) 2–5*/*HP1a* | Flies have at least five HP1 homologues (HP1a/b/c/d/e), see text for detail. *HP1a* KO flies die at 3rd instar larva (Eissenberg, Morris, Reuter, & Hartnett, 1992; Kellum, Raff, & Alberts, 1995). | *LHP1*/*TLF2* | Function with PRC1, see Table 8.3 and text for details (Gaudin et al., 2001; Turck et al., 2007). |

[a]In plants, *SUVR4* has been suggested to be most closely related to *G9a*.

of H3K9/HP1 components results in developmental defects in most organisms. *Suv39h1/2* double knockout (DKO) mice are viable although smaller in size and exhibit chromosomal defects during meiosis in the germline (Peters et al., 2001), partial loss of DNA methylation on PCH (Lehnertz et al., 2003), and increased tumor risks (Braig et al., 2005; Peters et al., 2001). Embryonic stem cells (ESCs) deficient for *Suv39h1/2* can maintain stemness, proliferate, and show enrichment of H3K27me3 on PCH, suggesting plasticity between Suv39h1/2 and PcG repressive pathways (Peters et al., 2003). Inactivation of *Suv39h* homologues in other organisms does not strongly impair their viability as it was shown in *Drosophila* (Schotta et al., 2002; Tschiersch et al., 1994), plants (Jackson et al., 2002; Naumann et al., 2005), or even yeast (Allshire, Nimmo, Ekwall, Javerzat, & Cranston, 1995), with the exception of the fungi *Neurospora crassa*, where the mutation of *Dim-5* results in growth defects and sterility (Tamaru & Selker, 2001). Interestingly, in *Dim-5* mutant, both H3K9 and DNA methylation are affected, suggesting an interdependency of these two chromatin marks in this organism (for more detail about the link between DNA and H3K9 methylation, see Box 8.1). This suggests that the Suv39h enzymes act principally as gatekeeper of genome integrity during development by regulating constitutive heterochromatin more than affecting gene transcription. Interestingly, a recent report demonstrated the requirement of transcription factors Pax3 and Pax9 for repression of pericentric transcripts and maintenance of PCH in mouse embryonic fibroblast (Bulut-Karslioglu et al., 2012). However, whether Pax transcription factors directly interact with Suv39h enzymes and how they could promote their targeting in a developmental manner remains an open question. Transcriptional regulation by Suv39h1 was recently pointed out in a report from Alder *et al.*, suggesting that developmental genes targeted by the PcG proteins in mouse ESCs are repressed in a Suv39h1-dependent manner in the trophoblast lineage (Alder et al., 2010). Knockdown of *Suv39h1* in cultured trophoblast stem cells (TSCs) resulted in decreased expression of the TSC key transcription factor *Cdx2*, and enhanced TSC differentiation. This suggests that, at least during the blastocyst stage of preimplantation development, Suv39h enzymes could be required for proper trophoblast formation through specific gene regulation. In plants, SUPERMAN (a C2H2-type zinc-finger protein), which is required for maintaining boundaries between floral organs in *Arabidopsis*, is controlled by the Su(var)3–9 homologue SUVH4/KRYPTONITE (KYP). KYP indirectly represses the *SUPERMAN* locus through the recruitment of the DNA methyltransferase CHROMETHYLASE3 (CMT3), allowing

## BOX 8.1 Interplay between H3K9 and DNA methylation

Generally, DNA methylation (Jones, 2012) and H3K9 methylation serve similar purposes in long-term silencing. However, a direct (inter-)dependence between these two silencing mechanisms is not always clear. DNA demethylation for the generation of induced pluripotent stem cell is critical, as it can be enhanced with inhibitors like azacytidine (Yamanaka & Blau, 2010). Interestingly, a screen for the identification of barriers against *in vitro* reprogramming identified *Suv39h1* (Onder et al., 2012), suggesting that losing DNA methylation and H3K9me3 is required for reprogramming. On the other hand, MEFs deficient for *Suv39h1/2* show a concomitant loss of H3K9me3 and DNAme at PCH (Lehnertz et al., 2003). This decrease could be a consequence of the loss of HP1$\alpha$ binding. Indeed, artificial targeting of HP1 to euchromatic genes allows recruitment of DNMT1 and transcriptional silencing (Smallwood, Estève, Pradhan, & Carey, 2007). G9a appears to act synergistically with DNMTs to mediate *de novo* epigenetic silencing. Dnmt3a/b associates with G9a through its ankyrin domains to silence key pluripotency factors like *Oct3/4* and suppresses proviruses (Epsztejn-Litman et al., 2008; Leung et al., 2011). G9a also interacts directly with DNMT1 to form a ternary complex with PCNA required for maintaining DNA and histone methylation on rDNA repeats throughout replication (Estève et al., 2006). However, DNA methylation and G9a-mediated H3K9 methylation seem to occur mostly as two parallel pathways (Tachibana et al., 2008), because most of the DNA methylation defects seen in *G9a*$^{-/-}$ ESCs can be rescued by a catalytically dead enzyme (Dong et al., 2008). DNMT1 associates with Np95 (also known as Uhrf1 or ICBP90) (Sharif et al., 2007), which has the ability to bind hemimethylated DNA through its SET and RING finger-associated (SRA) domain and H3K9me2/3 through its tandem tudor domain. Np95 can bind H3K9me2/3 regardless of the presence of H3S10P. Insensitive to the phosphomethyl switch (Fischle et al., 2005), Np95 provides a way of maintaining DNA methylation during mitosis (Rothbart et al., 2012). This bridging protein would then allow DNMT1 to prevent the loss of DNA methylation states during critical steps of the cell-cycle.

Such coordination between DNA replication and maintenance of histone and DNA methylation seems to apply for PCH through the association of the methyl-CpG binding protein MBD1 and SETDB1 to replication forks (Sarraf & Stancheva, 2004). Indeed, it has been suggested that SETDB1-mediated H3K9me1 would serve as a substrate for the Suv39h enzymes to restore H3K9me3 on PCH as replication is ongoing (Dambacher, Hahn, & Schotta, 2010). SETDB1 associates with the corepressor KAP1 (also known as Trim28/Tif1b) (Schultz et al., 2002). Biochemical studies showed that KAP1 is in a complex with remodeling enzymes (Mi2a), DNA methyltransferases (DNMT1, DNMT3a, DNMT3b), KMT (SETDB1), HP1, and KRAB-ZNF proteins (for review see Iyengar & Farnham, 2011). KAP1 repressive complex was shown to be implicated in the overall silencing of euchromatic genes, retrotransposons, and imprinted control regions (ICRs) during development

*Continued*

**BOX 8.1 Interplay between H3K9 and DNA methylation—cont'd**

(Quenneville et al., 2011; Rowe et al., 2010; Schultz et al., 2002). Targeting to specific genomic regions is mediated by KRAB-ZNF proteins, like Znf57, which can target this complex to ICRs when they are methylated, a mechanism suggested to protect these loci from demethylation during early embryo development (Quenneville et al., 2011, 2012; Zuo et al., 2012). KAP1 therefore acts as a scaffold linking DNA methylation, H3K9me3 and HP1 binding to specific loci which repressive state needs to be maintained for proper development. With the hundreds of KRAB-ZNF proteins encoded by the human genome (Huntley et al., 2006), it will be challenging to understand the contribution of these proteins for targeting KAP1-repressive complex (and then DNA and H3K9 methylation) to specific genomic sites during early embryo development.

HP1$\gamma$, which was recently mapped in HCT116 cells show a localization toward gene bodies and is thought to be involved in RNA processing (Smallwood et al., 2012). Strikingly, DNA methylation in gene bodies is often associated with transcriptional activity and it has even been suggested to participate in RNA splicing (Laurent et al., 2010); however, it is not known if there is a direct interplay between these two pathways for achieving this function. Recently, new evidence was brought for a direct link between H3K9 and DNA methylation, as PGC7 (also known as Dppa3/Stella) has been shown to bind H3K9me2. Pgc7 is involved in protecting from active demethylation the maternal genome, marked by H3K9me2, and several paternally imprinted genes in the zygote (Nakamura et al., 2012). In plants and mammals, DNA methylation can occur on CG, CHG, and CHH context (where H is either a C, T, or A). In *A. thaliana*, methylation on CHG is maintained by a positive feedback loop between the Su(var)3-9 homologue SUVH4/KYP and DNA methyltransferase CMT3 in order to silence genes and retrotransposons (Jackson et al., 2002, 2004). Indeed, H3K9me1/2 catalyzed by SUVH4/KYP (the major enzyme responsible for H3K9me1/2 in plant heterochromatin) allows recruitment of HP1, which then allows binding of CMT3 and DNA methylation. Recently, it was also shown that SUVH4/KYP is required for the chromatin remodeler DDM1 (decrease in DNA methylation 1) dependent *de novo* methylation (Sasaki, Kobayashi, Saze, & Kakutani, 2012). Furthermore, plant SUVH homologs contain a YDG/SRA domain in their N-terminus that can bind methylated DNA (Johnson et al., 2007), suggesting that methylated DNA can reinforce the silencing signal by enhancing the recruitment of SUVH enzymes (Ebbs & Bender, 2006).

However, the most striking evidence of a direct link between DNA and H3K9 methylation comes from studies in the fungi *N. crassa*. In this organism, mutation of *DIM-5*, a *Su(var)3-9* homologue, results in a global loss of both H3K9me3 and DNA methylation (Tamaru & Selker, 2001), similar to what is seen in the DNA methyltransferase *DIM-2* mutants (Kouzminova & Selker, 2001), although the latter does not affect H3K9 methylation. The link between these two marks is HP1,

**BOX 8.1 Interplay between H3K9 and DNA methylation—cont'd**

which binds to H3K9me3 through its CD and to DIM-2 through its CSD (Honda & Selker, 2008). Mutation of the *N. crassa HP1* gene, *hpo*, results in severe defects of DNA methylation without altering H3K9me3, like *DIM-2* mutants (Freitag, Hickey, Khlafallah, Read, & Selker, 2004). Strikingly, this suggests that DIM-5 and HP1 are upstream of DNA methylation in *N. crassa*. Moreover, H3K9me3, HP1, and DNA methylation colocalize almost perfectly on 44 defined heterochromatic domains on linkage group VII (Lewis et al., 2009). Interestingly, HP1 also prevents the spreading of heterochromatic domains by association with the jmjC domain containing KDM DMM-1/2 (DNA methylation modulator 1/2). DMM-1/2 removes H3K9me3 and then prevents further accumulation of HP1 and DNA methylation (Honda et al., 2010). Similarly, such antisilencing mechanisms also exist in *S. pombe* and *A. thaliana* (Miura et al., 2009; Saze, Shiraishi, Miura, & Kakutani, 2008; Zofall & Grewal, 2006). In the yeast *S. pombe*, there is clear evidence of the role played by the RNAi pathway in recruitment of the H3K9me3 KMT Clr4 and heterochromatin formation (Zhang, Mosch, Fischle, & Grewal, 2008) (although it lacks DNA methylation) and to a lesser extend in plants and *Drosophila* from the RNA-directed DNA methylation (RdDM) (Law & Jacobsen, 2010) and the piRNAs pathway, respectively (Pal-Bhadra et al., 2004). In *N. crassa*, targeting of DIM-5 to chromatin relies on its interaction with another factor, DIM-7 (Lewis, Adhvaryu, Honda, Shiver, & Selker, 2010). Therefore, despite the differences between the model organisms cited so far, it seems that methylation of DNA and H3K9 cooperate for mediating chromatin silencing.

development of flowering organs (Jackson et al., 2002; Yun, Weigel, & Lee, 2002). However, the principal role of plant Su(var)3–9 homologues SUVH and SUVR seems to be the control of transposons in heterochromatin during plant development (Kuhlmann & Mette, 2012; Naumann et al., 2005; Thorstensen, Grini, & Aalen, 2011).

### *2.1.2 SETDB1*

During mouse preimplantation development, transcriptional regulation seems to rely more on other classes of the H3K9 KMT including G9a and ESET/SETDB1 and on the H3K9 KDMs of the JMJD2 family. Indeed, murine ESCs inactivated for *Setdb1* or the two KDM *Jmjd2a* and *Jmjd2c* are unable to maintain self-renewal and differentiate (Loh, Zhang, Chen, George, & Ng, 2007). SETDB1 associates with the core pluripotency transcription factor Oct3/4 and regulates a specific set of developmental genes, most of them related to the trophectoderm lineage

(Bilodeau et al., 2009; Cho, Park, Kwon, & Kang, 2012; Lohmann et al., 2010; Yeap, Hayashi, & Surani, 2009; Yuan et al., 2009). This set of genes is also targeted by PcG proteins (discussed again in Section 3.2). This suggests cooperation between different epigenetic repressive pathways for maintaining the stemness state of ESC (Azuara et al., 2006; Boyer et al., 2006). Furthermore, SETDB1 association with the zinc-finger transcription factor ZNF274, allows repression of Zinc finger genes (Frietze, O'Geen, Blahnik, Jin, & Farnham, 2010), whereas its association with the serine/threonine kinase Akt/PKB mediates repression of certain transcription factors such as Forkhead family member (Gao et al., 2007). This suggests that SETDB1 mediates transcriptional silencing of specific sets of genes depending on its binding partners. Zygotic expression of *Setdb1* begins at the blastocyst stage, while the maternal transcript is present in the oocyte and persists throughout preimplantation development (Dodge et al., 2004). It would be of great interest to identify binding partners of SETDB1 that could modulate its targets and thereby its biological output during the earliest stages of embryo development. Despite this role in euchromatic regions, one striking feature of SETDB1 is the repression of transposable elements and repeats in mESC (Karimi et al., 2011; Matsui et al., 2010), a function achieved in cooperation with the corepressor KAP1/Trim28 and HP1 proteins (Schultz, Ayyanathan, Negorev, Maul, & Rauscher, 2002; Sripathy, Stevens, & Schultz, 2006), although the latter appears dispensable for this function (Maksakova et al., 2011). Indeed, even if heterochromatin is more compacted and silent than euchromatin, there is evidence that transcription of repeats is tightly controlled (Eymery, Callanan, & Vourc'h, 2009). SETDB1 is also implicated in the regulation of the structure of promyelocytic leukemia-nuclear body (PML-NBs) and the transcription of its associated genes (Cho, Park, & Kang, 2011). PML-NBs that have been linked to many cellular processes such as apoptosis, DNA damage responses, and transcriptional regulation (Torok, Ching, & Bazett-Jones, 2009), expanding the many potential biological roles of SETDB1 for proper embryo development. Altogether, these multiple functions achieved by SETDB1 could therefore explain in part the early phenotype seen in *Setdb1* knockout mice, which die around 3.5–5.5 dpc (Dodge et al., 2004).

### *2.1.3 G9a*

G9a takes part in the control of genes like *Mage-a* (Tachibana et al., 2002), *p21/waf1* (Nishio & Walsh, 2004), some imprinted genes in the trophoblast (Wagschal et al., 2008), interferon beta through its association with PRDM1 (human homologue of mouse Blimp1) (Gyory, Wu, Fejér, Seto, & Wright, 2004), and the key developmental regulators *Oct3/4* and *Nanog* (Epsztejn-Litman et al., 2008; Feldman et al., 2006; Yamamizu et al., 2012). G9a is targeted to the promoter of the transcription factors *Oct3/4* and *Nanog*, where it deposits H3K9me2 and allows the recruitment of HP1 and Dnmt3a/b. Gene silencing of *Oct3/4* and *Nanog* is tightly controlled by the joint action of APC/Cdh1-mediated degradation of G9a by the proteasome and removal of the H3K9me2 mark from their promoters by the JMJD2A and JMJD2C KDM (Loh et al., 2007; Wang et al., 2010; Whetstine et al., 2006). Indeed, *Jmjd2c* KO in mice results in down-regulation of *Oct3/4*, *Nanog*, and *Sox2* mRNA (Wang et al., 2010), whereas *G9a* KO aberrantly prolongs expression of these genes up to embryonic day 7.5 (Yamamizu et al., 2012). Both mutant mouse models show developmental defects, suggesting that proper control of G9a-mediated gene silencing is crucial for the embryo. In fly, *dG9a* was shown to be a suppressor of position effect variegation (Mis et al., 2006), and although *dG9a* mutant flies show minor developmental defects and are viable (Seum, Bontron, Reo, Delattre, & Spierer, 2007; Seum, Reo, et al., 2007; Stabell et al., 2006), such role in gene regulation in the embryo could apply. In fact, *dG9a* over-expression affects transcription of genes involved in the pupal eye formation (Kato et al., 2008). In plants, the SUVR proteins are the most closely related to G9a (Baumbusch et al., 2001; Thorstensen et al., 2006), but so far, they have been described to function essentially in the repression of transposons and ribosomal DNA (rDNA) (Thorstensen et al., 2006; Veiseth et al., 2011).

### 2.1.4 HP1

Given its affinity for H3K9 methylated residues, involvement of HP1 for gene silencing is anticipated (Bannister et al., 2001; Lachner et al., 2001). HP1 can induce compaction of targeted loci (Verschure et al., 2005) with heritability of the repressed state over mitotic division (Ayyanathan et al., 2003). Indeed, artificial targeting of exogenous HP1$\alpha$ to a specific locus induces increased H3K9me3 overtime which can be maintained through cell division even in absence of further exogenous *HP1$\alpha$* expression (Hathaway et al., 2012). However, it is now clear that HP1 isoforms, which show different localizations within the nucleus (Minc, Allory, Worman, Courvalin, & Buendia, 1999), do not completely share redundant function (Cammas, Janoshazi, Lerouge, & Losson, 2007). Some HP1 isoforms are required for either promoting or repressing transcription, or both (de Wit, Greil, & van Steensel, 2007; Vermaak & Malik, 2009). Indeed, there is now substantial evidence arguing that HP1 can promote transcription (Kwon & Workman, 2011). The mammalian HP1β has been found to play a role in the control of the expression of rDNA genes by RNA polymerase I (RNAPI) in a Suv39h-dependent manner (Horáková et al., 2010), whereas HP1γ regulates transcriptional elongation through its association with RNA polymerase II (RNAPII) (Vakoc, Mandat, Olenchock, & Blobel, 2005) and RNA processing (Smallwood et al., 2012). Moreover, HP1γ is part of a complex with the Polycomb group protein L3mbtl2, forming a PRC1-like complex involved in gene repression in mouse ESC and early embryogenesis (Qin et al., 2012; Trojer et al., 2011).

Interestingly, *Drosophila HP1a (Su(var)2–5),* which is required for fly development (Eissenberg et al., 1992; Kellum & Alberts, 1995; Kellum et al., 1995), may share some function in RNA processing as it was found to positively regulate euchromatic gene expression by interacting with the heterogeneous ribonucleoprotein DDP1, HRB87F, and PEP (Piacentini et al., 2009). In *Drosophila*, HP1 isoforms are also involved in gene regulation. Indeed, HP1a, positively regulates (indirectly or not) heterochromatic as well as euchromatic genes, like heat-shock (*Hsp70*) or cell-cycle related genes (*Mcms*, *ORC4*, *CAF-1*, *Cdc45*, and *Aurora B*). More generally, HP1a is a positive regulator of transcription by facilitating H3K36 demethylation via chromoshadow domain (CSD)-mediated recruitment of dKDM4A at active and/or heterochromatic regions (Lin, Paulson, Abmayr, & Workman, 2012; Lin et al., 2008). Either knockdown or overexpression of *HP1b* is lethal for fly development, where it seems to play an important role in transcription. HP1c associates with the transcription factors WOC (*without children*) and ROW (*related of WOC*) via their PxVxL motif to regulate a common set of genes involved in nervous system development (Abel, Eskeland, Raffa, Kremmer, & Imhof, 2009; Font-Burgada, Rossell, Auer, & Azorín, 2008). Moreover, HP1c appears also to be required for proper recruitment of FACT (facilitates chromatin transcription) to chromatin (Kwon et al., 2010).

In addition to fly and mammals, both isoforms of HP1 in *C. elegans* (HPL-1, HPL-2) show preferential euchromatic localization, *HPL-2* mutants showed that it is required for the regulation of germline genes, as well as sets of genes involved in lipid metabolism or innate immunity (Couteau, Guerry, Muller, & Palladino, 2002; Meister et al., 2011; Studencka et al., 2012). The only HP1 homologue identified in *Arabidopsis thaliana* (LHP1/TLF2) also localizes to euchromatic domains, where it associates with PcG proteins to represses genes involved in plant development

(Gaudin et al., 2001; Kotake, Takada, Nakahigashi, Ohto, & Goto, 2003; Nakahigashi, Jasencakova, Schubert, & Goto, 2005; Turck et al., 2007).

The role of HP1 isoforms for gene regulation during mouse preimplantation development is, however, still poorly understood. *HP1β* is predominantly expressed in oocytes and zygotes (Puschendorf et al., 2008), whereas *HP1α* appears at the 2-cell stage. *HP1γ* is expressed later and throughout development (Meglicki, Teperek-Tkacz, & Borsuk, 2012). Investigation of germline knockout mutant mice for either *HP1* isoform would help us to understand better their role, if any, for proper transcriptional regulation during early embryo development.

## 2.2. Function in the germline

The germline can be viewed as the immortal lineage of cells that gives rise to haploid gametes in sexually reproducing organisms. Germ cells undergo numerous DNA-directed events that must be tightly coordinated and controlled while these cells progress through their development. Several recent reports indicate that members of the H3K9/HP1 pathway have important biological functions in germline maintenance, differentiation, and possibly in the process of meiotic silencing of unpaired chromosomes and meiotic sex chromosome inactivation (MSCI). These last two phenomena, collectively called "meiotic silencing," target sex chromosomes in the heterogametic sex (the X chromosome in male nematodes and the XY body in male mice) and also any other chromosomes that fail to synapse due to mutation or chromosomal rearrangement. Meiotic silencing is of crucial importance as it is hypothesized to maintain genome integrity (Turner, 2007; Zamudio, Chong, & O'Bryan, 2008). Many members of the H3K9/HP1 pathway are highly expressed in germ cells (Khalil, Boyar, & Driscoll, 2004; Peters et al., 2001; van der Heijden et al., 2007). Mouse *Suv39h2* is abundant in testes and in oocytes compared to *Suv39h1* which is more ubiquitously expressed in somatic tissues (O'Carroll et al., 2000; Puschendorf et al., 2008). *Suv39h1/2* DKO mice are viable, but display impaired spermatogenesis. Spermatocytes undergo apoptosis at the pachytene stage as a consequence of incomplete homologue pairing and synapsis defects (Peters et al., 2001). Interestingly, *G9a* germline conditional knockout males are sterile, while fertility is severely impaired in females (Tachibana et al., 2007). Mutant spermatocytes fail to progress through the pachytene stage, show defects in double-strand break (DSB) repair, and undergo apoptosis. It can, however, not be excluded that these defects are a consequence of misregulation of G9a target genes.

In mammals, knockout of *HP1γ* induces a dramatic reduction in the number of PGCs due to cell cycle defects (Abe et al., 2011). These animals are sterile and exhibit defects in centromere clustering and synapsis in spermatocytes (Brown et al., 2010; Naruse et al., 2007; Takada et al., 2011). Takada and colleagues further show that *Suv39h1/2* H3K9-dependent methylation at PCH serves as a platform to recruit HP1γ, which then recruits G9a, highlighting a putative cooperative role between H3K9 KMTs and HP1 for meiotic progression. Strikingly, HP1γ and HP1β, but not HP1α associate with the transcriptionally silent XY body during male meiosis, suggesting possibly isoform-specific functions (Metzler-Guillemain, Luciani, Depetris, Guichaoua, & Mattei, 2003). Recently, it was shown that the SETDB1 homologue in worms, MET-2, is involved in MSCI and protects the germline from undergoing apoptosis (Checchi & Engebrecht, 2011). Such function has not yet been described for mammalian SETDB1 and SETDB2 proteins. In murine embryonic ovary and postnatal testis, *Setdb2* expression correlates with that of *Stra8*, a gene involved in the onset of meiosis in germ cells (Hogarth, Mitchell, Evanoff, Small, & Griswold, 2011), suggesting a potential role for *Setdb2* in mitotic to meiotic transition in germ cells.

In the mouse female germline, *HP1β* is the predominantly expressed isoform, and so far, no report indicates that its inactivation induces defects in this lineage (Aucott et al., 2008). In *Drosophila*, among the five isoforms of HP1, *HP1d* and *HP1e* show a germline-specific expression in female ovary and male testis, respectively (Vermaak, Henikoff, & Malik, 2005; Volpe, Horowitz, Grafer, Jackson, & Berg, 2001). *HP1d*, also named *Rhino*, was found in a screen for female sterile mutant flies. *HP1d/Rhino* mutant flies are characterized by defects in polytene chromosome structure of nurse cells and egg polarity defects. HP1d/Rhino is thought to act as a safeguard of the female germline against mobile elements through modulation of the piRNA pathway (Klattenhoff et al., 2009). Even though there is less information about HP1e, it is hypothesized to have a function similar to HP1d but in the male germline (Vermaak et al., 2005).

Among the H3K9 KMTs in *Drosophila*, only dSETDB1, has been shown to have a crucial role in the female germline (Clough, Moon, Wang, Smith, & Hazelrigg, 2007; Koch, Honemann-Capito, Egger-Adam, & Wodarz, 2009; Wang et al., 2011; Yoon et al., 2008). *dSETDB1* is expressed most strongly at early stages of oogenesis, in germ cells in the germarium. *dSETDB1* mutant ovaries primarily exhibit germ cell differentiation defects and apoptosis in young females, and gradually the niche of germ stem cells is lost, indicating that dSETDB1 regulates both germ cell maintenance and differentiation. It has been proposed that dSETDB1 and *SU(VAR)3–9* cooperate during germ cell maturation. As germ cells mature and differentiate, *dSETDB1* expression decreases and its function is gradually taken over by *Su(var)3–9* (Yoon et al., 2008). Some interplay between these two H3K9 KMTs is reinforced by the fact that the *dSETDB1* mutant phenotype is less severe in *Su(var)3–9*-deficient flies (Brower-Toland, Riddle, Jiang, Huisinga, & Elgin, 2009; Seum, Bontron, et al., 2007; Seum, Reo, et al., 2007). This puzzling phenotypic connection remains unresolved at the molecular level. In mammals, the exact function of SETDB1 in the germline is unknown and possible links between the different H3K9 KMT enzymes are not clear either. One study even suggested that Suv39h1, G9a, Glp, and SETDB1 are part of the same complex (Fritsch et al., 2010). However, since these enzymes show different expression pattern during development, the molecular composition and function of such putative H3K9 methylation complex is still an open question (Box 8.2).

Recently, PRDM3/16 were characterized as H3K9me1 KMTs that provide a template for Suv39-mediated H3K9me3 conversion at PCH (Pinheiro et al., 2012). Importantly, these proteins are essential for the clustering of pericentromeric regions into chromocenters. Although using different enzymes, this pathway seems to be by largely conserved in *C. elegans*, where it is required for maintenance of the anchoring of heterochromatic regions to the nuclear periphery (Towbin et al., 2012). The role of these enzymes, during germline and early embryonic development awaits further investigations.

## 2.3. New insights into the function of H3K9/HP1 pathway?

During the past decade, we gathered a better understanding of the function of H3K9 KMT and HP1 proteins. Although many points need further investigation, knockout and knockdown studies in various organisms allowed us to appreciate more their role for proper development. However, most of these experiments focus on the catalytic activity of KMT toward histones. Since its discovery, methylation of proteins involves many different substrates (Paik, Paik, & Kim, 2007). For instance, G9a has substrates other than the canonical histones (Chin et al., 2007; Sampath et al., 2007). Therefore, identification of nonhistone substrates for the other H3K9 KMT would be of great interest, as PTMs of proteins could lead to a variety of biological outputs.

**BOX 8.2 Removal of H3K9 methylation by KDMs during germ-cell development**

The balance between establishment and erasure of H3K9 methylation is also crucial for proper embryo development. The expression of the H3K9me1/2 JmjC-domain-containing histone demethylase 2a (*Jhdm2a*) enzyme partially overlaps with that of *G9a*, whereas *G9a* is continuously expressed in the germline until its downregulation during meiotic prophase, and *Jhdm2a* is found transcribed from late pachytene onward, with high expression levels in round and elongating spermatids (Okada, Tateishi, & Zhang, 2010). Although *Jhdm2a* knockout mice are viable, they display smaller testis, infertility, and obesity (Okada, Scott, Ray, Mishina, & Zhang, 2007, Okada et al., 2010). Disruption of this enzyme causes defects in postmeiotic chromatin condensation in elongating spermatids leading to impaired nuclear elongation. Interestingly, Jhdm2a was shown to bind to the promoter region of genes encoding Transition Protein 1 (*Tnp1*) and Protamine 1 (*Prm1*). Tnp1 and Prm1 are required for correct nuclear condensation during spermiogenesis (Cho et al., 2001; Zhao et al., 2004). Furthermore, deficiency for *Jhdm2a* impairs transcriptional activation of *Tnp1* and *Prm1* leading to infertility (Okada et al., 2010). Interestingly, the four members of the Jmjd2 family of KDMs in mammals show dual specificity for the removal of H3K9me3/2 and H3K36me3/2 *in vitro* (Fodor et al., 2006; Klose, Kallin, & Zhang, 2006; Whetstine et al., 2006). RNAi depletion of the *C. elegans* homologue *jmjd-2* induces increased levels of H3K9me3 (and H3K36me3 on one end of the X chromosome) in the germline. Germ cells depleted for jmjd-2 showed increased apoptosis and altered DSB repair although they do not harbor defects in pairing and synapsis (Whetstine et al., 2006). This dynamic interplay between the deposition and the erasure of H3K9 methylation is reinforced by the fact that the phenotype of *jmjd-2−/−* animals can be partially rescued by the deletion of HPL-2, the *C. elegans* homologue of HP1 (Black et al., 2010).

Indeed, HP1 is dynamically posttranslationally modified, which can affect its localization toward PCH (Maison et al., 2011) or following DNA damage (Ayoub, Jeyasekharan, Bernal, & Venkitaraman, 2008; Dinant & Luijsterburg, 2009). Moreover, the different HP1 isoforms do not show similar localization, associate with different proteins, and harbor noncompletely redundant functions in various organisms. Thus, HP1 isoforms and their PTMs may constitute additional layer of information, increasing the complexity of genome control by chromatin-modifying enzymes.

# 3. POLYCOMB REPRESSIVE PATHWAYS

## 3.1. Composition and developmental role of PcG proteins

In recent years, the composition and variety of Polycomb complexes in different cell types and organisms have received substantial attention. Many of the core components of PRCs have been duplicated during evolution, but instead of acting as "spare parts" serving redundant functions, these paralogs acquired different developmental roles during their divergence. In addition, there are other components that associate with Polycomb complexes (Polycomb cofactors), which further increase the variety and have an effect on the functionality and the targeting of the complexes in different cell types. Remarkably, the divergence of Polycomb complexes is not a rare event in evolution, since it occurred in plants, flies, and mammals. Below, we will focus on the composition and developmental roles of PRC1 and PRC2 in three model organisms.

### *3.1.1 PRC2*

#### 3.1.1.1 Drosophila

E(z), Su(z)12, Esc/Escl, and Nurf55 are the core PRC2 components in *Drosophila* (see Table 8.2), which together have been isolated as a 600-kDa complex (Müller et al., 2002; Tie, Furuyama, Prasad-Sinha, Jane, & Harte, 2001). Pcl has been identified as part of a bigger Pcl–PRC2 complex (Tie, Prasad-Sinha, Birve, Rasmuson-Lestander, & Harte, 2003), where Pcl functions as an enhancer of the H3K27 trimethylation *in vivo* (Nekrasov et al., 2007). In *Drosophila* larvae, Pcl facilitates the recruitment of PRC2 to chromosomes (Savla, Benes, Zhang, & Jones, 2008). Classical Polycomb phenotypes related to the misregulation of *Hox* genes have been described for all PRC2 members and with the exception of *esc* and *escl*, all homozygous null alleles show larval lethality (Anderson et al., 2011; Birve et al., 2001; Duncan, 1982; Phillips & Shearn, 1990; Struhl & Brower, 1982). According to the *Drosophila* developmental transcriptome project (Gelbart & Emmert, 2011; McQuilton, St Pierre, & Thurmond, 2012), mRNA levels for all *PRC2* genes peak in early embryos, then decline at larval stages and increase in female but not male adults. The only exceptions are the partially redundant paralogs *Esc* and *Escl* with *Esc* predominantly expressed in embryos and *Escl* in late larval/early pupal stages (Kurzhals et al., 2008). In addition, *Esc* and *Escl* mRNAs undergo splicing with different efficiencies and the two proteins

**Table 8.2** PRC2 components

| *M. musculus* | | | *D. melanogaster* | | *A. thaliana* | |
|---|---|---|---|---|---|---|
| *Ezh2*, *Ezh1* | Embryo: *Ezh2* KO is lethal around E5.5–8.5 and displays abnormal gastrulation (O'Carroll et al., 2001). *Ezh1* KO does not have an embryonic phenotype. No embryonic/germ line DKO has been reported yet (Ezhkova et al., 2011). | ESC: Single or double KO do not have a significant effect on the maintenance of ESC, although the H3K27 methylation mark is lost. An effect of the KO is seen upon differentiation of ESC due to failure in extinguishing expression of pluripotency genes (Shen et al., 2008). | *E(z)* | Loss of function of *E(z)* is 100% lethal around end of 3rd instar larva/puparium. Hypomorph mutations result in homeotic transformations and failure of germ-cell development. *E(z)* is a maternal effect gene (Phillips & Shearn, 1990). | *SWN*, *MEA*, *CLF* | Plants lacking *CLF* display pleiotropic effects on leafs and flowers and flower prematurely (Goodrich et al., 1997). *MEA* is a maternal effect gene and loss of function leads to overproliferation of the endosperm, resulting in embryo lethality later in seed development (Grossniklaus, Vielle-Calzada, Hoeppner, & Gagliano, 1998). Mutations in the *SWN* gene do not show significant phenotype but enhance the clf phenotype (Chanvivattana et al., 2004). |
| *Suz12* | Embryo: KO of *Suz12* is lethal around E7.5–E8.5 and gastrulation is not initiated (Pasini, Bracken, | ESC: In *Suz12* deficient cells proliferation and self-renewal are not affected but some lineage-specific genes are upregulated. | *Su(z)12* | Loss of function is lethal around end of 1st/beginning of 2nd instar larva and results in misexpression of *Hox* genes | *VRN2*, *FIS2*, *EMF2* | *FIS2* is a maternal effect gene, which phenocopies the *MEA loss of function* mutation (Luo et al., 1999). *EMF2* mutation leads to |

| | | | | | | |
|---|---|---|---|---|---|---|
| | Jensen, Lazzerini Denchi, & Helin, 2004). | Differentiation of ESC is disturbed in KO cells (Pasini, Bracken, Hansen, Capillo, & Helin, 2007). | | and strong homeotic phenotypes. *Su(z)12* is required in the germline and for suppression of PEV through heterochromatin silencing (Birve et al., 2001). | | premature flowering (Yoshida et al., 2001). *VRN2* is required for the maintenance of *FLC* repression in the process of vernalization (Gendall, Levy, Wilson, & Dean, 2001). |
| *Eed* | Embryo: KO of *Eed* is lethal around E8.5 and displays abnormal gastrulation (Faust, Lawson, Schork, Thiel, & Magnuson, 1998). | ESC: In *Eed*-deficient cells proliferation and self-renewal are mildly affected. Differentiation in the absence of *Eed* is biased toward endoderm (Leeb et al., 2010; Shen et al., 2008, 2009). | *Escl*, *Esc* | *Esc* is needed for segment establishment in the early embryo, but not for maintenance at later stages (Struhl & Brower, 1982). *Escl* mutation does not show a phenotype but enhances the phenotype of esc and leads to homeotic transformations and lethality at 3rd instar larva (Kurzhals, Tie, Stratton, & Harte, 2008; Ohno, McCabe, Czermin, Imhof, & Pirrotta, 2008). | *FIE* | A maternal effect gene phenocopying *MEA* and *EMF2* mutants (Ohad et al., 1999). *FIE* is required for embryo-to-seedling transition if maternal effect is rescued (Bouyer et al., 2011). |

control the enzymatic activity of PRC2 complexes differentially (Ohno et al., 2008). As a consequence, the maternal *Esc* contribution is required for development.

#### 3.1.1.2 Mammals

The mammalian PRC2 homologues are *Ezh1/Ezh2*, *Eed*, *Suz12*, *Rbbp4/Rbbp7*, and *Pcl1/Pcl2/Pcl3* (see Table 8.2). Homozygous mutations for *Ezh2*, *Eed*, or *Suz12* are lethal in early postimplantation development (Faust et al., 1998; O'Carroll et al., 2001; Pasini et al., 2004), while *Pcl2* mutant mice are viable but show posterior transformations as the result of *Hox* gene misregulation (Li et al., 2011). Gene expression data from mouse and human for the two KMT homologues *Ezh1* and *Ezh2* suggest that *Ezh2* is predominant in embryogenesis and in proliferating cells, while *Ezh1* might be more important for postnatal development (Margueron et al., 2008; Shen et al., 2008). Biochemical characterization has identified differences in the activity of Ezh1-PRC2 and Ezh2-PRC2, implying a strong chromatin compaction activity and a comparably weaker H3K27me2/3 KMT activity for Ezh1-PRC2 (Margueron et al., 2008; Shen et al., 2008). We could speculate that the enzymatic activity of PRC2 is more needed early in development, when the chromatin is still "young" and changing, so keeping the right set of genes silent in a dynamic environment would require a more active repressor. On the other hand, a number of studies using *Ezh2* conditional mutants revealed that Ezh2 is required for the terminal somatic cell differentiation and in some cases also for the maintenance of the multipotent or progenitor cell state (reviewed in Aldiri & Vetter, 2012). In most of these cells, *Ezh2* is coexpressed together with *Ezh1*, so it is difficult to assign roles to one or the other. The developmental role of Ezh1-PRC2 is therefore still unclear, as it is either masked by the presence of Ezh2 or is not critical, since homozygous *Ezh1* mutant mice are healthy, fertile, and do not show any transformations (Ezhkova et al., 2011). Another component of PRC2, *Eed*, does not have paralogs in mammals but instead has four isoforms, which are associated with three variants of the complex—PRC2, PRC3, and PRC4. These PRC2 variants show biochemical differences, as PRC2 and PRC4 are suggested to have the canonical H3K27—and an additional H1K26-KMT specificity, the biological significance of which remains unknown (Kuzmichev, Jenuwein, Tempst, & Reinberg, 2004, Kuzmichev et al., 2005).

#### 3.1.1.3 Plants

Homologues of all four core PRC2 members are found in *A. thaliana*: *MEA*, *SWN*, *CLF* (homologues of *Ezh2*); *VRN2*, *FIS2*, *EMF2* (homologues of *Suz12*); the *Rbbp4* homologue *MSI1*; and the *Eed* homologue *FIE* (see Table 8.2). Furthermore, there is a group of three PHD-finger proteins—*VRN5, VIN3*, and *VEL1*, which are considered as functionally related to Pcl because they enhance the KMT activity of the complex (Lucia, Crevillen, Jones, Greb, & Dean, 2008). As expected from the existence of numerous paralogs, there are multiple PRC2 complexes, which are involved in different processes. The development of the female gametophyte and the initiation of embryogenesis are controlled by the MEA–FIS2–MSI1–FIE complex and mutation in any of the four genes is embryonically lethal (Grossniklaus et al., 1998; Köhler et al., 2003; Luo et al., 1999; Ohad et al., 1999). The PRC2 maternal effect phenotype is caused by imprinting defects in the endosperm (functionally analogous to the mammalian placenta) and derepression of MADS box TFs, which leads to hyperproliferation of the endosperm and eventually seed abortion. Furthermore, by expressing a FIE transgene in *fie* mutant plants, it was possible to overcome the maternal requirement for PRC2 and identify pleiotropic phenotypes as the result of derepression of *KNOX* homeobox genes (Katz, Oliva, Mosquna, Hakim, & Ohad, 2004). Another PRC2 variant consisting of CLF/SWN-EMF2/VRN2-MSI1-FIE is involved in the regulation of several key transcription factors (AG, FLC), which control the transition from vegetative-to-reproductive development and the cold-induced flowering response (Coustham et al., 2012; reviewed in Holec & Berger, 2012; Song, Angel, Howard, & Dean, 2012).

### 3.1.2 PRC1

#### 3.1.2.1 Drosophila

The *Drosophila* PRC1 consists of Sce, Pc, Psc, and Ph and similar to PRC2 has been implicated in the *Hox* gene regulation (see Table 8.3). The expression of PRC1 genes throughout fly development resembles the one of PRC2 and mutations lead to classical homeotic phenotypes and embryonic lethality (Breen & Duncan, 1986; Dura, Brock, & Santamaria, 1985; Dura et al., 1987; Graveley et al., 2011; Jürgens, 1985; Lewis, 1947, 1978; McQuilton et al., 2012). More recently, Sce and Psc have been identified as members of a distinct complex, dRING-associated factors (dRAF), containing also the H3K36-demethylase dKDM2 (Lagarou et al., 2008). The *kdm2* mutant allele significantly enhances the *Pc* homeotic phenotype and in the same time rescues the *Trx* and *Ash1* mutations (ASH1 and TRX are H3K36- and H3K4-specific KMTs, respectively). Biochemical analysis revealed that dRAF but not PRC1 is the major complex involved in the ubiquitination of H2AK119, and this activity is directly linked to the removal of the H3K36me2 PTM. This is a striking example of how

**Table 8.3** PRC1 components

| *M. musculus* | | | *D. melanogaster* | | *A. thaliana* | |
|---|---|---|---|---|---|---|
| *Ring1a/Ring1*, *Ring1b/Rnf2* | Embryo: *Ring1b* KO is lethal around E6.5–9.5, gastrulation defects. *Ring1a* KO mice are viable and fertile with abnormalities in the axial skeleton. Conditional DKO in oogenesis has a strong maternal effect and leads to 2-cell arrest (del Mar Lorente et al., 2000; Posfai et al., 2012; Voncken et al., 2003). | ESC: Single *Ring1a* or *Ring1b* KO ESC does not show significant phenotype, while DKO leads to derepression of lineage genes and loss of ESC identity. H2AK119ub is globally absent in DKO (Endoh et al., 2008; Leeb & Wutz, 2007, 2010; Van der Stoop et al., 2008). | *Sce/ dRING* | Loss of function is lethal around 1st instar larva. *Hox* genes are derepressed in Sce mutants (Breen & Duncan, 1986). | *AtRING1a*, *AtRING1b* | Single KO plants do not exhibit significant phenotypes, while DKO leads to derepression of *KNOX* genes, homeotic transformations and ectopic meristemic formation. AtRING1a/b interacts with the plant HP1 homologue LHP1 (Xu & Shen, 2008). |
| *Pcgf6*, *Pcgf5*, *Pcgf4/Bmi1*, *Pcgf3*, *Pcgf2/ Mel18*, *Pcgf1* | Embryo: Only *Pcgf2* and *Pcgf4* have been functionally studied and single KO of either *Pcgf2* or *Pcgf4* results in posterior transformations of axial skeleton. DKO is lethal around E9.5 due to failure of somite formation and | | *Psc* | *Psc* KO is embryonic lethal, affecting the embryonic body pattern and most severely the head formation. Derepression of *Hox* genes in Psc mutants (Beh, Colwell, & Francis, 2012; Jürgens 1985). | *AtBMI1c*, *AtBMI1b*, *AtBMI1a* | Single KO mutants do not exhibit significant phenotypes. DKO for *AtBMI1a* and *AtBMI1b* leads to aberrant cell differentiation and early embryonic arrest, as well as upregulation of the stem-cell regulators *WUS*, *STM*, *FUS3*, *LEC1*, *and* |

| | | | | | | |
|---|---|---|---|---|---|---|
| | organogenesis (Akasaka et al., 2001). | | | | | *WOX5* (Bratzel, López-Torrejón, Koch, Del Pozo, & Calonje, 2010). |
| *Cbx8/Pc3*, *Cbx7*, *Cbx6*, *Cbx4/Pc2*, *Cbx2/M33* | Embryo: From the five homologues, only *Cbx2* has been functionally studied and loss of function results in perinatal lethality, as well as male-to-female sex reversal, meiotic defects and chromosome instability (Baumann & De La Fuente, 2011; Katoh-Fukui et al., 1998). | | *Pc* | *Pc* loss of function is embryonic lethal and giving a strong homeotic phenotype, which is enhanced by overexpression of *BX-C*. Derepression of *Hox* genes in Pc mutants (Lewis, 1978). | *LHP1/TLF2* | A homologue of fly and mouse *HP1*, but localizing to euchromatin and interacting with the other PRC1 members in plants. Mutant *lhp1* plants show pleiotropic phenotypes with overall size reduction, changes in leaf morphology, early flowering. *LHP1* is also required for the repression of *FLC* in the process of vernalization (Gaudin et al., 2001; Sung et al., 2006). |

changing the components of the complex can dramatically alter the enzymatic activity of Sce. In fact, the discovery of the Polycomb repressive deubiquitinase complex "PR-DUB" revealed that the ubiquitination activity of dRAF/PRC1 needs to be counterbalanced in order to prevent repression of unintended targets (Scheuermann et al., 2010). PR-DUB consists of the deubiquitinating enzyme Calypso and the PcG protein ASX that co-occupy Polycomb target genes. The functional combination of dRAF and PR-DUB allows precise regulation of the dosage of Polycomb repression (Scheuermann et al., 2010). Genetic, genome-wide expression, and chromatin analyses further show that SCE, PSC, and PR-DUB, regulating H2A monoubiquitination levels, are only required for repression of a subset of PRC1 target genes. Repression of other targets depends on the function of the PSC paralog Su(z)2 and the Ph protein, possibly by mediating chromatin compaction (Gutiérrez et al., 2012).

#### 3.1.2.2 Mammals

The PRC1 members in mammals have undergone multiple duplications during evolution, and there are six homologues of *Drosophila Psc* (*Pcgf1/2/3/4/5/6*), five homologues of *Pc* (*Cbx2/4/6/7/8*), three homologues of *Ph* (*Phc1/2/3*), and two homologues of *Sce* (*Ring1/Rnf2*). The only gene that has been shown to be embryonic lethal at early postimplantation is *Rnf2* (see Table 8.3), while *Pcgf2*, *Pcgf4*, *Cbx2*, and *Phc1* show perinatal lethality and/or homeotic transformations (Akasaka et al., 1996; Coré et al., 1997; Katoh-Fukui et al., 1998; Takihara et al., 1997; van der Lugt et al., 1994). *Phc2* and *Ring1* mutant mice are healthy and fertile with minor homeotic transformations in the anterior–posterior axis (del Mar Lorente et al., 2000; Isono et al., 2005). Furthermore, two studies analyzing the role of *Cbx2* point to a function of PRC1 in sex determination and meiotic regulation (Baumann & De La Fuente, 2011; Katoh-Fukui et al., 1998). The fact that only *Rnf2* is indispensable for embryonic development could be explained with the presence of redundant paralogs. This is evident from several studies of DKO mice, which show dramatically enhanced phenotypes compared to the single mutations. In the first study, Akasaka and colleagues showed that double deficiency for *Pcgf2* and *Pcgf4* results in postimplantation lethality during somite formation and organogenesis as a result of misregulated *Hox* gene expression (Akasaka et al., 2001). In another study, Posfai and colleagues removed both *Ring1* and *Rnf2* in the female germline and observed a strong maternal effect leading to a 2-cell embryonic arrest, that is, the stage before zygotic genome activation (Posfai et al., 2012). The authors profiled the transcriptome of

*Ring1/Rnf2*-deficient fully-grown oocytes and revealed massive gene misregulation. Furthermore, they presented evidence that Ring1/Rnf2 are responsible for global H2AK119 ubiquitination in mammalian oocytes. With a similar genetic approach, Lapthanasupkul et al. examined the function of *Ring1* and *Rnf2* in mesenchymal stem cells and revealed that upon DKO, the proliferation of the stem cells and the differentiation process are severely affected (Lapthanasupkul et al., 2012). A number of developmental regulator genes that show a restricted expression pattern in normal tissue were broadly expressed in the mutant tissue. Interestingly, these authors reported that the removal of Ring1/Rnf2 resulted in a massive depletion of Kdm2b protein. The latter has been identified as a member of the human BCOR complex, which is the homologue of dRAF (Gearhart, Corcoran, Wamstad, & Bardwell, 2006). It remains to be shown whether mammalian BCOR has specialized as the major H2AK119 ubiquitin ligase and whether the H3K36 demethylase activity is coupled to Polycomb-mediated repression. A hint into this direction presents the recently published large-scale genomics data from the human ENCODE project, which shows that H3K27me3 and H3K36me3 are largely mutually exclusive (Consortium T.E.P., 2012; Voigt et al., 2012). The study that identified the fly PR-DUB complex also showed that the human homologues of *Calypso* and *ASX* (*BAP1* and *Asxl1*, respectively) form a complex and have deubiquitinase activity *in vitro*, suggesting a conserved molecular function for this Polycomb-related complex (Scheuermann et al., 2010). Indeed, homozygous *Asxl1*-deficient mice have homeotic transformations and die shortly after birth (Fisher et al., 2010). Finally, Gao and colleagues addressed the variety of mammalian PRC1-like complexes in a systematic way and identified four major PRC1 subtypes, which differ by the presence of different Pcgf and Cbx homologues and target sequences (Gao et al., 2012). Tavares and colleagues identified another PRC1 complex containing RYBP, Rnf2, and Mel18/Pcgf2 that is highly catalytically active and is targeted to chromatin in an unknown manner, independently of *Eed* function and H3K27me3 (Tavares et al., 2012). *Rybp* is essential for gastrulation (Pirity, Locker, & Schreiber-Agus, 2005) and *in vitro* differentiation of ESCs. Rybp is not required for self-renewal of ESCs and mediates repression of certain endogenous retroviruses and preimplantation and germline genes (Hisada et al., 2012). The extensive variety in complexes potentially underlies different biological roles in specific developmental contexts. To address this issue, future studies will have to target individual Pcgf, Cbx, and Rybp/Yaf2 components rather than the common Ring1 and Rnf2.

#### 3.1.2.3 Plants

The composition of PRC1 complexes in *Arabidopsis* has not been studied in a systematic way and to the depth as in mammalian systems or in *Drosophila*, mostly because the PRC1 homologues have been identified only recently in *Arabidopsis* (Sanchez-Pulido, Devos, Sung, & Calonje, 2008). A plant PRC1-like complex has been characterized by Xu and Shen (2008), showing interaction between AtRING1a/AtRING1b and the H3K27me3-binding CD protein LHP1 (Turck et al., 2007). While single *AtRING1a* or *AtRING1b* homozygous mutants did not show any abnormal phenotype, double mutant plants displayed homeotic transformations and meristem stem cell phenotypes similar to the ones observed in *lhp1*−/− and *clf*−/− plants (Gaudin et al., 2001; Goodrich et al., 1997; Larsson, Landberg, & Meeks-Wagner, 1998).The loss of *AtRING1a/b* caused derepression of *KNOX* genes promoting meristem proliferation, whereas H3K27me3 levels at the promoters of affected genes did not change, which shows that PRC2 function was still intact but transcriptional repression requires AtRING1a/b (Xu & Shen, 2008). The latter observation was confirmed in a study by Bratzel and colleagues who examined the role of *AtBMI1a* and *AtBMI1b*, two of the three identified Pcgf homologues in *Arabidopsis* (Bratzel et al., 2010; Sanchez-Pulido et al., 2008). Plants deficient for both *AtBMI1a* and *AtBMI1b* show a variety of phenotypes, ranging from complete developmental arrest of early seedlings to more mild effects related to cell differentiation and formation of callus-like structures. This phenotypes correspond to the observed upregulation of stem cell regulators such as *WUS, STM, FUS3, LEC1*, and *WOX5*. Furthermore, Bratzel and coworkers identified for the first time that *Arabidopsis* PRC1 proteins AtBMI1a/B and AtRING1a/b are involved in monoubiquitination of H2A.1 (the *Arabidopsis* H2A homologue that retained the lysine substrate at position 121) and are associated in a complex with LHP1 and a nonconserved plant-specific protein EMF1. Interestingly, EMF1 has been implicated with a dual role as being part of both PRC1 and PRC2 complexes in *Arabidopsis* and repressing two independent sets of genes (Kim, Lee, Eshed-Williams, Zilberman, & Sung, 2012). Plants depleted of *emf1* show homeotic and flowering phenotypes (Aubert et al., 2001), as well as derepression of PRC2 targets like AG, FLC, and SEP1-3 but not the imprinted *PHE* genes.

## 3.2. Polycomb regulation

As evident from the many examples given above, PcG proteins are required throughout the life cycle of eukaryotes: during embryogenesis, for somatic cell differentiation, in germline development, as well as in disease (Albert & Helin, 2010; Richly, Aloia, & Di Croce, 2011). A common feature of all these biological processes is the transition from one developmental state to another, which is accompanied by major changes in gene expression. Original genetic studies in various species demonstrated major roles of Polycomb group proteins in the maintenance of stable repression during differentiation (e.g., *Hox* genes in flies and mammals and *MADS* box genes in plants). More recently, however, multiple chromatin immunoprecipitation (ChIP) experiments in different organisms have revealed thousands of putative Polycomb targets that are controlled in a cell-type specific manner (Bouyer et al., 2011; Consortium T.E.P., 2012; Endoh et al., 2012; Gerstein et al., 2010; Gutiérrez et al., 2012; Kwong et al., 2008; Mohn et al., 2008; Roy et al., 2010). This data allows us to address the divergent developmental roles of Polycomb by classifying the Polycomb targets and considering the specificity of distinct PRC complexes. Finally, we will discuss the various mechanisms for targeting Polycomb to the chromatin.

### 3.2.1 Dynamics of core and specific polycomb target sets

On the basis of chromatin profiling experiments, Polycomb targets in mammals can be classified into several classes: (I) core Polycomb target loci, co-occupied by PRC1 and PRC2 proteins and labeled by H3K27me3 and H2AK119ub1; (II) PRC2-only targets, marked by H3K27 methylation and sometimes co-occupied by core PRC2 members; (III) PRC1-only targets, bound by one or more PRC1 members and harboring H2AK119ub1. There is a fourth class that includes targets bound by PRC1 or PRC2 members acting in a Polycomb-independent manner (for instance, as part of other complexes), which we will not consider here. Although classification may in part relate to differences in immunoprecipitation efficiencies and definitions of enrichment threshold values, genes in classes I and II do encode for different biological functions and respond differently in ESCs upon induction of differentiation (Ku et al., 2008). H2AK119ub1 mediated by Ring1 and Rnf2 is required for repression of class I genes as well as for the maintenance of ESC identity. PRC1 activity is, however, not required for binding of PRC1 proteins to target genes, neither for compaction of, for example, the HoxB cluster in ESCs (Endoh et al., 2012). Generally, these recent results are in accordance with the first genome-wide study that compared PRC1/PRC2 co-occupancy in ESCs and identified common 512 targets (Boyer et al., 2006), from which 291 overlap with the 510 class I targets found by Endoh and coworkers (Endoh et al., 2012). In summary, PRC1 and particularly the E3-ligase activity of its Ring finger proteins are required in ESCs and during mouse development for silencing of a rather small but important set of evolutionary conserved genes encoding for developmental regulators.

Interestingly, from the thousands of Polycomb targets described, only several hundred genes belong to class I marked by PRC1 and PRC2 loci. A similar observation was made in *Drosophila*. Several independent genome-wide experiments identified around 200–400 Polycomb targets co-occupied by several PcG members (Gutiérrez et al., 2012; Schuettengruber et al., 2009; Schwartz et al., 2006), which roughly correspond to 20% of all H3K27me3 target sites. In a systematic genome-wide ChIP study, analyzing the chromatin localization of 53 proteins in Kc167 embryonic cells, Filion and colleagues identified five distinct types of chromatin (Filion et al., 2010). Class I target genes were represented by one chromatin type. H3K27me3 only states were found in one of the two actively transcribed chromatin types, suggesting that class II genes might be partially associated with active chromatin. Furthermore, upon removal of Polycomb, the number of upregulated genes (derepressed targets) is significantly lower than the number of targets bound, and most of the misexpressed genes are "classical" Polycomb targets or class I targets (e.g., *Hox* genes, Wnt-, Fgf-, Tgf-signaling genes, and other developmental regulators) (Bracken, Dietrich, Pasini, Hansen, & Helin, 2006; Ezhkova et al., 2011; Posfai et al., 2012). Therefore, the role of class II and III target loci remains to be determined. Are they serving as a buffer of regulation or perhaps as a structural component of the chromatin landscape?

### 3.2.2 Mechanisms of polycomb recruitment

Over the years, a lot of effort has been put into revealing the targeting mechanism of PRCs and propagation of the modified state. The classical model of epigenetic inheritance of the H3K9 methylated state by HP1 proteins recognizing the methylated histone as well as interacting with H3K9 HMTs fueled the field (Jenuwein & Allis, 2001). Seminal work by Margueron et al. demonstrated that binding of the PRC2 component ESC/EED via its WD40 propeller to H3K27me3 stimulates E(Z)/EZH2 to catalyze trimethylation of the unmodified substrate (Margueron et al., 2009). Together with work by Hansen et al. (2008), these data provided, in principle, a mechanism for epigenetic inheritance. More recent work indicates that reestablishment of H3K27me3 levels (and also for H3K9me3) does not occur during replication but gradually during subsequent cell cycle stages (Xu, Wang, Chen, & Zhu, 2012). Importantly, instead of newly incorporated histones, parental histones with intermediate methylation states are preferentially used as substrate. Together, these data suggest that number of modified nucleosomes within a region likely affects the efficiency of propagation (Brykczynska et al., 2010; Xu et al., 2012). The catalytic activity of the PRC2 complex is inhibited by H3K4 and H3K36 trimethylation, when residing on the same histone tail in the nucleosome (Schmitges et al., 2011; Voigt et al., 2012; Yuan et al., 2011), potentially providing means for inhibition of spreading of the PRC2 repressed state by Trithorax group proteins. Furthermore, independent of the stimulatory effect of preexisting H3K27me3, PRC2 activity is stimulated by high nucleosomal density that is sensed by the VEFS-box domain of the Su(z)12 protein interacting with amino acids 35–42 of H3 protruding from the nucleosome core. For PRC1, Psc interacts with nucleosomes and self-interacts in cell-fee replication systems thereby forming oligomeric structures. Since some Psc-chromatin contacts are dynamic while others are stable, Psc may enable inheritance of PRC1 on chromatin during replication (Lo et al., 2012).

*De novo* targeting and propagation may also be in part mediated by interactions of PRC proteins with the underlying DNA. Indeed, fly PcG proteins interact with specific DNA-binding factors such as PHO that associate with complex DNA elements termed Polycomb response elements (PREs) (Ringrose & Paro, 2007). Regions around such sites are marked by H3K27me3 and are co-occupied by Pc, likely due to its CD that has a high binding affinity to H3K27me3 (Fischle et al., 2003; Schuettengruber et al., 2009). In *Drosophila* embryos, E(z) and Pc proteins, but not H3K27me3, have been reported to be associated with PREs on recently replicated DNA, suggesting that these proteins may be directly involved in epigenetic heritability. In mammals, the role of transcription factors is less understood and only two PRE-like sequences have been identified up to date (Sing et al., 2009; Woo, Kharchenko, Daheron, Park, & Kingston, 2010). The role of PHO in targeting is probably not conserved in mammals since YY1, the mammalian orthologue, is not localized at PRC target genes in ESCs (Ku et al., 2008; Mendenhall et al., 2010). PRC proteins generally localize at CpG-rich sequences suggesting a possible function of transcription factors binding within such elements (Ku et al., 2008; Zheng, Zhao, & Mehler, 2009). Arnold and colleagues revealed a role for Rest in H3K27me3 establishment at specific target sequences in neuronal progenitor cells during differentiation of ESCs (Arnold et al., 2012). Sequences containing Rest and Snail transcription factors are sufficient for the recruitment of H3K27me3 at targeted transgenic insertion sites suggesting that transcription factors can target PRC2 for gene repression which is consistent with reported biochemical interactions between REST and PcG proteins (Dietrich et al., 2012; Ren & Kerppola, 2011).

Finally, both PRC2 and PRC1 complexes are independently required for contraction of the *Kcnq1* imprint cluster and imprinted gene silencing during early mouse development (Terranova et al., 2008), as well for the formation of facultative heterochromatin at one of the two X-chromosomes in female mammalian cells (Plath et al., 2003). In these processes, noncoding RNAs (ncRNAs) may target PRC complexes to chromatin. Indeed, over the past few years an ever-increasing amount of data has been accumulating on the link between ncRNA and Polycomb repression. A prerequisite for ncRNA-mediated targeting is RNA-binding affinity by PRC members, and this has been (so far) reported for the PRC2 members Ezh2 and Suz12 (Guil et al., 2012; Kanhere et al., 2010; Ng, Johnson, & Stanton, 2012), as well as for PRC1 member Cbx7 (Yap et al., 2010). Until recently, only a few of the known ncRNAs have been functionally analyzed and prominent examples linked to PcG targeting are the HOTAIR ncRNA responsible for the silencing of the HoxD cluster in mammals (Rinn et al., 2007; Tsai et al., 2010), COLDAIR/COOLAIR required for the cold-induced silencing of the flowering repressor FLC in *Arabidopsis* (reviewed in Song et al., 2012), as well as Polycomb/Trithorax-related ncRNAs in flies (Hekimoglu & Ringrose, 2009). A recent study in mouse ESC identified the "Polycomb transcriptome" consisting of almost 10,000 PRC2-bound RNAs (Zhao et al., 2010). Another report focused on a subclass of ncRNAs and found 24 of the 226 lincRNAs (large intergenic ncRNAs) in ESC to be bound by PcG proteins (Guttman et al., 2011). Ng et al. described lincRNA-dependent PcG recruitment in human ESC (Ng et al., 2012). The molecular mechanisms underlying target selection in *cis* and in *trans* in relation to timing of ncRNA expression remain little understood. Interestingly, HOTAIR is able to bind to PRC2 and a REST complex, also containing LSD1 and CoREST (Tsai et al., 2010), suggesting scaffold functions for ncRNAs bridging DNA-binding factors and PRC2. In all, future work is required to determine the relative contributions of transcription factors, ncRNAs, and recognition of existing chromatin states in the *de novo* formation versus maintenance of Polycomb gene repression.

## 4. CONCLUSION

One role of chromatin silencing mechanisms is in the maintenance of repressed states through cell division. But this does not mean that this is some rigid regulatory mechanism, on the contrary, there is a need of plasticity to support the development of multicellular organisms. One way of achieving plasticity is through the interconnection of pathways. The two major chromatin-based silencing mechanisms that we reviewed above share many common ways of targeting based on protein interaction, ncRNA, or recognition of PTM on histones (Fig. 8.1). However, they are often considered as distinct and mutually exclusive in regard to their distribution in the nucleus and genome wide (de Wit et al., 2007). As mentioned in the beginning, studies in *T. thermophila* would argue that these pathways have common ancestral functions (Liu et al., 2007). Could these two silencing mechanisms be independent of each other? HP1 is found in an atypical PRC1 complex

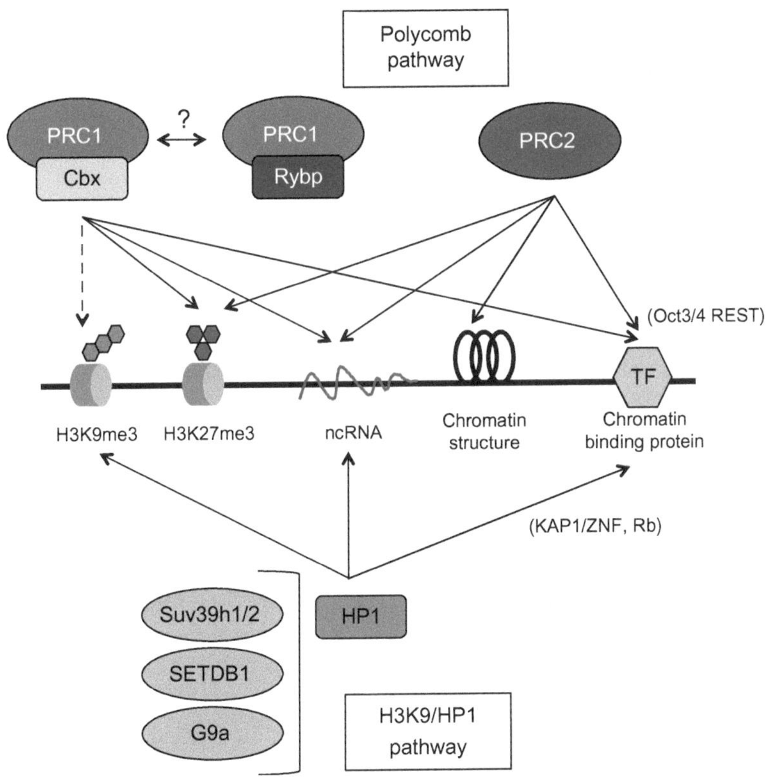


**Figure 8.1** Overview of targeting of Polycomb complexes and H3K9 KMT/HP1 proteins. Polycomb and H3K9 KMT/HP1 complexes can be targeted to chromatin through many diverse ways implicating direct interaction with modified histones, noncoding RNA (ncRNA), chromatin structure, and DNA-binding protein like transcription factors or zinc-finger-containing protein (ZNF). Solid lines indicate interactions. Dotted lines indicate potential interactions. Some examples are indicated between brackets. See text for details. (See Color Insert.)

(Gao et al., 2012; Qin et al., 2012), and SETDB1 represses some sets of developmental genes also targeted by the PcG pathway (Bilodeau et al., 2009). The PRC1 subunits Cbx2 and Cbx7 have affinity for H3K27me3 and H3K9me3 *in vitro* (Bernstein et al., 2006; Kaustov et al., 2011) and were suggested to associate with Suv39h KMT (Li et al., 2010; Sewalt et al., 2002). Suv39h KMT could then influence PRC1 targeting as suggested by Yang and colleagues (Yang et al., 2011), who showed that Cbx4/hPC2 localization in PcG is dependent on Suv39h1 in quiescent cells. Interestingly, in the zygote, PRC1 is prevented from binding to maternal PCH in a Suv39h2-dependent manner (Puschendorf et al., 2008). Many of these enzymes and their relative marks are present in germ cells of both sex and could therefore influence the outcome of embryogenesis (Brykczynska et al., 2010; Hammoud et al., 2009; Posfai et al., 2012). Though the concept of transgenerational inheritance is still under debate (Gill, Erkek, & Peters, 2012), the interplay between these two major chromatin silencing pathway would be an interesting way to accommodate chromatin plasticity during developmental transition within germ cell maturation and the developing embryo.

## ACKNOWLEDGMENTS

M. T. greatly acknowledges his EMBO long-term fellowship and P. N. his Boehringer Ingelheim Fonds PhD fellowship. Research in the Peters group is supported by the Novartis Research Foundation, the Swiss National Science Foundation (NRP 63—Stem Cells and Regenerative Medicine), SystemsX.ch (Cell plasticity), the Japanese Swiss Science and Technology Cooperation Program, and the EMBO Young Investigator Program.

## REFERENCES

Abe, K., Naruse, C., Kato, T., Nishiuchi, T., Saitou, M., & Asano, M. (2011). Loss of heterochromatin protein 1 gamma reduces the number of primordial germ cells via impaired cell cycle progression in mice. *Biology of Reproduction, 85,* 1013–1024.

Abel, J., Eskeland, R., Raffa, G. D., Kremmer, E., & Imhof, A. (2009). Drosophila HP1c is regulated by an auto-regulatory feedback loop through its binding partner Woc. *PLoS One, 4,* e5089.

Akasaka, T., Kanno, M., Balling, R., Mieza, M. A., Taniguchi, M., & Koseki, H. (1996). A role for mel-18, a Polycomb group-related vertebrate gene, during theanteroposterior specification of the axial skeleton. *Development, 122,* 1513–1522.

Akasaka, T., van Lohuizen, M., van der Lugt, N., Mizutani-Koseki, Y., Kanno, M., Taniguchi, M., et al. (2001). Mice doubly deficient for the Polycomb Group genes Mel18 and Bmi1 reveal synergy and requirement for maintenance but not initiation of Hox gene expression. *Development, 128,* 1587–1597.

Albert, M., & Helin, K. (2010). Histone methyltransferases in cancer. *Seminars in Cell and Developmental Biology, 21,* 209–220.

Alder, O., Lavial, F., Helness, A., Brookes, E., Pinho, S., Chandrashekran, A., et al. (2010). Ring1B and Suv39h1 delineate distinct chromatin states at bivalent genes during early mouse lineage commitment. *Development, 137*, 2483–2492.

Aldiri, I., & Vetter, M. L. (2012). PRC2 during vertebrate organogenesis: A complex in transition. *Developmental Biology, 367*, 91–99.

Allshire, R. C., Nimmo, E. R., Ekwall, K., Javerzat, J. P., & Cranston, G. (1995). Mutations derepressing silent centromeric domains in fission yeast disrupt chromosome segregation. *Genes and Development, 9*, 218–233.

Anderson, A. E., Karandikar, U. C., Pepple, K. L., Chen, Z., Bergmann, A., & Mardon, G. (2011). The enhancer of trithorax and polycomb gene Caf1/p55 is essential for cell survival and patterning in Drosophila development. *Development, 138*, 1957–1966.

Arnold, P., Schöler, A., Pachkov, M., Balwierz, P., Jørgensen, H., Stadler, M. B., et al. (2012). Modeling of epigenome dynamics identifies transcription factors that mediate Polycomb targeting. *Genome Research, 23*, 60–73.

Aubert, D., Chen, L., Moon, Y. H., Martin, D., Castle, L. A., Yang, C. H., et al. (2001). EMF1, a novel protein involved in the control of shoot architecture and flowering in Arabidopsis. *The Plant Cell, 13*, 1865–1875.

Aucott, R., Bullwinkel, J., Yu, Y., Shi, W., Billur, M., Brown, J. P., et al. (2008). HP1-beta is required for development of the cerebral neocortex and neuromuscular junctions. *The Journal of Cell Biology, 183*, 597–606.

Ayoub, N., Jeyasekharan, A. D., Bernal, J. A., & Venkitaraman, A. R. (2008). HP1-beta mobilization promotes chromatin changes that initiate the DNA damage response. *Nature, 453*, 682–686.

Ayyanathan, K., Lechner, M. S., Bell, P., Maul, G. G., Schultz, D. C., Yamada, Y., et al. (2003). Regulated recruitment of HP1 to a euchromatic gene induces mitotically heritable, epigenetic gene silencing: A mammalian cell culture model of gene variegation. *Genes and Development, 17*, 1855–1869.

Azuara, V., Perry, P., Sauer, S., Spivakov, M., Jørgensen, H. F., John, R. M., et al. (2006). Chromatin signatures of pluripotent cell lines. *Nature Cell Biology, 8*, 532–538.

Bannister, A. J., Zegerman, P., Partridge, J. F., Miska, E. A., Thomas, J. O., Allshire, R. C., et al. (2001). Selective recognition of methylated lysine 9 on histone H3 by the HP1 chromo domain. *Nature, 410*, 120–124.

Baumann, C., & De La Fuente, R. (2011). Role of Polycomb group protein Cbx2/M33 in meiosis onset and maintenance of chromosome stability in the mammalian germline. *Genes, 2*, 59–80.

Baumbusch, L. O., Thorstensen, T., Krauss, V., Fischer, A., Naumann, K., Assalkhou, R., et al. (2001). The *Arabidopsis thaliana* genome contains at least 29 active genes encoding SET domain proteins that can be assigned to four evolutionarily conserved classes. *Nucleic Acids Research, 29*, 4319–4333.

Beh, L. Y., Colwell, L. J., & Francis, N. J. (2012). A core subunit of Polycomb repressive complex 1 is broadly conserved in function but not primary sequence. *Proceedings of the National Academy of Sciences of the United States of America, 109*(18), E1063–71.

Bernstein, E., Duncan, E. M., Masui, O., Gil, J., Heard, E., & Allis, C. D. (2006). Mouse polycomb proteins bind differentially to methylated histone H3 and RNA and are enriched in facultative heterochromatin. *Molecular and Cellular Biology, 26*, 2560–2569.

Bilodeau, S., Kagey, M. H., Frampton, G. M., Rahl, P. B., & Young, R. A. (2009). SetDB1 contributes to repression of genes encoding developmental regulators and maintenance of ES cell state. *Genes and Development, 23*, 2484–2489.

Birve, A., Sengupta, A. K., Beuchle, D., Larsson, J., Kennison, J. A., Rasmuson-Lestander, A., et al. (2001). Su(z)12, a novel Drosophila Polycomb group gene that is conserved in vertebrates and plants. *Development, 128*, 3371–3379.

Black, J. C., Allen, A., Van Rechem, C., Forbes, E., Longworth, M., Tschöp, K., et al. (2010). Conserved antagonism between JMJD2A/KDM4A and HP1γ during cell cycle progression. *Molecular Cell*, *40*, 736–748.

Bouyer, D., Roudier, F., Heese, M., Andersen, E. D., Gey, D., Nowack, M. K., et al. (2011). Polycomb repressive complex 2 controls the embryo-to-seedling phase transition. *PLoS Genetics*, 7, e1002014.

Boyer, L. A., Plath, K., Zeitlinger, J., Brambrink, T., Medeiros, L. A., Lee, T. I., et al. (2006). Polycomb complexes repress developmental regulators in murine embryonic stem cells. *Nature*, *441*, 349–353.

Bracken, A. P., Dietrich, N., Pasini, D., Hansen, K. H., & Helin, K. (2006). Genome-wide mapping of Polycomb target genes unravels their roles in cell fate transitions. *Genes and Development*, *20*, 1123–1136.

Braig, M., Lee, S., Loddenkemper, C., Rudolph, C., Peters, A. H. F. M., Schlegelberger, B., et al. (2005). Oncogene-induced senescence as an initial barrier in lymphoma development. *Nature*, *436*, 660–665.

Bratzel, F., López-Torrejón, G., Koch, M., Del Pozo, J. C., & Calonje, M. (2010). Keeping cell identity in Arabidopsis requires PRC1 RING-finger homologs that catalyze H2A monoubiquitination. *Current Biology*, *20*, 1853–1859.

Breen, T. R., & Duncan, I. M. (1986). Maternal expression of genes that regulate the bithorax complex of *Drosophila melanogaster*. *Developmental Biology*, *118*, 442–456.

Brower-Toland, B., Riddle, N. C., Jiang, H., Huisinga, K. L., & Elgin, S. C. R. (2009). Multiple SET methyltransferases are required to maintain normal heterochromatin domains in the genome of *Drosophila melanogaster*. *Genetics*, *181*, 1303–1319.

Brown, J. P., Bullwinkel, J., Baron-Lühr, B., Billur, M., Schneider, P., Winking, H., et al. (2010). HP1gamma function is required for male germ cell survival and spermatogenesis. *Epigenetics and Chromatin*, *3*, 9.

Brykczynska, U., Hisano, M., Erkek, S., Ramos, L., Oakeley, E. J., Roloff, T. C., et al. (2010). Repressive and active histone methylation mark distinct promoters in human and mouse spermatozoa. *Nature Structural and Molecular Biology*, *17*, 679–687.

Bulut-Karslioglu, A., Perrera, V., Scaranaro, M., de la Rosa-Velazquez, I. A., van de Nobelen, S., Shukeir, N., et al. (2012). A transcription factor-based mechanism for mouse heterochromatin formation. *Nature Structural and Molecular Biology*, *19*, 1023–1030.

Cammas, F., Janoshazi, A., Lerouge, T., & Losson, R. (2007). Dynamic and selective interactions of the transcriptional corepressor TIF1 beta with the heterochromatin protein HP1 isotypes during cell differentiation. *Differentiation*, *75*, 627–637.

Campos, E. I., & Reinberg, D. (2009). Histones: Annotating chromatin. *Annual Review of Genetics*, *43*, 559–599.

Chanvivattana, Y., Bishopp, A., Schubert, D., Stock, C., Moon, Y.-H., Sung, Z. R., & Goodrich, J. (2004). Interaction of Polycomb-group proteins controlling flowering in *Arabidopsis*. *Development*, *131*, 5263–5276.

Checchi, P. M., & Engebrecht, J. (2011). *Caenorhabditis elegans* histone methyltransferase MET-2 shields the male X chromosome from checkpoint machinery and mediates meiotic sex chromosome inactivation. *PLoS Genetics*, 7, e1002267.

Cheutin, T., McNairn, A. J., Jenuwein, T., Gilbert, D. M., Singh, P. B., & Misteli, T. (2003). Maintenance of stable heterochromatin domains by dynamic HP1 binding. *Science*, *299*, 721–725.

Chin, H. G., Estève, P.-O., Pradhan, M., Benner, J., Patnaik, D., Carey, M. F., et al. (2007). Automethylation of G9a and its implication in wider substrate specificity and HP1 binding. *Nucleic Acids Research*, *35*, 7313–7323.

Cho, S., Park, J. S., & Kang, Y.-K. (2011). Dual functions of histone-lysine N-methyltransferase Setdb1 protein at promyelocytic leukemia-nuclear body (PML-NB):

Maintaining PML-NB structure and regulating the expression of its associated genes. *The Journal of Biological Chemistry, 286*, 41115–41124.
Cho, S., Park, J. S., Kwon, S., & Kang, Y.-K. (2012). Dynamics of Setdb1 expression in early mouse development. *Gene Expression Patterns, 12*, 213–218.
Cho, C., Willis, W. D., Goulding, E. H., Jung-Ha, H., Choi, Y. C., Hecht, N. B., et al. (2001). Haploinsufficiency of protamine-1 or -2 causes infertility in mice. *Nature Genetics, 28*, 82–86.
Clough, E., Moon, W., Wang, S., Smith, K., & Hazelrigg, T. (2007). Histone methylation is required for oogenesis in Drosophila. *Development, 134*, 157–165.
Consortium T.E.P., (2012). An integrated encyclopedia of DNA elements in the human genome. *Nature, 489*, 57–74.
Coré, N., Bel, S., Gaunt, S. J., Aurrand-Lions, M., Pearce, J., Fisher, A., et al. (1997). Altered cellular proliferation and mesoderm patterning in Polycomb-M33-deficient mice. *Development, 124*, 721–729.
Coustham, V., Li, P., Strange, A., Lister, C., Song, J., & Dean, C. (2012). Quantitative modulation of polycomb silencing underlies natural variation in vernalization. *Science, 337*, 584–587.
Couteau, F., Guerry, F., Muller, F., & Palladino, F. (2002). A heterochromatin protein 1 homologue in *Caenorhabditis elegans* acts in germline and vulval development. *EMBO Reports, 3*, 235–241.
Dambacher, S., Hahn, M., & Schotta, G. (2010). Epigenetic regulation of development by histone lysine methylation. *Heredity, 105*, 24–37.
de Wit, E., Greil, F., & van Steensel, B. (2007). High-resolution mapping reveals links of HP1 with active and inactive chromatin components. *PLoS Genetics, 3*, e38.
del Mar Lorente, M., Marcos-Gutiérrez, C., Pérez, C., Schoorlemmer, J., Ramírez, A., Magin, T., et al. (2000). Loss- and gain-of-function mutations show a polycomb group function for Ring1A in mice. *Development, 127*, 5093–5100.
Dietrich, N., Lerdrup, M., Landt, E., Agrawal-Singh, S., Bak, M., Tommerup, N., et al. (2012). REST-mediated recruitment of polycomb repressor complexes in mammalian cells. *PLoS Genetics, 8*, e1002494.
Dinant, C., & Luijsterburg, M. S. (2009). The emerging role of HP1 in the DNA damage response. *Molecular and Cellular Biology, 29*, 6335–6340.
Dodge, J. E., Kang, Y.-K., Beppu, H., Lei, H., & Li, E. (2004). Histone H3-K9 methyltransferase ESET is essential for early development. *Molecular and Cellular Biology, 24*, 2478–2486.
Dong, K. B., Maksakova, I. A., Mohn, F., Leung, D., Appanah, R., Lee, S., et al. (2008). DNA methylation in ES cells requires the lysine methyltransferase G9a but not its catalytic activity. *The EMBO Journal, 27*, 2691–2701.
Duncan, I. M. (1982). Polycomblike: A gene that appears to be required for the normal expression of the bithorax and antennapedia gene complexes of *Drosophila melanogaster*. *Genetics, 102*, 49–70.
Dura, J. M., Brock, H. W., & Santamaria, P. (1985). Polyhomeotic: A gene of *Drosophila melanogaster* required for correct expression of segmental identity. *Molecular and General Genetics, 198*, 213–220.
Dura, J.-M., Randsholt, N. B., Deatrick, J., Erk, I., Santamaria, P., Freeman, J. D., et al. (1987). A complex genetic locus, polyhomeotic, is required for segmental specification and epidermal development in *D. melanogaster*. *Cell, 51*, 829–839.
Ebbs, M. L., & Bender, J. (2006). Locus-specific control of DNA methylation by the Arabidopsis SUVH5 histone methyltransferase. *The Plant Cell, 18*, 1166–1176.
Eissenberg, J. C., Morris, G. D., Reuter, G., & Hartnett, T. (1992). The heterochromatin-associated protein HP-1 is an essential protein in Drosophila with dosage-dependent effects on position-effect variegation. *Genetics, 131*, 345–352.

Endoh, M., Endo, T. A., Endoh, T., Fujimura, Y., Ohara, O., Toyoda, T., et al. (2008). Polycomb group proteins Ring1A/B are functionally linked to the core transcriptional regulatory circuitry to maintain ES cell identity. *Development, 135*, 1513–1524.
Endoh, M., Endo, T. A., Endoh, T., Isono, K.-I., Sharif, J., Ohara, O., et al. (2012). Histone H2A mono-ubiquitination is a crucial step to mediate PRC1-dependent repression of developmental genes to maintain ES cell identity. *PLoS Genetics, 8*, e1002774.
Epsztejn-Litman, S., Feldman, N., Abu-Remaileh, M., Shufaro, Y., Gerson, A., Ueda, J., et al. (2008). De novo DNA methylation promoted by G9a prevents reprogramming of embryonically silenced genes. *Nature Structural and Molecular Biology, 15*, 1176–1183.
Estève, P.-O., Chin, H. G., Smallwood, A., Feehery, G. R., Gangisetty, O., Karpf, A. R., et al. (2006). Direct interaction between DNMT1 and G9a coordinates DNA and histone methylation during replication. *Genes and Development, 20*, 3089–3103.
Eymery, A., Callanan, M., & Vourc'h, C. (2009). The secret message of heterochromatin: New insights into the mechanisms and function of centromeric and pericentric repeat sequence transcription. *The International Journal of Developmental Biology, 53*, 259–268.
Ezhkova, E., Lien, W.-H., Stokes, N., Pasolli, H. A., Silva, J. M., & Fuchs, E. (2011). EZH1 and EZH2 cogovern histone H3K27 trimethylation and are essential for hair follicle homeostasis and wound repair. *Genes and Development, 25*, 485–498.
Faust, C., Lawson, K. A., Schork, N. J., Thiel, B., & Magnuson, T. (1998). The Polycomb-group gene eed is required for normal morphogenetic movements during gastrulation in the mouse embryo. *Development, 125*, 4495–4506.
Feldman, N., Gerson, A., Fang, J., Li, E., Zhang, Y., Shinkai, Y., et al. (2006). G9a-mediated irreversible epigenetic inactivation of Oct-3/4 during early embryogenesis. *Nature Cell Biology, 8*, 188–194.
Filion, G. J., van Bemmel, J. G., Braunschweig, U., Talhout, W., Kind, J., Ward, L. D., et al. (2010). Systematic protein location mapping reveals five principal chromatin types in Drosophila cells. *Cell, 143*, 212–224.
Fischle, W., Tseng, B. S., Dormann, H. L., Ueberheide, B. M., Garcia, B. A., Shabanowitz, J., et al. (2005). Regulation of HP1-chromatin binding by histone H3 methylation and phosphorylation. *Nature, 438*, 1116–1122.
Fischle, W., Wang, Y., Jacobs, S. A., Kim, Y., Allis, C. D., & Khorasanizadeh, S. (2003). Molecular basis for the discrimination of repressive methyl-lysine marks in histone H3 by Polycomb and HP1 chromodomains. *Genes and Development, 17*, 1870–1881.
Fisher, C. L., Lee, I., Bloyer, S., Bozza, S., Chevalier, J., Dahl, A., et al. (2010). Additional sex combs-like 1 belongs to the enhancer of trithorax and polycomb group and genetically interacts with Cbx2 in mice. *Developmental Biology, 337*, 9–15.
Fodor, B. D., Kubicek, S., Yonezawa, M., O'Sullivan, R. J., Sengupta, R., Perez-Burgos, L., et al. (2006). Jmjd2b antagonizes H3K9 trimethylation at pericentric heterochromatin in mammalian cells. *Genes and Development, 20*, 1557–1562.
Font-Burgada, J., Rossell, D., Auer, H., & Azorín, F. (2008). Drosophila HP1c isoform interacts with the zinc-finger proteins WOC and Relative-of-WOC to regulate gene expression. *Genes and Development, 22*, 3007–3023.
Freitag, M., Hickey, P. C., Khlafallah, T. K., Read, N. D., & Selker, E. U. (2004). HP1 is essential for DNA methylation in neurospora. *Molecular Cell, 13*, 427–434.
Frietze, S., O'Geen, H., Blahnik, K. R., Jin, V. X., & Farnham, P. J. (2010). ZNF274 recruits the histone methyltransferase SETDB1 to the 3' ends of ZNF genes. *PLoS One, 5*, e15082.
Fritsch, L., Robin, P., Mathieu, J. R. R., Souidi, M., Hinaux, H., Rougeulle, C., et al. (2010). A subset of the histone H3 lysine 9 methyltransferases Suv39h1, G9a, GLP, and SETDB1 participate in a multimeric complex. *Molecular Cell, 37*, 46–56.

Gao, H., Yu, Z., Bi, D., Jiang, L., Cui, Y., Sun, J., et al. (2007). Akt/PKB interacts with the histone H3 methyltransferase SETDB1 and coordinates to silence gene expression. *Molecular and Cellular Biochemistry, 305*, 35–44.

Gao, Z., Zhang, J., Bonasio, R., Strino, F., Sawai, A., Parisi, F., et al. (2012). PCGF homologs, CBX proteins, and RYBP define functionally distinct PRC1 family complexes. *Molecular Cell, 45*, 344–356.

Garcia, B. A., Hake, S. B., Diaz, R. L., Kauer, M., Morris, S. A., Recht, J., et al. (2007). Organismal differences in post-translational modifications in histones H3 and H4. *The Journal of Biological Chemistry, 282*, 7641–7655.

Gaudin, V., Libault, M., Pouteau, S., Juul, T., Zhao, G., Lefebvre, D., et al. (2001). Mutations in LIKE HETEROCHROMATIN PROTEIN 1 affect flowering time and plant architecture in Arabidopsis. *Development, 128*, 4847–4858.

Gearhart, M. D., Corcoran, C. M., Wamstad, J. A., & Bardwell, V. J. (2006). Polycomb group and SCF ubiquitin ligases are found in a novel BCOR complex that is recruited to BCL6 targets. *Molecular and Cellular Biology, 26*, 6880–6889.

Gelbart, W., & Emmert, D. (2011). *Calculation of RPKM to generate quantitative expression data: Read-length values for modENCODE developmental timecourse RNA-Seq data.* [WWW Document]. FBrf0214760. http://flybase.org/reports/FBrf0214760.html.

Gendall, A. R., Levy, Y. Y., Wilson, A., & Dean, C. (2001). The VERNALIZATION 2 Gene Mediates the Epigenetic Regulation of Vernalization in *Arabidopsis*. *Cell, 107*, 525–535.

Gerstein, M. B., Lu, Z. J., Nostrand, E. L. V., Cheng, C., Arshinoff, B. I., et al. (2010). Integrative analysis of the *Caenorhabditis elegans* genome by the modENCODE project. *Science, 330*, 1775–1787.

Gill, M. E., Erkek, S., & Peters, A. H. (2012). Parental epigenetic control of embryogenesis: A balance between inheritance and reprogramming? *Current Opinion in Cell Biology, 24*, 387–396.

Goodrich, J., Puangsomlee, P., Martin, M., Long, D., Meyerowitz, E. M., & Coupland, G. (1997). A Polycomb-group gene regulates homeotic gene expression in Arabidopsis. *Nature, 386*, 44–51.

Graveley, B. R., Brooks, A. N., Carlson, J. W., Duff, M. O., Landolin, J. M., Yang, L., et al. (2011). The developmental transcriptome of *Drosophila melanogaster*. *Nature, 471*, 473–479.

Grossniklaus, U., Vielle-Calzada, J.-P., Hoeppner, M. A., & Gagliano, W. B. (1998). Maternal control of embryogenesis by MEDEA, a Polycomb group gene in Arabidopsis. *Science, 280*, 446–450.

Guil, S., Soler, M., Portela, A., Carrère, J., Fonalleras, E., Gómez, A., et al. (2012). Intronic RNAs mediate EZH2 regulation of epigenetic targets. *Nature Structural and Molecular Biology, 19*, 664–670.

Gutiérrez, L., Oktaba, K., Scheuermann, J. C., Gambetta, M. C., Ly-Hartig, N., & Müller, J. (2012). The role of the histone H2A ubiquitinase Sce in Polycomb repression. *Development, 139*, 117–127.

Guttman, M., Donaghey, J., Carey, B. W., Garber, M., Grenier, J. K., Munson, G., et al. (2011). lincRNAs act in the circuitry controlling pluripotency and differentiation. *Nature, 477*, 295–300.

Gyory, I., Wu, J., Fejér, G., Seto, E., & Wright, K. L. (2004). PRDI-BF1 recruits the histone H3 methyltransferase G9a in transcriptional silencing. *Nature Immunology, 5*, 299–308.

Hammoud, S. S., Nix, D. A., Zhang, H., Purwar, J., Carrell, D. T., & Cairns, B. R. (2009). Distinctive chromatin in human sperm packages genes for embryo development. *Nature, 460*, 473–478.

Hansen, K. H., Bracken, A. P., Pasini, D., Dietrich, N., Gehani, S. S., Monrad, A., et al. (2008). A model for transmission of the H3K27me3 epigenetic mark. *Nature Cell Biology, 10*, 1291–1300.

Hathaway, N. A., Bell, O., Hodges, C., Miller, E. L., Neel, D. S., & Crabtree, G. R. (2012). Dynamics and memory of heterochromatin in living cells. *Cell, 149*, 1447–1460.

Hayashi, M. T., Takahashi, T. S., Nakagawa, T., Nakayama, J., & Masukata, H. (2009). The heterochromatin protein Swi6/HP1 activates replication origins at the pericentromeric region and silent mating-type locus. *Nature Cell Biology, 11*, 357–362.

Hekimoglu, B., & Ringrose, L. (2009). Non-coding RNAs in polycomb/trithorax regulation. *RNA Biology, 6*, 129–137.

Hisada, K., Sánchez, C., Endo, T. A., Endoh, M., Román-Trufero, M., Sharif, J., et al. (2012). RYBP represses endogenous retroviruses and preimplantation- and germ line-specific genes in mouse embryonic stem cells. *Molecular and Cellular Biology, 32*, 1139–1149.

Hogarth, C. A., Mitchell, D., Evanoff, R., Small, C., & Griswold, M. (2011). Identification and expression of potential regulators of the mammalian mitotic-to-meiotic transition. *Biology of Reproduction, 84*, 34–42.

Holec, S., & Berger, F. (2012). Polycomb group complexes mediate developmental transitions in plants. *Plant Physiology, 158*, 35–43.

Honda, S., Lewis, Z. A., Huarte, M., Cho, L. Y., David, L. L., Shi, Y., et al. (2010). The DMM complex prevents spreading of DNA methylation from transposons to nearby genes in *Neurospora crassa*. *Genes and Development, 24*, 443–454.

Honda, S., & Selker, E. U. (2008). Direct interaction between DNA methyltransferase DIM-2 and HP1 is required for DNA methylation in *Neurospora crassa*. *Molecular and Cellular Biology, 28*, 6044–6055.

Horáková, A. H., Bártová, E., Galiová, G., Uhlírová, R., Matula, P., & Kozubek, S. (2010). SUV39h-independent association of HP1 beta with fibrillarin-positive nucleolar regions. *Chromosoma, 119*, 227–241.

Huntley, S., Baggott, D. M., Hamilton, A. T., Tran-Gyamfi, M., Yang, S., Kim, J., et al. (2006). A comprehensive catalog of human KRAB-associated zinc finger genes: Insights into the evolutionary history of a large family of transcriptional repressors. *Genome Research, 16*, 669–677.

Isono, K.-I., Fujimura, Y.-I., Shinga, J., Yamaki, M., O-Wang, J., Takihara, Y., et al. (2005). Mammalian polyhomeotic homologues Phc2 and Phc1 act in synergy to mediate polycomb repression of Hox genes. *Molecular and Cellular Biology, 25*, 6694–6706.

Iyengar, S., & Farnham, P. J. (2011). KAP1 protein: An enigmatic master regulator of the genome. *The Journal of Biological Chemistry, 286*, 26267–26276.

Jackson, J. P., Johnson, L., Jasencakova, Z., Zhang, X., PerezBurgos, L., Singh, P. B., et al. (2004). Dimethylation of histone H3 lysine 9 is a critical mark for DNA methylation and gene silencing in *Arabidopsis thaliana*. *Chromosoma, 112*, 308–315.

Jackson, J. P., Lindroth, A. M., Cao, X., & Jacobsen, S. E. (2002). Control of CpNpG DNA methylation by the KRYPTONITE histone H3 methyltransferase. *Nature, 416*, 556–560.

Jenuwein, T., & Allis, C. D. (2001). Translating the histone code. *Science, 293*, 1074–1080.

Johnson, L. M., Bostick, M., Zhang, X., Kraft, E., Henderson, I., Callis, J., et al. (2007). The SRA methyl-cytosine-binding domain links DNA and histone methylation. *Current Biology, 17*, 379–384.

Jones, P. A. (2012). Functions of DNA methylation: Islands, start sites, gene bodies and beyond. *Nature Reviews Genetics, 13*, 484–492.

Jürgens, G. (1985). A group of genes controlling the spatial expression of the bithorax complex in Drosophila. *Nature, 316*, 153–155.

Kanhere, A., Viiri, K., Araújo, C. C., Rasaiyaah, J., Bouwman, R. D., Whyte, W. A., et al. (2010). Short RNAs are transcribed from repressed Polycomb target genes and interact with Polycomb Repressive Complex-2. *Molecular Cell, 38*, 675–688.

Karimi, M. M., Goyal, P., Maksakova, I. A., Bilenky, M., Leung, D., Tang, J. X., et al. (2011). DNA methylation and SETDB1/H3K9me3 regulate predominantly distinct sets of genes, retroelements, and chimeric transcripts in mESCs. *Cell Stem Cell, 8*, 676–687.

Kato, Y., Kato, M., Tachibana, M., Shinkai, Y., & Yamaguchi, M. (2008). Characterization of Drosophila G9a in vivo and identification of genetic interactants. *Genes to Cells, 13*, 703–722.

Katoh-Fukui, Y., Tsuchiya, R., Shiroishi, T., Nakahara, Y., Hashimoto, N., Noguchi, K., et al. (1998). Male-to-female sex reversal in M33 mutant mice. *Nature, 393*, 688–692.

Katz, A., Oliva, M., Mosquna, A., Hakim, O., & Ohad, N. (2004). FIE and CURLY LEAF polycomb proteins interact in the regulation of homeobox gene expression during sporophyte development. *The Plant Journal, 37*, 707–719.

Kaustov, L., Ouyang, H., Amaya, M., Lemak, A., Nady, N., Duan, S., et al. (2011). Recognition and specificity determinants of the human cbx chromodomains. *The Journal of Biological Chemistry, 286*, 521–529.

Kellum, R., & Alberts, B. M. (1995). Heterochromatin protein 1 is required for correct chromosome segregation in Drosophila embryos. *Journal of Cell Science, 108*(Pt. 4), 1419–1431.

Kellum, R., Raff, J. W., & Alberts, B. M. (1995). Heterochromatin protein 1 distribution during development and during the cell cycle in Drosophila embryos. *Journal of Cell Science, 108*(Pt. 4), 1407–1418.

Khalil, A. M., Boyar, F. Z., & Driscoll, D. J. (2004). Dynamic histone modifications mark sex chromosome inactivation and reactivation during mammalian spermatogenesis. *Proceedings of the National Academy of Sciences of the United States of America, 101*, 16583–16587.

Kim, S. Y., Lee, J., Eshed-Williams, L., Zilberman, D., & Sung, Z. R. (2012). EMF1 and PRC2 cooperate to repress key regulators of Arabidopsis development. *PLoS Genetics, 8*, e1002512.

Klattenhoff, C., Xi, H., Li, C., Lee, S., Xu, J., Khurana, J. S., et al. (2009). The Drosophila HP1 homolog Rhino is required for transposon silencing and piRNA production by dual-strand clusters. *Cell, 138*, 1137–1149.

Klose, R. J., Kallin, E. M., & Zhang, Y. (2006). JmjC-domain-containing proteins and histone demethylation. *Nature Reviews Genetics, 7*, 715–727.

Koch, C. M., Honemann-Capito, M., Egger-Adam, D., & Wodarz, A. (2009). Windei, the Drosophila homolog of mAM/MCAF1, is an essential cofactor of the H3K9 methyl transferase dSETDB1/eggless in germ line development. *PLoS Genetics, 5*, e1000644.

Köhler, C., Hennig, L., Bouveret, R., Gheyselinck, J., Grossniklaus, U., & Gruissem, W. (2003). Arabidopsis MSI1 is a component of the MEA/FIE Polycomb group complex and required for seed development. *The EMBO Journal, 22*, 4804–4814.

Kotake, T., Takada, S., Nakahigashi, K., Ohto, M., & Goto, K. (2003). Arabidopsis TERMINAL FLOWER 2 gene encodes a heterochromatin protein 1 homolog and represses both FLOWERING LOCUS T to regulate flowering time and several floral homeotic genes. *Plant and Cell Physiology, 44*, 555–564.

Kouzarides, T. (2007). Chromatin modifications and their function. *Cell, 128*, 693–705.

Kouzminova, E., & Selker, E. U. (2001). dim-2 encodes a DNA methyltransferase responsible for all known cytosine methylation in Neurospora. *The EMBO Journal, 20*, 4309–4323.

Krauss, V. (2008). Glimpses of evolution: Heterochromatic histone H3K9 methyltransferases left its marks behind. *Genetica, 133*, 93–106.

Ku, M., Koche, R. P., Rheinbay, E., Mendenhall, E. M., Endoh, M., Mikkelsen, T. S., et al. (2008). Genomewide analysis of PRC1 and PRC2 occupancy identifies two classes of bivalent domains. *PLoS Genetics, 4,* e1000242.

Kuhlmann, M., & Mette, M. F. (2012). Developmentally non-redundant SET domain proteins SUVH2 and SUVH9 are required for transcriptional gene silencing in *Arabidopsis thaliana. Plant Molecular Biology, 79,* 623–633.

Kurzhals, R. L., Tie, F., Stratton, C. A., & Harte, P. J. (2008). Drosophila ESC-Like can substitute for ESC and becomes required for Polycomb silencing if ESC is absent. *Developmental Biology, 313,* 293–306.

Kuzmichev, A., Jenuwein, T., Tempst, P., & Reinberg, D. (2004). Different Ezh2-containing complexes target methylation of histone H1 or nucleosomal histone H3. *Molecular Cell, 14,* 183–193.

Kuzmichev, A., Margueron, R., Vaquero, A., Preissner, T. S., Scher, M., Kirmizis, A., et al. (2005). Composition and histone substrates of polycomb repressive group complexes change during cellular differentiation. *PNAS, 102,* 1859–1864.

Kwon, S. H., Florens, L., Swanson, S. K., Washburn, M. P., Abmayr, S. M., & Workman, J. L. (2010). Heterochromatin protein 1 (HP1) connects the FACT histone chaperone complex to the phosphorylated CTD of RNA polymerase II. *Genes and Development, 24,* 2133–2145.

Kwon, S. H., & Workman, J. L. (2011). The changing faces of HP1: From heterochromatin formation and gene silencing to euchromatic gene expression. *Bioessays, 33,* 280–289.

Kwong, C., Adryan, B., Bell, I., Meadows, L., Russell, S., Manak, J. R., et al. (2008). Stability and dynamics of polycomb target sites in Drosophila development. *PLoS Genetics, 4,* e1000178.

Lachner, M., O'Carroll, D., Rea, S., Mechtler, K., & Jenuwein, T. (2001). Methylation of histone H3 lysine 9 creates a binding site for HP1 proteins. *Nature, 410,* 116–120.

Lagarou, A., Mohd-Sarip, A., Moshkin, Y. M., Chalkley, G. E., Bezstarosti, K., Demmers, J. A. A., et al. (2008). dKDM2 couples histone H2A ubiquitylation to histone H3 demethylation during Polycomb group silencing. *Genes and Development, 22,* 2799–2810.

Lapthanasupkul, P., Feng, J., Mantesso, A., Takada-Horisawa, Y., Vidal, M., Koseki, H., et al. (2012). Ring1a/b polycomb proteins regulate the mesenchymal stem cell niche in continuously growing incisors. *Developmental Biology, 367,* 140–153.

Larsson, A. S., Landberg, K., & Meeks-Wagner, D. R. (1998). The TERMINAL FLOWER2 (TFL2) gene controls the reproductive transition and meristem identity in *Arabidopsis thaliana. Genetics, 149,* 597–605.

Laurent, L., Wong, E., Li, G., Huynh, T., Tsirigos, A., Ong, C. T., et al. (2010). Dynamic changes in the human methylome during differentiation. *Genome Research, 20,* 320–331.

Law, J. A., & Jacobsen, S. E. (2010). Establishing, maintaining and modifying DNA methylation patterns in plants and animals. *Nature Reviews Genetics, 11,* 204–220.

Leeb, M., & Wutz, A. (2007). Ring1B is crucial for the regulation of developmental control genes and PRC1 proteins but not X inactivation in embryonic cells. *Journal of Cell Biology, 178,* 219–229.

Leeb, M., Pasini, D., Novatchkova, M., Jaritz, M., Helin, K., & Wutz, A. (2010). Polycomb complexes act redundantly to repress genomic repeats and genes. *Genes Development, 24,* 265–276.

Lehnertz, B., Ueda, Y., Derijck, A. A. H. A., Braunschweig, U., Perez-Burgos, L., Kubicek, S., et al. (2003). Suv39h-mediated histone H3 lysine 9 methylation directs DNA methylation to major satellite repeats at pericentric heterochromatin. *Current Biology, 13,* 1192–1200.

Leung, D. C., Dong, K. B., Maksakova, I. A., Goyal, P., Appanah, R., Lee, S., et al. (2011). Lysine methyltransferase G9a is required for de novo DNA methylation and the establishment, but not the maintenance, of proviral silencing. *Proceedings of the National Academy of Sciences of the United States of America, 108*, 5718–5723.

Lewis, P. H. (1947). Polycomb:PC. *Drosophila Information Service, 21*, 69.

Lewis, E. B. (1978). A gene complex controlling segmentation in Drosophila. *Nature, 276*, 565–570.

Lewis, Z. A., Adhvaryu, K. K., Honda, S., Shiver, A. L., & Selker, E. U. (2010). Identification ofDIM-7, a protein required to target the DIM-5 H3 methyltransferase to chromatin. *Proceedings of the National Academy of Sciences of the United States of America, 107*, 8310–8315.

Lewis, Z. A., Honda, S., Khlafallah, T. K., Jeffress, J. K., Freitag, M., Mohn, F., et al. (2009). Relics of repeat-induced point mutation direct heterochromatin formation in *Neurospora crassa. Genome Research, 19*, 427–437.

Li, X., Isono, K.-I., Yamada, D., Endo, T. A., Endoh, M., Shinga, J., et al. (2011). Mammalian polycomb-like Pcl2/Mtf2 is a novel regulatory component of PRC2 that can differentially modulate polycomb activity both at the Hox gene cluster and at Cdkn2a genes. *Molecular and Cellular Biology, 31*, 351–364.

Li, Q., Wang, X., Lu, Z., Zhang, B., Guan, Z., Liu, Z., et al. (2010). Polycomb CBX7 directly controls trimethylation of histone H3 at lysine 9 at the p16 locus. *PLoS One, 5*, e13732.

Lin, C.-H., Li, B., Swanson, S., Zhang, Y., Florens, L., Washburn, M. P., et al. (2008). Heterochromatin protein 1a stimulates histone H3 lysine 36 demethylation by the Drosophila KDM4A demethylase. *Molecular Cell, 32*, 696–706.

Lin, C.-H., Paulson, A., Abmayr, S. M., & Workman, J. L. (2012). HP1a targets the Drosophila KDM4A demethylase to a subset of heterochromatic genes to regulate H3K36me3 levels. *PLoS One, 7*, e39758.

Liu, Y., Taverna, S. D., Muratore, T. L., Shabanowitz, J., Hunt, D. F., & Allis, C. D. (2007). RNAi-dependent H3K27 methylation is required for heterochromatin formation and DNA elimination in Tetrahymena. *Genes and Development, 21*, 1530–1545.

Lo, S. M., Follmer, N. E., Lengsfeld, B. M., Madamba, E. V., Seong, S., Grau, D. J., et al. (2012). A bridging model for persistence of a polycomb group protein complex through DNA replication in vitro. *Molecular Cell, 46*, 784–796.

Loh, Y.-H., Zhang, W., Chen, X., George, J., & Ng, H.-H. (2007). Jmjd1a and Jmjd2c histone H3 Lys 9 demethylases regulate self-renewal in embryonic stem cells. *Genes and Development, 21*, 2545–2557.

Lohmann, F., Loureiro, J., Su, H., Fang, Q., Lei, H., Lewis, T., et al. (2010). KMT1E mediated H3K9 methylation is required for the maintenance of embryonic stem cells by repressing trophectoderm differentiation. *Stem Cells, 28*, 201–212.

Lucia, F. D., Crevillen, P., Jones, A. M. E., Greb, T., & Dean, C. (2008). A PHD-Polycomb Repressive Complex 2 triggers the epigenetic silencing of FLC during vernalization. *PNAS, 105*, 16831–16836.

Luo, M., Bilodeau, P., Koltunow, A., Dennis, E. S., Peacock, W. J., & Chaudhury, A. M. (1999). Genes controlling fertilization-independent seed development in *Arabidopsis thaliana. Proceedings of the National Academy of Sciences of the United States of America, 96*, 296–301.

Maison, C., Bailly, D., Roche, D., Montes de Oca, R., Probst, A. V., Vassias, I., et al. (2011). SUMOylation promotes de novo targeting of HP1α to pericentric heterochromatin. *Nature Genetics, 43*, 220–227.

Maksakova, I. A., Goyal, P., Bullwinkel, J., Brown, J. P., Bilenky, M., Mager, D. L., et al. (2011). H3K9me3-binding proteins are dispensable for SETDB1/H3K9me3-dependent retroviral silencing. *Epigenetics and Chromatin, 4*, 12.

Margueron, R., Justin, N., Ohno, K., Sharpe, M. L., Son, J., Drury, W. J., 3rd., et al. (2009). Role of the polycomb protein EED in the propagation of repressive histone marks. *Nature, 461*, 762–767.

Margueron, R., Li, G., Sarma, K., Blais, A., Zavadil, J., Woodcock, C. L., et al. (2008). Ezh1 and Ezh2 maintain repressive chromatin through different mechanisms. *Molecular Cell, 32*, 503–518.

Matsui, T., Leung, D., Miyashita, H., Maksakova, I. A., Miyachi, H., Kimura, H., et al. (2010). Proviral silencing in embryonic stem cells requires the histone methyltransferase ESET. *Nature, 464*, 927–931.

McQuilton, P., St Pierre, S. E., & Thurmond, J. (2012). FlyBase 101—The basics of navigating FlyBase. *Nucleic Acids Research, 40*, D706–D714.

Meglicki, M., Teperek-Tkacz, M., & Borsuk, E. (2012). Appearance and heterochromatin localization of HP1$\alpha$ in early mouse embryos depends on cytoplasmic clock and H3S10 phosphorylation. *Cell Cycle, 11*, 2189–2205.

Meister, P., Schott, S., Bedet, C., Xiao, Y., Rohner, S., Bodennec, S., et al. (2011). *Caenorhabditis elegans* Heterochromatin protein 1 (HPL-2) links developmental plasticity, longevity and lipid metabolism. *Genome Biology, 12*, R123.

Mendenhall, E. M., Koche, R. P., Truong, T., Zhou, V. W., Issac, B., Chi, A. S., et al. (2010). GC-rich sequence elements recruit PRC2 in mammalian ES cells. *PLoS Genetics, 6*, e1001244.

Metzler-Guillemain, C., Luciani, J., Depetris, D., Guichaoua, M. R., & Mattei, M. G. (2003). HP1beta and HP1gamma, but not HP1alpha, decorate the entire XY body during human male meiosis. *Chromosome Research, 11*, 73–81.

Minc, E., Allory, Y., Worman, H. J., Courvalin, J. C., & Buendia, B. (1999). Localization and phosphorylation of HP1 proteins during the cell cycle in mammalian cells. *Chromosoma, 108*, 220–234.

Mis, J., Ner, S. S., & Grigliatti, T. A. (2006). Identification of three histone methyltransferases in Drosophila: dG9a is a suppressor of PEV and is required for gene silencing. *Molecular Genetics and Genomics, 275*, 513–526.

Miura, A., Nakamura, M., Inagaki, S., Kobayashi, A., Saze, H., & Kakutani, T. (2009). An Arabidopsis jmjC domain protein protects transcribed genes from DNA methylation at CHG sites. *The EMBO Journal, 28*, 1078–1086.

Mohn, F., Weber, M., Rebhan, M., Roloff, T. C., Richter, J., Stadler, M. B., et al. (2008). Lineage-specific Polycomb targets and de novo DNA methylation define restriction and potential of neuronal progenitors. *Molecular Cell, 30*, 755–766.

Müller, J., Hart, C. M., Francis, N. J., Vargas, M. L., Sengupta, A., Wild, B., et al. (2002). Histone methyltransferase activity of a Drosophila Polycomb group repressor complex. *Cell, 111*, 197–208.

Nakahigashi, K., Jasencakova, Z., Schubert, I., & Goto, K. (2005). The Arabidopsis heterochromatin protein1 homolog (TERMINAL FLOWER2) silences genes within the euchromatic region but not genes positioned in heterochromatin. *Plant and Cell Physiology, 46*, 1747–1756.

Nakamura, T., Liu, Y.-J., Nakashima, H., Umehara, H., Inoue, K., Matoba, S., et al. (2012). PGC7 binds histone H3K9me2 to protect against conversion of 5mC to 5hmC in early embryos. *Nature, 486*, 415–419.

Naruse, C., Fukusumi, Y., Kakiuchi, D., & Asano, M. (2007). A novel gene trapping for identifying genes expressed under the control of specific transcription factors. *Biochemical and Biophysical Research Communications, 361*, 109–115.

Naumann, K., Fischer, A., Hofmann, I., Krauss, V., Phalke, S., Irmler, K., et al. (2005). Pivotal role of AtSUVH2 in heterochromatic histone methylation and gene silencing in Arabidopsis. *The EMBO Journal, 24*, 1418–1429.

Nekrasov, M., Klymenko, T., Fraterman, S., Papp, B., Oktaba, K., Köcher, T., et al. (2007). Pcl-PRC2 is needed to generate high levels of H3-K27 trimethylation at Polycomb target genes. *The EMBO Journal, 26*, 4078–4088.

Ng, S.-Y., Johnson, R., & Stanton, L. W. (2012). Human long non-coding RNAs promote pluripotency and neuronal differentiation by association with chromatin modifiers and transcription factors. *The EMBO Journal, 31*, 522–533.

Nishio, H., & Walsh, M. J. (2004). CCAAT displacement protein/cut homolog recruits G9a histone lysine methyltransferase to repress transcription. *Proceedings of the National Academy of Sciences of the United States of America, 101*, 11257–11262.

O'Carroll, D., Erhardt, S., Pagani, M., Barton, S. C., Surani, M. A., & Jenuwein, T. (2001). The polycomb-group gene Ezh2 is required for early mouse development. *Molecular and Cellular Biology, 21*, 4330–4336.

O'Carroll, D., Scherthan, H., Peters, A. H., Opravil, S., Haynes, A. R., Laible, G., et al. (2000). Isolation and characterization of Suv39h2, a second histone H3 methyltransferase gene that displays testis-specific expression. *Molecular and Cellular Biology, 20*, 9423–9433.

Ohad, N., Yadegari, R., Margossian, L., Hannon, M., Michaeli, D., Harada, J. J., et al. (1999). Mutations in FIE, a WD polycomb group gene, allow endosperm development without fertilization. *The Plant Cell, 11*, 407–416.

Ohno, K., McCabe, D., Czermin, B., Imhof, A., & Pirrotta, V. (2008). ESC, ESCL and their roles in Polycomb group mechanisms. *Mechanisms of Development, 125*, 527–541.

Okada, Y., Scott, G., Ray, M. K., Mishina, Y., & Zhang, Y. (2007). Histone demethylase JHDM2A is critical for Tnp1 and Prm1 transcription and spermatogenesis. *Nature, 450*, 119–123.

Okada, Y., Tateishi, K., & Zhang, Y. (2010). Histone demethylase JHDM2A is involved in male infertility and obesity. *Journal of Andrology, 31*, 75–78.

Onder, T. T., Kara, N., Cherry, A., Sinha, A. U., Zhu, N., Bernt, K. M., et al. (2012). Chromatin-modifying enzymes as modulators of reprogramming. *Nature, 483*, 598–602.

Paik, W. K., Paik, D. C., & Kim, S. (2007). Historical review: The field of protein methylation. *Trends in Biochemical Sciences, 32*, 146–152.

Pal-Bhadra, M., Leibovitch, B. A., Gandhi, S. G., Rao, M., Bhadra, U., Birchler, J. A., et al. (2004). Heterochromatic silencing and HP1 localization in Drosophila are dependent on the RNAi machinery. *Science, 303*, 669–672.

Pasini, D., Bracken, A. P., Jensen, M. R., Lazzerini Denchi, E., & Helin, K. (2004). Suz12 is essential for mouse development and for EZH2 histone methyltransferase activity. *The EMBO Journal, 23*, 4061–4071.

Pasini, D., Bracken, A. P., Hansen, J. B., Capillo, M., & Helin, K. (2007). The polycomb group protein Suz12 is required for embryonic stem cell differentiation. *Molecular and Cellular Biology, 27*, 3769–3779.

Peters, A. H. F. M., Kubicek, S., Mechtler, K., O'Sullivan, R. J., Derijck, A. A. H. A., Perez-Burgos, L., et al. (2003). Partitioning and plasticity of repressive histone methylation states in mammalian chromatin. *Molecular Cell, 12*, 1577–1589.

Peters, A. H., O'Carroll, D., Scherthan, H., Mechtler, K., Sauer, S., Schöfer, C., et al. (2001). Loss of the Suv39h histone methyltransferases impairs mammalian heterochromatin and genome stability. *Cell, 107*, 323–337.

Phillips, M. D., & Shearn, A. (1990). Mutations in polycombeotic, a Drosophila polycomb-group gene, cause a wide range of maternal and zygotic phenotypes. *Genetics, 125*, 91–101.

Piacentini, L., Fanti, L., Negri, R., Del Vescovo, V., Fatica, A., Altieri, F., et al. (2009). Heterochromatin protein 1 (HP1a) positively regulates euchromatic gene expression through RNA transcript association and interaction with hnRNPs in Drosophila. *PLoS Genetics, 5*, e1000670.

Pinheiro, I., Margueron, R., Shukeir, N., Eisold, M., Fritzsch, C., Richter, F. M., et al. (2012). Prdm3 and Prdm16 are H3K9me1 methyltransferases required for mammalian heterochromatin integrity. *Cell, 150*, 948–960.

Pirity, M. K., Locker, J., & Schreiber-Agus, N. (2005). Rybp/DEDAF is required for early postimplantation and for central nervous system development. *Molecular and Cellular Biology, 25*, 7193–7202.

Plath, K., Fang, J., Mlynarczyk-Evans, S. K., Cao, R., Worringer, K. A., Wang, H., et al. (2003). Role of histone H3 lysine 27 methylation in X inactivation. *Science, 300*, 131–135.

Posfai, E., Kunzmann, R., Brochard, V., Salvaing, J., Cabuy, E., Roloff, T. C., et al. (2012). Polycomb function during oogenesis is required for mouse embryonic development. *Genes and Development, 26*, 920–932.

Probst, A. V., Dunleavy, E., & Almouzni, G. (2009). Epigenetic inheritance during the cell cycle. *Natura Reviews. Molecular Cell Biology, 10*, 192–206.

Puschendorf, M., Terranova, R., Boutsma, E., Mao, X., Isono, K., Brykczynska, U., et al. (2008). PRC1 and Suv39h specify parental asymmetry at constitutive heterochromatin in early mouse embryos. *Nature Genetics, 40*, 411–420.

Qin, J., Whyte, W. A., Anderssen, E., Apostolou, E., Chen, H.-H., Akbarian, S., et al. (2012). The Polycomb group protein L3mbtl2 assembles an atypical PRC1-family complex that is essential in pluripotent stem cells and early development. *Cell Stem Cell, 11*, 319–332.

Quenneville, S., Turelli, P., Bojkowska, K., Raclot, C., Offner, S., Kapopoulou, A., et al. (2012). The KRAB-ZFP/KAP1 system contributes to the early embryonic establishment of site-specific DNA methylation patterns maintained during development. *Cell Reports, 2*, 766–773.

Quenneville, S., Verde, G., Corsinotti, A., Kapopoulou, A., Jakobsson, J., Offner, S., et al. (2011). In embryonic stem cells, ZFP57/KAP1 recognize a methylated hexanucleotide to affect chromatin and DNA methylation of imprinting control regions. *Molecular Cell, 44*, 361–372.

Rea, S., Eisenhaber, F., O'Carroll, D., Strahl, B. D., Sun, Z. W., Schmid, M., et al. (2000). Regulation of chromatin structure by site-specific histone H3 methyltransferases. *Nature, 406*, 593–599.

Ren, X., & Kerppola, T. K. (2011). REST interacts with Cbx proteins and regulates polycomb repressive complex 1 occupancy at RE1 elements. *Molecular and Cellular Biology, 31*, 2100–2110.

Richly, H., Aloia, L., & Di Croce, L. (2011). Roles of the Polycomb group proteins in stem cells and cancer. *Cell Death & Disease, 2*, e204.

Ringrose, L., & Paro, R. (2007). Polycomb/Trithorax response elements and epigenetic memory of cell identity. *Development, 134*, 223–232.

Rinn, J. L., Kertesz, M., Wang, J. K., Squazzo, S. L., Xu, X., Brugmann, S. A., et al. (2007). Functional demarcation of active and silent chromatin domains in human HOX loci by noncoding RNAs. *Cell, 129*, 1311–1323.

Rosnoblet, C., Vandamme, J., Völkel, P., & Angrand, P.-O. (2011). Analysis of the human HP1 interactome reveals novel binding partners. *Biochemical and Biophysical Research Communications, 413*, 206–211.

Rothbart, S. B., Krajewski, K., Nady, N., Tempel, W., Xue, S., Badeaux, A. I., et al. (2012). Association of UHRF1 with methylated H3K9 directs the maintenance of DNA methylation. *Nature Structural and Molecular Biology, 19*, 1155–1160.

Rowe, H. M., Jakobsson, J., Mesnard, D., Rougemont, J., Reynard, S., Aktas, T., et al. (2010). KAP1 controls endogenous retroviruses in embryonic stem cells. *Nature, 463*, 237–240.

Roy, S., Ernst, J., Kharchenko, P. V., Kheradpour, P., Negre, N., et al. (2010). Identification of functional elements and regulatory circuits by Drosophila modENCODE. *Science, 330*, 1787–1797.

Sampath, S. C., Marazzi, I., Yap, K. L., Sampath, S. C., Krutchinsky, A. N., Mecklenbräuker, I., et al. (2007). Methylation of a histone mimic within the histone methyltransferase G9a regulates protein complex assembly. *Molecular Cell, 27*, 596–608.

Sanchez-Pulido, L., Devos, D., Sung, Z. R., & Calonje, M. (2008). RAWUL: A new ubiquitin-like domain in PRC1 Ring finger proteins that unveils putative plant and worm PRC1 orthologs. *BMC Genomics*, *9*, 308.

Sarraf, S. A., & Stancheva, I. (2004). Methyl-CpG binding protein MBD1 couples histone H3 methylation at lysine 9 by SETDB1 to DNA replication and chromatin assembly. *Molecular Cell*, *15*, 595–605.

Sasaki, T., Kobayashi, A., Saze, H., & Kakutani, T. (2012). RNAi-independent de novo DNA methylation revealed in Arabidopsis mutants of chromatin remodeling gene DDM1. *The Plant Journal*, *70*, 750–758.

Savla, U., Benes, J., Zhang, J., & Jones, R. S. (2008). Recruitment of Drosophila Polycomb-group proteins by Polycomblike, a component of a novel protein complex in larvae. *Development*, *135*, 813–817.

Saze, H., Shiraishi, A., Miura, A., & Kakutani, T. (2008). Control of genic DNA methylation by a jmjC domain-containing protein in *Arabidopsis thaliana*. *Science*, *319*, 462–465.

Scheuermann, J. C., de Ayala Alonso, A. G., Oktaba, K., Ly-Hartig, N., McGinty, R. K., Fraterman, S., et al. (2010). Histone H2A deubiquitinase activity of the Polycomb repressive complex PR-DUB. *Nature*, *465*, 243–247.

Schmitges, F. W., Prusty, A. B., Faty, M., Stützer, A., Lingaraju, G. M., Aiwazian, J., et al. (2011). Histone methylation by PRC2 is inhibited by active chromatin marks. *Molecular Cell*, *42*, 330–341.

Schotta, G., Ebert, A., Krauss, V., Fischer, A., Hoffmann, J., Rea, S., et al. (2002). Central role of Drosophila SU(VAR)3-9 in histone H3-K9 methylation and heterochromatic gene silencing. *The EMBO Journal*, *21*, 1121–1131.

Schotta, G., Ebert, A., & Reuter, G. (2003). SU(VAR)3-9 is a conserved key function in heterochromatic gene silencing. *Genetica*, *117*, 149–158.

Schuettengruber, B., Chourrout, D., Vervoort, M., Leblanc, B., & Cavalli, G. (2007). Genome regulation by polycomb and trithorax proteins. *Cell*, *128*, 735–745.

Schuettengruber, B., Ganapathi, M., Leblanc, B., Portoso, M., Jaschek, R., Tolhuis, B., et al. (2009). Functional anatomy of Polycomb and Trithorax chromatin landscapes in Drosophila embryos. *PLoS Biology*, 7, e1000013.

Schultz, D. C., Ayyanathan, K., Negorev, D., Maul, G. G., & Rauscher, F. J., 3rd. (2002). SETDB1: A novel KAP-1-associated histone H3, lysine 9-specific methyltransferase that contributes to HP1-mediated silencing of euchromatic genes by KRAB zinc-finger proteins. *Genes and Development*, *16*, 919–932.

Schwaiger, M., Kohler, H., Oakeley, E. J., Stadler, M. B., & Schübeler, D. (2010). Heterochromatin protein 1 (HP1) modulates replication timing of the Drosophila genome. *Genome Research*, *20*, 771–780.

Schwartz, Y. B., Kahn, T. G., Nix, D. A., Li, X.-Y., Bourgon, R., Biggin, M., et al. (2006). Genome-wide analysis of Polycomb targets in *Drosophila melanogaster*. *Nature Genetics*, *38*, 700–705.

Seum, C., Bontron, S., Reo, E., Delattre, M., & Spierer, P. (2007). Drosophila G9a is a nonessential gene. *Genetics*, *177*, 1955–1957.

Seum, C., Reo, E., Peng, H., Rauscher, F. J., 3rd., Spierer, P., & Bontron, S. (2007). Drosophila SETDB1 is required for chromosome 4 silencing. *PLoS Genetics*, *3*, e76.

Sewalt, R. G. A. B., Lachner, M., Vargas, M., Hamer, K. M., den Blaauwen, J. L., Hendrix, T., et al. (2002). Selective interactions between vertebrate polycomb homologs and the SUV39H1 histone lysine methyltransferase suggest that histone H3-K9 methylation contributes to chromosomal targeting of Polycomb group proteins. *Molecular and Cellular Biology*, *22*, 5539–5553.

Sharif, J., Muto, M., Takebayashi, S., Suetake, I., Iwamatsu, A., Endo, T. A., et al. (2007). The SRA protein Np95 mediates epigenetic inheritance by recruiting Dnmt1 to methylated DNA. *Nature*, *450*, 908–912.

Shaver, S., Casas-Mollano, J. A., Cerny, R. L., & Cerutti, H. (2010). Origin of the polycomb repressive complex 2 and gene silencing by an E(z) homolog in the unicellular alga Chlamydomonas. *Epigenetics, 5*, 301–312.

Shen, X., Liu, Y., Hsu, Y.-J., Fujiwara, Y., Kim, J., Mao, X., et al. (2008). EZH1 mediates methylation on histone H3 lysine 27 and complements EZH2 in maintaining stem cell identity and executing pluripotency. *Molecular Cell, 32*, 491–502.

Sing, A., Pannell, D., Karaiskakis, A., Sturgeon, K., Djabali, M., Ellis, J., et al. (2009). A vertebrate Polycomb response element governs segmentation of the posterior hindbrain. *Cell, 138*, 885–897.

Smallwood, A., Estève, P.-O., Pradhan, S., & Carey, M. (2007). Functional cooperation between HP1 and DNMT1 mediates gene silencing. *Genes and Development, 21*, 1169–1178.

Smallwood, A., Hon, G. C., Jin, F., Henry, R. E., Espinosa, J. M., & Ren, B. (2012). CBX3 regulates efficient RNA processing genome-wide. *Genome Research, 22*, 1426–1436.

Song, J., Angel, A., Howard, M., & Dean, C. (2012). Vernalization—A cold-induced epigenetic switch. *Journal of Cell Science, 125*, 3723–3731.

Sripathy, S. P., Stevens, J., & Schultz, D. C. (2006). The KAP1 corepressor functions to coordinate the assembly of de novo HP1-demarcated microenvironments of heterochromatin required for KRAB zinc finger protein-mediated transcriptional repression. *Molecular and Cellular Biology, 26*, 8623–8638.

Stabell, M., Eskeland, R., Bjørkmo, M., Larsson, J., Aalen, R. B., Imhof, A., et al. (2006). The Drosophila G9a gene encodes a multi-catalytic histone methyltransferase required for normal development. *Nucleic Acids Research, 34*, 4609–4621.

Struhl, G., & Brower, D. (1982). Early role of the esc+ gene product in the determination of segments in Drosophila. *Cell, 31*, 285–292.

Studencka, M., Konzer, A., Moneron, G., Wenzel, D., Opitz, L., Salinas-Riester, G., et al. (2012). Novel roles of *Caenorhabditis elegans* heterochromatin protein HP1 and linker histone in the regulation of innate immune gene expression. *Molecular and Cellular Biology, 32*, 251–265.

Sung, S., He, Y., Eshoo, T. W., Tamada, Y., Johnson, L., Nakahigashi, K., et al. (2006). Epigenetic maintenance of the vernalized state in *Arabidopsis thaliana* requires LIKE HETEROCHROMATIN PROTEIN 1. *Nature Genetics, 38*, 706–710.

Tachibana, M., Matsumura, Y., Fukuda, M., Kimura, H., & Shinkai, Y. (2008). G9a/GLP complexes independently mediate H3K9 and DNA methylation to silence transcription. *The EMBO Journal, 27*, 2681–2690.

Tachibana, M., Nozaki, M., Takeda, N., & Shinkai, Y. (2007). Functional dynamics of H3K9 methylation during meiotic prophase progression. *The EMBO Journal, 26*, 3346–3359.

Tachibana, M., Sugimoto, K., Nozaki, M., Ueda, J., Ohta, T., Ohki, M., et al. (2002). G9a histone methyltransferase plays a dominant role in euchromatic histone H3 lysine 9 methylation and is essential for early embryogenesis. *Genes and Development, 16*, 1779–1791.

Takada, Y., Naruse, C., Costa, Y., Shirakawa, T., Tachibana, M., Sharif, J., et al. (2011). HP1γ links histone methylation marks to meiotic synapsis in mice. *Development, 138*, 4207–4217.

Takihara, Y., Tomotsune, D., Shirai, M., Katoh-Fukui, Y., Nishii, K., Motaleb, M. A., et al. (1997). Targeted disruption of the mouse homologue of the Drosophila polyhomeotic gene leads to altered anteroposterior patterning and neural crest defects. *Development, 124*, 3673–3682.

Tamaru, H., & Selker, E. U. (2001). A histone H3 methyltransferase controls DNA methylation in *Neurospora crassa*. *Nature, 414*, 277–283.

Tavares, L., Dimitrova, E., Oxley, D., Webster, J., Poot, R., Demmers, J., et al. (2012). RYBP-PRC1 complexes mediate H2A ubiquitylation at polycomb target sites independently of PRC2 and H3K27me3. *Cell*, *148*, 664–678.

Terranova, R., Yokobayashi, S., Stadler, M. B., Otte, A. P., van Lohuizen, M., Orkin, S. H., et al. (2008). Polycomb group proteins Ezh2 and Rnf2 direct genomic contraction and imprinted repression in early mouse embryos. *Developmental Cell*, *15*, 668–679.

Thorstensen, T., Fischer, A., Sandvik, S. V., Johnsen, S. S., Grini, P. E., Reuter, G., et al. (2006). The Arabidopsis SUVR4 protein is a nucleolar histone methyltransferase with preference for monomethylated H3K9. *Nucleic Acids Research*, *34*, 5461–5470.

Thorstensen, T., Grini, P. E., & Aalen, R. B. (2011). SET domain proteins in plant development. *Biochimica et Biophysica Acta*, *1809*, 407–420.

Tie, F., Furuyama, T., Prasad-Sinha, J., Jane, E., & Harte, P. J. (2001). The Drosophila Polycomb Group proteins ESC and E(Z) are present in a complex containing the histone-binding protein p55 and the histone deacetylase RPD3. *Development*, *128*, 275–286.

Tie, F., Prasad-Sinha, J., Birve, A., Rasmuson-Lestander, Å., & Harte, P. J. (2003). A 1-megadalton ESC/E(Z) complex from Drosophila that contains polycomblike and RPD3. *Molecular and Cellular Biology*, *23*, 3352–3362.

Torok, D., Ching, R. W., & Bazett-Jones, D. P. (2009). PML nuclear bodies as sites of epigenetic regulation. *Frontiers in Bioscience*, *14*, 1325–1336.

Towbin, B. D., González-Aguilera, C., Sack, R., Gaidatzis, D., Kalck, V., Meister, P., et al. (2012). Step-wise methylation of histone H3K9 positions heterochromatin at the nuclear periphery. *Cell*, *150*, 934–947.

Trojer, P., Cao, A. R., Gao, Z., Li, Y., Zhang, J., Xu, X., et al. (2011). L3MBTL2 protein acts in concert with PcG protein-mediated monoubiquitination of H2A to establish a repressive chromatin structure. *Molecular Cell*, *42*, 438–450.

Tsai, M.-C., Manor, O., Wan, Y., Mosammaparast, N., Wang, J. K., Lan, F., et al. (2010). Long noncoding RNA as modular scaffold of histone modification complexes. *Science*, *329*, 689–693.

Tschiersch, B., Hofmann, A., Krauss, V., Dorn, R., Korge, G., & Reuter, G. (1994). The protein encoded by the Drosophila position-effect variegation suppressor gene Su(var)3-9 combines domains of antagonistic regulators of homeotic gene complexes. *The EMBO Journal*, *13*, 3822–3831.

Turck, F., Roudier, F., Farrona, S., Martin-Magniette, M.-L., Guillaume, E., Buisine, N., et al. (2007). Arabidopsis TFL2/LHP1 specifically associates with genes marked by trimethylation of histone H3 Lysine 27. *PLoS Genetics*, *3*, e86.

Turner, J. M. A. (2007). Meiotic sex chromosome inactivation. *Development*, *134*, 1823–1831.

Vakoc, C. R., Mandat, S. A., Olenchock, B. A., & Blobel, G. A. (2005). Histone H3 lysine 9 methylation and HP1gamma are associated with transcription elongation through mammalian chromatin. *Molecular Cell*, *19*, 381–391.

van der Heijden, G. W., Derijck, A. A. H. A., Pósfai, E., Giele, M., Pelczar, P., Ramos, L., et al. (2007). Chromosome-wide nucleosome replacement and H3.3 incorporation during mammalian meiotic sex chromosome inactivation. *Nature Genetics*, *39*, 251–258.

van der Lugt, N. M., Domen, J., Linders, K., van Roon, M., Robanus-Maandag, E., te Riele, H., et al. (1994). Posterior transformation, neurological abnormalities, and severe hematopoietic defects in mice with a targeted deletion of the bmi-1 proto-oncogene. *Genes and Development*, *8*, 757–769.

Van der Stoop, P., Boutsma, E.A., Hulsman, D., Noback, S., Heimerikx, M., Kerkhoven, R.M., et al., (2008). Ubiquitin E3 ligase Ring1b/Rnf2 of polycomb repressive complex 1 contributes to stable maintenance of mouse embryonic stem cells. PLoS ONE 3, e2235.

Veiseth, S. V., Rahman, M. A., Yap, K. L., Fischer, A., Egge-Jacobsen, W., Reuter, G., et al. (2011). The SUVR4 histone lysine methyltransferase binds ubiquitin and converts H3K9me1 to H3K9me3 on transposon chromatin in Arabidopsis. *PLoS Genetics*, *7*, e1001325.

Vermaak, D., Henikoff, S., & Malik, H. S. (2005). Positive selection drives the evolution of rhino, a member of the heterochromatin protein 1 family in Drosophila. *PLoS Genetics*, *1*, 96–108.

Vermaak, D., & Malik, H. S. (2009). Multiple roles for heterochromatin protein 1 genes in Drosophila. *Annual Review of Genetics*, *43*, 467–492.

Verschure, P. J., van der Kraan, I., de Leeuw, W., van der Vlag, J., Carpenter, A. E., Belmont, A. S., et al. (2005). In vivo HP1 targeting causes large-scale chromatin condensation and enhanced histone lysine methylation. *Molecular and Cellular Biology*, *25*, 4552–4564.

Voigt, P., LeRoy, G., Drury, W. J., III., Zee, B. M., Son, J., Beck, D. B., et al. (2012). Asymmetrically modified nucleosomes. *Cell*, *151*, 181–193.

Volpe, A. M., Horowitz, H., Grafer, C. M., Jackson, S. M., & Berg, C. A. (2001). Drosophila rhino encodes a female-specific chromo-domain protein that affects chromosome structure and egg polarity. *Genetics*, *159*, 1117–1134.

Voncken, J. W., Roelen, B. A. J., Roefs, M., De Vries, S., Verhoeven, E., Marino, S., et al. (2003). Rnf2 (Ring1b) deficiency causes gastrulation arrest and cell cycle inhibition. *Proceedings of the National Academy of Sciences of the United States of America*, *100*, 2468–2473.

Wagschal, A., Sutherland, H. G., Woodfine, K., Henckel, A., Chebli, K., Schulz, R., et al. (2008). G9a histone methyltransferase contributes to imprinting in the mouse placenta. *Molecular and Cellular Biology*, *28*, 1104–1113.

Wang, X., Pan, L., Wang, S., Zhou, J., McDowell, W., Park, J., et al. (2011). Histone H3K9 trimethylase Eggless controls germline stem cell maintenance and differentiation. *PLoS Genetics*, 7, e1002426.

Wang, J., Zhang, M., Zhang, Y., Kou, Z., Han, Z., Chen, D.-Y., et al. (2010). The histone demethylase JMJD2C is stage-specifically expressed in preimplantation mouse embryos and is required for embryonic development. *Biology of Reproduction*, *82*, 105–111.

Whetstine, J. R., Nottke, A., Lan, F., Huarte, M., Smolikov, S., Chen, Z., et al. (2006). Reversal of histone lysine trimethylation by the JMJD2 family of histone demethylases. *Cell*, *125*, 467–481.

Woo, C. J., Kharchenko, P. V., Daheron, L., Park, P. J., & Kingston, R. E. (2010). A region of the human HOXD cluster that confers polycomb-group responsiveness. *Cell*, *140*, 99–110.

Xu, L., & Shen, W.-H. (2008). Polycomb silencing of KNOX genes confines shoot stem cell niches in Arabidopsis. *Current Biology*, *18*, 1966–1971.

Xu, M., Wang, W., Chen, S., & Zhu, B. (2012). A model for mitotic inheritance of histone lysine methylation. *EMBO Reports*, *13*, 60–67.

Yamamizu, K., Fujihara, M., Tachibana, M., Katayama, S., Takahashi, A., Hara, E., et al. (2012). Protein kinase A determines timing of early differentiation through epigenetic regulation with G9a. *Cell Stem Cell*, *10*, 759–770.

Yamanaka, S., & Blau, H. M. (2010). Nuclear reprogramming to a pluripotent state by three approaches. *Nature*, *465*, 704–712.

Yang, L., Lin, C., Liu, W., Zhang, J., Ohgi, K. A., Grinstein, J. D., et al. (2011). ncRNA- and Pc2 methylation-dependent gene relocation between nuclear structures mediates gene activation programs. *Cell*, *147*, 773–788.

Yap, K. L., Li, S., Muñoz-Cabello, A. M., Raguz, S., Zeng, L., Mujtaba, S., et al. (2010). Molecular interplay of the noncoding RNA ANRIL and methylated histone H3 lysine 27 by polycomb CBX7 in transcriptional silencing of INK4a. *Molecular Cell*, *38*, 662–674.

Yeap, L.-S., Hayashi, K., & Surani, M. A. (2009). ERG-associated protein with SET domain (ESET)-Oct4 interaction regulates pluripotency and represses the trophectoderm lineage. *Epigenetics and Chromatin, 2*, 12.

Yoon, J., Lee, K.-S., Park, J. S., Yu, K., Paik, S.-G., & Kang, Y.-K. (2008). dSETDB1 and SU(VAR)3-9 sequentially function during germline-stem cell differentiation in *Drosophila melanogaster*. *PLoS One, 3*, e2234.

Yoshida, N., Yanai, Y., Chen, L., Kato, Y., Hiratsuka, J., Miwa, T., Sung, Z. R., & Takahashi, S. (2001). EMBRYONIC FLOWER2, a Novel Polycomb Group Protein Homolog, Mediates Shoot Development and Flowering in *Arabidopsis*. *Plant Cell, 13*, 2471–2481.

Yuan, P., Han, J., Guo, G., Orlov, Y. L., Huss, M., Loh, Y.-H., et al. (2009). Eset partners with Oct4 to restrict extraembryonic trophoblast lineage potential in embryonic stem cells. *Genes and Development, 23*, 2507–2520.

Yuan, W., Xu, M., Huang, C., Liu, N., Chen, S., & Zhu, B. (2011). H3K36 methylation antagonizes PRC2-mediated H3K27 methylation. *The Journal of Biological Chemistry, 286*, 7983–7989.

Yun, J.-Y., Weigel, D., & Lee, I. (2002). Ectopic expression of SUPERMAN suppresses development of petals and stamens. *Plant and Cell Physiology, 43*, 52–57.

Zamudio, N. M., Chong, S., & O'Bryan, M. K. (2008). Epigenetic regulation in male germ cells. *Reproduction, 136*, 131–146.

Zeng, W., Ball, A. R., Jr., & Yokomori, K. (2010). HP1: Heterochromatin binding proteins working the genome. *Epigenetics, 5*, 287–292.

Zhang, K., Mosch, K., Fischle, W., & Grewal, S. I. S. (2008). Roles of the Clr4 methyltransferase complex in nucleation, spreading and maintenance of heterochromatin. *Nature Structural and Molecular Biology, 15*, 381–388.

Zhao, J., Ohsumi, T. K., Kung, J. T., Ogawa, Y., Grau, D. J., Sarma, K., et al. (2010). Genome-wide identification of Polycomb-associated RNAs by RIP-seq. *Molecular Cell, 40*, 939–953.

Zhao, M., Shirley, C. R., Hayashi, S., Marcon, L., Mohapatra, B., Suganuma, R., et al. (2004). Transition nuclear proteins are required for normal chromatin condensation and functional sperm development. *Genesis, 38*, 200–213.

Zheng, D., Zhao, K., & Mehler, M. F. (2009). Profiling RE1/REST-mediated histone modifications in the human genome. *Genome Biology, 10*, R9.

Zofall, M., & Grewal, S. I. S. (2006). Swi6/HP1 recruits a JmjC domain protein to facilitate transcription of heterochromatic repeats. *Molecular Cell, 22*, 681–692.

Zuo, X., Sheng, J., Lau, H.-T., McDonald, C. M., Andrade, M., Cullen, D. E., et al. (2012). Zinc finger protein ZFP57 requires its co-factor to recruit DNA methyltransferases and maintains DNA methylation imprint in embryonic stem cells via its transcriptional repression domain. *The Journal of Biological Chemistry, 287*, 2107–2118.

CHAPTER NINE

# Parental Epigenetic Asymmetry in Mammals

**Rachel Duffié, Déborah Bourc'his**[1]

Unité Génétique Biologie du Développement, Institut Curie, UMR3215/INSERMU394, Paris, France
[1]Corresponding author: e-mail address: deborah.bourchis@curie.fr

## Contents



## Abstract

The early mammalian embryo is marked by genome-wide parental epigenetic asymmetries, which are directly inherited from the sperm and the oocyte, but are also amplified a few hours after fertilization. The yin–yang of these complementary parental programs is essential for proper development, as uniparental embryos are not viable. The majority of these parental asymmetries are erased, as the embryonic genome assumes its own chromatin signature toward pluripotency and then differentiation,

*Current Topics in Developmental Biology*, Volume 104
ISSN 0070-2153
http://dx.doi.org/10.1016/B978-0-12-416027-9.00009-7

reducing the risk for haploinsufficiency. At a few loci, however, parent-of-origin information persists through development, via maintenance and protective complexes. In this review, we discuss the parental asymmetries that are inherited from the gametes, the forces involved in their elimination, reinforcement or protection, and how this influences the embryonic program. We highlight the gradual loss of all parental asymmetries occurring throughout development, except at imprinted loci, which maintain distinct parent-of-origin chromatin and transcriptional characteristics for life. A deeper understanding of the nongenetic contributions of each germline is important to provide insight into the origin of non-Mendelian inheritance of phenotypic traits, as well as the risk of incompatibilities between parental genomes.

# 1. INTRODUCTION

The reproductive cells, the oocyte and the spermatozoon, are specialized to transmit genetic material across generations through the process of fertilization. While this function is equally shared between the two parental gametes, essential pools of proteins and mitochondria are uniquely inherited from the oocyte. The maternal ooplasm is notably required for supporting early development before the activation of the embryonic genome. Because the oocyte and the sperm harbor vastly different types of chromatin and carry different sets of RNA populations (Bourc'his & Voinnet, 2010; Gill, Erkek, & Peters, 2012), the potential for additional parental asymmetry in the early embryo is tremendous. In this chapter, we focus on nongenetically encoded information contained in the nucleus in the form of biochemical modifications imposed on the DNA molecule itself and on its supporting proteins, the histones. Because this information is intrinsically tied to the genetic material, it is more likely to persist longer after fertilization and to be maintained in an epigenetic manner throughout cell divisions in the newly developing individual and potentially to adulthood. Unraveling the origin, targets, and consequences of parent-specific information is key to understanding complex phenotypic traits in mammals, in particular, forms of non-Mendelian inheritance. Important insights can also be gained from deciphering the duration of parental epigenetic asymmetries and the forces exerted to reduce or protect their occurrence throughout life.

## 1.1. Flexibility of histone-associated information

Histone-borne information encompasses a very complex system of various redundant, exclusive, and complementary post-translational modifications mainly localized to the N-terminal tails, which freely extrude from the nucleosomes. These modifications can alter gene expression via direct effects on chromatin condensation and accessibility. This is the case specifically for acetylation, which loosens histone and DNA interactions by modifying histone charge and in turn enhances transcription. In most cases, however, histone modifications and, in particular, methylation indirectly influence gene expression through the recruitment of specialized proteins, the histone readers, which commonly act in association with large chromatin-remodeling complexes (Musselman, Lalonde, Cote, & Kutateladze, 2012). Depending on the methylation mark and the subsequent binding of different effector proteins, the impact on transcription can be positive (H3K4 methylation) or negative (H3K9, H3K27, and H4K20 methylation).

From a conceptual point of view, histone modifications are thought to confer a form of flexible regulation. The existence of specialized enzymes, known as histone modifiers, which catalyze both the establishment and the removal of histone marks, is key to allow highly dynamic patterns of histone modifications and, consequently, rapid switches between gene expression programs. This property enables temporal silencing, poising of developmentally important genes, and immediate response to signaling pathways (Bernstein et al., 2006; Song, Angel, Howard, & Dean, 2012). Most importantly, the current lack of a compelling mechanistic model for mitotic inheritance of histone modifications has forged the notion that this type of information cannot be autonomously propagated in dividing cells and instead requires continuous input from transcription factors and perhaps, in some cases, from DNA methylation. The extrapolation of this notion to the context of intergenerational inheritance has led to the prevailing view that if histone modifications can be passed from parents to progeny in mammals, they are likely to provide only transient, short-term information limited to immediate postfertilization stages. Nonetheless, the markedly divergent chromatin composition of the oocyte and sperm, resulting from dramatically disparate differentiation programs, is a major source of parental asymmetry after fertilization.

## 1.2. Long-term potential of DNA methylation-associated information

In contrast, DNA methylation is thought to be extremely stable and associated with long, and sometimes definitive, locked-in silent states. This view notably emanates from the existence of a faithful and autonomous system of DNA methylation maintenance, which is intimately coupled to DNA replication, and has been extensively studied at the biochemical, genetic, and structural levels (Bestor & Ingram, 1983; Bostick et al., 2007; Li, Bestor, & Jaenisch, 1992; Song, Teplova, Ishibe-Murakami, & Patel, 2012). The sequence symmetry of CpG motifs, which are predominantly targeted for cytosine methylation in mammals, allows the exact carbon copy of methylation patterns from the parental to the newly synthesized strand of DNA by a maintenance DNA methyltransferase enzyme, DNMT1, and thus the propagation of DNA methylation patterns to subsequent cell generations. The intrinsic robustness of this system has been demonstrated *in vitro*, as predetermined states of DNA methylation are conserved over a hundred cell divisions, provided that the initial methylation density is high (Lorincz, Schubeler, Hutchinson, Dickerson, & Groudine, 2002; Stein, Gruenbaum, Pollack, Razin, & Cedar, 1982). Because of the truly epigenetic nature of DNA methylation, this mark is thought to be more likely to be passed to the next generation upon fertilization and to contribute to parent-inherited long-term silencing in the progeny, possibly extending to adulthood. Genomic imprinting is the quintessence of DNA methylation-based transmission of gametic information occurring in a parent-of-origin manner with life-long consequences on the next generation (Ferguson-Smith, 2011).

The long-term potential for inheritance of DNA methylation marks has historically been supported by the lack of DNA demethylase activity, able to direct the rapid removal of methylated cytosines in mammalian genomes (Ooi & Bestor, 2008). The reversion from methylated to unmethylated states only seemed possible through passive dilution, requiring multiple cell divisions without replication-coupled maintenance, a mechanism incompatible with rapid changes in DNA methylation. However, the recent identification of the ten-eleven translocation (TET) enzymes has provided an avenue by which demethylation could occur on a short- time scale (Kriaucionis & Heintz, 2009; Tahiliani et al., 2009). TET enzymes catalyze the oxidation of 5-methylcytosines (5mC) into 5-hydroxymethylcytosines (5hmC), which could represent an intermediate stage toward unmethylated cytosines (see Chapter 2). In direct relevance to the question of gametic inheritance, such a mechanism permitting active alteration of methylation profiles has been shown to be potent just after fertilization and to reinforce parental asymmetry (Gu et al., 2011; Iqbal, Jin, Pfeifer, & Szabo, 2011; Wossidlo et al., 2011). This implies the existence of protective factors, which allow maintenance of gamete-inherited cytosine methylation at certain loci and the persistence of genomic imprinting beyond this period.

### 1.3. Forces exerted to reduce parental epigenetic inheritance

The extent of gametic histone and DNA modification inheritance is limited by different forces driving the extensive reprogramming of the early developing mammalian embryo (Morgan, Santos, Green, Dean, & Reik, 2005). Maternal to zygotic transition begins within the first divisions of the newly formed embryo, as early as the two-cell stage in mice and around the four- to eight-cell stage in humans. Previously inherited silent states must be overcome rapidly to set up chromatin states permissive for the activation of embryonic genes. Next, the fully differentiated gametic state must be reset between fertilization and the blastocyst stage to ensure the acquisition of pluripotency, the ability to generate all cell lineages. This is followed by embryonic differentiation programs around the time of implantation of the embryo in the uterus, concomitantly with loss of pluripotency and cell lineage diversification. These early events leave few opportunities for inheritance of gametic information. Moreover, gametic information that occurs in a parent-specific manner is naturally counter-selected; allelic epigenetic specificities tend to be reduced to allow functional equivalence of the two parental chromosomes and to limit haploinsufficiency, where normal function is lost due to the availability of one active copy of a gene instead of two.

However, functional non-equivalence of the two parental alleles of a gene may also be useful, in certain developmental and physiological circumstances, where it may be crucial to fine-tune expression levels. Early embryonic development may be particularly prone to this type of single-dose regulation, as minute differences in gene expression may be key for promoting differential cell fates (Guo et al., 2010; Miyanari & Torres-Padilla, 2012; Tang et al., 2011). Interactions through the placenta between the mother and the fetus also seem to be especially sensitive to single-dose regulation, as dysregulation of the monoallelic expression of imprinted genes is particularly deleterious in biological pathways involved in maternal–fetal exchanges (Schulz et al., 2010). Here, we provide a survey of parental epigenetic asymmetries in mammals, organized in a developmental time frame, from their establishment in gametes, to their selective consolidation or erasure immediately after fertilization, to their further restriction during embryonic development.

## 2. GENERATING ASYMMETRIC NUCLEAR-BASED INFORMATION IN PARENTAL GAMETES

The cartography of the histone and DNA modifications that are differentially carried by the mature oocyte and spermatozoa must first be described to address the potential for long-lasting parental asymmetry in the progeny. Gametogenesis provides a convenient period for parental genomes to be distinctly marked, as they are physically separated in different cell types and even different organisms. Profound differences exist between males and females regarding the hormonal, temporal, structural, and functional control of germ-cell development, which shape gametic epigenomes in a highly sexually dimorphic manner. The various chromatin components and the histone and DNA modifications that define the sperm and oocyte genomes have just started to be characterized on the genome-wide level.

## 2.1. Sex-specific chromatin organization and histone modifications

Enzymes involved in histone modification and chromatin organization are strictly required for proper gametogenesis (Gill et al., 2012). Changes in chromatin modifications and expression of specific histone modifiers are among the first markers of emerging germ-cell precursors, amid surrounding somatic cells of the early embryo (Kota & Feil, 2010). At this early stage, global histone modification patterns are indistinguishable in male and female germ cells. At the end of gametogenesis, however, mature oocyte and sperm chromatin profiles are dramatically different, reflecting the specific constraints exerted on the two parental germlines during their differentiation.

To reach the oocyte, the sperm must travel along both the male and female reproductive tracts. To enhance motility and protection from DNA damage, the sperm nucleus adopts an extreme state of condensation, made possible via the replacement of canonical histone proteins by smaller basic structural proteins, termed protamines. This transformation in DNA packaging occurs in a sequential manner, starting in transcriptionally inactive spermatids with the incorporation of testis-specific histone variants, followed by a phase of histone hyperacetylation which may provide an accessible environment for the final histone eviction and substitution with protamines (Gaucher et al., 2010). It was believed that in this way the sperm could contribute little to no histone-based information to the progeny.

However, biochemical and microscopy methods suggest that 10–15% of the paternal genome can persist in the form of histone-containing nucleosomes in human sperm (Brykczynska et al., 2010; Gusse & Chevaillier, 1980; Tanphaichitr, Sobhon, Taluppeth, & Chalermisarachai, 1978), a feature confirmed by the detection of several H3 histone variants by immunofluorescence (van der Heijden et al., 2008). Recent genome-wide mapping of nucleosome distribution indicated that retention occurs nonrandomly, with a twofold enrichment at regulatory sequences around transcription start sites, and a significant overrepresentation at the promoters of transcription factors and signaling molecules that guide embryonic development (Brykczynska et al., 2010; Hammoud et al., 2009). Loci enriched in histones are not completely devoid of protamines; instead, histones are diluted among protamines, forming heterogeneous histone/protamine mixtures. Further cartography of histone modifications by chromatin immunoprecipitation (ChIP) followed by array analysis or high-throughput sequencing has thus far been focused on permissive H3K4me2/me3 marks and repressive H3K27me3 marks (Brykczynska et al., 2010; Hammoud et al., 2009). While transcriptionally inert, the mature sperm contains H3K4me2/3 marks, which map specifically to promoters of genes involved in spermatogenesis and basic cellular processes, potentially reflecting expression and requirement of these gene functions at prior stages of male germ-cell development (Brykczynska et al., 2010). A distinct set of genes is marked by H3K27me3 modifications. These are mostly developmentally important genes such as the *Hox* genes and early embryonic transcription factors, such as *Sox2*, *Cdx2*, and *Gata6* (Brykczynska et al., 2010; Hammoud et al., 2009). In patterns reminiscent of what was previously reported as bivalent chromatin domains in ES cells, dual enrichment of H3K4me2/3 and H3K27me3 was also observed at the promoters of some development-related genes, which may be important for poising future expression states (Bernstein et al., 2006; Pan et al., 2007). Programmed histone retention and modifications in sperm could be important for fertility, as genome-wide analyses have highlighted altered sperm histone composition in subfertile men with various reproductive dysfunctions (Hammoud et al., 2011).

In comparison with humans, protamination occurs more extensively in mouse sperm cells, with an estimation of only 1% of total DNA retained in nucleosomes (Brykczynska et al., 2010; van der Heijden et al., 2006). Nonetheless, high conservation exists between mouse and human with respect to H3K4me2 marks at housekeeping and testis-specific genes, as well as H3K27me3 enrichment at developmentally important loci (Brykczynska et al., 2010; van der Heijden et al., 2006). However, some discrepancies exist, such as the dual occupancy of H3K4me3 and H3K27me3 marks at some testis-specific genes observed in mice only. This specificity may reflect the need to prevent spurious expression of these genes after fertilization, a risk that could be incurred due to the relatively early activation of the mouse embryonic genome as compared to humans. Genome-wide mapping of these and other histone modifications remains to be conducted; however, these data suggest that the paternal genome could transmit a certain level of histone-based information, provided that these are not immediately replaced by oocyte-inherited histones in the zygote. Of note, H3K27me3-marked regions exhibited higher histone retention in nucleosome/protamine mixtures than H3K4me3-marked regions, arguing for a greater likelihood that sperm H3K27me3 marks play a role in paternal epigenetic transmission. In support of this, a correlation was found between high enrichment of H3K27me3 marks in spermatozoa and repression of associated genes in early embryogenesis (Brykczynska et al., 2010; van der Heijden et al., 2006).

The major sex-specific constraint exerted on the oocyte is to produce stores of RNAs and proteins to support embryonic development immediately after fertilization. Accordingly, a state of acute transcriptional activity characterizes the oocyte during the growing phase of development. Our current knowledge about histone variants and modifications in oocytes comes mainly from immunofluorescence studies and genetic deletions of histone-modifying enzymes (Abe, Tsai, Jin, Pfeifer, & Szabo, 2011; Ciccone et al., 2009; Gu, Wang, & Sun, 2010; Liu, Kim, & Aoki, 2004). Growing oocytes gain increasing levels of histone acetylation, in agreement with a transcriptional burst, which are then reversed into hypoacetylated states as meiosis is resumed (Kim, Liu, Tazaki, Nagata, & Aoki, 2003). Methylation marks associated with active transcription such as H3K4me2/3 and H3K79me2/3, and those associated with repression such as H3K9me2/3 and H3K27me2/3, are also globally detected in mouse oocyte nuclei (Mayer, Smith, Fundele, & Haaf, 2000; Ooga et al., 2008; Sarmento et al., 2004). Due to the difficulty in obtaining large numbers of oocytes for ChIP assays, studies addressing the modification status of specific loci have not been amenable so far. However, recent advances in techniques applicable to small quantities of material should soon allow the genome-wide analysis of the histone modification landscape of the mammalian oocyte (Dahl & Collas, 2008; Goren et al., 2010). Currently, it is not known which specific loci carry differential histone marks in oocyte and sperm and could thus impart asymmetric information to the embryo. Importantly, not only are oocyte nucleosomes passed to the zygote, both in the form of incorporated and free histones, but also the enzymes necessary to posttranslationally modify histones in the early embryo. Further studies in early embryos are required to determine how stably information contained in oocyte and sperm nucleosomes is transmitted during preimplantation development, and most importantly, whether it can impact the embryonic program.

## 2.2. Sex-specific DNA methylation patterns

The acquisition of DNA methylation patterns in the gametes has been more extensively studied than histone modifications, partly because of the larger availability of complementary techniques to study DNA methylation. Details on the timing and drivers of gametic DNA methylation are reported in Chapters 2 and 5. To summarize, erasure of somatic DNA methylation patterns in the future gametes occurs in two waves during embryonic life, first during the migration of the primordial germ cells (PGCs) toward the genital ridges and secondly at the time of PGC entry into these future gonads (Guibert, Forne, & Weber, 2012; Hackett et al., 2012; Popp et al., 2010). Next, establishment of DNA methylation profiles is specified separately in the differentiating female and male germlines under the control of sex-specific signaling pathways and kinetics (Schaefer, Ooi, Bestor, & Bourc'his, 2007). Maternal methylation patterns are established postnatally during the growing phase of meiotically arrested oocytes. In contrast, paternal methylation patterns are acquired very early, during fetal life, and will be propagated until the release of mature sperm, which require successive cell divisions and can span decades in long-lived mammalian species such as humans.

As a result of these sex-specific programs, the mature oocyte and sperm harbor very different genomic methylation patterns. This was shown on a global level in early studies, where quantification of DNA methylation content revealed that the oocyte genome is less methylated than the sperm genome (Monk, Boubelik, & Lehnert, 1987). Evidence for gametic differences in DNA methylation also came from the identification of imprinting control regions (ICRs), which are *cis*-regulatory elements that control the expression of imprinted genes and which carry, by definition, parent-specific DNA methylation marks inherited from the gametes. Around 23 ICRs have been identified in the mouse, which in turn control the parent-specific monoallelic expression of some 130 imprinted genes. Maternal ICRs are specifically methylated in the oocyte but not in sperm, while paternal ICRs are methylated in sperm but not in the oocyte. The sexual dimorphism of the ICRs not only arises from exclusive parental methylation but also from the number of ICRs methylated in each parental gamete: only three ICRs are methylated in sperm and all the remaining ones are methylated in the oocyte, illustrating a strong bias toward the maternal genome in imprinting control (Proudhon et al., 2012; Reik & Walter, 2001; Schulz et al., 2010). Moreover, while maternally methylated ICRs coincide systematically with CpG-rich promoters, the three known paternally methylated ICRs map to intergenic regions with a relatively poor CpG content.

It was generally assumed that sex-specific DNA methylation was preferentially targeted to these ICRs. However, recent advances in techniques that permit genome-wide methylation analysis of small quantities of DNA have demonstrated that sex-specific DNA methylation is much more prevalent in mouse gametic genomes (Kobayashi et al., 2012; Smallwood et al., 2011; Smith et al., 2012). Genome coverage differed between the studies due to the different techniques employed. All are based on the chemical method of bisulfite conversion, but on the one hand, the reduced representation bisulfite sequencing (RRBS) technique enriches for CpG-rich regions such as CpG islands (CGIs) and in reducing genome size allows deeper coverage in high-throughput sequencing (Smallwood et al., 2011; Smith et al., 2012). On the other hand, whole-genome bisulfite sequencing (Bisul-Seq) enables a less biased assessment but at the expense of lower sequencing coverage (Kobayashi et al., 2012). Different cutoffs were used for minimum number of reads per CpG for subsequent analyses. Moreover, two studies used a recent cartography of 23,021 CGIs in mouse (Illingworth et al., 2010), while the other one adopted a 100 bp sliding window approach (Smith et al., 2012). These differences in sequencing coverage, thresholds for concluding whether a CpG is methylated or not, and genomic region classification lead to different numbers of CpGs, CGIs, or germline differentially methylated regions (gDMRs) that were counted as methylated in oocyte and/or sperm (Table 9.1).

Globally, these studies confirmed that the sperm genome is more methylated than the oocyte genome, although the percentage of total CpG methylation was variably reported as 25%, 83%, or 90% in sperm and as 15%, 32%, or 40% in the oocyte, depending on the study (Kobayashi et al., 2012; Smallwood et al., 2011; Smith et al., 2012). Genomic methylation in sperm is relatively enriched at regions of low to intermediate CpG density, corresponding to intergenic regions and repeats, especially Intracisternal A Particle (IAP) and Long Interspersed Nuclear Elements, group 1 (LINE-1) retrotransposons (Kobayashi et al., 2012; Smallwood et al., 2011; Smith et al., 2012). The repetitive nature of these sequences is sufficient to explain quantitative dominance of CpG methylation in the sperm versus oocyte genome, as retrotransposons account for half of mammalian genomes (Rollins et al., 2006). In contrast, CpG-rich CGIs, genomic elements typically associated with gene promoters, are more likely to be methylated in the oocyte than in the sperm. Around a thousand CGIs are methylated only in the oocyte genome, while between 3- and 17-fold fewer CGIs may be specifically methylated in the sperm genome, depending on the study (Kobayashi et al., 2012; Smallwood et al., 2011). These numbers exceed the current estimates for ICRs by far (23) and demonstrate that sex-specific methylation is a general outcome of the divergent programs

**Table 9.1** Genome-wide studies of oocyte and sperm DNA methylation patterns in mouse

| Study | Technique | Coverage | Global % methylation | Genomic distribution | Oocyte gDMRs in CGIs | Oocyte CGI gDMRs in blastocysts | % Oocyte CGI in blastocysts |
|---|---|---|---|---|---|---|---|
| *Oocyte methylation* | | | | | | | |
| Smallwood et al. (2011) | RRBS | >5 × | 15 | 40% promoter; 50% intragenic; 10% intergenic | 962 | 207 | 21.5 |
| Kobayashi et al. (2012) | Bisul-Seq | 70% at 1 ×; 40% at 5 × (avg. 18.8 ×) | 40 | Intragenic | 1329 | 817 | 61.5 |
| Smith et al. (2012) | RRBS | > 10 × for a single CpG; > 5 × for 100 bp tiles | 32 | Intragenic CGI promoters | 376 (100 bp tiles) | NP | NP |

| Study | Technique | Coverage | Global % methylation | Genomic distribution | Sperm gDMRs in CGIs | Sperm CGI gDMRs in blastocysts | % Sperm CGI in blastocysts |
|---|---|---|---|---|---|---|---|
| *Sperm methylation* | | | | | | | |
| Smallwood et al. (2011) | RRBS | >5 × | 25 | 14% promoter; 74% intragenic; 12% intergenic | 58 | 35 | 60.3 |
| Kobayashi et al. (2012) | Bisul-Seq | 80% at 1 ×; 63% at 5 × (avg. 12.5 ×) | 90 | Intergenic; intragenic | 349 | 34 | 9.7 |
| Smith et al. (2012) | RRBS | > 10 × for a single CpG; > 5 × for 100 bp tiles | 83 | Intergenic Repeat elements | 4894 (100 bp tiles) | NP | NP |

undergone by female and male germ cells. Interestingly, oocyte- and sperm-specific methylation of distinct genomic regions according to their CpG content and localization with respect to genes is consistent with the different nature of maternal and paternal ICRs. This divergence may reflect the different DNA methylation-guiding pathways in the two parental germlines.

The acquisition of methylation patterns in gametes depends on the *de novo* methyltransferase Dnmt3A and its cofactor Dnmt3L in both sexes (Fig. 9.1A; Bourc'his & Bestor, 2004; Bourc'his, Xu, Lin, Bollman, & Bestor, 2001; Kaneda et al., 2004). It is also believed that active H3K4me2/3 marks could repel DNA methyltransferases at defined genomic sequences in the two parental germlines. The strongest evidence for this comes from *in vitro* biochemical studies, which revealed that DNMT3A and DNMT3L bind lysine 4 of histone H3 directly, but that these interactions are antagonized when the lysine residue is methylated (Ooi et al., 2007; Otani et al., 2009). Moreover, the histone H3K4 demethylase KDM1B (also known as LSD2 or AOF1) is required for the establishment of DNA methylation for at least four maternal ICRs in the oocyte (Ciccone et al., 2009). In agreement with this, a general depletion of DNA methylation occurs at loci with H3K4me3 enrichment in mouse oocytes (Smallwood et al., 2011). Conversely, enrichment of H3K4me3 marks is observed at maternal ICRs during male germ-cell development, potentially underlying their exclusion from the paternal *de novo* methylation program (Henckel, Chebli, Kota, Arnaud, & Feil, 2012).

An additional component that shapes the oocyte and sperm methylomes lies in transcription-dependent mechanisms. Pioneering genetic studies demonstrated that transcription traversing the imprinted *Gnas* maternal ICRs is essential for oocyte methylation establishment at this locus (Chotalia et al., 2009). This requirement for transcription appears to be a genome-wide phenomenon, as RNA-seq studies in oocytes link DNA methylation in gene units with active transcription of these units (Kobayashi et al., 2012). Interestingly, alternative oocyte-specific promoters are commonly located upstream of canonical, somatically used promoters. These downstream promoters frequently become methylated in the oocyte and not in the sperm (Smallwood et al., 2011), such as genes encoding members of the DNMT family, *Dnmt1*, *Dnmt3b*, and *Dnmt3L* (Guenatri, Duffie, Iranzo, Fauque, & Bourc'his, 2013; Smallwood et al., 2011; Smith et al., 2012). The mechanism for transcription-associated DNA methylation in the oocyte has yet to be elucidated. A similar mechanism to gene-body methylation in somatic cells could be at play. Gene-body methylation has been proposed to occur through the recruitment of DNA methyltransferases along the path of the elongating transcript (Ball et al.,

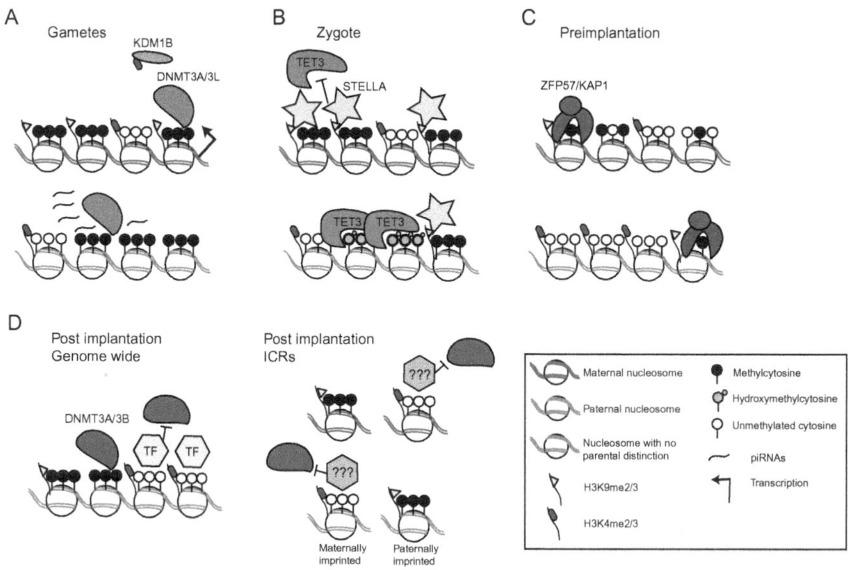


**Figure 9.1** *Forces acting toward the establishment and the protection of gametically inherited DNA methylation marks.* (A) DNA methylation establishment in the gametes. DNMT3A and DNMT3L work in the female and male germlines to establish DNA methylation profiles genome-wide. In oocytes (upper part, red nucleosomes), this process is guided, at least in part, by traversing transcription and by the removal of H3K4me2/me3 marks, which can be mediated by the KDM1B histone demethylase. In the developing sperm (lower part, blue nucleosomes), DNMT3A/3L-dependent DNA methylation can be guided by piRNAs. Major remodeling of sperm chromatin will occur in the last stages of spermatogenesis by histone to protamine exchange (not shown). (B) DNA methylation loss and protection in the one-cell zygote. Before the onset of the first S-phase, H3K9me2, which is globally associated with maternally methylated DNA, recruits STELLA thus preventing TET3 binding. On paternal chromatin, TET3 binds 5mC and converts it to 5hmC, except at certain loci such as paternal ICRs, which are protected by H3K9me2-dependent STELLA binding. (C) DNA methylation loss and protection during preimplantation development. DNA methylation is passively lost during this period, following embryonic cell division. Methylated sequences that harbor TGCCGC motifs escape this passive demethylation through the recruitment of the heterochromatic ZFP57/KAP1 complex. (D) DNA methylation gain and protection in the early postimplantation embryo. While *de novo* DNA methylation is set up genome wide, mainly via the activity of DNMT3B and to some extent of DNMT3A, some loci are protected with an unmethylated state. Transcription factor binding and/or H3K4 methylation marks could protect these regions from the action of DNMT3A/3B. This shielding mechanism is likely to be operant genome wide and at ICRs, in particular, leading in this case to the maintenance of parent-specific DNA methylation. (See Color Insert.)

2009; Feng et al., 2010; Zemach, McDaniel, Silva, & Zilberman, 2010). Alternatively, the mechanism for transcription-associated DNA methylation in the oocyte could rely on *trans*-acting small RNA species, such as endo-siRNAs (small interfering RNAs), which are abundant in oocytes (Watanabe et al., 2008). This mode would mirror the small RNA-dependent pathway of *de novo* methylation observed in the sperm, where other small RNA populations are involved, the piRNAs (piwi-interacting RNAs) (Fig. 9.1A). In this model, retrotransposon transcripts produced in gamete precursors as DNA methylation is erased would serve as substrates for the production of piRNAs via the slicing activity of PIWI proteins. PIWIs loaded with piRNAs can in turn feedback into the nucleus to promote DNA methylation at the promoters of retrotransposons by homology-dependent recognition (Aravin & Bourc'his, 2008; Aravin et al., 2008). The mechanism of action by which piRNA/PIWI complexes target *de novo* DNA methyltransferases to retrotransposons in the male germline is still a matter of intense investigation. Recently, one of the three paternal ICRs, *Rasgrf1*, was shown to acquire DNA methylation via the piRNA pathway, owing to the presence of a retrotransposon in close proximity to this ICR (Watanabe et al., 2011). However, the two other paternal ICRs acquire DNA methylation in a piRNA-independent manner, leaving the possibility open that other DNA methylation targeting systems act at these loci.

Finally, in addition to unequal distribution of methylation, there are also differences in the sequence motifs enriched for cytosine methylation in the two mature parental gametes. Indeed, while DNA methylation was thought to concern only symmetric CG dinucleotide contexts, recent reports have provided evidence for asymmetric non-CG methylation in the oocyte, ES cells, and brain (Arand et al., 2012; Guenatri et al., 2013; Lister et al., 2009; Proudhon et al., 2012; Smith et al., 2012; Tomizawa et al., 2011; Xie et al., 2012). This non-CG methylation is believed to reflect the recent activity of *de novo* DNA methyltransferases, which methylate cytosines in all contexts. However, asymmetric cytosine methylation cannot be maintained by DNMT1 upon cell divisions. Therefore, non-CG methylation probably only exists in nondividing cells, such as oocytes, or in dividing cell types containing large amounts of *de novo* DNA methyltransferases, such as ES cells, which will reiteratively establish non-CG methylation after each replication. Non-CG methylation is present in the oocyte genome, but not in the sperm genome, which has lost this methylation type upon the multiple cell divisions that separate prenatal *de novo* DNA methylation from mature sperm (Ichiyanagi, Ichiyanagi, Miyake, & Sasaki, 2012). The function, if any, of non-CG methylation is still elusive.

To summarize, at fertilization, the oocyte methylome includes 1000 methylated single-copy CGIs, which represent only a minor fraction of the total 23,000 CGIs. The sperm is hypermethylated as compared to the oocyte, and intergenic repeats are the main targets of this methylation. Oocyte-specific methylation of somatic promoters may regulate levels, kinetics, or alternative splicing of genes during early embryonic development. In contrast, sperm-specific methylation could prevent retrotransposon expression following fertilization and at the time of zygotic genome activation, ensuring protection of the embryonic genome, as it does for the sperm genome (Bourc'his & Bestor, 2004). However, as developed in Section 3, the sperm-derived methylome is rapidly remodeled after fertilization, leaving several families of retrotransposons methylation free, while the maternal genome progressively loses DNA methylation through embryonic cell divisions. The field will now need to dissect exactly which marks at which sequences are only byproducts of gamete differentiation and which marks can directly impact embryonic development.

## 3. IMMEDIATE INHERITANCE AND REINFORCEMENT OF PARENTAL ASYMMETRY IN THE ONE-CELL ZYGOTE

As the immediate product of gametic fusion, the one-cell zygote represents the developmental stage at which parental differences are the most apparent. Some of these are directly inherited from the gametes, but new differences are also generated by maternally provided stores of proteins that catalyze histone and DNA modifications, resulting in the amplification of parental asymmetry. Although the two sets of parental chromosomes share the same ooplasm environment, they maintain striking topological separation after fertilization, which remains even when they unite within the same nucleus after the first mitosis and at least until the four-cell stage (Mayer et al., 2000). This physical dissociation may be more than just a consequence of the different developmental histories of the oocyte- and the sperm-derived chromatin; it may also play a crucial role in the differential treatment of the two parental pronuclei. Indeed, despite exposure to the same maternally provided factors, the parental pronuclei undergo very different programs during the few hours following fertilization, as a prelude to zygotic genome activation. The maternal pronucleus must be released from the metaphase II arrest to complete meiosis and to reduce maternal genome content to one set of haploid chromosomes. Meanwhile, the paternal pronucleus must be converted into an accessible, transcriptionally permissive chromatin state. An important time point to consider is the first DNA replication phase (Fig. 9.2), which starts around 5–6 h after fertilization and is completed before 9 h in the mouse, with strain-to-strain variability. Any histone or DNA modification preceding the first S-phase is replication-independent and likely to involve active enzymatic mechanisms; changes occurring at replication are likely to be passive, resulting from the synthesis of new unmethylated DNA strands and the incorporation of new replication-dependent histones.

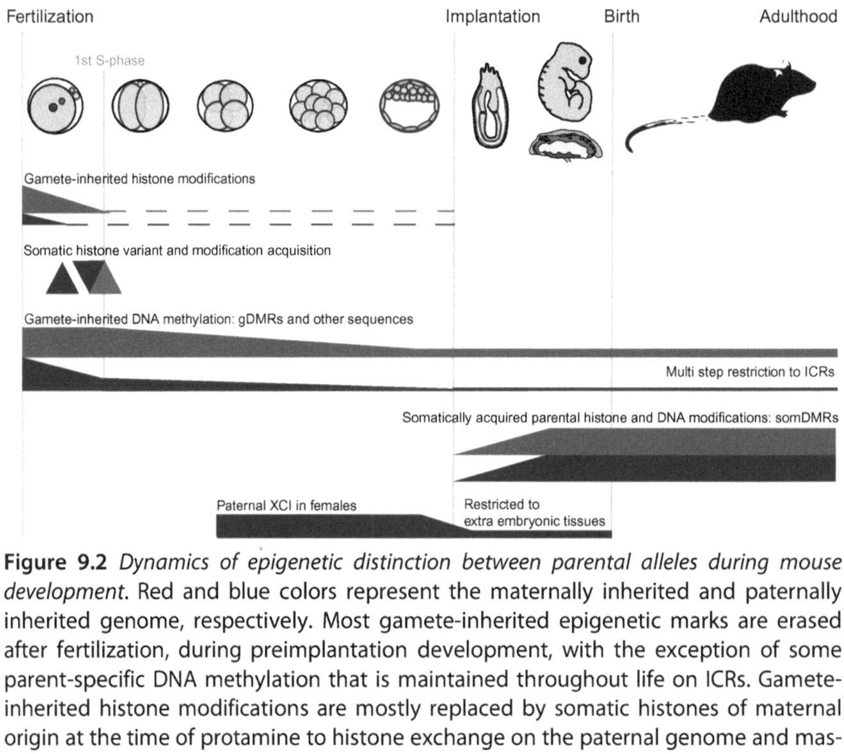


**Figure 9.2** *Dynamics of epigenetic distinction between parental alleles during mouse development.* Red and blue colors represent the maternally inherited and paternally inherited genome, respectively. Most gamete-inherited epigenetic marks are erased after fertilization, during preimplantation development, with the exception of some parent-specific DNA methylation that is maintained throughout life on ICRs. Gamete-inherited histone modifications are mostly replaced by somatic histones of maternal origin at the time of protamine to histone exchange on the paternal genome and massively on the two parental genomes of the one-cell zygote by replication-dependent incorporation at the first S-phase. Gametic DNA methylation is lost in part on the paternal genome before the first S-phase by an active mechanism and then progressively on the two parental genomes by a passive mechanism of replication-dependent dilution during preimplantation development. Only ICRs maintain gamete and parent-specific DNA methylation patterns throughout life. At implantation, new loci acquire parent-specific histone and DNA modifications, under the control of ICRs, the somatic DMRs (somDMRs). Finally, paternal XCI occurs in female preimplantation embryos but is reset at implantation to allow random XCI, except in extraembryonic tissues where paternal XCI continues. (See Color Insert.)

### 3.1. Reshuffling of histone variants and modifications

Very early on, within the first 3 hours after entry into the oocyte, the condensed sperm chromatin is submitted to extensive, active remodeling, while the maternal chromatin is relatively stable. Paternal protamines are stripped and replaced with oocyte-supplied histones, using H3.3 histone variants, which can be incorporated in a DNA replication-independent manner, through the control of the histone assembly chaperone HIRA (Loppin et al., 2005; Torres-Padilla, Bannister, Hurd, Kouzarides, & Zernicka-Goetz, 2006). However, some H3.1 and H3.2 variants, which can be assembled only through DNA replication, are still detected on paternal chromatin by immunofluorescence at this stage. This observation suggests a possible participation of the limited amount of sperm-derived histones to zygotic chromatin, at least transiently before the first S-phase (Fig. 9.2; van der Heijden et al., 2008). The maternal pronucleus is, on the other hand, exclusively loaded with canonical H3 histones, H3.1 and H3.2, in direct continuation with the oocyte chromatin state (van der Heijden et al., 2005).

The parental pronuclei also display different histone modifications before the onset of the first S-phase (for review, see Burton & Torres-Padilla, 2010). Compared to maternal chromatin, the newly formed paternal chromatin is hyperacetylated, a state that could participate in the global decondensation of the paternal pronucleus (Adenot, Mercier, Renard, & Thompson, 1997; Santos, Peters, Otte, Reik, & Dean, 2005). Acetylated histones H4 may be directly incorporated in this modified form at the time of protamine to histone exchange (Sobel, Cook, Perry, Annunziato, & Allis, 1995). In terms of histone methylation, however, the paternal pronucleus is globally depleted, while the maternal pronucleus is enriched, as was the oocyte, for H3K4me2/3, H3K9me2/3, and H3K27me1/2/3 marks (van der Heijden et al., 2005). By the final stage of the zygote (PN5), after replication and incorporation of replication-dependent histones, the asymmetry between the two parental genomes in terms of histone variants and the majority of histone acetylation and methylation marks mostly disappears.

H3K9me2/3 marks represent an exception, as these modifications continue to specifically decorate the maternal genome, but not the paternal one after the first S-phase (Liu et al., 2004). The combination of histone hyperacetylation and lack of repressive H3K9 methylation may participate in the more precocious transcriptional activation of the paternal pronucleus, which can be observed starting from the S-phase of the one-cell zygote and onward, while maternal pronucleus activity is not detectable until the two-cell stage (Aoki, Worrad, & Schultz, 1997). As developed later in this chapter, one function of maternal H3K9me2 retention could be the protection of the maternal pronucleus from the removal of 5mC marks, which occurs on the paternal pronucleus during the early phase of zygotic development (Nakamura et al., 2006, 2012; Szabo & Pfeifer, 2012). Heterochromatic domains formed by pericentric satellite DNA show some other striking differences in histone modifications. Maternal pericentric regions exhibit classical constitutive heterochromatic marks, such as H3K9me3 and the associated heterochromatin protein 1 Beta (HP1β), inherited from the oocyte (Probst & Almouzni, 2011; Santos et al., 2005). These features are not present on paternal pericentric regions, which are originally devoid of heterochromatic marks after fertilization. At the time of DNA replication, they acquire a form of facultative heterochromatin, H3K27me3 marks, under the control of the Polycomb repressive complex 2 (Puschendorf et al., 2008). Accordingly, maternally inherited Polycomb repressive complex 1 proteins specifically label paternal heterochromatin (Puschendorf et al., 2008). Remarkably, abolition of parental asymmetry occurs when the H3K9 methyltransferase Suv39h2 is inactivated, leading to the acquisition of a paternal-like status by the maternal heterochromatin, which becomes similarly coated by H3K27me3 marks (Puschendorf et al., 2008).

### 3.2. Parent-specific DNA methylation fates

Pioneering immunofluorescence studies using antibodies directed against 5mC showed a rapid loss of methylation from the paternal pronucleus within 5 hours postfertilization, creating a dramatic asymmetry between the two parental pronuclei before the first S-phase (Fig. 9.2; Mayer et al., 2000; Rougier et al., 1998; Santos, Hendrich, Reik, & Dean, 2002). Site-specific bisulfite analysis, at LINE-1 elements, in particular, suggested a two-step requirement for this loss, with an active initiation phase, reinforced by a replication-dependent mechanism of passive demethylation (Wossidlo et al., 2011). Recent studies have demonstrated that the first step is concomitant with gain of 5hmC on the paternal pronucleus, mediated by a store of maternally provided TET3 protein (Iqbal et al., 2011; Wossidlo et al., 2011). Zygotic TET3 deficiency leads to the retention of 5mC on the paternal pronucleus, abolishing global parental differences in cytosine methylation levels. Transition from methylated to unmethylated cytosines on the paternal pronucleus seems thus to require an intermediate conversion step toward hydroxymethylation. How conversion from hydroxymethylation to demethylation is achieved is the subject of intense investigation. Leading hypotheses, including further oxidation and DNA repair pathways, can be found in Chapter 2 and W. Reik's chapter.

The differential treatment of the two parental pronuclei probably arises through a combination of mechanisms. Notably, H3K9me2 enrichment was shown to protect the maternal pronucleus from global TET3-induced conversion to hydroxymethylation via the recruitment of oocyte-provided STELLA/PGC7 protein (Fig. 9.1B), a protein involved in chromatin condensation and transcriptional repression (Liu, Nakamura, & Nakano, 2012; Nakamura et al., 2006, 2012; Szabo & Pfeifer, 2012). On the paternal pronucleus, lack of H3K9me2, combined with hyperacetylation, may contribute to enhanced accessibility to TET enzymes. However, remarkably, STELLA specifically protects two paternal ICRs from active paternal DNA demethylation, which allows their propagation beyond the one-cell stage (Nakamura et al., 2006).

A recent stage-by-stage RRBS methylome analysis provided the first genome-wide sequence resolution map of DNA methylation changes during this important transition period (Smith et al., 2012). In agreement with the cytological studies, a 10–60% drop was observed between gametic and one-cell zygotic DNA methylation levels, such that the sequences which are intermediately methylated in the zygote correspond to the sequences that are highly methylated in the oocyte and sequences with nearly no methylation in the zygote are highly methylated in the sperm. By tracking parent-specific information at several loci using single nucleotide polymorphisms, the specific loss of paternal methylation was demonstrated. As the sperm mostly provides DNA methylation at repeats, the methylation loss was confirmed to occur at specific families of LINE-1 elements, in particular, the Tf and Gf classes, as well as at Long Terminal Repeat (LTR) retrotransposons, which resemble infectious retroviruses in their genomic organization and sequences. One notable exception occurs at the LTR family of IAP elements, which are particularly aggressive in the mouse genome and exhibit intrinsic resistance to DNA demethylation in several developmental contexts (Lane et al., 2003). As hydroxymethylcytosines and methylcytosines are indistinguishable in bisulfite-based assays (Huang et al., 2010), the lack of methylation-associated reads on the paternal pronucleus implies that complete conversion toward demethylation has already occurred. This is in agreement with the fact that the one-cell embryo pools used in this RRBS study were collected 10 hours postfertilization.

The biological significance of such an abrupt and early demethylation of the paternal genome is unknown. Allowing demethylation of paternal retrotransposons could pose a danger to the early embryo, as reactivation of these elements may lead to mobilization and induction of mutagenic insertions. However, certain elements may serve beneficial functions, and a particular class of LTR elements, the murine endogenous retrovirus-like (MuERV-L), was recently involved in promoting pluripotent properties in embryonic cells (Macfarlan et al., 2012). Expression of these elements exclusively peaks at the time of major embryonic genome activation, at the two-cell stage. It is not known whether MuERV-L elements are methylated in sperm, but if so, their autonomous activation in the embryonic genome would require prior removal of DNA methylation from their promoters. Still in relation with pluripotency, TET3 deficiency greatly impairs embryonic development potential (Gu et al., 2011). This was linked to retention of DNA methylation at the promoter of *Nanog* and *Oct4* genes, which may delay the expression of these pluripotency factors from the paternal genome.

## 4. GLOBAL LOSS AND SPECIFIC RETENTION OF PARENTAL ASYMMETRIES DURING PREIMPLANTATION DEVELOPMENT

At the two-cell stage of mouse development, the two parental genomes become equally involved in transcription to promote full embryonic genome activation. Moreover, at that time, most of the parent-specific histone-based differences have been reconciled due to the incorporation of new replication-dependent histones at the first S-phase. Gametic inheritance of DNA methylation patterns was reduced during the first-cell stage by an active paternal-specific demethylation process. Patterns are further reduced by a progressive loss of DNA methylation occurring equally on the two parental genomes as the embryo progresses toward the acquisition of pluripotency (Fig. 9.2). However, site-specific mechanisms of protection become fully operant during this period for the retention of a subset of germline-inherited differentially methylated regions, which include autosomal imprinted loci. Another form of parental imprinting coexists, which occurs on a chromosome-wide level, leading to the systematic silencing of the paternally inherited X chromosome in females, at least in the mouse model.

### 4.1. Selective protection of parent-specific DNA methylation patterns

Early immunofluorescence studies originally revealed the occurrence of hemimethylated chromosomes at the two-cell stage and their progressive replacement by unmethylated chromosomes toward the blastocyst stage, as a signature of replication-dependent loss of DNA methylation (Rougier et al., 1998). More recent genome-wide molecular studies confirmed a gradual decrease in both CG and non-CG methylation toward implantation, reaching a minimal level by E3.5 (embryonic day 3.5 postfertilization) blastocysts, for both oocyte- and sperm-specific gDMRs (Smith et al., 2012). However, while this phenomenon is prominent at repeated sequences, a significant proportion of CpG-rich CGI sequences are resistant to DNA demethylation. In particular, 20–60% of oocyte-methylated CGIs maintain methylation in the blastocyst, which represent a variable estimation between 200 and 800 loci, depending on studies (Table 9.1; Kobayashi et al., 2012; Smallwood et al., 2011). As discussed earlier, fewer sperm-specific gDMRs exist as CGIs and about 35 paternal CGIs are maintained from sperm to blastocysts. These numbers indicate that protection of parent-specific methylation extends beyond the 23 known ICRs.

Key molecular players responsible for the maintenance and protection of specific loci against passive demethylation have recently been identified. The Krüppel-associated box-containing zinc-finger protein ZFP57 was the first component shown to be involved in maintenance of ICR methylation during early embryonic development (Li et al., 2008). ZFP57 enables DNA binding of a heterochromatin protein complex organized around KAP1 (Krüppel-associated protein, also known as TRIM28), which also contains the H3K9 methyltransferase SETB1, HP1, and the nucleosome remodeling NuRD complex. Accordingly, loss of maternally inherited KAP1 also results in stochastic loss of DNA methylation at known maternal and paternal ICRs as well as embryonic lethality at midgestation (Li et al., 2008; Messerschmidt et al., 2012). In ES cells, KAP1 partners include the different DNA methyltransferases, DNMT1, DNMT3A, and DNMT3B (Quenneville et al., 2011; Zuo et al., 2012). KAP1 could therefore promote sequence-specific conservation of DNA methylation during cell divisions either by targeting DNMT1 maintenance activity or by targeting reiterative *de novo* methylation via DNMT3A and DNMT3B at these loci.

Two recent ChIP-Seq studies in ES cells describe binding sites for the KAP1 maintenance complex genome wide (Quenneville et al., 2011; Zuo et al., 2012). Three members of the KAP1 repressive complex (ZFP57, KAP1, and SETDB1) colocalize at 91 discrete loci, including all known ICRs. ZFP57 was moreover shown to be required for proper KAP1 binding and maintenance of DNA methylation at these loci (Quenneville et al., 2011). Sequence analysis of ZFP57 binding sites revealed (TGCCGC) hexameric motifs, which can only be bound by

ZFP57 when the internal CG motif is methylated. Known ICRs notably harbor these motifs in multiple copies. By superimposing genetic and epigenetic information, ZFP57 allows the exclusive recognition of genomic regions, which are CG-rich and are inherited in the zygote as methylated, allowing their selective protection against the passive wave of DNA demethylation. Among the 91 sequences bound by ZFP57, possible novel ICRs are present, provided that they acquire parent-specific methylation patterns in gametes and subsist throughout life. Some of them were indeed confirmed to be new genuine parent-specific gDMRs, persisting throughout life, like the *Cdh15* ICR. But for a number of ZFP57-bound gDMRs, parent-specific methylation appeared to be lost after implantation during embryonic differentiation (Proudhon et al., 2012).

Taken altogether, key advances in understanding how DNA methylation is maintained during preimplantation, from the one-cell zygote to implantation, could support a two-step model whereby genome-wide protection against active DNA demethylation occurs on the maternal genome and specific paternal ICRs via H3K9me2 and STELLA. Subsequent protection against passive DNA demethylation of distinct loci of maternal and paternal origin would then occur via ZFP57/KAP1 (Fig. 9.1C). Considering the link between the KAP1 complex and the H3K9 methyltransferase SETB1 protein, H3K9me2 marks could also be mechanistically linked to this second protective process. These players appear as key determinants for the conservation of gametic inherited epigenetic information and, more specifically, for allowing perpetuation of some parental asymmetries beyond preimplantation development. However, the greater number of blastocyst gDMRs compared to the number of ZFP57 binding sites suggests that additional proteins may be at play. The plethora of zinc-finger proteins containing Krüppel-associated box represent attractive candidates for conveying additional sequence specificities to the KAP1-mediated retention of gamete-inherited DNA methylation.

## 4.2. Imprinted X-chromosome inactivation

In addition to gene-specific imprinting, chromosome-wide imprinting occurs in early preimplantation embryos, specifically in females. In mammals, X-linked dosage compensation occurs between XX females and XY males through the silencing of one of the X chromosome in females. This process is achieved differently among therian mammals, with some species exhibiting systematic, imprinted inactivation of the paternal X chromosome, like the kangaroo, and others exhibiting random X inactivation, like human women (Deakin, Chaumeil, Hore, & Marshall Graves, 2009; Okamoto, Otte, Allis, Reinberg, & Heard, 2004; Okamoto et al., 2011). In the mouse, these two forms of X-chromosome inactivation (XCI) exist and occur in non-overlapping developmental stages and tissues. While random XCI is the prominent form of inactivation in female somatic tissues, paternal XCI is specific to preimplantation embryos, and this choice is maintained in extraembryonic lineages, such as the placenta, after implantation (Fig. 9.2). Parental X asymmetry is lost in embryonic tissues just before implantation, as the inactive paternal X is reactivated in the inner cell mass of the blastocyst to allow subsequent random choice to inactivate the maternal or the paternal chromosome (Okamoto et al., 2004). Both imprinted and random XCI in mouse depend upon a genetically defined region, named the X-inactivation center, which notably encodes the *cis*-acting determinant of X inactivation, the *Xist* long noncoding RNA (Augui, Nora, & Heard, 2011). In imprinted XCI, the *Xist* gene is only expressed from the paternal X chromosome, while in random XCI, clonal cell populations express *Xist* from either the maternal or the paternal allele.

Despite avid search for an epigenetic feature inherited from the gametes, the nature of the imprint responsible for the differential treatment of the two parental X chromosomes, while they unite within the same ooplasm after fertilization, is yet to be discovered. The oogenesis-derived imprint prevents *Xist* expression from the maternal allele, while the spermatogenesis-derived state promotes, actively or by default, paternal *Xist* expression. Interestingly, the refractory mark of the maternal X chromosome is acquired during oocyte growth at the same time as autosomal ICRs acquire their DNA methylation-dependent imprint (Bourc'his et al., 2001; Tada et al., 2000). However, no perturbation of imprinted XCI is observed in embryos derived from the fertilization of DNA methylation-free oocytes even when *Xist* transcription is genetically abolished on the paternal X chromosome (Chiba et al., 2008). Of note, a maternally provided *trans*-acting factor, RNF12/RLIM, was shown to be required for paternal *Xist* upregulation and paternal XCI in preimplantation embryos and extraembryonic tissues (Shin et al., 2010). However, while it may act as a facilitating factor for XCI, it does not explain why the maternal X chromosome, which is exposed to the same maternal factors, is unable to express *Xist*.

Transcriptional asymmetry between the parental X chromosomes could play a role in the initial choice to inactivate the paternally derived X chromosome. One hypothesis proposes inheritance of a predetermined state of transcriptional inertia of the paternal X chromosome from the sperm to the female embryo (Huynh & Lee, 2003). During a process called meiotic sex chromosome inactivation (MSCI) (reviewed in Turner, 2007), the sex body formed by the partially paired X and Y chromosomes becomes transcriptionally inactive in male spermatocytes, via the accumulation of repressive histone marks, but in a *Xist*-independent manner (McCarrey et al., 2002). Delivery of this silent paternal X chromosome may be responsible for its preferential inactive state in the embryo. A contrasting hypothesis proposes *de novo* inactivation of the paternal X chromosome after fertilization (Okamoto et al., 2005; Patrat et al., 2009). Using RNA FISH, X-linked genes were shown to be active on both alleles in the two-cell embryo, followed by reduction to monoallelic expression from the four-cell embryo, as *Xist* becomes upregulated on the paternal X chromosome. Moreover, using an autosomal paternally inherited *Xist* transgene, which inherently does not undergo MSCI, imprinted *de novo Xist* expression and *cis* silencing of the autosome carrying the transgene was observed in preimplantation embryos when paternally inherited (Okamoto et al., 2005). This study argues for *de novo* imprinted XCI in the embryo and suggests that paternal inheritance of *Xist* but not paternal inheritance of a silenced X chromosome is a determining factor. Imprinted X chromosome inactivation could thus occur in the mouse as a result of the intense chromatin remodeling, active DNA demethylation, and precocious timing of transcriptional activation which characterize the paternal genome. Combined with gamete-inherited resistance of the maternal X chromosome to express *Xist*, earlier expression of *Xist* or other activators of XCI from the paternal X copy likely explains the origin of imprinted X inactivation in preimplantation mouse embryos.

## 5. SCARCE MAINTENANCE OF PARENTAL EPIGENETIC ASYMMETRY AFTER EMBRYO IMPLANTATION AND THROUGHOUT ADULTHOOD

At the blastocyst stage, after the extensive reprogramming that drives the embryonic genome toward the acquisition of pluripotency, most of the parental asymmetries characteristic of the one-cell stage have disappeared. Limited parent-specific information is selectively protected, so that a few hundred gDMRs maintain oocyte-specific or sperm-specific DNA methylation, which notably include the ICRs. The wave of *de novo* methylation that occurs at implantation eliminates most of these gDMRs. The ICRs represent a small and unique set of gDMRs that retain parent-specificity throughout life.

However, additional, *de novo* parental asymmetry is known to arise in the postimplantation embryo, shaping the somatic or secondary DMRs (somDMRs), under the indirect control of gDMRs. As differentiation and expression programs are established, persisting ICRs and somDMRs accumulate differential posttranslational histone modifications on the two parental alleles. However, even the seemingly permanent state of the ICR can be lost, and this is known to occur in a tissue- and cell type-specific way.

## 5.1. Reduction of germline DMRs

During preimplantation, active and passive mechanisms of demethylation are the main forces that reduce the number of parent-specific gDMRs. The parental allele that is inherited in a methylated state is the one at risk during this period. Among the 100 gDMRs that were protected until this stage, very few will proceed through the early postimplantation stages. During this period, the main leveling force is the massive *de novo* methylation accompanying pluripotency loss and embryonic differentiation. To allow further conservation of gDMRs, this time protective mechanisms must be deployed on the parental allele that is inherited in an unmethylated state.

Embryonic *de novo* DNA methylation has been shown to be mostly DNMT3B dependent, although some loci also require the alternative *de novo* DNA methyltransferase, DNMT3A (Borgel et al., 2010; Okano, Bell, Haber, & Li, 1999). This process is initiated around E4.5 and completed by E9.5 (Borgel et al., 2010). On a genome-wide scale, RRBS analysis revealed a transition from 20% global methylation in the inner cell mass at E3.5 to almost 60% in the E7.5 embryo, which is in range with final somatic genomic methylation levels (Smith et al., 2012). According to this study, sperm-derived gDMRs disappear by DNA methylation gain on the formerly unmethylated allele, while most oocyte-derived gDMRs are resolved to unmethylated states in adult tissues. However, numerous cases of oocyte gDMRs were clearly demonstrated to be lost by methylation gain on the formerly unmethylated paternal allele in early postimplantation development, so that these so-called transient gDMRs exhibit fully methylated patterns for the rest of life, with no sign of prior imprinted status (Borgel et al., 2010; Proudhon et al., 2012). The function of these preimplantation-specific gDMRs is of great interest, as they could very transiently regulate monoallelism and levels of expression of transcripts during a critical window of embryonic development, in relationship with pluripotency acquisition and early lineage specification.

While several protective mechanisms against gDMR demethylation have been uncovered, the regulation of gDMR protection against DNA methylation is not fully understood. CGIs are usually methylation free in mid-gestation embryos and in adult tissues (Borgel et al., 2010; Weber et al., 2007). Because most of the ICRs are CGIs (the 20 maternal ICRs), the origin of the intrinsic resistance of CGIs to DNA methylation is likely to be shared by the unmethylated allele of imprinted CGIs. Genome-wide studies have excluded the mere CpG enrichment as sufficient to predict unmethylated states (Straussman et al., 2009). This observation is further confirmed by the fact that transient gDMRs and permanent ICRs do not differ in terms of CpG content, while they have very different fates in the wake of embryonic *de novo* methylation (Proudhon et al., 2012). In contrast, sequence specificities embedded within CGIs play an important role in the exclusion of DNA methylation from most CGIs. Methylation-resistant CGIs harbor significant strand asymmetry in C and G distribution, leading to the formation of long R-loop structures, which repel DNMT3B (Ginno, Lott, Christensen, Korf, & Chedin, 2012). In direct relevance to the conservation of ICRs, R-loop structures were detected at the unmethylated allele of the human *SNRPN* ICR. In parallel, studies performed with environment-controlled transgenes have demonstrated that the presence of DNA binding sites for transcription factors is sufficient *per se* to recapitulate hypomethylated states, at CGIs but also at CpG-poor sequences, without the requirement of active transcription (Lienert et al., 2011; Stadler et al., 2011). Binding of transcription factors may thus have a shielding effect against DNMTs and be involved in the protection of unmethylated alleles of both CpG-rich maternal ICRs and CpG-poor paternal ICRs at implantation. As we discussed earlier, active chromatin profiles linked to H3K4 methylation could also play a role in repelling DNMTs during this stage of development (Fig. 9.1D).

A specific survey of transcription factor binding sites at ICRs may provide important cues toward unraveling the origin of ICR protection at implantation. Interestingly, the same rule of protection by binding may apply later in life during tissue formation. A recently identified new maternal ICR, associated with the *Cdh15* M-Cadherin gene, revealed the unexpected existence of tissue-specific ICRs, arguing against the prevailing view that ICR parent-specific methylation patterns are permanent and unalterable throughout life (Proudhon et al., 2012). While being resolved to biallelic methylation in all investigated somatic tissues, maternal-specific methylation is specifically conserved at the *Cdh15* ICR in some cell types of the hypothalamus, where it acts as an alternative promoter of the *Cdh15* transcript, only active on the unmethylated paternal allele. A hypothalamus-specific transcription factor may be involved in this tissue-specific maintenance of parental asymmetry. Systematic studies of different tissues and at different ages must be conducted to truly determine how permanently maintained the ICRs are. In this regard, although a genome-wide survey of maternal-specific DNA methylation uncovered around 30 new maternal ICR candidates in the midgestation mouse embryo (Proudhon et al., 2012), only nine ICR candidates were identified in the adult cortex specifically (Xie et al., 2012). As a whole, embryonic differentiation at implantation and tissue differentiation throughout life gradually restrain the number of ICRs, reducing the occurrence of parental epigenetic asymmetry step by step to just a few dozen loci.

## 5.2. Emergence of somatic DMRs

Most germline DMRs disappear in the wake of the embryonic *de novo* DNA methylation associated with peri-implantation development. However, as a counterbalance, new loci acquire parent-specific DNA methylation at this stage, outside the germline context. These are called the secondary, somDMRs (Fig. 9.2). Distinction of the parental genomes occurs within the same nuclear environment, via germline, primary ICRs located in the vicinity and acting *in cis*. somDMRs are associated with the monoallelic, parent-specific expression of surrounding genes and, like ICRs, harbor opposite states of histone modifications (see Section 5.3). Postimplantation emergence of somDMRs is a very common theme, found at many imprinted loci including *Dlk1–Gtl2*, *Gnas*, *Gpr1–Zdbf2*, *H19–Igf2*, *Airn–Igf2r*, and *Kcnq1* (reviewed in John & Lefebvre, 2011). Acquisition of methylation at somDMRs is associated with transcription *in cis*, of either protein-coding genes or noncoding RNAs, whose promoter is regulated by an ICR. The act of transcription rather than the transcript product is usually required for the acquisition of somDMRs on the same allele (Latos et al., 2012). The precise molecular mechanisms for transcription-dependent somDMR emergence merit further attention.

## 5.3. Parent-specific histone modifications at imprinted loci

ICRs and somDMRs typically harbor parent-specific histone modifications, with the unmethylated allele marked by permissive H3K4me2/3 and H3K9ac modifications, and the methylated allele marked by a combination of H3K9me3 and H3K20me3 (Carr, Yevtodiyenko, Schmidt, & Schmidt, 2007; Regha et al., 2007). Histone modifications have been shown to play a role in the modulation of tissue-specific imprinted expression. In the placenta, histone modifications are even thought to be more important than DNA methylation in governing imprinted expression; roles for Polycomb complexes and the G9a methyltransferase have been assigned to monoallelic expression of several imprinted genes in this tissue (Umlauf et al., 2004; Wagschal et al., 2008).

Parent-specific histone profiles seem to occur in a DNA methylation-dependent manner, as midgestation *Dnmt3L−/+* maternal methylation-free embryos show a lack of parent-specific histone modifications (Henckel et al., 2012). Interestingly, two ICRs, which can sporadically regain methylation in *Dnmt3L−/+* embryos, displayed the correct parent-specific histone modification profile when the ICR was properly methylated. Transcription is also an important component in proper histone modification establishment at imprinted loci. Even in the absence of DNA methylation of the *H19* ICR, transcription is sufficient to establish histone modifications in *cis* (Verona, Thorvaldsen, Reese, & Bartolomei, 2007). DNA methylation of the ICRs, parent-specific histone modifications, and monoallelic transcription are all tightly interconnected. More studies will be required to determine whether histone marks control the transcription of imprinted loci or whether transcription itself induces histone marks, and these mechanisms could differ in different developmental contexts.

## 6. CONCLUSIONS

At fertilization, the sexually dimorphic oocyte and sperm unite to form the zygote. They deliver their respective haploid genomes to the embryo, which dramatically differ in terms of chromatin arrangement, histone modifications, and DNA methylation. Delineating which marks are only byproducts of oocyte and sperm differentiation and which marks are also important for further embryonic development will be the next important task for the community to undertake. Nevertheless, despite intense embryonic genome reprogramming and natural elimination of haploinsufficiency, some parental asymmetries are maintained and play an important role in development. As exemplified by genomic imprinting, DNA methylation is the most stable, if not only, vector for life-long persistence of parental asymmetries inherited from the gametes. Maintenance of gamete-inherited parent-specific DNA methylation occurs very scarcely in the mammalian genome at 23 ICRs identified so far. Despite the paucity of these elements, they require the intervention of highly specialized protein complexes, which alternatively protect against DNA methylation loss or against DNA methylation gain.

While they are completely different in nature, the preimplantation ZFP57/KAP1 system and the postimplantation transcription factor-based system similarly rely on DNA sequence features. This coincidence illustrates that genomic imprinting, which is considered as one of the most important paradigms of epigenetic regulation, relies in part on genetic information for its selective maintenance during development. This genetic determinism further implies that sequence variation during evolution may act to shape the number of ICRs and the extent of parental epigenetic asymmetry.

## ACKNOWLEDGMENTS

Research in the Bourc'his lab is supported by the Schlumberger Foundation, the Agence Nationale pour la Recherche (ANR), and a EURYI Award. R. D. acknowledges the Institut Curie for her International PhD program fellowship.

## REFERENCES

Abe, M., Tsai, S. Y., Jin, S. G., Pfeifer, G. P., & Szabo, P. E. (2011). Sex-specific dynamics of global chromatin changes in fetal mouse germ cells. *PLoS One*, *6*, e23848.

Adenot, P. G., Mercier, Y., Renard, J. P., & Thompson, E. M. (1997). Differential H4 acetylation of paternal and maternal chromatin precedes DNA replication and differential transcriptional activity in pronuclei of 1-cell mouse embryos. *Development*, *124*, 4615–4625.

Aoki, F., Worrad, D. M., & Schultz, R. M. (1997). Regulation of transcriptional activity during the first and second cell cycles in the preimplantation mouse embryo. *Developmental Biology*, *181*, 296–307.

Arand, J., Spieler, D., Karius, T., Branco, M. R., Meilinger, D., Meissner, A., et al. (2012). In vivo control of CpG and non-CpG DNA methylation by DNA methyltransferases. *PLoS Genetics*, *8*, e1002750.

Aravin, A. A., & Bourc'his, D. (2008). Small RNA guides for de novo DNA methylation in mammalian germ cells. *Genes and Development*, *22*, 970–975.

Aravin, A. A., Sachidanandam, R., Bourc'his, D., Schaefer, C., Pezic, D., Toth, K. F., et al. (2008). A piRNA pathway primed by individual transposons is linked to de novo DNA methylation in mice. *Molecular Cell*, *31*, 785–799.

Augui, S., Nora, E. P., & Heard, E. (2011). Regulation of X-chromosome inactivation by the X-inactivation centre. *Nature Reviews Genetics*, *12*, 429–442.

Ball, M. P., Li, J. B., Gao, Y., Lee, J. H., LeProust, E. M., Park, I. H., et al. (2009). Targeted and genome-scale strategies reveal gene-body methylation signatures in human cells. *Nature Biotechnology*, *27*, 361–368.

Bernstein, B. E., Mikkelsen, T. S., Xie, X., Kamal, M., Huebert, D. J., Cuff, J., et al. (2006). A bivalent chromatin structure marks key developmental genes in embryonic stem cells. *Cell*, *125*, 315–326.

Bestor, T. H., & Ingram, V. M. (1983). Two DNA methyltransferases from murine erythroleukemia cells: Purification, sequence specificity, and mode of interaction with

DNA. *Proceedings of the National Academy of Sciences of the United States of America*, *80*, 5559–5563.

Borgel, J., Guibert, S., Li, Y., Chiba, H., Schübeler, D., Sasaki, H., et al. (2010). Targets and dynamics of promoter DNA methylation during early mouse development. *Nature Genetics*, *42*, 1093–1100.

Bostick, M., Kim, J. K., Esteve, P. O., Clark, A., Pradhan, S., & Jacobsen, S. E. (2007). UHRF1 plays a role in maintaining DNA methylation in mammalian cells. *Science*, *317*, 1760–1764.

Bourc'his, D., & Bestor, T. H. (2004). Meiotic catastrophe and retrotransposon reactivation in male germ cells lacking Dnmt3L. *Nature*, *431*, 96–99.

Bourc'his, D., & Voinnet, O. (2010). A small-RNA perspective on gametogenesis, fertilization, and early zygotic development. *Science*, *330*, 617–622.

Bourc'his, D., Xu, G. L., Lin, C. S., Bollman, B., & Bestor, T. H. (2001). Dnmt3L and the establishment of maternal genomic imprints. *Science*, *294*, 2536–2539.

Brykczynska, U., Hisano, M., Erkek, S., Ramos, L., Oakeley, E. J., Roloff, T. C., et al. (2010). Repressive and active histone methylation mark distinct promoters in human and mouse spermatozoa. *Nature Structural and Molecular Biology*, *17*, 679–687.

Burton, A., & Torres-Padilla, M. E. (2010). Epigenetic reprogramming and development: A unique heterochromatin organization in the preimplantation mouse embryo. *Briefings in Functional Genomics*, *9*, 444–454.

Carr, M. S., Yevtodiyenko, A., Schmidt, C. L., & Schmidt, J. V. (2007). Allele-specific histone modifications regulate expression of the Dlk1-Gtl2 imprinted domain. *Genomics*, *89*, 280–290.

Chiba, H., Hirasawa, R., Kaneda, M., Amakawa, Y., Li, E., Sado, T., et al. (2008). De novo DNA methylation independent establishment of maternal imprint on X chromosome in mouse oocytes. *Genesis*, *46*, 768–774.

Chotalia, M., Smallwood, S. A., Ruf, N., Dawson, C., Lucifero, D., Frontera, M., et al. (2009). Transcription is required for establishment of germline methylation marks at imprinted genes. *Genes and Development*, *23*, 105–117.

Ciccone, D. N., Su, H., Hevi, S., Gay, F., Lei, H., Bajko, J., et al. (2009). KDM1B is a histone H3K4 demethylase required to establish maternal genomic imprints. *Nature*, *461*, 415–418.

Dahl, J. A., & Collas, P. (2008). A rapid micro chromatin immunoprecipitation assay (microChIP). *Nature Protocols*, *3*, 1032–1045.

Deakin, J. E., Chaumeil, J., Hore, T. A., & Marshall Graves, J. A. (2009). Unravelling the evolutionary origins of X chromosome inactivation in mammals: Insights from marsupials and monotremes. *Chromosome Research*, *17*, 671–685.

Feng, S., Cokus, S. J., Zhang, X., Chen, P. Y., Bostick, M., Goll, M. G., et al. (2010). Conservation and divergence of methylation patterning in plants and animals. *Proceedings of the National Academy of Sciences of the United States of America*, *107*, 8689–8694.

Ferguson-Smith, A. C. (2011). Genomic imprinting: The emergence of an epigenetic paradigm. *Nature Reviews Genetics*, *12*, 565–575.

Gaucher, J., Reynoird, N., Montellier, E., Boussouar, F., Rousseaux, S., & Khochbin, S. (2010). From meiosis to postmeiotic events: The secrets of histone disappearance. *The FEBS Journal*, *277*, 599–604.

Gill, M. E., Erkek, S., & Peters, A. H. (2012). Parental epigenetic control of embryogenesis: A balance between inheritance and reprogramming? *Current Opinion in Cell Biology*, *24*, 387–396.

Ginno, P. A., Lott, P. L., Christensen, H. C., Korf, I., & Chedin, F. (2012). R-loop formation is a distinctive characteristic of unmethylated human CpG island promoters. *Molecular Cell*, *45*, 814–825.

Goren, A., Ozsolak, F., Shoresh, N., Ku, M., Adli, M., Hart, C., et al. (2010). Chromatin profiling by directly sequencing small quantities of immunoprecipitated DNA. *Nature Methods*, *7*, 47–49.

Gu, T. P., Guo, F., Yang, H., Wu, H. P., Xu, G. F., Liu, W., et al. (2011). The role of Tet3 DNA dioxygenase in epigenetic reprogramming by oocytes. *Nature*, *477*, 606–610.

Gu, L., Wang, Q., & Sun, Q. Y. (2010). Histone modifications during mammalian oocyte maturation: Dynamics, regulation and functions. *Cell Cycle*, *9*, 1942–1950.

Guenatri, M., Duffie, R., Iranzo, J., Fauque, P., & Bourc'his, D. (2013). Plasticity in Dnmt3L-dependent and -independent modes of de novo methylation in the developing mouse embryo. *Development*, *140*, 562–572.

Guibert, S., Forne, T., & Weber, M. (2012). Global profiling of DNA methylation erasure in mouse primordial germ cells. *Genome Research*, *22*, 633–641.

Guo, G., Huss, M., Tong, G. Q., Wang, C., Li Sun, L., Clarke, N. D., et al. (2010). Resolution of cell fate decisions revealed by single-cell gene expression analysis from zygote to blastocyst. *Developmental Cell*, *18*, 675–685.

Gusse, M., & Chevaillier, P. (1980). Electron microscope evidence for the presence of globular structures in different sperm chromatins. *The Journal of Cell Biology*, *87*, 280–284.

Hackett, J. A., Sengupta, R., Zylicz, J. J., Murakami, K., Lee, C., Down, T. A., et al. (2012). Germline DNA demethylation dynamics and imprint erasure through 5-hydroxymethylcytosine. *Science*, *339*, 448–452.

Hammoud, S. S., Nix, D. A., Hammoud, A. O., Gibson, M., Cairns, B. R., & Carrell, D. T. (2011). Genome-wide analysis identifies changes in histone retention and epigenetic modifications at developmental and imprinted gene loci in the sperm of infertile men. *Human Reproduction*, *26*, 2558–2569.

Hammoud, S. S., Nix, D. A., Zhang, H., Purwar, J., Carrell, D. T., & Cairns, B. R. (2009). Distinctive chromatin in human sperm packages genes for embryo development. *Nature*, *460*, 473–478.

Henckel, A., Chebli, K., Kota, S. K., Arnaud, P., & Feil, R. (2012). Transcription and histone methylation changes correlate with imprint acquisition in male germ cells. *The EMBO Journal*, *31*, 606–615.

Huang, Y., Pastor, W. A., Shen, Y., Tahiliani, M., Liu, D. R., & Rao, A. (2010). The behaviour of 5-hydroxymethylcytosine in bisulfite sequencing. *PLoS One*, *5*, e8888.

Huynh, K. D., & Lee, J. T. (2003). Inheritance of a pre-inactivated paternal X chromosome in early mouse embryos. *Nature*, *426*, 857–862.

Ichiyanagi, T., Ichiyanagi, K., Miyake, M., & Sasaki, H. (2012). Accumulation and loss of asymmetric non-CpG methylation during male germ-cell development. *Nucleic Acids Research*, *41*, 738–745.

Illingworth, R. S., Gruenewald-Schneider, U., Webb, S., Kerr, A. R. W., James, K. D., Turner, D. J., et al. (2010). Orphan CpG Islands identify numerous conserved promoters in the mammalian genome. *PLoS Genetics*, *6*, e1001134.

Iqbal, K., Jin, S. G., Pfeifer, G. P., & Szabo, P. E. (2011). Reprogramming of the paternal genome upon fertilization involves genome-wide oxidation of 5-methylcytosine. *Proceedings of the National Academy of Sciences of the United States of America*, *108*, 3642–3647.

John, R. M., & Lefebvre, L. (2011). Developmental regulation of somatic imprints. *Differentiation*, *81*, 270–280.

Kaneda, M., Okano, M., Hata, K., Sado, T., Tsujimoto, N., Li, E., et al. (2004). Essential role for de novo DNA methyltransferase Dnmt3a in paternal and maternal imprinting. *Nature*, *429*, 900–903.

Kim, J. M., Liu, H., Tazaki, M., Nagata, M., & Aoki, F. (2003). Changes in histone acetylation during mouse oocyte meiosis. *The Journal of Cell Biology*, *162*, 37–46.

Kobayashi, H., Sakurai, T., Imai, M., Takahashi, N., Fukuda, A., Yayoi, O., et al. (2012). Contribution of intragenic DNA methylation in mouse gametic DNA methylomes to establish oocyte-specific heritable marks. *PLoS Genetics*, *8*, e1002440.

Kota, S. K., & Feil, R. (2010). Epigenetic transitions in germ cell development and meiosis. *Developmental Cell*, *19*, 675–686.

Kriaucionis, S., & Heintz, N. (2009). The nuclear DNA base 5-hydroxymethylcytosine is present in Purkinje neurons and the brain. *Science*, *324*, 929–930.

Lane, N., Dean, W., Erhardt, S., Hajkova, P., Surani, A., Walter, J., et al. (2003). Resistance of IAPs to methylation reprogramming may provide a mechanism for epigenetic inheritance in the mouse. *Genesis*, *35*, 88–93.

Latos, P. A., Pauler, F. M., Koerner, M. V., Senergin, H. B., Hudson, Q. J., Stocsits, R. R., et al. (2012). Airn transcriptional overlap, but not its lncRNA products, induces imprinted Igf2r silencing. *Science*, *338*, 1469–1472.

Li, E., Bestor, T. H., & Jaenisch, R. (1992). Targeted mutation of the DNA methyltransferase gene results in embryonic lethality. *Cell*, *69*, 915–926.

Li, X., Ito, M., Zhou, F., Youngson, N., Zuo, X., Leder, P., et al. (2008). A maternal-zygotic effect gene, Zfp57, maintains both maternal and paternal imprints. *Developmental Cell*, *15*, 547–557.

Lienert, F., Wirbelauer, C., Som, I., Dean, A., Mohn, F., & Schubeler, D. (2011). Identification of genetic elements that autonomously determine DNA methylation states. *Nature Genetics*, *43*, 1091–1097.

Lister, R., Pelizzola, M., Dowen, R. H., Hawkins, R. D., Hon, G., Tonti-Filippini, J., et al. (2009). Human DNA methylomes at base resolution show widespread epigenomic differences. *Nature*, *462*, 315–322.

Liu, H., Kim, J. M., & Aoki, F. (2004). Regulation of histone H3 lysine 9 methylation in oocytes and early pre-implantation embryos. *Development*, *131*, 2269–2280.

Liu, Y. J., Nakamura, T., & Nakano, T. (2012). Essential role of DPPA3 for chromatin condensation in mouse oocytogenesis. *Biology of Reproduction*, *86*, 40.

Loppin, B., Bonnefoy, E., Anselme, C., Laurencon, A., Karr, T. L., & Couble, P. (2005). The histone H3.3 chaperone HIRA is essential for chromatin assembly in the male pronucleus. *Nature*, *437*, 1386–1390.

Lorincz, M. C., Schubeler, D., Hutchinson, S. R., Dickerson, D. R., & Groudine, M. (2002). DNA methylation density influences the stability of an epigenetic imprint and Dnmt3a/b-independent de novo methylation. *Molecular and Cellular Biology*, *22*, 7572–7580.

Macfarlan, T. S., Gifford, W. D., Driscoll, S., Lettieri, K., Rowe, H. M., Bonanomi, D., et al. (2012). Embryonic stem cell potency fluctuates with endogenous retrovirus activity. *Nature*, *487*, 57–63.

Mayer, W., Smith, A., Fundele, R., & Haaf, T. (2000). Spatial separation of parental genomes in preimplantation mouse embryos. *The Journal of Cell Biology*, *148*, 629–634.

McCarrey, J. R., Watson, C., Atencio, J., Ostermeier, G. C., Marahrens, Y., Jaenisch, R., et al. (2002). X-chromosome inactivation during spermatogenesis is regulated by an Xist/Tsix-independent mechanism in the mouse. *Genesis*, *34*, 257–266.

Messerschmidt, D. M., de Vries, W., Ito, M., Solter, D., Ferguson-Smith, A., & Knowles, B. B. (2012). Trim28 is required for epigenetic stability during mouse oocyte to embryo transition. *Science*, *335*, 1499–1502.

Miyanari, Y., & Torres-Padilla, M.-E. (2012). Control of ground-state pluripotency by allelic regulation of Nanog. *Nature*, *483*, 470–473.

Monk, M., Boubelik, M., & Lehnert, S. (1987). Temporal and regional changes in DNA methylation in the embryonic, extraembryonic and germ cell lineages during mouse embryo development. *Development*, *99*, 371–382.

Morgan, H. D., Santos, F., Green, K., Dean, W., & Reik, W. (2005). Epigenetic reprogramming in mammals. *Human Molecular Genetics, 14*(Spec. No.), R47–R58.
Musselman, C. A., Lalonde, M. E., Cote, J., & Kutateladze, T. G. (2012). Perceiving the epigenetic landscape through histone readers. *Nature Structural and Molecular Biology, 19*, 1218–1227.
Nakamura, T., Arai, Y., Umehara, H., Masuhara, M., Kimura, T., Taniguchi, H., et al. (2006). PGC7/Stella protects against DNA demethylation in early embryogenesis. *Nature Cell Biology, 9*, 64–71.
Nakamura, T., Liu, Y. J., Nakashima, H., Umehara, H., Inoue, K., Matoba, S., et al. (2012). PGC7 binds histone H3K9me2 to protect against conversion of 5mC to 5hmC in early embryos. *Nature, 486*, 415–419.
Okamoto, I., Arnaud, D., Le Baccon, P., Otte, A. P., Disteche, C. M., Avner, P., et al. (2005). Evidence for de novo imprinted X-chromosome inactivation independent of meiotic inactivation in mice. *Nature, 438*, 369–373.
Okamoto, I., Otte, A. P., Allis, C. D., Reinberg, D., & Heard, E. (2004). Epigenetic dynamics of imprinted X inactivation during early mouse development. *Science, 303*, 644–649.
Okamoto, I., Patrat, C., Thepot, D., Peynot, N., Fauque, P., Daniel, N., et al. (2011). Eutherian mammals use diverse strategies to initiate X-chromosome inactivation during development. *Nature, 472*, 370–374.
Okano, M., Bell, D. W., Haber, D. A., & Li, E. (1999). DNA methyltransferases Dnmt3a and Dnmt3b are essential for de novo methylation and mammalian development. *Cell, 99*, 247–257.
Ooga, M., Inoue, A., Kageyama, S., Akiyama, T., Nagata, M., & Aoki, F. (2008). Changes in H3K79 methylation during preimplantation development in mice. *Biology of Reproduction, 78*, 413–424.
Ooi, S. K., & Bestor, T. H. (2008). The colorful history of active DNA demethylation. *Cell, 133*, 1145–1148.
Ooi, S. K., Qiu, C., Bernstein, E., Li, K., Jia, D., Yang, Z., et al. (2007). DNMT3L connects unmethylated lysine 4 of histone H3 to de novo methylation of DNA. *Nature, 448*, 714–717.
Otani, J., Nankumo, T., Arita, K., Inamoto, S., Ariyoshi, M., & Shirakawa, M. (2009). Structural basis for recognition of H3K4 methylation status by the DNA methyltransferase 3A ATRX-DNMT3-DNMT3L domain. *EMBO Reports, 10*, 1235–1241.
Pan, G., Tian, S., Nie, J., Yang, C., Ruotti, V., Wei, H., et al. (2007). Whole-genome analysis of histone H3 lysine 4 and lysine 27 methylation in human embryonic stem cells. *Cell Stem Cell, 1*, 299–312.
Patrat, C., Okamoto, I., Diabangouaya, P., Vialon, V., Le Baccon, P., Chow, J., et al. (2009). Dynamic changes in paternal X-chromosome activity during imprinted X-chromosome inactivation in mice. *Proceedings of the National Academy of Sciences of the United States of America, 106*, 5198–5203.
Popp, C., Dean, W., Feng, S., Cokus, S. J., Andrews, S., Pellegrini, M., et al. (2010). Genome-wide erasure of DNA methylation in mouse primordial germ cells is affected by AID deficiency. *Nature, 463*, 1101–1105.
Probst, A. V., & Almouzni, G. (2011). Heterochromatin establishment in the context of genome-wide epigenetic reprogramming. *Trends in Genetics, 27*, 177–185.
Proudhon, C., Duffie, R., Ajjan, S., Cowley, M., Iranzo, J., Carbajosa, G., et al. (2012). Protection against de novo methylation is instrumental in maintaining parent-of-origin methylation inherited from the gametes. *Molecular Cell, 47*, 909–920.
Puschendorf, M., Terranova, R., Boutsma, E., Mao, X., Isono, K., Brykczynska, U., et al. (2008). PRC1 and Suv39h specify parental asymmetry at constitutive heterochromatin in early mouse embryos. *Nature Genetics, 40*, 411–420.

Quenneville, S., Verde, G., Corsinotti, A., Kapopoulou, A., Jakobsson, J., Offner, S., et al. (2011). In embryonic stem cells, ZFP57/KAP1 recognize a methylated hexanucleotide to affect chromatin and DNA methylation of imprinting control regions. *Molecular Cell, 44*, 361–372.

Regha, K., Sloane, M. A., Huang, R., Pauler, F. M., Warczok, K. E., Melikant, B., et al. (2007). Active and repressive chromatin are interspersed without spreading in an imprinted gene cluster in the mammalian genome. *Molecular Cell, 27*, 353–366.

Reik, W., & Walter, J. (2001). Evolution of imprinting mechanisms: The battle of the sexes begins in the zygote. *Nature Genetics, 27*, 255–256.

Rollins, R. A., Haghighi, F., Edwards, J. R., Das, R., Zhang, M. Q., Ju, J., et al. (2006). Large-scale structure of genomic methylation patterns. *Genome Research, 16*, 157–163.

Rougier, N., Bourc'his, D., Gomes, D. M., Niveleau, A., Plachot, M., Paldi, A., et al. (1998). Chromosome methylation patterns during mammalian preimplantation development. *Genes and Development, 12*, 2108–2113.

Santos, F., Hendrich, B., Reik, W., & Dean, W. (2002). Dynamic reprogramming of DNA methylation in the early mouse embryo. *Developmental Biology, 241*, 172–182.

Santos, F., Peters, A. H., Otte, A. P., Reik, W., & Dean, W. (2005). Dynamic chromatin modifications characterise the first cell cycle in mouse embryos. *Developmental Biology, 280*, 225–236.

Sarmento, O. F., Digilio, L. C., Wang, Y., Perlin, J., Herr, J. C., Allis, C. D., et al. (2004). Dynamic alterations of specific histone modifications during early murine development. *Journal of Cell Science, 117*, 4449–4459.

Schaefer, C. B., Ooi, S. K., Bestor, T. H., & Bourc'his, D. (2007). Epigenetic decisions in mammalian germ cells. *Science, 316*, 398–399.

Schulz, R., Proudhon, C., Bestor, T. H., Woodfine, K., Lin, C. S., Lin, S. P., et al. (2010). The parental non-equivalence of imprinting control regions during mammalian development and evolution. *PLoS Genetics, 6*, e1001214.

Shin, J., Bossenz, M., Chung, Y., Ma, H., Byron, M., Taniguchi-Ishigaki, N., et al. (2010). Maternal Rnf12/RLIM is required for imprinted X-chromosome inactivation in mice. *Nature, 467*, 977–981.

Smallwood, S. A., Tomizawa, S., Krueger, F., Ruf, N., Carli, N., Segonds-Pichon, A., et al. (2011). Dynamic CpG island methylation landscape in oocytes and preimplantation embryos. *Nature Genetics, 43*, 811–814.

Smith, Z. D., Chan, M. M., Mikkelsen, T. S., Gu, H., Gnirke, A., Regev, A., et al. (2012). A unique regulatory phase of DNA methylation in the early mammalian embryo. *Nature, 484*, 339–344.

Sobel, R. E., Cook, R. G., Perry, C. A., Annunziato, A. T., & Allis, C. D. (1995). Conservation of deposition-related acetylation sites in newly synthesized histones H3 and H4. *Proceedings of the National Academy of Sciences of the United States of America, 92*, 1237–1241.

Song, J., Angel, A., Howard, M., & Dean, C. (2012). Vernalization—A cold-induced epigenetic switch. *Journal of Cell Science, 125*, 3723–3731.

Song, J., Teplova, M., Ishibe-Murakami, S., & Patel, D. J. (2012). Structure-based mechanistic insights into DNMT1-mediated maintenance DNA methylation. *Science, 335*, 709–712.

Stadler, M. B., Murr, R., Burger, L., Ivanek, R., Lienert, F., Scholer, A., et al. (2011). DNA-binding factors shape the mouse methylome at distal regulatory regions. *Nature, 480*, 490–495.

Stein, R., Gruenbaum, Y., Pollack, Y., Razin, A., & Cedar, H. (1982). Clonal inheritance of the pattern of DNA methylation in mouse cells. *Proceedings of the National Academy of Sciences of the United States of America, 79*, 61–65.

Straussman, R., Nejman, D., Roberts, D., Steinfeld, I., Blum, B., Benvenisty, N., et al. (2009). Developmental programming of CpG island methylation profiles in the human genome. *Nature Structural and Molecular Biology, 16*, 564–571.

Szabo, P. E., & Pfeifer, G. P. (2012). H3K9me2 attracts PGC7 in the zygote to prevent Tet3-mediated oxidation of 5-methylcytosine. *Journal of Molecular Cell Biology*, *4*, 427–429.

Tada, T., Obata, Y., Tada, M., Goto, Y., Nakatsuji, N., Tan, S., et al. (2000). Imprint switching for non-random X-chromosome inactivation during mouse oocyte growth. *Development*, *127*, 3101–3105.

Tahiliani, M., Koh, K. P., Shen, Y., Pastor, W. A., Bandukwala, H., Brudno, Y., et al. (2009). Conversion of 5-methylcytosine to 5-hydroxymethylcytosine in mammalian DNA by MLL partner TET1. *Science*, *324*, 930–935.

Tang, F., Barbacioru, C., Nordman, E., Bao, S., Lee, C., Wang, X., et al. (2011). Deterministic and stochastic allele specific gene expression in single mouse blastomeres. *PLoS One*, *6*, e21208.

Tanphaichitr, N., Sobhon, P., Taluppeth, N., & Chalermisarachai, P. (1978). Basic nuclear proteins in testicular cells and ejaculated spermatozoa in man. *Experimental Cell Research*, *117*, 347–356.

Tomizawa, S., Kobayashi, H., Watanabe, T., Andrews, S., Hata, K., Kelsey, G., et al. (2011). Dynamic stage-specific changes in imprinted differentially methylated regions during early mammalian development and prevalence of non-CpG methylation in oocytes. *Development*, *138*, 811–820.

Torres-Padilla, M. E., Bannister, A. J., Hurd, P. J., Kouzarides, T., & Zernicka-Goetz, M. (2006). Dynamic distribution of the replacement histone variant H3.3 in the mouse oocyte and preimplantation embryos. *The International Journal of Developmental Biology*, *50*, 455–461.

Turner, J. M. (2007). Meiotic sex chromosome inactivation. *Development*, *134*, 1823–1831.

Umlauf, D., Goto, Y., Cao, R., Cerqueira, F., Wagschal, A., Zhang, Y., et al. (2004). Imprinting along the Kcnq1 domain on mouse chromosome 7 involves repressive histone methylation and recruitment of Polycomb group complexes. *Nature Genetics*, *36*, 1296–1300.

van der Heijden, G. W., Derijck, A. A., Ramos, L., Giele, M., van der Vlag, J., & de Boer, P. (2006). Transmission of modified nucleosomes from the mouse male germline to the zygote and subsequent remodeling of paternal chromatin. *Developmental Biology*, *298*, 458–469.

van der Heijden, G. W., Dieker, J. W., Derijck, A. A., Muller, S., Berden, J. H., Braat, D. D., et al. (2005). Asymmetry in histone H3 variants and lysine methylation between paternal and maternal chromatin of the early mouse zygote. *Mechanisms of Development*, *122*, 1008–1022.

van der Heijden, G. W., Ramos, L., Baart, E. B., van den Berg, I. M., Derijck, A. A., van der Vlag, J., et al. (2008). Sperm-derived histones contribute to zygotic chromatin in humans. *BMC Developmental Biology*, *8*, 34.

Verona, R. I., Thorvaldsen, J. L., Reese, K. J., & Bartolomei, M. S. (2007). The transcriptional status but not the imprinting control region determines allele-specific histone modifications at the imprinted H19 locus. *Molecular and Cellular Biology*, *28*, 71–82.

Wagschal, A., Sutherland, H. G., Woodfine, K., Henckel, A., Chebli, K., Schulz, R., et al. (2008). G9a histone methyltransferase contributes to imprinting in the mouse placenta. *Molecular and Cellular Biology*, *28*, 1104–1113.

Watanabe, T., Tomizawa, S., Mitsuya, K., Totoki, Y., Yamamoto, Y., Kuramochi-Miyagawa, S., et al. (2011). Role for piRNAs and noncoding RNA in de novo DNA methylation of the imprinted mouse Rasgrf1 locus. *Science*, *332*, 848–852.

Watanabe, T., Totoki, Y., Toyoda, A., Kaneda, M., Kuramochi-Miyagawa, S., Obata, Y., et al. (2008). Endogenous siRNAs from naturally formed dsRNAs regulate transcripts in mouse oocytes. *Nature*, *453*, 539–543.

Weber, M., Hellmann, I., Stadler, M. B., Ramos, L., Paabo, S., Rebhan, M., et al. (2007). Distribution, silencing potential and evolutionary impact of promoter DNA methylation in the human genome. *Nature Genetics*, *39*, 457–466.

Wossidlo, M., Nakamura, T., Lepikhov, K., Marques, C. J., Zakhartchenko, V., Boiani, M., et al. (2011). 5-Hydroxymethylcytosine in the mammalian zygote is linked with epigenetic reprogramming. *Nature Communications*, *2*, 241.

Xie, W., Barr, C. L., Kim, A., Yue, F., Lee, A. Y., Eubanks, J., et al. (2012). Base-resolution analyses of sequence and parent-of-origin dependent DNA methylation in the mouse genome. *Cell*, *148*, 816–831.

Zemach, A., McDaniel, I. E., Silva, P., & Zilberman, D. (2010). Genome-wide evolutionary analysis of eukaryotic DNA methylation. *Science*, *328*, 916–919.

Zuo, X., Sheng, J., Lau, H. T., McDonald, C. M., Andrade, M., Cullen, D. E., et al. (2012). Zinc finger protein ZFP57 requires its co-factor to recruit DNA methyltransferases and maintains DNA methylation imprint in embryonic stem cells via its transcriptional repression domain. *The Journal of Biological Chemistry*, *287*, 2107–2118.

# INDEX

Note: Page numbers followed by "*f*" indicate figures, and "*t*" indicate tables.

## R



## S



## T

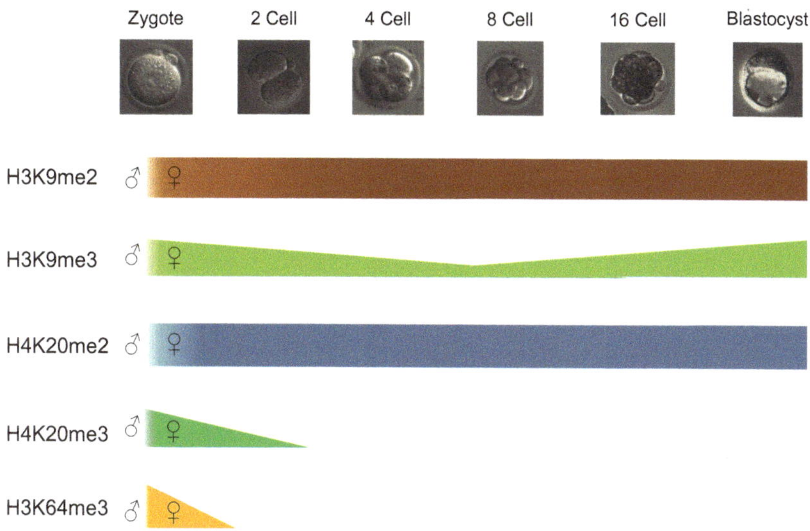


**Anas Fadloun *et al.*, Figure 1.2** Dynamics of the main constitutive heterochromatic marks during preimplantation development. All heterochromatin marks analysed to date are asymmetrically localized on the maternal chromatin during the zygote stage. The levels of H3K9me2 and H4K20me2 are maintained relatively stable during the early stages of development, whereas global H3K9me3 levels decrease between the 4-cell and the 16-cell stage. H4K20me3 is undetectable on embryonic chromatin by the end of the 2-cell stage and similar kinetics have been described for H3K64me3, which is only present during the zygote stage and is absent afterward.

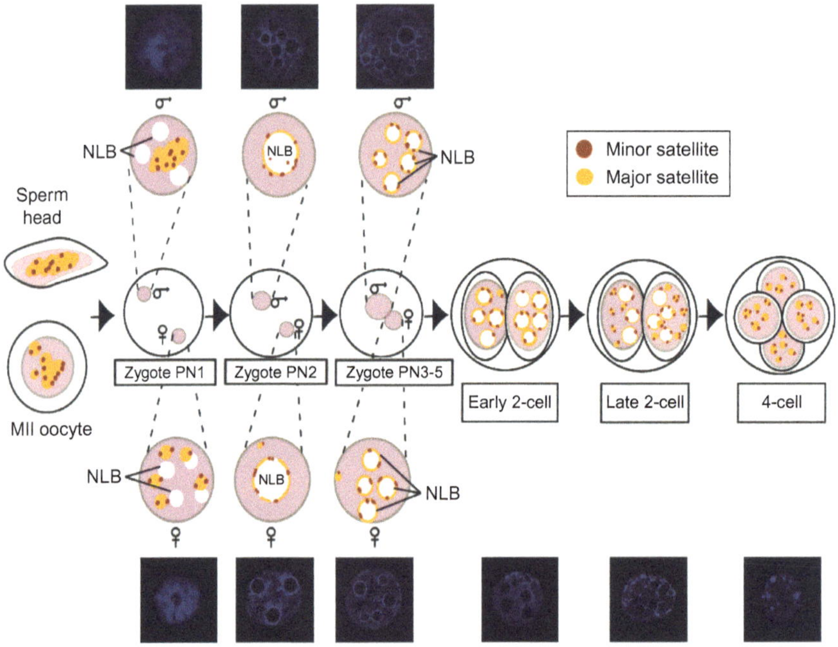


**Anas Fadloun *et al.*, Figure 1.3** Nuclear architecture changes during preimplantation development in the mouse. The nuclear organization changes dynamically after fertilization. The nucleolar-like bodies (NLBs) are established after fertilization at the pronuclear stage (PN) 1 in the zygote. At the PN2 stage, the minor and major satellites (MMS) reorganize to form a ring-like structure around the NLBs. Between the PN3 and PN5 an increase in the number of NLBs occurs, but the position of the MMS remains the same. At the early 2-cell stage (E2C), the nuclear organization starts to change and by the end of the 2-cell stage (L2C) many MMS start to form new chromocenters independently of the NLBs structure. By the 4-cell stage, most of the NLBs have disappeared and instead the majority of MMS are clustered into "somatic"-like chromocenters. Note that nuclei are not at the same scale.

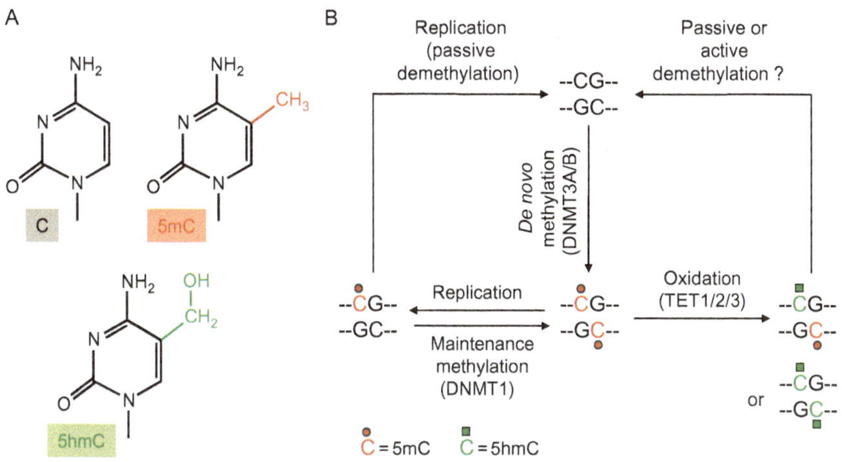


**Sylvain Guibert and Michael Weber, Figure 2.1** DNA methylation and hydroxymethylation occur on cytosines. (A) Chemical structure of cytosine (C), 5-methylcytosine (5mC), and 5-hydroxymethylcytosine (5hmC). (B) Cytosine methylation occurs symmetrically at CpG dinucleotides. *De novo* methylation is mostly catalyzed by DNMT3A/B enzymes, whereas DNMT1 methylates hemimethylated sites generated after DNA replication. Lack of maintenance methylation results in passive loss of DNA methylation with DNA replication. TET proteins are capable of oxidizing 5mC into 5hmC, which acts as an intermediate in demethylation pathways.

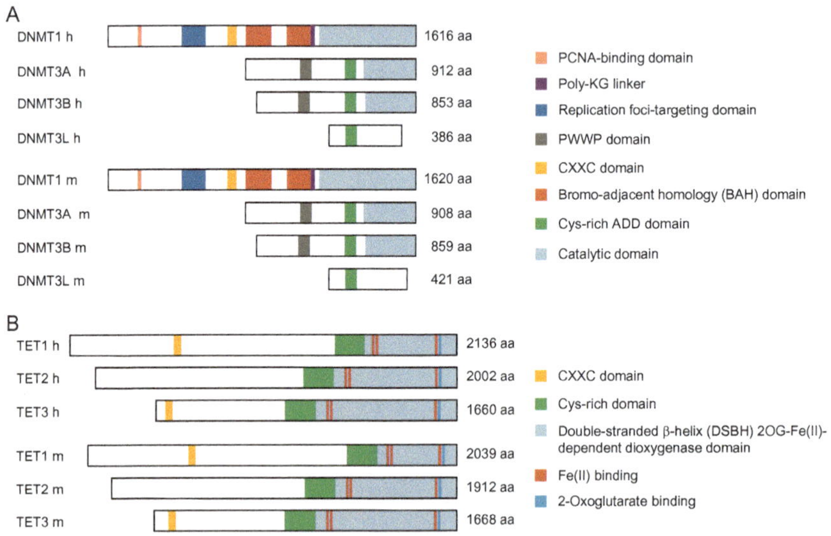


**Sylvain Guibert and Michael Weber, Figure 2.2** Structure of mammalian DNMT and TET enzymes. (A) DNMTs share a C-terminal catalytic domain, except DNMT3L which does not contain a functional catalytic domain. The N-terminal part of DNMT1 contains interaction domains with PCNA, replication foci, and unmethylated DNA (CXXC domain). DNMT1 also contains two BAH domains. DNMT3A and DNMT3B comprise a PWWP domain recognizing H3K36me3 and an ADD domain mediating interaction with histone H3 unmethylated at H3K4. (B) TET proteins contain in their C terminus a cysteine-rich domain adjacent to a DSBH domain typical of 2-oxoglutarate-Fe(II)-dependent dioxygenases. TET1 and TET3 also contain a CXXC domain. The size of the proteins is indicated by the number of amino acids (aa).

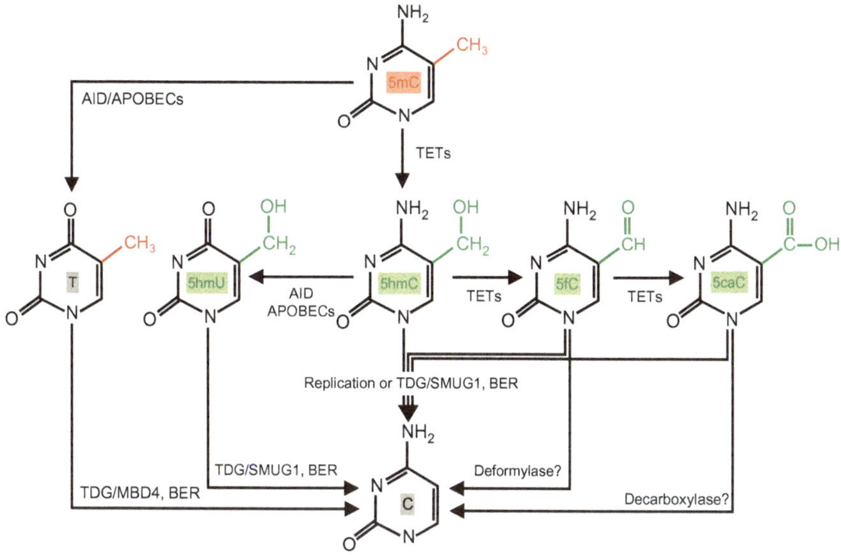


**Sylvain Guibert and Michael Weber, Figure 2.6** Potential chemical pathways for active cytosine demethylation. One potential mechanism for active replication-independent demethylation is deamination of 5mC into thymine (T) by cytosine deaminases AID or APOBEC, followed by excision of the thymine by DNA glycosylases TDG or MBD4 and base excision repair (BER). 5mC can also be converted into 5hmC by TET enzymes, which can further be oxidized into 5-formylcytosine (5fC) and 5-carboxylcytosine (5caC), or deaminated into 5-hydroxymethyluracil (5hmU). All these modified bases could lead to replication-dependent passive demethylation or active demethylation via excision by DNA glycosylases (TDG, SMUG1) followed by BER. In addition, a putative decarboxylase or deformylase could convert 5caC or 5fC directly back to cytosine, but such enzymes have not yet been discovered.

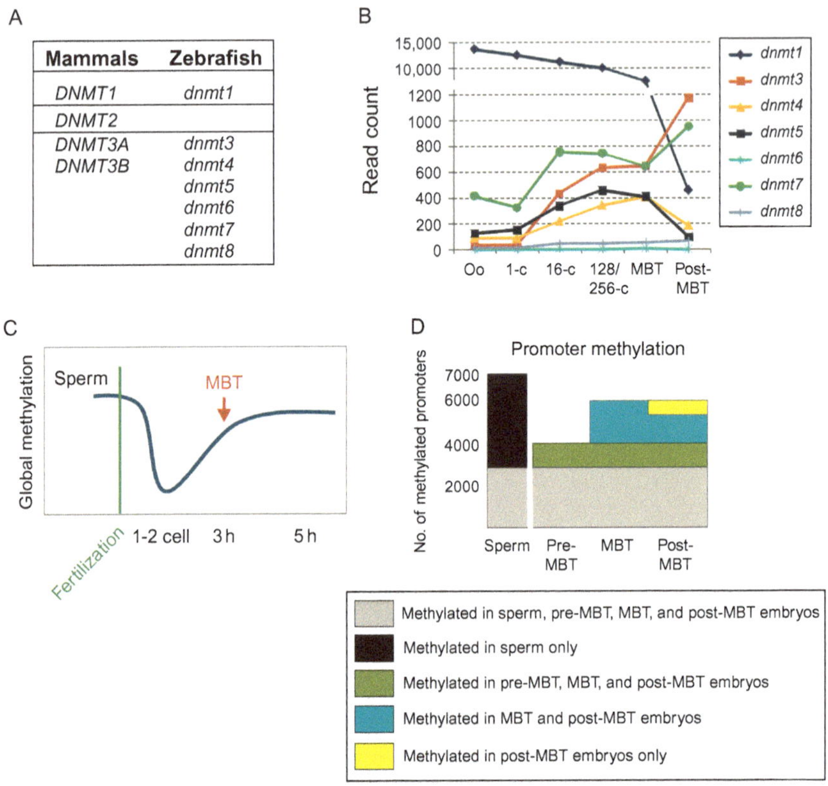


**Ingrid S. Andersen *et al.*, Figure 3.2** DNA methylation patterns in the zebrafish embryo. (A) *dnmt* gene orthologs in mammals and zebrafish. (B) Expression patterns of zebrafish *dnmt* genes during pre-MBT, MBT, and post-MBT development. Graph depicts mRNA detection levels from our RNA-seq data (Aanes et al., 2011) in unfertilized eggs (Oo), and 1-cell (fertilized), 16-cell, 128-/256-cell, MBT, and post-MBT (4.5-h postfertilization) embryos. (C) Global DNA methylation changes during early development. (D) Changes in promoter DNA methylation levels in sperm and at pre-MBT (256-cell stage), MBT (1- to 2000-cell stage), and post-MBT (4.5-h postfertilization) stages. Distinct shadings indicate developmental stage-specific methylation.

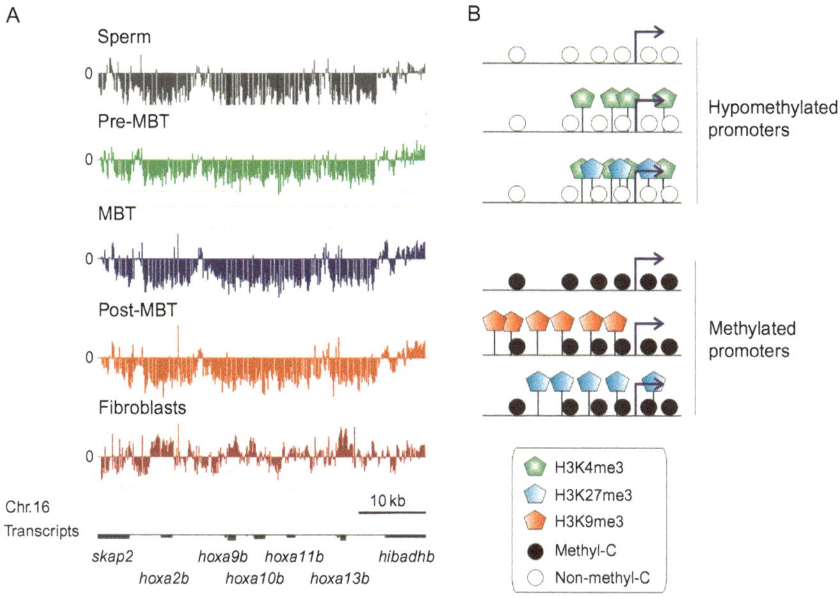


**Ingrid S. Andersen *et al.*, Figure 3.3** Differential marking of hypomethylated and methylated DNA regions by modified histones in zebrafish embryos. (A) DNA hypomethylation of a developmentally regulated gene cluster. Browser view of DNA methylation profiles over a 54-kb fragment of the zebrafish hoxa locus on chromosome 16. Note the hypomethylated state of the locus in sperm and embryos, and local recurrent methylation in fibroblasts. (B) Differential association of H3K4me3 and H3K9me3 or H3K27me3 with hypomethylated and methylated promoters, respectively, in pre-MBT embryos.

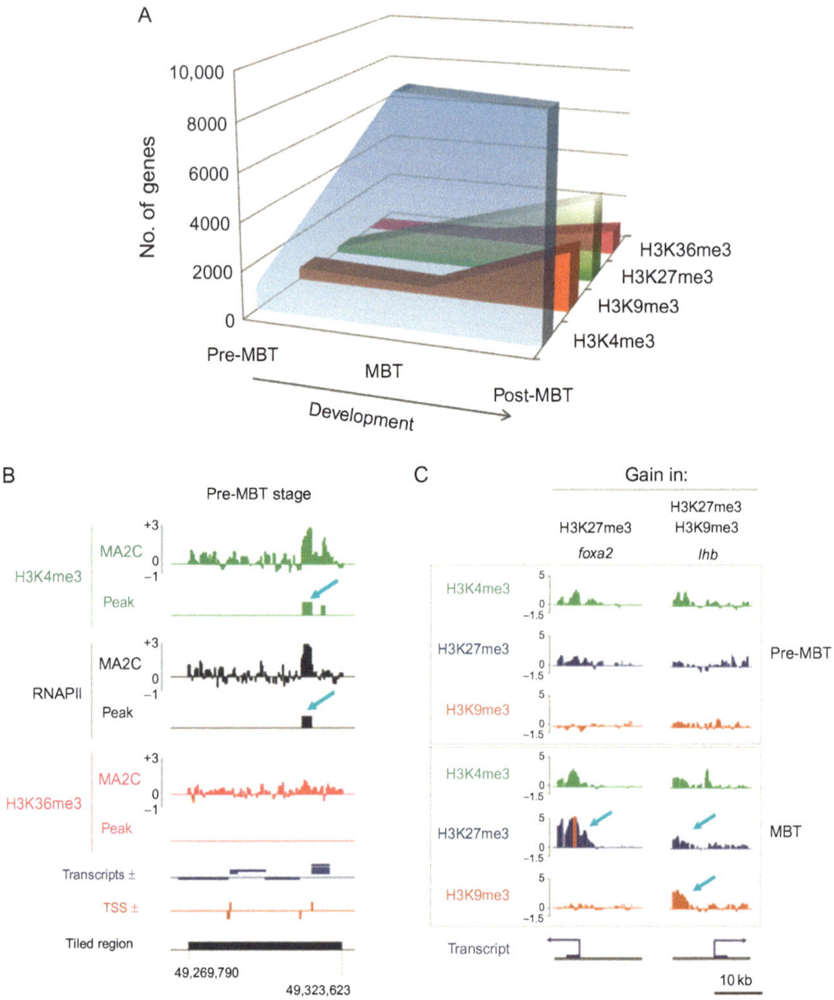


**Ingrid S. Andersen *et al.*, Figure 3.4** Marking of the zebrafish embryonic genome by posttranslationally modified histones. (A) Developmental enrichment in H3K4me3, H3K9me3, H3K27me3, and H3K36 me3 during pre-MBT, MBT, and post-MBT development. (B) Example of pre-MBT promoter occupancy by H3K4me3 and RNAPII on an inactive locus (ChIP-chip browser shot showing enrichment levels and peaks (arrow)); the tiled region covers nucleotides 49,269,790–49,323,623 of chromosome 1. (C) Examples of gain of pre-MBT H3K4me3-marked genes (upper track) in H3K27me3 (foxa2) and in H3K27me3 and H3K9me3 (lhb). *(B and C) Modified from Lindeman et al. (2011).*

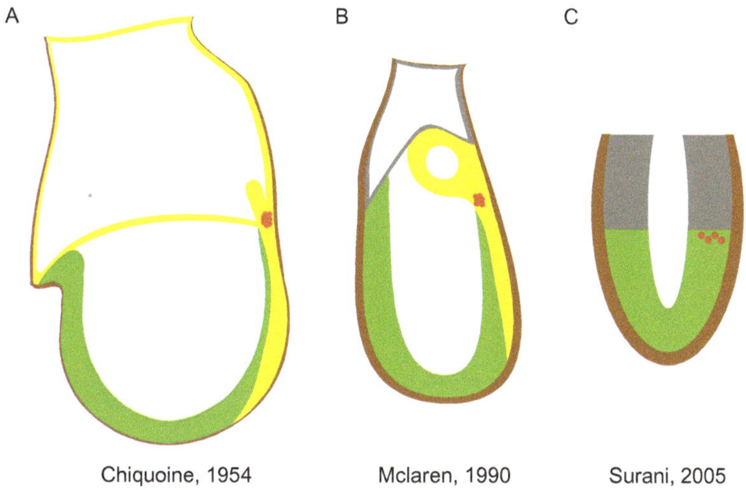


**Harry G. Leitch *et al.*, Figure 5.1** History of mouse PGC specification. Schematic diagrams of embryos depicting the earliest developmental stage at which PGCs were observed in landmark studies. (A) Chiquoine observed PGCs in early E8.0 embryos in the allantoic and yolk sac mesoderm (Chiquoine, 1954). (B) Modifications to the AP staining protocol allowed PGCs to be detected a day earlier emerging from the posterior primitive streak, in extraembryonic mesoderm (Ginsburg, Snow, & McLaren, 1990). (C) Using Blimp1-trangenic reporter mice, four to eight PGCs were observed in the E6.25 prestreak epiblast (Ohinata et al., 2005).

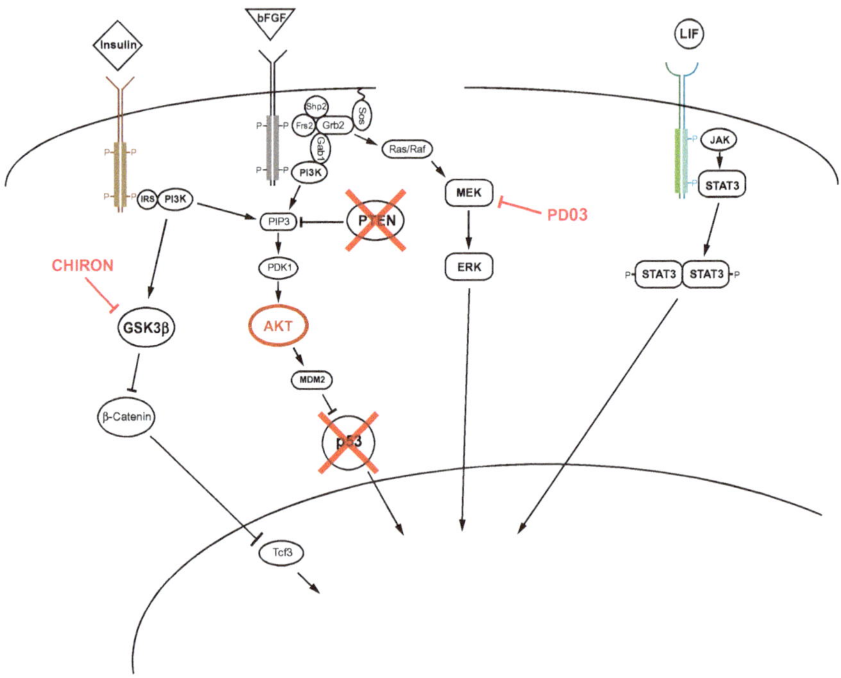


**Harry G. Leitch *et al.*, Figure 5.4** PGC signaling pathways.

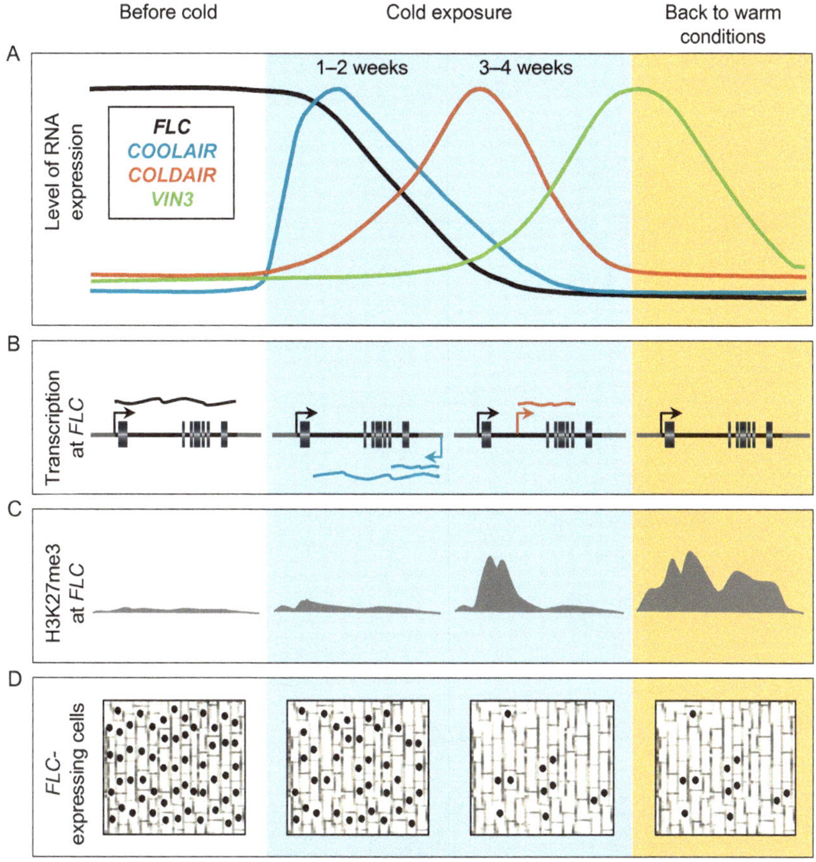


**Daniel Grimanelli and François Roudier, Figure 6.1** Chromatin-based repression of *FLC* during vernalization. The following features are depicted at different stages before, during and after prolonged exposure to cold. (A) Expression levels of *FLC* mRNA, *COOLAIR* noncoding antisense RNAs, *COLDAIR* noncoding sense RNA, and *VIN3* mRNA. (B) Location of the different transcripts over the *FLC* locus. The transcription start site of the three types of transcripts is indicated by colored arrows. (C) H3K27me3 enrichment over the *FLC* locus. (D) Bistable expression of *FLC* in individual cells within a tissue. Black dots represent nuclei of cells expressing *FLC*. Before cold exposure, *FLC* sense mRNA levels are high in all cells. Soon after cold exposure, *FLC* expression is downregulated in some cells concomitantly with the transcription of COOLAIR. Longer exposure to cold triggers the expression of COLDAIR that recruits a catalytically active PRC2 complex at the nucleation region around the TSS of *FLC*, which results in the local enrichment of H3K27me3. This leads to an increasing number of cells in which *FLC* switches to a repressed state. Return to warm temperature after at least 4 weeks of cold exposure results in stable *FLC* silencing associated with the spreading of VIN3-PRC2 and H3K27me3 over the whole locus.

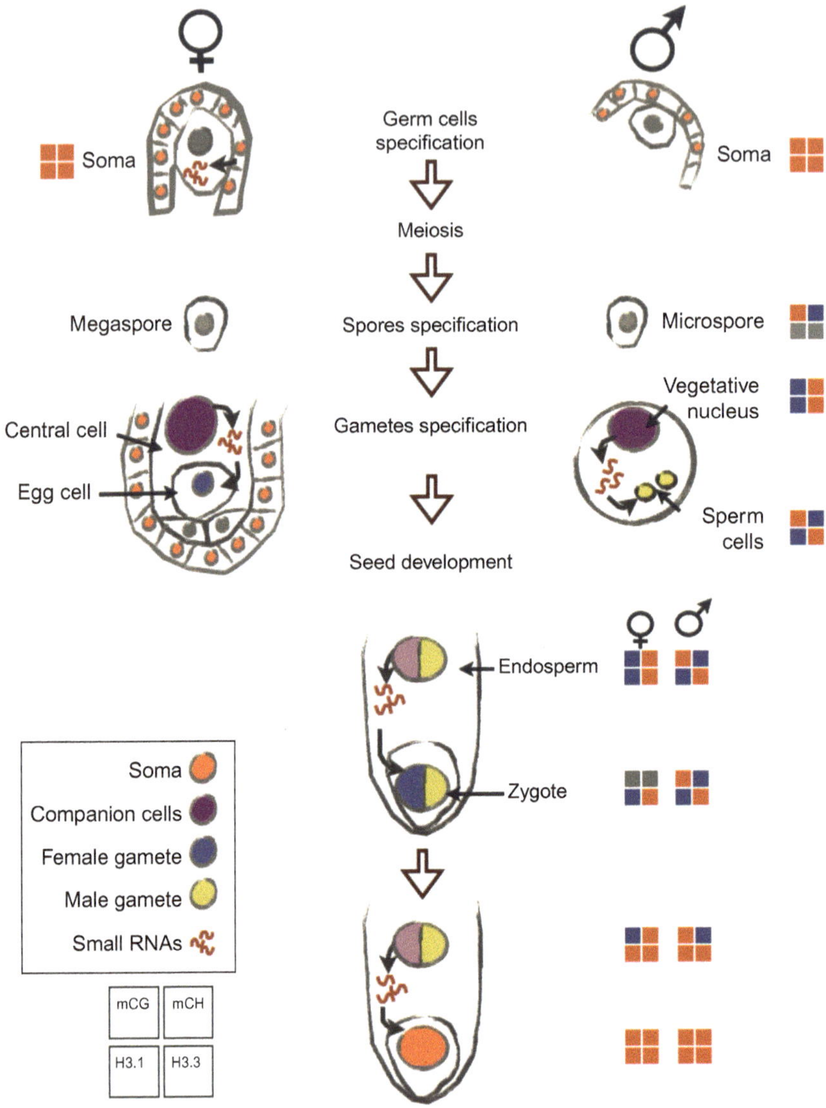


**Daniel Grimanelli and François Roudier, Figure 6.2**—See legend on next page.

**Daniel Grimanelli and François Roudier, Figure 6.2** Chromatin dynamics during germ cells development and reprogramming. While much remains to be determined, existing data suggest highly dynamic chromatin changes starting with the specification of germ cells. Known patterns of CG methylation (mCG), non-CG methylation (mCH), canonical H3 (H3.1), and variant H3 (H3.3) deposition are indicated in the squares for the different cell types. Red indicates enrichment, while blue indicates depletion. Gray boxes indicate missing data. Virtually nothing is known with respect to male or female sporogenesis. Small RNA molecules have been identified at different time points, originating potentially from both somatic and companion cells, but their biological role remains unclear. A non cell-autonomous mode of action, involving movement of the small RNAs from the soma, has been proposed during germ cell specification but may also act at later stages of gamete development (not indicated). In the mature gametophytes, small RNAs originating from companion cells target gametic chromatin in the egg and the sperm cells. The gametes are characterized by distinct epigenomes indicated by different colors. The epigenetic states of the paternally (♂) and maternally (♀) inherited genomes following fertilization are indicated separately. The extent of epigenome reprogramming in the embryo remains a matter of debate, but the picture assumes a phase of reprogramming before the first division of the zygote, resulting in the reestablishment of a mostly "somatic" genome where parental origins have been largely erased.

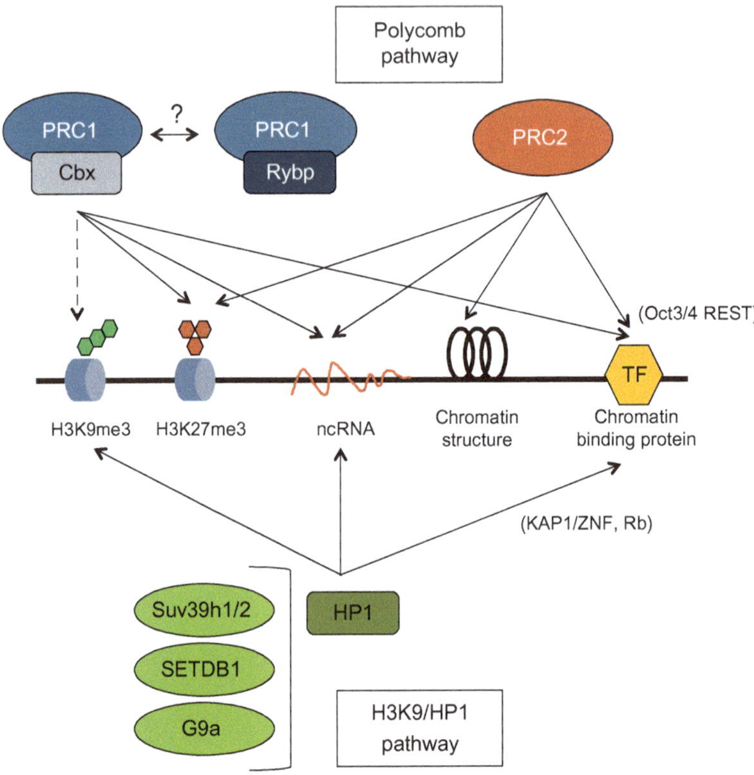


**Peter Nestorov *et al.*, Figure 8.1** Overview of targeting of Polycomb complexes and H3K9 KMT/HP1 proteins. Polycomb and H3K9 KMT/HP1 complexes can be targeted to chromatin through many diverse ways implicating direct interaction with modified histones, noncoding RNA (ncRNA), chromatin structure, and DNA-binding protein like transcription factors or zinc-finger-containing protein (ZNF). Solid lines indicate interactions. Dotted lines indicate potential interactions. Some examples are indicated between brackets. See text for details.

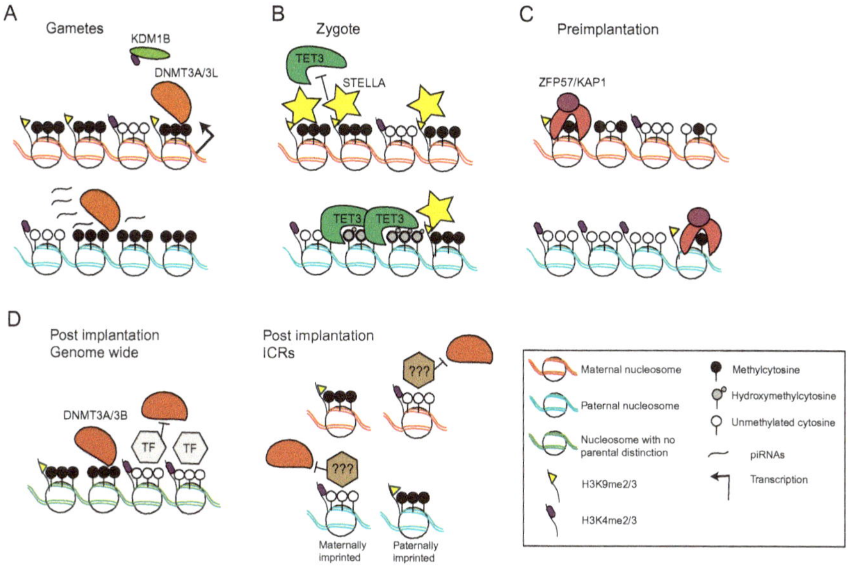


**Rachel Duffié and Déborah Bourc'his, Figure 9.1** *Forces acting toward the establishment and the protection of gametically inherited DNA methylation marks.* (A) DNA methylation establishment in the gametes. DNMT3A and DNMT3L work in the female and male germlines to establish DNA methylation profiles genome-wide. In oocytes (upper part, red nucleosomes), this process is guided, at least in part, by traversing transcription and by the removal of H3K4me2/me3 marks, which can be mediated by the KDM1B histone demethylase. In the developing sperm (lower part, blue nucleosomes), DNMT3A/3L-dependent DNA methylation can be guided by piRNAs. Major remodeling of sperm chromatin will occur in the last stages of spermatogenesis by histone to protamine exchange (not shown). (B) DNA methylation loss and protection in the one-cell zygote. Before the onset of the first S-phase, H3K9me2, which is globally associated with maternally methylated DNA, recruits STELLA thus preventing TET3 binding. On paternal chromatin, TET3 binds 5mC and converts it to 5hmC, except at certain loci such as paternal ICRs, which are protected by H3K9me2-dependent STELLA binding. (C) DNA methylation loss and protection during preimplantation development. DNA methylation is passively lost during this period, following embryonic cell division. Methylated sequences that harbor TGCCGC motifs escape this passive demethylation through the recruitment of the heterochromatic ZFP57/KAP1 complex. (D) DNA methylation gain and protection in the early postimplantation embryo. While *de novo* DNA methylation is set up genome wide, mainly via the activity of DNMT3B and to some extent of DNMT3A, some loci are protected with an unmethylated state. Transcription factor binding and/or H3K4 methylation marks could protect these regions from the action of DNMT3A/3B. This shielding mechanism is likely to be operant genome wide and at ICRs, in particular, leading in this case to the maintenance of parent-specific DNA methylation.

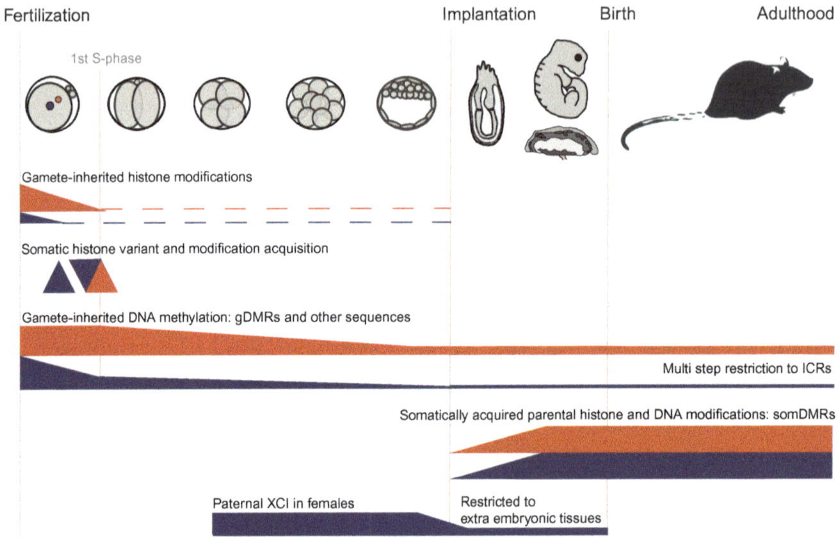


**Rachel Duffié and Déborah Bourc'his, Figure 9.2** *Dynamics of epigenetic distinction between parental alleles during mouse development*. Red and blue colors represent the maternally inherited and paternally inherited genome, respectively. Most gamete-inherited epigenetic marks are erased after fertilization, during preimplantation development, with the exception of some parent-specific DNA methylation that is maintained throughout life on ICRs. Gamete-inherited histone modifications are mostly replaced by somatic histones of maternal origin at the time of protamine to histone exchange on the paternal genome and massively on the two parental genomes of the one-cell zygote by replication-dependent incorporation at the first S-phase. Gametic DNA methylation is lost in part on the paternal genome before the first S-phase by an active mechanism and then progressively on the two parental genomes by a passive mechanism of replication-dependent dilution during preimplantation development. Only ICRs maintain gamete and parent-specific DNA methylation patterns throughout life. At implantation, new loci acquire parent-specific histone and DNA modifications, under the control of ICRs, the somatic DMRs (somDMRs). Finally, paternal XCI occurs in female preimplantation embryos but is reset at implantation to allow random XCI, except in extraembryonic tissues where paternal XCI continues.

MIX
Papier | Fördert
gute Waldnutzung
FSC® C083411

Kolibri 360 GmbH
Ferdinand-Jühlke-Straße 7
99095 Erfurt, Deutschland
produktsicherheit@kolibri360.de

# Law Beyond Code

This book explores one of the most pressing questions of our time: What happens to law when technological systems—algorithms, smart contracts, digital platforms—begin to regulate our lives as effectively as, or even more effectively than, traditional legal rules?

While many scholars have framed this debate in terms of efficiency, compliance, and enforcement, this book takes a step back to ask a deeper question: What is law, and what distinguishes it from mere technological control? Through three interconnected chapters, the book moves across theory, contemporary issues, and history. It critiques the popular thesis of the "death of law" under conditions of algorithmic regulation, arguing that law itself can be seen as a technology of a special kind. The book's central argument is that technological regulation does not "perfect" the law by guaranteeing compliance. Instead, it abolishes the normative dimension that defines law: the distinction between what is and what ought to be. When algorithms collapse this gap—by making non-compliance impossible—the unique reality that law creates disappears. Accessible, provocative, and interdisciplinary, this is not simply a book about legal theory. It is a book about the future of human freedom, responsibility, and political life in a world increasingly governed by technology.

This book will appeal primarily to scholars and high-level students working in the areas of legal theory, the philosophy of law, and law and technology.

**Tommaso Gazzolo** is Associate Professor of Philosophy of Law at the University of Sassari. His work is situated within general legal theory and focuses in particular on the concept of law and the relationship between being and ought-to-be. His research also addresses the ontology of legal institutions, the nature of legal normativity, and the relationship between law and technology. Alongside these theoretical concerns, he has published extensively on the history of modern legal thought, with particular reference to authors such as Hans Kelsen, Carl Schmitt, G.W.F. Hegel, Immanuel Kant, Thomas Hobbes, and John Locke. More recently, his work has focused on the transformation of law in the context of digital technologies, algorithmic regulation, and artificial intelligence, exploring the conditions under which law can retain its normative autonomy in technologically mediated societies. His research combines systematic legal philosophy with close engagement with the history of legal and political thought.

# Law Beyond Code

## Technology and the Future of Normativity

**Tommaso Gazzolo**

Routledge
Taylor & Francis Group
LONDON AND NEW YORK

First published 2026
by Routledge
4 Park Square, Milton Park, Abingdon, Oxon OX14 4RN

and by Routledge
605 Third Avenue, New York, NY 10158

*Routledge is an imprint of the Taylor & Francis Group, an informa business*



*British Library Cataloguing-in-Publication Data*
A catalogue record for this book is available from the British Library

*Library of Congress Cataloging-in-Publication Data*
Names: Gazzolo, Tommaso, 1984- author
Title: Law beyond code : technology and the future of normativity / Tomasso Gazzolo.
Description: 1. | New York : Routledge, 2026. |
Includes bibliographical references and index.
Identifiers: LCCN 2026010740 (print) | LCCN 2026010741 (ebook) | ISBN 9781041307150 hardback | ISBN 9781041307181 paperback | ISBN 9781003775850 ebook
Subjects: LCSH: Jurisprudence | Law--Philosophy | Technology and law | Normativity (Ethics) | Artificial intelligence–Law and legislation | Algorithms–Social aspects
Classification: LCC K237 .G39 2026 (print) | LCC K237 (ebook)
LC record available at https://lccn.loc.gov/2026010740
LC ebook record available at https://lccn.loc.gov/2026010741

ISBN: 978-1-041-30715-0 (hbk)
ISBN: 978-1-041-30718-1 (pbk)
ISBN: 978-1-003-77585-0 (ebk)

DOI: 10.4324/9781003775850

Typeset in Times New Roman
by Deanta Global Publishing Services, Chennai, India

# Contents

# Introduction

## Law and Technology

Over the past decade, the intersection of law and technology has moved to the centre of both legal theory and philosophy. This does not mean that jurists had never addressed such questions before, or that the field admits of a single definition.

Established labels such as *Cyber-Law*, *Techno-Law*, and the broader *Law and Technology* encompass widely different domains—from cybercrime and digital intellectual property to bioinformatics and drone regulation. They also signal contrasting understandings of law itself rather than of technology alone.

The now-canonical phrases "*code is law*" and "*law is code*" capture this divergence. The first points to the normative capacity of computer code and asks whether technological regulation merely imitates, or instead transforms, traditional legal norms. The second suggests the automation of legal reasoning and the translation of norms into algorithms, as in projects of *computational law*. Some scholars take these views as mutually reinforcing: if code is inherently normative, legal norms should, in principle, be expressible as code. Others maintain that moving from one to the other alters the very meaning of law, because the normativity of code differs qualitatively from that of legal rules and resists such translation.

Discussing "law and technology," therefore, is never only a matter of their interaction. In analyzing technology's implications, jurists also advance implicit conceptions of what law *is* and what it *ought to be*.

As H.L.A. Hart observed, the question "What is law?" remains a constant in legal discourse—one that the digital revolution and developments in artificial intelligence have once again brought into sharp relief.

This book approaches that question from a different angle. Phrases such as "law and technology" or "law and technique" assume that law and technology exist as distinct spheres whose points of contact we then attempt to map. The chapters that follow challenge this assumption. They argue that law is not external to technique but a form of it: a distinctive human technology whose mode of operation differs fundamentally from what is usually meant when we speak of "technological regulation."

DOI: 10.4324/9781003775850-1

## Law, Technique, and Technology

It is necessary to clarify from the outset what is meant here by "technique" and "technology." Although the two terms are often used interchangeably, especially in the United States, it is useful to draw a distinction between them:

(a) by *technique*, I refer to a set of procedures capable of producing a certain kind of object—namely, *artefacts*, that is, artificial objects. When I claim that law is a technique, I therefore mean that law is, first and foremost, a set of procedures whose outcome is the production of artificial "objects," that is, entities that do not exist in nature (for example, property, contract, or even paternity). *Technique*, in this sense, refers to what is *done*—to activity itself, to the fact that law is *used* for something.

(b) by *technology*, on the other hand, I refer to the body of knowledge concerning the technical processes that enable the production of certain artificial objects. It concerns, therefore, less what is *done* than the reflection on what allows one to *conceive* and *design* what is done.

To speak of law as a technology thus means to refer to the fact that law is a form of knowledge that *conceives* and *designs* technical solutions to the specific problems it identifies as relevant.

This does not mean, of course, that these two dimensions do not overlap or fail to form a single continuum. It is therefore not inappropriate to use them interchangeably at times, and to speak of law both as *technique* and as *technology*.

Nevertheless, drawing this distinction remains crucial, for it makes clear that law is always *praxis* as well as *episteme*—a productive activity grounded in its own distinctive forms of knowledge, and in this sense, law both acts upon and reflects the world it helps to construct.

Let us clarify this point with an example. Running can be defined as a *technique*—that is, a form of know-how, or a way of doing. Yet, as Don Ihde observes, what enables one, through a series of analyses and instruments, to understand how to produce new materials for shoes or other innovations capable of enhancing its efficiency—and thereby altering the very gestures through which running is performed—is *technology*.

A similar distinction can be drawn in the field of law. We might say, for instance, that the possibility of acquiring ownership of a thing through the payment of a price is a *technical* possibility, in the sense of a "can-do" that only law makes possible. Yet this occurs on the basis of a specific *technology*—the juridical one—which is what allows the *price* to become that which *ought to be paid*, that is, to constitute in a particular way the artificial "object" we call *price*.

The thesis advanced throughout this book is that the specific form of technology proper to law is defined by that distinctive kind of normativity expressed by the *ought-to-be* (in German: *Sollen*). This observation already points toward what will become one of the central theses of this work. Technological regulation does not simply "complete" or "perfect" the law by ensuring compliance more effectively; instead, it threatens to *negate* the normative dimension that is constitutive of law itself. Law is not reducible to a tool for guiding behaviour or enforcing compliance; it institutes a distinct normative reality—an "*ought-to-be*"—that technological devices, by collapsing the distinction between "ought" and "is," effectively abolish.

## Structure of the Book

This study aims to fill a crucial gap in the current literature. While leading scholars such as Brownsword, Lucy, and Hildebrandt have approached the question through the lenses of enforcement, efficiency, and governance, the present work steps back to examine the ontology of legal norms themselves. In doing so, it offers a renewed conceptual framework that invites future debate to move beyond pragmatic concerns of effectiveness toward the deeper terrain of legal philosophy. The argument unfolds across three chapters, following the itinerary outlined below:

(a) In the first chapter, we will seek to clarify in what sense law is not, as previously suggested, something distinct from "technique," but rather a particular kind of technique and technology in its own right. In particular, we will focus on how law may be understood as a *cybernetic art*—that is, as a form of "governing" technique that operates through the creation of connections and interactions between the natural and the artificial. We will attempt to show how legal concepts such as *property*, *obligations*, and *rights* can be conceived and interpreted within this framework.

Highlighting law's capacity to bridge and integrate the natural and the artificial helps reveal in what sense it may be understood as an *anthropotechnics*—a human-making technology through which humanity continuously shapes and redefines itself. Rights, in this perspective, can be seen as frameworks that allow individuals to become what they are not "by nature." They are not merely tools or instruments, but techniques that act back upon those who use them, transforming the very subjects who exercise them.

Let us clarify this point with the simplest possible example. The legal norms that establish, regulate, and modify the attribution of paternity can, in this sense, be understood as genuine frameworks that enable individuals to subjectify themselves respectively as "fathers" and "sons." Becoming a father, in legal terms, does not mean having a certain "quality" merely recognized or verified—a quality that one would supposedly possess "by nature."

In fact, one can be a "father" in the eyes of the law even without being so biologically. Rather, it means being able to rely on a set of rights that allow an individual to assume the role and status of a father—to become, therefore, a subject who, in and of himself, that is, in the absence of law's operation, would not exist as such.

(b) The second chapter will illustrate how this *cybernetics* operates in a concrete case: the contemporary debate on so-called *neuro-rights*. Understanding law as a form of *anthropotechnics* also makes it possible to clarify how the question of legal protection for the brain can be approached. Those who have argued for the creation of so-called *neuro-rights* start from a fundamentally *naturalistic* conception of law—one that assumes that, in order to determine the most appropriate legal forms of protection for the products of our mental activity, we must first answer the question of what the brain and the mind "naturally" are.

In contrast to this view, the argument advanced here is that law is effective precisely to the extent that it can make the brain into something other than what it "naturally" is.

From this perspective, supporting the idea that property law may offer the most appropriate framework for protecting our mental activities does not entail reifying the brain. On the contrary, it affirms law's unique capacity to depart from mere "reality" and to forge new connections between nature and artifice whenever such reconnections better serve the ends of justice that law seeks to achieve.

Law, in this sense, grants individuals the capacity to engage with their own brains and mental activities *as if* they were objects of property. Not, of course, because they truly are, but because this very legal construction allows what the brain and the mind *are by nature* to be protected more effectively.

(c) Finally, the third chapter will address the relationship between the technology proper to law and the form of normativity characteristic of algorithmic "technologies" emerging from the digital revolution. In doing so, we will take up the question—raised by several scholars—of the possible "death of law," and reconsider what it might mean, in the digital age, for law to preserve its own normative autonomy.

Together, the three chapters constitute a trajectory through which the book progressively deepens its inquiry—from the foundational conceptualization of law as technique/technology, through a concrete case study, to the normative and ontological implications of digital governance. In so doing, the work not only unifies these elements into an integrated whole, but also challenges the reader to reconsider the conditions under which law retains its normative autonomy in an age of technological mediation.

## Conceptual Clarifications

Although this book is a work of general legal theory rather than of positive law, it inevitably bears the imprint of the author's own intellectual formation within the civil law tradition. The conceptual vocabulary and style of reasoning adopted here belong to the broader European-continental legal culture, whose categories and methods continue to shape the way law, normativity, and technology are understood throughout this study.

For this reason, it may be helpful to offer the reader, at the outset, some guidance on the meaning of the key terms and central concepts that will be used throughout the book.

Since two of the central concepts employed in this book are drawn directly from Hans Kelsen's legal theory—particularly from the phase preceding his emigration to the United States—it is essential to begin by clarifying them.

The German term *Sollen* (rendered in Italian as *dover-essere* and in French as *devoir-être*) will here be translated as *ought-to-be*. The reader should be aware, however, that within a Kelsenian framework to say that something *soll*—that it "*ought-to-be*"—does not mean that it *soll sein*, that it "*ought to become being*." In other words, *ought-to-be* does not denote a duty aimed at bringing about a state of affairs in the realm of being (*Sein*); rather, it refers to a mode of existence entirely distinct from, and irreducible to, any form of empirical or factual being. This usage follows the convention already established in the English translations of Kelsen's works. Occasionally, however, *Sollen* has also been rendered simply as *ought*, and in such cases, this term is used here interchangeably with *ought-to-be*.

What is crucial is that the reader should avoid conflating two concepts that Kelsen carefully distinguishes: *Sollen*, on the one hand, and *Pflicht* (or *legal duty*), on the other. The latter refers to the obligation to perform a certain action—more precisely, to the duty to refrain from the conduct that a legal norm defines as unlawful.

A further clarification concerns the examples occasionally used throughout the book, many of which are drawn from the civil law tradition. This choice makes it necessary to prevent possible conceptual misunderstandings. In particular, this issue arises in Chapter 2, where the discussion of so-called "neuro-rights" relies on the distinction between rights of the person and property rights.

Some ambiguity may result, especially with regard to the meaning of "property." Not only does the English language distinguish between *ownership* and *property*, but there are also significant differences between the understanding of property rights in legal systems such as the Italian one and their meaning within the common law tradition. Since it is not possible to examine this matter in detail here, I will simply note that the expression *property right* is used in the sense derived from the French codification tradition—namely, as the right to use and dispose of a thing fully and exclusively, within the limits established by law.

A final remark concerns the distinction between *reasons* and *causes*. Throughout this book, I argue that to claim that legal norms constitute "reasons for action" ultimately amounts to adopting a causal conception of normativity, insofar as norms are thereby understood as rules designed to *influence* or determine behaviour. As will be shown, some of Frederick Schauer's theses lend support to this line of argument.

At the same time, however, certain passages in the text suggest that norms may also be understood as "reasons" in a different sense (see especially chapter 3). Let me briefly clarify this point.

If we take a norm of the form "If x, then ought y," it seems to me that two mechanisms are at work, which must be kept distinct: (a) on the one hand, the norm states that *x* is the *cause* of the duty *y*; (b) on the other hand, it also indicates that *x* is the *reason* for that duty—that is, what *justifies* why *y* is owed.

It is well known that, particularly within the Anglo-American tradition, the distinction between *causes* and *reasons* is sometimes rejected on grounds that vary widely among authors. Yet I believe the distinction can, and should, be maintained in the present context.

There is, in my view, a significant difference between saying that the contract I have entered into is the *cause* of my duty to pay the price and saying that it is the *reason* why I *ought to* pay it.

Certainly, the contract is both the cause and the reason for that duty—but in the first case, it explains, at the causal level, why the duty to pay exists, whereas in the second, it provides the *normative justification* for why I must fulfil it.

These clarifications, of course, will not suffice to dispel all the difficulties that a reader coming from a civil law background may encounter. Moreover, although the traditional divide between the analytic and the continental styles of legal thought has, over time, become considerably less pronounced, it is undeniable that certain habits of reasoning remain characteristic of one tradition rather than the other.

It is my hope, nonetheless, that these differences will not discourage the reader, and that the effort towards synthesis and clarity that has guided this work may render its argument accessible, even across the boundaries of our diverse legal cultures.

# 1 Cyberlaw

## A New Approach

### Cyber

The expression "cyberlaw" has already entered legal usage (in American English, *Cyber-law*, and French, *Cyberdroit*), even though its meaning remains fragmented and multifaceted.

Its derivation, however, is clear: it belongs to that set of neologisms that refer, directly or indirectly, to the cyberspace imagined by Gibson in *Neuromancer*.

For Gibson, terms compounded with the prefix *cyber-* point to a double connection: first, to the Greek root of the word—*κυβερνάω*, "to govern," "to steer the helm," and thus also "to navigate" (hence *κυβερνήτης*, the helmsman); and second, to the *cyborg*, the synthesis of the connection between human and machine.

Speaking of "cyberlaw" therefore means, for us, understanding law from within this dual articulation: as a technique, a cybernetic art, insofar as it is concerned with governing, and governing in turn concerns interactions between organs.

We must explain this last point. As Stiegler observed, if governing means—following the Platonic strategy—"holding the helm," the "helm," however, is nothing other than an (artificial) organ of another equally artificial organ, the boat, such that the governor is one who "governs a set of artificial organs through that organ of command that is the helm" (Stiegler 2019, p. 48).

This clarifies the connection that is proper to what we call "cyber-law:" it is an art of government, of navigation, through rules that are nonetheless rules for the production, coordination, command, and functioning of organs. But what kind of organs? Artificial or "natural"? And what does it mean to say that law has to do with them?

For now, let us say that *cyber* marks the root of that field of concepts constituted by a certain relationship between the organic and the technological.

It is the field that, again borrowing from Stiegler, we may call *organological*—a field in which there is nothing, no thought, without the organic being what it is only insofar as it is modified, situated, transformed by specific techniques internal to a technology.

DOI: 10.4324/9781003775850-2

A somatic organ like the hand does not exist *as such*—as, precisely, "my" hand, or "this" brain—except insofar as it has always already been shaped and formed by a technique that has enabled it to externalize itself, to "free itself" through the tool. The hand is not, in other words, an organ that, thanks to an instrument, is liberated—it is, rather, the organ that results from its own liberation through the tool.

*Cyber* says this: that nothing exists, on earth or in the heavens, that is not technological. Technology has always embedded itself in the skin. And for this reason, it has always tended to render itself invisible, to disappear, to become nature.

"I will answer very simply that the Internet will disappear," was Eric Schmidt's reply, when he was Google's CEO, to a question about the future of the web. The Internet will disappear because, from being a dimension separate from our real life, it will eventually become part of it, that is, it will become part of us, integrated into our body through sensors, wearable devices, and instruments through which we will *be* on the Internet.

As has been noted, this is nothing more than the reprise of what Mark Weiser had already theorized in 1991 when speaking of so-called ubiquitous computing (*ubicomp*), starting from the principle that the deepest technologies are those that disappear, merging with "the fabric of everyday life until they are indistinguishable from it."

To disappear, to become invisible, indistinguishable from life, is the way technology fulfils itself. More precisely: it is by becoming life—by being nothing other than life—that technology achieves itself as what it is. The Walkman fulfils itself in the AirPods, and these in something else, until the ear and its prosthesis, its supplement, are no longer distinguishable.

More generally, we can say that everything—every "individual," every object—exists only insofar as it has always already been "technical:"

(a) "man," the human, is nothing other than his becoming human through technological, anthropotechnical manipulations (the "athletic" techniques through which the human becomes human).
(b) "things" are not *res*, goods, except in their being rendered objects of exchange; animals are not such except through a series of operations that separate them from the human.

We can thus begin to see in what sense law is always cyberlaw, always a cybernetics: it is one of the techniques through which reality is rendered legal reality—and such realization passes essentially through exosomatization, that is, technical externalization, the production of "artificial" organs that are nonetheless the very condition for the constitution of what is "natural."

This is the *paradox of exteriorization* (*paradoxe de l'extériorisation*), namely, that the exteriorization of man is the condition for his very constitution as such, that going outside oneself is the condition for the constitution of the self. These "feedback modifications"—how technique constitutes the life that, logically, precedes it—make clear how life is thus always already technical, *vie technique* (Stiegler 1998).

Law, as cyberlaw, belongs among these techniques, and it is from this perspective that we must analyze it here. To study law as a cybernetics, that is, means to study it insofar as it constitutes a technique of "government" realized through interactions and integrations between the natural and the artificial—or better, between the organic and the technological.

## Hybridization

To understand in what sense law is essentially a cybernetics, one can begin with various examples. But if we wish, first of all, to preserve its closest connection with its "organological" function, we may certainly start by observing how the legal is, above all, a technology of the body as *juridical*—that is, of the production of "organs," faculties, that constitute the human being as a subject capable of acting in a certain way.

I believe it is worth returning here, to explain the point, to a passage from Savigny that can serve as a useful entry point into the problem:

> It can and must be taken as beyond doubt," writes Savigny in a page of his *System des heutigen römischen Rechts*, "that man has the right to dispose of himself and his powers (*rechtmäßige Macht des Menschen über sich selbst und seine Kräfte*); and furthermore, that this capacity (*diese Macht*) is precisely the basis and presupposition (*die Grundlage und Voraussetzung*) of all other genuine rights (*allen wahren Rechte*), inasmuch as, for example, property and obligations have for us value and meaning only as an artificial extension of our own personal powers, as new organs, artificially added to our natural essence (*Eigenthum und Obligationen nur Bedeutung und Werth für uns haben als künstliche Erweiterung unsrer eigenen persönlichen Kräfte, als neue Organe, die unserm Naturwesen künstlich hinzugefügt werden*).
>
> (Savigny 1840, p. 336)

I will avoid here a full commentary on the passage, which would require a necessary clarification regarding how Savigny redefines the relationship between the natural and the artificial, through his critique of any idea that rights—starting with property—can be conceived as the reflection or outward manifestation of some "original" legal relation internal to the person (their right over themselves, their being "owner" of themselves).

For our purposes, it is enough to note this: Savigny is thinking in terms of a *logic of the supplement*: property and obligations are artificial organs through which something is "externalized" (the "natural" faculty of man to have rights, we might say) which, however, is nothing other than what this externalization itself produces as its own presupposition, as that which it claims to supplement.

Let us now limit ourselves to *obligation*. We may here set aside the way Savigny constructs the concept of obligation, understood as "dominion" over the person (or, more precisely, over the conduct of the debtor). As has been rightly observed, the structures of obligation differ from those of liability, which historically precede them:

> in obligation, payment is seen as the satisfaction of a creditor's interest—that is, as an event apt to bring about a planned future benefit in favor of a subject who thereby comes to have a right to it; in the system of liability, by contrast, payment was seen only as an act of release for the debtor, because the system itself was nothing more than a means of punishing a subject for failing to do certain things.
>
> (Cannata 1995, p. 416)

The invention of the concept of obligation is what brings about a radical transformation of the earlier legal structure of subjection—characteristic of the *legatum per damnationem*—in which payment was not the execution of a duty of performance, but how one could free oneself from subjection to the power, the *manus*, of the legatee.

Through *obligatio*, it now becomes possible to externalize a will capable of binding itself to what it now wills—a will that, through obligation, becomes able to commit itself for the future.

From this point of view, *obligatio* must be understood as a mechanism that "creates" a new form—already from the outset "technical," therefore—of human action, which we might define as an action oriented and bound towards the future, in which what the debtor "must" do is such by a planned and anticipated future satisfaction to which a subject initially without rights has now become entitled.

When Nietzsche conceives memory as a "faculty"—an "organ," we might say—artificial because shaped through education, as the deactivation and suspension of "natural" forgetfulness, he rightly links it to the very structures of *obligatio*: to a mechanism through which the individual "is answerable for his own future" (Nietzsche 1994, p. 36).

More than the "long history of the origin of responsibility," we should see here a genealogy of obligation as an artificial organ. It is no accident that the "contractual relationship between creditor and debtor" is precisely what Nietzsche identifies—though he then tends to reread it as a relation of "subjection"—as the relation through which it becomes possible to manufacture a memory for the one who promises.

What is at stake in the legal technology of obligation is the creation of an organ: memory. That is, a memory understood here as a faculty—an entirely artificial, wholly constructed "organ," with nothing "natural" about it—through which man becomes human, is "reared," is constituted as that subject who can bind himself, commit himself to willing what he has willed, to anticipate the future and dispose of it.

*Obligatio* is what constitutes "memory" by externalizing it through the bond the debtor assumes, according to the mechanism on which we have already insisted.

Legal technology is not the only one among the techniques that function in this way: indeed, one might say that in general, technique is always a vector of memory, since memory itself exists only through its externalization—through the invention of instruments that allow us to record, preserve, and recall our thoughts and knowledge.

But *hypomnesis* retroacts upon *anamnesis*: what we can remember depends on the way we have objectified and externalized our memory, on the techniques through which this occurs, on the procedures for calling back into presence what has been entrusted to the support.

Let us, however, remain for a moment longer with the legal technology of *obligatio*. If obligations are—as Savigny puts it—"artificial organs," prosthetic extensions of our "natural" faculty, this must be understood in this precise sense: the structures of obligation constitute—according to the movement of exteriorization/interiorization—the techniques through which "a memory is manufactured" in the one who promises, binding him to answer for his future conduct because he has declared that he "wishes to go on willing" what he has promised to another.

## The Roman "Venter"

I would now like to attempt another example—drawing on what Yan Thomas has written about the mother's womb—since it points precisely to the intersection, the connection (certainly a highly problematic one) between the organic and the technological (Thomas 1986).

In Roman legal logic, *venter* ("womb") is an ambiguous term, or rather a "double" one. On the one hand, it denotes the child conceived but not yet born: it is not a "reality," not anything that exists, but a juridical construction designating nothing other than the posthumous son of the father.

Here the name of the uterus does not refer either to a part of the mother's body or to what is physically contained in that body (since, for the Romans, *partus* was the term used—naturalistically—to indicate one who is to be born or ready to be born; the embryo, we might say, was entirely alien and irrelevant to law: "it cannot be considered that, before being born, the embryo was a human being," affirms Papinian).

*Venter*, therefore, is nothing other than a reality constructed entirely by law.

Yet, on the other hand, the womb is always also the "mother's womb:" it is, that is, also always a part of the woman's body, to which the law itself grants no "autonomy." This is shown by the fact that, for the Romans, a woman who procures an abortion never endangers anyone other than herself: there is, in other words, no "subject," no individuality, that lives within her, in her uterus, when she is pregnant. Hence, the ambivalence of a term that denotes both the unborn child—as something that exists only for the law—and the uterus.

This is why, as Thomas writes, posthumous children are at once and contradictorily a physical part of the mother's viscera and, themselves, "venters"—in the singular sense of abstractions identified solely through their relationship to a dead father of whom they are already legitimate heirs.

Here, then, we find the space for a connection, for a possible articulation, between the organic and the technological: the "womb" is at once a part of the pregnant woman, of her body, of her viscera (*mulieris portio est vel viscerum*), and, at the same time, a subject having no relation to her, as the posthumous heir of a deceased man, and one that has no existence except within the law.

What is interesting is that, in this case, what is defined is a field of possibility, of potentiality, in which the relationship between the organic and the technological remains open. *Venter* indicates, as we have seen, a splitting between nature and artifice, such that the conceived being exists at once as a future person and as (part of) the maternal body.

It is worth pausing here to consider a further issue. As is well known, with Christianity, there would eventually prevail—though not without initial difficulty—the idea that abortion was a form of anticipatory homicide, since he who will be born is already a man.

Here it is "nature" that grounds and binds the law: since the fetus is already life, already an individual, abortion amounts to homicide.

But, as Thomas has noted, for a time, the debate over the reasons for prohibiting abortion was also articulated according to the opposite logic—namely, that it was law that "demonstrated" or gave proof of nature.

In one of the treatises of Pseudo-Galen, for instance, devoted to the question of whether the embryo is a living being, one of the answers given is that, since it is already an heir, it must consequently be considered "alive." It is not that the embryo is legally an heir because it is alive; rather, it is because it is juridically an heir that it is alive—the law, in other words, produces the "naturalization" of itself, of its own artifice.

Regardless of the specific solutions adopted, the example of *venter* further clarifies the cybernetic function of law.

Law does not prohibit or permit abortion based on the observation of something belonging to the order of the "natural," something given to it as a fact—such as whether or not the embryo is "life," whether it is part of the woman's body or an individuality distinct from it.

On the contrary, we must say that what the womb is, for law, depends on the juridical operations that are connected to it.

And, whatever direction is taken, the mother's "womb" is always, juridically speaking, something that belongs neither to the side of nature, the organic, nor to that of pure "fiction," as the logic of the Roman jurists demonstrates: if it is at once uterus and child, organ and subject, this is because it is the product of a technology—of a "cybernetics," we might say—that operates by constituting connections between these two levels, by creating reciprocal interactions and integrations between the organic and the artificial, between the human and the machinic.

The "womb" has always been, one might say, essentially cyborg—just as cyborg is Max's womb in *Videodrome*, the womb into which he inserts the videocassette and from which he later draws a pistol (which allows us to frame correctly the problems that today's so-called "new technologies" pose in matters of reproduction, procreation, and gestation). Law does not hinder or promote a given solution based on whether it is or is not in conformity with some "nature." Rather, it constructs nature in function of the solution it deems correct—that is, for which it considers there to be reasons to promote or discourage it.

## Naturalization and Legal Function

Law is always "cyber." This could be demonstrated simply by what Italian jurist Carnelutti reminds us—namely, that, juridically, there are things that were once human beings, and human beings who were once things (Carnelutti 1938).

He is here referring to the phenomena of separation and incorporation so often discussed by jurists: consider hair which, once cut, ceases to be part of the body and becomes *res*, a thing; but also those artificial objects—such as a wooden arm or a gold tooth—which, once "incorporated" or "inserted" into the human body, can no longer be treated as things, becoming instead *pars hominis*.

Thus, Carnelutti observes, a part of the human being can become a thing when it is definitively separated from the human itself: blood drawn, or a braid of cut hair, becomes juridically a "thing" when originally it was not. Similarly, a thing "becomes man, that is, part of man, when it is incorporated into him"—and this, for Carnelutti, is a juridical phenomenon that implies the capacity of law to "transform a thing into part of a man."

Far from being founded on the distinction between persons and things, law in reality proves itself to be a "cybernetics:" that is, a technique which proceeds by separations and incorporations, by continual transformations of things into human beings and of human beings into things.

Carnelutti does not, however, take the definitive step; he seeks to conceal it and to re-inscribe the logic he has discovered within the principle that what occurs according to law is "because so it is according to nature."

Incorporation, that is, would not consist for Carnelutti in the "introduction of something into the organism," but only "in the insertion of such a thing as is destined to restore it:" while blood is properly incorporated, becoming part of the human body, this should not, in his view, apply to the bullet lodged within it or to the precious stone swallowed by a thief.

It is evident here that Carnelutti attempts to retrace his steps with respect to his own reasoning, proposing a "naturalistic" criterion to distinguish between what can and cannot be "incorporated."

In reality, however, what he discovers here is precisely the cybernetic function of law—its capacity to create zones of indistinction, of reciprocal articulation, of grafting, between the organic and the technological.

Certainly, Carnelutti raises an essential question, that of the "boundaries" of the body: when can I say, juridically, that a thing—a prosthesis, but also a smartphone or some other device that may be installed in the body—has become human, part of the human body?

Yet it is difficult to maintain that there truly exists a criterion external to law which could bind its choices. Is it, for example, the fact that I can experience that thing as part of myself—that I cease to perceive it as a thing within my body, and instead perceive it as my body itself? But—even granting that such a "fact" could be ascertained—for what reason should law decide on that basis?

In reality, in principle, nothing prevents law, according to its own logics, from effecting the transformation of something into a part of the body or, conversely, of a part of the body into a thing.

The point that Carnelutti in fact fails to grasp is that these devices of separation/incorporation that constitute law demonstrate that in it there is no given criterion for distinguishing the person from the *res*—or, more precisely, for distinguishing what would form part of the self of the individual, what would be the individual itself, from that which could only belong to the individual as property or some other right over the thing.

And it is precisely this that makes law a cybernetics: its ability to transform things into organs and organs into things, to constitute the body beyond any "naturalistic" constraint. I can be the bullet lodged in my brain, just as, in principle, nothing prevents my hand from being considered a thing belonging to me. In itself, the fact that such a human-thing integration is permanent or not, separable or not in nature, is irrelevant for the purposes of what law may decide.

One must, however, avoid a fundamental mistake—one that, it seems to me, is rather widespread. Let me explain the point. The legal affirmation, to return to our earlier example, that the uterus "is" the child should in no way be understood as if it were a recognition of some "reality," a statement of fact.

In other words, the law does not hold that the *venter* is a person because it assumes that this is so in reality, in some reality to which it must, therefore, conform (in our case, as we have seen, it is in fact the exact opposite). I believe this is an essential point, because it clarifies the function proper to rights—that is, to those mechanisms that constitute subjective legal positions.

This function is not to recognize how things are in reality, but—on the contrary—to provide schemes that enable those who follow them to act legally in certain ways.

To give a simple example: the fact that the law "ascertains" (that is: constitutes me as) the "father" of my child does not in any way mean that it recognizes some fact or state of affairs (such as that I am biologically the child's father). Rather, its meaning is to confer upon me the power to act legally as a father.

This, after all, is precisely what makes law a "cybernetics:" a technique not of recognition/ascertainment of reality, but of the constitution of a reality autonomous and independent from that which is "given" or assumed as "natural."

What could be more "cyber" than a law which—here I am referring to the striking case of the *mulatto pisano* (Court of Appeal of Florence, 15 July 1949)[1]—considers a child of colour born to a white woman married to a white man to be the son of that man, despite recognizing, of course, that it is physiologically impossible for him to be the father?

The Roman distinction between *persona* and *homo* is, in this sense, a "cybernetic" distinction, in that it separates the role—or, more precisely, the subjective position constituted by the attribution of a certain bundle of rights—from the concrete individual, the juridical reality from the factual reality.

And this separation is what makes rights nothing other than prosthetic devices: mechanisms through which the individual can externalize himself, act, regardless of what he "naturally" is.

Certainly, this will also entail effects of internalization, of "subjectivation:" it is in acting, on the legal plane, as the father of my child that I strive

1 This was a legal dispute that, in Italy, received wide coverage in the press. It concerned an action to deny paternity brought by the husband of a woman who, shortly after the end of the Second World War, had given birth to a child of colour. The court rejected the action to deny paternity, for reasons that will be examined in the course of the text. In connection with the case, a bill was introduced in the Chamber of Deputies aimed at extending "the cases of denial of paternity provided for by Article 235 of the Civil Code" so as to allow such an action "also in the case in which, on the basis of manifest and indisputable somatic characteristics—scientifically established—it is determined that the child conceived during the marriage is of a different ethnic origin from that of the presumed father."

to become a father, to be the father that, through law, I become. But—and this is the essential point—my "being" a father here has nothing to do with what I "naturally" am or am not, but exclusively with a technology—namely, the juridical technology—that enables me to become one. One must therefore not confuse "nature" with the effects of "naturalization" that law can produce. To do so would be to lose sight of legal logic.

To remain with our example: it would mean assuming that the reason law "recognizes" me as the father, gives me the right to act as the father of my child, is the fact that I am one. Nothing could be more mistaken.

No fact can constitute a reason to assign or recognize a right (to think otherwise would be to fall into that "naturalism" which claims to recognize normative force in facts). Whether or not I am recognized by law as the father does not depend on how things are, on factual "reality," but, rather, on reasons always internal to law—that is, on the operations to which such recognition is directed. In the *Code civil*, for example, it is the logic of marriage that grounds paternity: the father is the husband of the woman who has given birth to a child.

Ultimately, this is merely a case of applying the logical principle according to which a norm can never be derived from a fact, but always and only from another norm—a principle that risks being obscured, if not violated, whenever one claims that a certain right must be assigned based on some fact (one or more qualities of the subjects who should hold it, a given historical or political circumstance, etc.).

Logically, I cannot claim, for example, that a certain class of subjects *x* must have the right to do *y* on the basis of the fact that, otherwise, in the absence of such recognition, it would suffer a certain harm. The argument is not valid unless one introduces, or makes explicit, the norm according to which it is right or obligatory that the class of subjects *x* should not suffer that harm. Only if this norm is admitted as valid does it become possible to justify the right being claimed.

To speak of "cybernetics" about that specific technology which is law thus means recognizing that, in juridical reality, there is nothing that is not juridical—nothing that is not legally constructed. And it also means recognizing that the attribution of a certain right can never be justified by a fact, by the observation of a given state of affairs, but always and only by a norm (by which that state of affairs is considered as a reason to attribute that right).

To return, once again, to our example: even under the current Italian Civil Code, it is not the mere fact, ascertained, that biologically *x* is the father of *y* that constitutes the reason for declaring paternity; rather, it is the norm—Article 269 of the Italian Civil Code, which provides that proof of paternity may be given by any means, that justifies the legal effects that such a fact may produce. Put differently: even when law says that it is nature that decides, it is always and in every case law that decides—and thus it is always law, and not nature, that determines the outcome.

## Property

To clarify in what sense law constitutes a specific technology, a good example can be found in property law. I will pause on this point, because clarifying the meaning of property will also help us to better frame the discussion that, in the next chapter, will be devoted to the issue of neuro-rights.

We have already referred to Savigny's view of property as an "artificial organ." It is now necessary to take a further step and show that property is not simply a right over things, but a technique through which, by means of our relation to things, it becomes possible to *individualize* or *subjectivize* ourselves in a certain way.

If we were to answer the question "what is property?", I believe the answer, in this sense, could be the following: property is a *technical form* that relates an individual to a thing in such a way that, through the powers exercised over the thing (to use it, enjoy it, dispose of it, and so on), the individual can *appropriate*—that is, make their own—the qualities, knowledges, and skills that have been externalized in the thing itself, and thereby constitute themselves as the individual they become.

In part, this definition is similar to the one proposed by Margaret Radin and to her understanding of property as a relation with objects—as the way in which individuals interact with and relate to things as a condition for their own physical and psychological individuation.

Radin speaks, in this sense, of a *self-identification through objects* (Radin 1982. See Austin 2010). Her theory of *property for personhood* responds to the need to emphasize that the act of *appropriating* things constitutes a mode through which individuals constitute themselves as *persons*.

In contrast to Radin, however, I do not consider it necessary to ground property in *personality*, since this would seem to presuppose the existence of the "person" prior to their constitution through the relation with things held in property.

We could, then, offer a slightly different explanation. Let us take the simplest possible example: my smartphone. I am not asking what it means for it to be "mine"—the obvious answer here would be that I purchased it, or that it was given to me, and so on, according to the rules governing the acquisition of ownership. Nor am I asking what powers or faculties follow from the fact that it is "mine." Rather, I am asking: what does this *being mine* consist in? What does property mean here, and what function does it perform?

Property is not merely—as an approach in the style of Ross would suggest—a way of presenting the connection between a series of facts (its modes of acquisition) and a series of legal consequences (the powers and faculties it gives rise to).

It is also, in fact, what provides its holder with a certain schema—a particular form through which they can enter into a relation with the thing; that is, they can engage with it in ways that would not be possible without property.

It is, in fact, precisely because the smartphone is "mine"—because I am its owner—that I will decide to store my personal notes in it, to keep a record of the conversations I have had with my friends, to save numbers, appointments, thoughts, images, and so forth.

In this sense, the right of property is what allows me to enter into a relation with the smartphone such that, by externalizing within it a series of knowledges, pieces of information, and so on, I can later re-internalize them—that is, appropriate them, make them *mine*, make them become *my* knowledges and *my* information.

Through this relation, therefore, I can constitute myself as the person who will remember those conversations, who will recall what the saved photographs evoke in them—*a self that exists through the technical form of law itself.*

In claiming here the autonomy of law as a technology, my point is precisely that only legal knowledge is capable of operating technically without falling into errors that could otherwise be easily made. A clear example of this can be found in the way philosophers such as David Chalmers, starting from theoretical premises similar to those expressed here, reach conclusions about law that, in our view, are not entirely coherent.

According to Chalmers—who, together with Andy Clark, formulated the so-called "extended mind" hypothesis (Clark – Chalmers 1988)—the tools of our environment can literally become "parts of the mind." If this is the case, he concludes, then *obviously* my smartphone is part of my mind. This, however, entails a series of normative consequences, which Chalmers himself consistently develops:

> The extended-mind hypothesis may also reconfigure how we think of morality and the self. If someone steals my smartphone, we typically think of this as theft. But if the extended-mind hypothesis is right, its more akin to assault. If the phone is part of me, then interfering with the phone is interfering with my person. This trend is likely to increase as our reliance on augmenting technology grows [...]. At some point, our social and legal norms may need to change to acknowledge the extension of our minds.
>
> (Chalmers 2022)

It is interesting to note that this conclusion is, in part, sometimes shared by the courts themselves. Consider the 2014 case of *Riley v. California* before the United States Supreme Court. The case concerned the exclusion of evidence obtained by the police after the defendant's arrest, when officers seized the smartphone he had in his pocket and accessed its contents.

In the opinion written by Chief Justice Roberts, it is noted, among other things, that the question of whether the so-called *search incident to arrest* doctrine—that is, the possibility of conducting a warrantless search at the time of an arrest—should apply, had to be assessed in light of the fact that smartphones are so pervasively and constantly present in people's lives that "a visitor from Mars might conclude they were an important feature of human anatomy."

Although the issue in that case concerned privacy protection, the Court's reasoning seems to suggest something remarkably close to what Chalmers himself proposes.

Let us return, however, to the argument developed by Chalmers, which, as can be seen, takes the following form:

(a) if the hypothesis of the extended mind is true, then my smartphone is literally a part of me—I do not *own* a smartphone; I *am* my smartphone.
(b) if someone "steals" my smartphone, they are therefore not depriving me of a piece of property but committing a violence against my own person.

In my view, this reasoning is fallacious, insofar as it attempts to derive normative consequences from a descriptive premise. Let us concede, for the sake of argument, that Chalmers is right on the descriptive level: the mind is extended, in the sense that there is no convincing argument proving that it is internal, located in the brain, while there are several arguments suggesting that it extends into the environment—into external processes that perform the same functions as internal ones (the so-called *parity principle*).

From this, however, nothing follows about how norms *ought* to regulate the relationship between myself and things. Consider, moreover, the substance of the argument: on the normative level, it results in transforming the right of property into a right over one's own person. But what kind of right over one's person would that be?

Nothing, in principle, excludes the possibility that such a right might indeed be conceived according to the logic of property. Chalmers' argument could, at least in the abstract, even be used to defend the opposite move—that is, the reduction of all rights of the person to rights of property.

Yet the real objection lies elsewhere, though it remains connected to the previous ones. As we have seen, even if the extended mind hypothesis were true, it would still tell us nothing about how individuals relate themselves to things.

Granted that—ontologically—my smartphone is part of me, part of my mind, this still says nothing about the *way* I live or experience that part of myself. I can, for instance, experience my body or my hand as something that I *am*, but I can also, at the same time, observe my hand as an *object*.

The same would hold, even under the assumption of an extended mind, for the smartphone: it will always be possible to experience it as a *thing*, regardless of the fact that, ontologically, it is part of myself. After all, rights such as property precisely fulfil this function: they make it possible to experience objects as *things*—as *res*—and thus as something distinct from the subject of that very experience.

What Chalmers fails to grasp is that law is a technology that *constitutes* reality, not one that merely reflects it. Law is a discursive order that enables us to experience reality in a certain way—to manipulate and to experience "objects" that do not exist in nature, and so on. From this perspective, whether the extended mind hypothesis is "true" or not has no relevance for the law or its consequences.

Juridically, the right of property constitutes a device whose function is to establish and maintain—at the level of legal relations, and partly therefore also of social relations—the separation between persons and things.

Yet, as we have seen, this separation is not assumed as given; on the contrary, it is the outcome of a process that requires the internalization of what is originally externalized. From this perspective, to institute the smartphone as an object *in property* means to decide that, regardless of what it "is" in reality, my relation to it will be treated, legally, as if the smartphone were *not me*.

## Three Paradigms

From what we have seen, law has always already been a technology, one that works by constituting technical forms of action—of which "subjective rights" and procedural "actions" are today the paradigm. At this point, the fundamental question arises: what kind of "cybernetics" is the legal one—or, in other words, what kind of "government" or "navigation" does it entail? This means asking: what are "rights" for, what are they, and what function do they serve, along with the technical forms proper to legal action?

Here, I would distinguish three different "paradigms."

(1) the first is what I would call, in the strict sense, the "technical" paradigm, in the classical legal meaning: one that conceives the forms of legal action in terms of a means-end relationship.

Rights and actions are here understood as means provided by the legal order for the achievement and realization of certain ends ("the protection of rights"), which the system predetermines.

Legal norms would ultimately define "technical rules" (of the type: "if you want x, then you must do y"). In my view, this conception is unsatisfactory, for it tends to think of the "legal" as the means, while the end or purpose would be, in principle, independent of law. Let me explain.

If I say that a will is a "means" for achieving a certain end—such as leaving my property to a particular person—I am reasoning as follows: since I have the desire, interest, or will to leave my property to x, law then intervenes, ex post, granting me the "means" to do so.

This, however, is false, for it is only because of the norm that provides the means that such an end can exist in the first place: without the norm on wills, it is impossible even to want to appoint an heir. Likewise, without the norm that institutes marriage, it is logically impossible to desire to marry. The end, in fact, is nothing other than the legal effect that exists only through the means, and therefore does not precede it.

(2) the second "paradigm" is what I would call "narcissistic," or voluntaristic.

Here, rights—and more generally the legal forms of action—are understood as the outward expression of what an individual is, of the qualities they possess. For example, it is because I am the child of x that I have the right to be supported by them until I reach economic independence.

As we have already said, this is not true from a legal standpoint, since such a conception presupposes that the ought depends on the is—that what is owed to me depends on what I "am" (or want or feel myself to be).

From this perspective, it is not hard to see how the new wave of "identity-based" claims marking the struggle for our civil rights has been articulated along this logic. While this would require a separate discussion, here we are interested in it only to clarify the sense of this paradigm, according to which the is founds the ought—in other words, what I am (or believe or want to be) is taken to ground my right to be it. We have already stressed how such logic is essentially anti-legal, since in law, only one ought can ground another ought.

(3) the last paradigm—the one I wish to defend here—is what I call the "cybernetic" paradigm.

The legal forms of action are nothing other than mechanisms of exteriorization-interiorization—according to the recursive capacity proper to cybernetic feedback. They are mechanisms through which I, by acting as if I were a certain "subject," a certain person, can constitute myself, become the subject that I am not.

In cybernetics, as is well known, feedback refers to the mechanism by which part of the output returns as input—or, more generally, to retroaction, whereby the effect acts upon its causes or presuppositions, modifying them.

This is, in fact, what Sloterdijk calls the "fundamental anthropotechnical law," which consists in the autoplastic repercussion of all actions and movements upon the actor. The general principle is that every activity retroacts upon the operator who performs it, but what retroacts also acts in advance.

Action produces actors, reflection produces the reflective, emotion produces the sensitive, and examination of conscience produces conscience itself.

Law—and its forms of action (most exemplarily: subjective rights)—operates in exactly this "cybernetic" way. If we return to our earlier example: I am not a "father" before assuming—through the norm—the possibility of acting as a father, of "exteriorizing" myself, that is, as an artificial centre of rights and duties; but it is precisely by acting as a father that I can become one, that I can constitute myself as such, be what I become.

Put otherwise: the "subject" does not stand before the action, is not its "author;" rather, it is the result of the action itself—it is the subject that constitutes us in the action, through its performance.

Jurists, moreover, are thoroughly familiar with this mechanism, even if they are not always aware of it—and one need only recall the entire history of the problem of the relationship between substantive and procedural law to prove it.

Strictly speaking, it is not because I am the "owner" of a thing that I exercise my right to reclaim it; on the contrary, it is in exercising that right (which presupposes only that I constitute myself as if I were the owner, asserting title to the right) that I can constitute myself as owner, that I become the owner of that thing.

Suppose we shift this logic to the plane of the relationship between norm and reality. In that case, we can then say: "my" right is not the expression, the outward manifestation, of what I am, but rather a scheme that allows me to become what—naturally—I am not.

## Conclusions

If law is a cybernetics, then it is because it provides artificial techniques and action-schemes—modes of "governing" one's conduct—such that we act within it as "persons" (as fathers, children, creditors, owners, men, women, etc.), that is, as beings radically distinct from the individuals we "naturally" are.

Thus—to return to Savigny—obligations are schemes through which I can act as a being capable of memory and of the will to bind myself for the future. And this—note well—regardless of whether such "memory" and "will" are things I actually possess, or even whether they exist naturally as "attributes" of the human being.

Law does not presuppose human faculties—it manufactures them. Through obligation, law produces the very memory and will it requires.

Indeed, one might say that if the human being is an individual capable of memory, this is only because it has been able to constitute itself as such—has been able to become what it is—through the instrument of obligations.

Similarly, real rights must be considered schemes through which the individual constitutes themselves in a certain relation to things (using relations with others), and so on.

If, then, this is the logic of law—its being a cybernetics—law loses more and more of its consistency and its function, the more it is forced into becoming a sort of instrument for "recognizing" what already exists, for "reflecting" reality as it is.

To reclaim the cybernetic character of law is, therefore, to reclaim its specific technique of "government," founded on the separation between artifice and "nature."

I want to return to the already mentioned case of the *mulatto pisano*, to revisit certain points. In 1949, the court denied a man "of the white race" the possibility of contesting paternity of a child of colour born to his wife, also white. The reason was that Article 235 of the Italian Civil Code, in its wording at the time, required proof of the husband's impotence to conceive to grant a paternity challenge—a circumstance which, in this case, did not exist.

The judges, of course, admitted that the child whose paternity was disputed was certainly not the man's biological son. But this simple fact in no way sufficed to admit the action for disavowal.

As they wrote:

> To laypeople this may appear—and indeed will certainly appear—to be a logical incongruity, but this is because they are inclined to reason simply as follows: 'It is proven that the child is illegitimate, therefore disavowal must be allowed;' not realizing that they are inverting the terms of the problem.

The logic proper to law is here clear: whether the husband is to be recognized as the father of the child does not depend on whether he is so in fact, but on how the norms regulate the attribution, and possible disavowal, of paternity.

We are not concerned here with whether this solution is "unjust," or whether it serves the real interests of the child. Rather, we wish to emphasize another point: law has no particular interest in determining whether the man is the father of the child—he "is" not, as the judgement itself has no problem explicitly admitting.

If, nevertheless, it attributes paternity to the wife's husband, it is because such an operation is justified by its way of "governing" conduct: for law, the husband must henceforth act as a father. Is this a burden or an opportunity? We cannot know; it will certainly vary from case to case.

It is possible that in this way the husband may come to become the father he is not, or not. The outcome, in other words, is not predetermined, and precisely for this reason, "rights" are action-schemes which do not, however, guarantee the result of the action itself.

When law tells me I am the father of my child, it is neither recognizing a pre-existing reality, as it exists, nor, note well, is it "realizing" it: it is only telling me that I must (and may) now become the father that I am not already (for otherwise I could not become him), providing me with the forms that might, perhaps, allow me to do so (the duties of care and education, of maintenance, the succession rights that will now govern my relations with the child, etc.).

If law is *cyberlaw*, it is precisely because it is not an "instrument" to ensure that we can be what we are, or to fulfil the "desires" or interests we already have.

It does not serve the human being for what they are. It serves the human being, rather, for what they may become. In this sense, it is part of anthropogenetic techniques: techniques that preside over the becoming-human of the human being, which constitute the means through which we can become what we are not yet.

## References

Austin L.M. (2010), "Person, Place, or Thing? Property and the Structuring of Social Relations." *The University of Toronto Law Journal*, 2: 445–465.

Cannata C.A. (1995), "Obbligazioni nel diritto romano, medievale e moderno." *Digesto delle Discipline Privatistiche. Sezione Civile*, XII. Torino: Utet: 408–457.

Carnelutti F. (1938), "Problema giuridico della trasfusione del sangue." *Il foro italiano*, 63, IV: 90–103.

Chalmers D. (2022), *Reality+: Virtual Worlds and the Problems of Philosophy*. New York: Norton.

Clark A. – Chalmers D. (1998), "The Extended Mind." *Analysis*, 1: 7–19.

Nietzsche F. (1994), *On the Genealogy of Morality*. Cambridge: Cambridge University Press.

Radin M.J. (1982), "Property and Personhood." *Stanford Law Review*, 5: 957–1015.

Savigny F.C. (1840), *System des heutigen römischen Rechts*, I. Berlin: Veit.

Stiegler B. (1998), *Technics and Time, 1. The Fault of Epimetheus*. Stanford: Stanford University Press.

Stiegler B. (2019), "Differire l'ingovernabile in direzione del Negantropocene." In R.Ronchi and Stiegler B. (eds.), *L'ingovernabile*. Genova: Il Melangolo.

Thomas Y. (1986), "Le 'ventre'. Corps maternel, droit paternel." *La valeur*, 1: 211–236.

# 2 Law and Neuropolitics

## The Question of Neuro-Rights

As mentioned in the Introduction, reinterpreting law as a "cybernetic" technique—in the specific sense developed throughout this book—makes it possible to address a number of contemporary legal issues from what we hope is a new and original perspective.

In this chapter, we turn to a concrete case concerning the ways in which innovations in neurotechnology and the growing hybridization between human and machine have raised questions and problems to which jurists have sought to provide answers and solutions.

We will begin by introducing the general problem—what has come to be known as "neuropolitics"—in order to identify what, in my view, constitutes the fundamental legal issue it entails.

To this end, I shall revisit and critically develop the meaning that the expression originally had when it was coined by Timothy Leary, the psychologist and writer who became a well-known and controversial figure in the 1960s for his theories on psychedelics.

Leary was among the first to observe that advances in the neurosciences would have, as a fundamental consequence, the emergence of a new right over one's own brain—concerning not only the freedom to modify it, but also the need to preserve it from external forms of control and manipulation.

For this reason, I will first draw on Leary's reflections to clarify what is at stake in today's debates surrounding the so-called "connected brain."

This will then allow us, in the second part of the chapter, to turn to the question of neuro-rights, recently presented as a legal innovation capable of offering effective protection for the person and for mental activity itself.

The notion of *neuro-rights*, however, presents a number of difficulties and weaknesses that we shall attempt to analyze. This will allow us to propose an alternative solution—namely, the recovery of the traditional right of property as the legal foundation for the protection of the brain and its activities. Yet, as we shall see, this recovery must be understood and interpreted precisely in light of a proper understanding of the kind of technology that law itself represents.

DOI: 10.4324/9781003775850-3

To conceive of the brain as a "thing owned" does not, of course, mean to "reify" it or to assume that it is, in any real or natural sense, an "object," something that can literally be possessed. One can perfectly well accept—without contradiction—that we do not *have* a brain, but rather *are* our brain. Yet this has nothing to do with the way in which we may choose, legally, to define it.

To say that the brain is a good held in ownership means, instead, to institute a specific form of connection and integration between the individual and their brain—one that makes it possible to create certain forms of legal protection and safeguard. In this sense, law operates as a technology that enables individuals to experience their own brain *as if* it were an object of property; and it is precisely this possibility that allows legal protection to be effectively extended to the challenges raised by neurotechnological innovation.

## The Liberation of the Brain

The expression *neuropolitics* has now become established, in scientific contexts, to refer to a field with still uncertain boundaries and still in the process of developing theories focused on the relationship between brain functioning and people's political choices.

It usually designates both the studies in the field of cognitive neuroscience that attempt to clarify the cerebral mechanisms underlying political behaviour (Benedikter – Giordano 2024; Qvortrup 2024; Schreiber 2017), and those theories concerned, instead, with examining how the "governing of the brain" has today become the main stake in political control (Rose – Abi-Rached 2014; Connolly 2022).

Often, however, it is forgotten that the term *neuropolitics* was introduced, with a more precise and defined meaning, in the mid-1970s in the work of Timothy Leary (1920–1996).

I do not intend here to dwell on the controversial figure of Leary, on the role he played for the "acid generation" and in the research and experimentation on psychedelics (Devonis 2021). What interests me, rather, is to recover the sense he gave to *neuropolitics*, by way of which he articulated a series of issues that perhaps only today have become truly pressing and decisive.

Let us follow, first, a few of Leary's pages. If there is such a thing as "neuropolitics," it is because the brain, for Leary, is essentially "a bio-chemical-electrical network capable of receiving and creating a changing series of adaptive realities," and therefore, as such, modifiable (Leary 1991, p. 5).

The point is this: the brain can be programmed, integrated, controlled—which immediately raises the political question of whether and how this should be done, and of who should be entitled to (re)programme it.

What Leary hoped to achieve through LSD experimentation was the separation—the liberation—of the brain from the limits of the conscious mind, from the "normal mind," which for him was essentially a "source of illness." What was required was that each of us might finally acquire an understanding of—and, above all, control over—our own nervous system, that we might develop new neural circuits (new minds) of reception, integration, and transmission.

The first—decisive—consequence that Leary drew from this was that it implied a definitive passage from philosophy to cybernetics. Or, more precisely, it meant that philosophy might finally be brought to completion and realized as cybernetics—as the real possibility of constructing a new mind able to free itself from social and political conditioning, from "programmed fears," and to reach a human nature responsible for itself.

Here Leary stands far from Heidegger, who had seen in cybernetics the end of philosophy. For Leary, in fact, the point was not to determine man's freedom as something planned, that is, controllable; on the contrary, cybernetics became, for him, the practice by which each individual learns to change station, tune in, focus, programme and reprogram their own brain, accessing the freedom to choose how to become the human they are.

For this reason, the stakes were immediately political: the liberation of the brain—which is always also liberation from the mind (from the "normal mind," programmed by power, in Leary's view)—could only be liberation from existing social and political relations. Without grasping this, one cannot understand the 1960s, the Beatles, the "hippie beads," or the Summer of Love. The revolution that came from America in the 1960s was, indeed, political, but not in the sense in which we, especially in Europe, understood it at the time.

Leary's *neuropolitics*—the lysergic revolution—was not rooted, except indirectly, in the terrain of class struggle, in the contradictions inherent to the economic-social structure, or in the critique of the control, by the ruling classes, of the apparatus of production. Ultimately, what American "counter-culture" was the first to grasp was that "politics"—the decision on the kind of society we would live in during the years of the rise of "cognitive" capitalism—would concern only secondarily the level of wealth production or that of the political-institutional order: the real stake was the brain, the way we would think, the way we would reorganize our minds:

> The real revolution of the sixties was neurological. The emergence of a new philosophy occurs at rare, crossroads points in history. Political, sodal, economic changes follow changes in the conceptions of human nature. The consciousness revolution of the sixties initiated what is now called the Information, Communication, Cybernetic Age and challenged all of the institutions and moral principles upon which industrial society was based.
>
> (Leary 1991, pp. 9–10)

This is Leary's idea of *neuropolitics*: to provide each person with the ability to finally understand and control their own nervous system, to learn how to program and reprogram their own brain.

We must therefore begin to distinguish the core theme of *neuropolitics* from the means and techniques—certainly variable over time—used to modify the brain. In the early 1960s, Leary was thinking essentially of psychedelic drugs, or more precisely of a shift from drugs that "affect mood," whose daily consumption was already widespread (alcohol, tobacco, amphetamines, barbiturates, etc.), to those that "affect the brain," such as LSD. For Leary, it was these latter substances that could activate those cerebral circuits capable of enabling individuals to control, experiment with, and freely programme their own perception, their relationship with themselves, and with reality.

What Leary outlines is a "neurological space" characterized by eight circuits, or potential "gears," or "mini-brains," which may or may not be activated under certain circumstances and through different kinds of alterations. Without entering here into the full detail of Leary's descriptions of each circuit, we should pause on a few essential points in his reasoning. As mentioned, Leary believed in the 1960s that drugs were the main tools capable of activating new neurological levels, enabling the transition from one circuit to another. Yet he had also perceived that what he considered the truly decisive leap—between the fifth and the sixth circuit—was in fact closely tied to the electronic revolution.

It is true that, for Leary, "cyber-chemical psychedelic" drugs such as LSD were designed "to achieve results, not to escape from reality," namely, to open "the nervous system to possibilities of a future post-terrestrial evolution" (today we would say: post-human). But he also warned that the evolution of drug policy, and of people's experience of drugs, constantly ran the risk of reducing them to nothing more than tools of the "hedonics of the fifth circuit," a mere detachment of the brain from reality.

For Leary, however, the real goal remained another: to enable the brain to reach the sixth circuit, in which the nervous system meta-programmes itself, creating serial re-imprinting upon itself. This is the level that, for Leary, would allow individuals to communicate and share their thoughts without the need for language or speech—and to do so, more than through drugs, via computer technology:

> The evolutionary function of the sixth circuit is to enable us to communicate in Einsteinian relativities and neuroelectric accelerations, not using third circuit laryngeal manual symbols but directly via feedback, telepathy and computer link-up. Neuro-electric signals will increasingly replace "speech" (hominid grunts) as personal computers become standard home appliances.
>
> (Leary 1991, p. 94)

This was written by Leary in 1973. In 2016, Elon Musk founded Neuralink, which quickly became a reference point for research on the "connected brain," that is, the possibility of achieving a direct link between an individual's mental processes and an artificial intelligence capable of reading them, encoding them, and transposing them directly into reality (Musk 2019).

To take a well-known example: I think of something like "turning on the air conditioner," the computer deciphers my thought and activates the air conditioner itself. The brain-computer interface (BCI) will thus be able to capture brain signals (or thoughts) and translate them into commands, sending them to devices capable of performing the desired actions.

Even if cautiously and aware of the risks, Musk himself has also made clear that, in his view, the development of artificial intelligence will lead, in the coming years, to what has been called the *Singularity*—a term that, as we have seen, already appears in Leary, but which would only be popularized years later by the works of Vernor Vinge and Ray Kurzweil—to refer to the point at which artificial intelligence will surpass "human" intelligence, and a permanent symbiosis, a connection between human and machine, will be achieved (Shananahan 2015; Krüger 2021; Kurzweil 2024).

For Kurzweil, as for Leary, the human brain is something that can be integrated, modified, freed from its limits. For this reason, the Singularity will make it possible, he writes, to overcome those conditions, to gain power over our own destiny, when the non-biological portion of our intelligence will be billions of billions of times more powerful than unaided human intelligence.

What concerns us here is this tendency towards the creation of a connected brain, a *digital brain*. Here the Singularity, as Žižek has observed, refers to the idea that through the direct sharing of my thoughts and experiences with others (a machine that reads my mental processes can also transpose them into another mind), a sphere of mental experience will be born, shared on a global level:

> "Wired brain" refers to a direct link between our mental processes and a digital machine, a link which, while it enables me to directly trigger events in reality with a mere thought (I think about something like starting my air conditioner, the computer deciphers my thought and triggers the air conditioner), it also enables the digital machine to control my thoughts. "Singularity" refers to the idea that, by way of directly sharing my thoughts and experiences with others (a machine which reads my mental processes can also transpose it to another mind), a domain of global shared mental experience will emerge which will function as a new form of divinity—my thoughts will be directly immersed in a global Thought of the universe itself.
>
> (Žižek 2020, p. 21)

An integrated brain, in this sense, will be a brain in which each person's thoughts can be immediately and directly shared with others—this is the "telepathy" Leary spoke of, as the activation of the sixth circuit, the cyber-electronic brain. It will no longer be necessary, then, to speak, to make use of language, since every experience and thought could be transmitted directly from one brain to another, and thus shared. Physicist Freeman Dyson coined, in this context, the term "radiotelepathy" to describe this shift towards a collective mind:

> Well one thing I imagine quite a lot is sort of radical changes in human consciousness, like something I think of called radio telepathy. Old-fashioned telepathy was supposed to be something supernatural: I dream a dream and somebody hundreds of miles away can communicate magically mind to mind. That probably doesn't exist but we can imagine doing that by mechanical means, putting little radio transmitters and receivers into people's heads so that they can actually communicate thoughts directly. I don't think that's impossible. It's a problem of understanding how the brain actually works, how the mental processes work. It might even be feasible in a hundred years be that two people can communicate directly mind to mind and if that's possible of course it'll change the whole nature of life profoundly, much more so than any of the other things I've been talking about.
>
> (Dyson 1999)

Leary's idea has today been taken up by those working toward the possibility of direct neural communication. Musk has often argued, in various interviews, that "human" language will soon become obsolete, and that the "connected" brain will allow us to read each other's minds, thus overcoming the difficulties and misunderstandings typical of verbal communication.

Neuroscientist Miguel Nicolelis thinks along similar lines: the "liberation of the brain," which for him means the incorporation into it of artificial intelligence devices, will allow us to reduce, or eliminate, the once unsurpassable limits that define the individual human being, so that one day, in our remote future, we may experience what it is like to be part of a conscious network of brains, a true cerebral framework of collective thought.

Although he frames the issue as a question, he seems to have little doubt that it will be possible, and thus that it will happen:

> Assuming for a moment that somehow, through some amazing and harmless future technology, this brain-net become real, could the individuals participating in it not only communicate back and forth with one another just by thinking, but also vividly experience what their counterparts feel and perceive, as they seamlessly adhere to this true "mind meld"?
>
> (Nicolelis 2012)

The *mind-meld*, or the *collective brain*, is what Leary already had in mind when speaking of the cyber-atomic circuit. But the point here is not, of course, to ascribe to him the authorship or foresight of these ideas. Rather, the reference to Leary must serve for us as a way to locate the field within which such questions—now discussed in light of technological and neuroscientific developments—ought to be investigated.

## Neuropolitics

It could reasonably be objected that, in order to clarify the issues relating to the development of BCI mechanisms, the "Singularity," and the transition to the digital and collective brain, Leary's concepts are not particularly rigorous and lack "scientific" grounding: his definition of brain circuits, of "electronic" or "cybernetic" culture, is not only dated but inseparable from his use and fusion of New Age assumptions, terms, and suggestions, along with contaminations from Eastern thought (one might think of *The Tibetan Book of the Dead*) and cyberpunk literary culture.

This is undeniable. Leary himself, after all, recalled that during the early 1960s experimentation with psilocybin and LSD, his research group was confronted with the problem of describing the effects of such drugs without having at its disposal "any scientific language capable of communicating or even acknowledging the vast scope and the 'strange' effects of these chaotic phenomena."

Still, the reason why the reference to Leary must be preserved is that it allows us to maintain "neuropolitics" as a question that, as such, cannot be thought simply through the language, methodologies, and problems specific to neuroscience, medicine, and their technological developments. Neuropolitics, in other words, is not merely the set of ethical, legal-political, and institutional issues that would follow from discoveries and problems belonging exclusively to science.

It is therefore worth holding on to certain of Leary's insights: from this perspective, neuropolitics is not what derives from the institutional and political implications of neuroscientific research, but is the *fulfilment of philosophy as cybernetics*—that is, the move to a certain idea of the relationship between brain and power (on which neuroscientific research and technological developments themselves depend).

For this reason, neuropolitics must invent its own concepts: these are not meant to "describe" brain function or neuronal networks as they are studied in scientific research, but instead to give form to, and work upon, the problems—philosophical in nature (or, as Leary would say, "cybernetic")—to which the brain-power relationship gives rise.

From this standpoint, a notion such as that of the "cyber-atomic" brain is neither precise nor imprecise, correct nor incorrect, precisely because it is not to be understood as the description of a fact, but as the identification of a field of problems.

For this reason, Leary's *topica*—his distinction between eight brain circuits, four of which he places in the right hemisphere (those yet to be activated) and the first four in the left—must also be understood as a framework for organizing two kinds of political practice, which he respectively calls: (a) *cybernetics* (management of the left brain), tasked with building "common info-structures" and protecting each person's cyberspaces "from invasion and exploitation by others;" and (b) *psibernetics* (management of the right brain), concerning the creation of "info-environments in one's own neuroworld, linking one's neurospace with that of others," and "protecting one's brain from invasion and exploitation from outside."

Without necessarily adopting Leary's categories or vocabulary, it is essential to retain and develop the meaning he gave to the term *neuropolitics*. If we were to summarize its key characteristics schematically, they could be outlined as follows:

(1) neuropolitics assumes that the brain has become the object, the stake, and the very field within which power relations—always, by definition, "molecular," to borrow Foucault's term, and plural—are defined, structured, and articulated in the age of artificial intelligence (or, if one prefers, cognitive capitalism).

We might say that the fundamental political technology of our time is no longer simply the governance of life and biological processes—as in biopolitics—but the governance of the brain, namely the regulation, production, and control of processes relating to the "connected brain" and the sharing of thoughts.

Sociologist Nikolas Rose has put it clearly, noting that since the 1990s, the governance of populations has made knowledge of the brain and its functioning one of the central issues of political and economic intervention (Rose 2006).

Neuroscientist Marcello Ienca observes that human beings "are today in an unprecedented historical transition," and that:

> our generation will be remembered either as the one that steered the last great historical transition of human bio-cultural evolution for the benefit of all; or, conversely, as the one that abused the most complex organ ever produced in the history of the known Universe—the human brain—to destabilize human societies, amplify inequalities, definitively erode the private sphere, and abolish the basic principles of social organisation.
>
> (Ienca 2019, p. 149)

(2) the central question of neuropolitics is not to probe the ethical or political implications of neuroscientific research or the development of digital technologies, nor to assess their feasibility.

Rather, it concerns the techniques and strategies through which the brain will be redefined by virtue of being embedded within relations of power—through procedures aimed at governing, influencing, and directly "programming" it, in other words, through its immediate connection to machines.

(3) Neuropolitics further implies—as Leary intuited—that any transformation (including the very possibility of revolutionary activity) capable of fundamentally altering our social, economic, and moral arrangements will pass through, and depend primarily upon, the "programming" of our brain.

This refers to the ways in which we will experience our own thinking, and our modes of perceiving and interacting with reality. Any change that might occur in the coming decades—in economic structures, social relations, or the organization of our societies—will carry a meaning strictly dependent upon, and therefore subordinate to, what is happening in the meantime to the way our brains are being configured and programmed.

To give just one example: we might imagine that the introduction of a universal basic income will finally bring liberation from work. Yet its significance will be entirely different if this occurs in a context where intelligence and consciousness have been separated—in which case it will be the machines that know our thoughts and desires better than we ourselves do, and thus the machines that tell us how to spend our liberated time, what to desire, and so on.

Although *neuropolitics* shares more than one feature with what Stiegler calls *psychopower*, it identifies a specific moment corresponding to the technological phase we are now entering. Stiegler had already identified in *psychotechnologies*—"technologies that make it possible to control the mental activity of individuals in increasingly calculable ways"—the factor that marked, at the start of the twenty-first century, a shift away from the series of questions Foucault had examined in his analysis of biopower.

We will therefore need to return to Stiegler's strategy of analysis and critique, which cannot be said to have been superseded simply because here we are insisting on a further step—namely, the direct intervention on the brain by new neurotechnologies.

This is all the more so since Stiegler himself sometimes spoke of "psycho-neuro-politics" and repeatedly emphasized, as we will also see later, how cognitive capitalism advances by modifying the synaptic organization of the brain—especially that of young people. Indeed, the transductive relationship between technology and the brain is one of the fundamental axes of Stiegler's thought, and one that neuropolitics cannot but take up—albeit starting from the problems that define the new and additional stage represented by *neurocapitalism*.

## A Jurisprudence of the Brain

My perspective is that of a jurist, and for this reason, it is first and foremost oriented towards thinking about the problems of neuropolitics through the grammar and techniques proper to legal systems.

Since 2017, Ienca and Andorno have introduced the notion of *neuro-rights* to refer to a set of strategies aimed at safeguarding and protecting the brain against the possibility that—through advances in neural engineering and pervasive neurotechnologies—certain components of mental information might become the object of control, manipulation, or interference by third parties. This approach echoes what Richard Glen Boire had earlier suggested when speaking of the need to begin sketching a "jurisprudence of the mind:"

> It is time to develop a jurisprudence of the mind; one that takes account of the latest understandings of the brain, the advancing powers of psychopharmacology, and which situates these within our country's tradition of embracing individual freedom, self-determination, and limited government. Cognitive liberty is the human right most in need of immediate elaboration and defense. Nothing is more private, more intimate, more properly within the sphere of each individual's sovereignty than the interior environment of his or her own mind and intellect. The right of a person to liberty, autonomy and privacy over his or her own intellect is situated at the core of what it means to be a free person. Yet as the law currently stands there is no explicit recognition of, let alone elaboration upon, a basic right to mental sovereignty. The "war on drugs" has taken a new turn. No longer is it just an issue of the government telling you which drugs are off-limits, it's now also about which drugs the government can force a person to take. Welcome to the age of chemical coercion and cognitive censorship.
>
> (Glen Boire 2003, pp. 9–10)

Andorno and Ienca have especially stressed the need to reconceptualize existing individual rights and to create new, specific neuro-rights, which they identify as the *right to mental privacy*, *mental integrity*, and *psychological continuity*. Their perspective is undoubtedly valuable, as it makes a fundamental contribution to rethinking the problem of rights in light of the particular nature of brain data and the possibilities for "mind reading" opened up by new technologies.

Regarding the first of these, the information to be protected is such that it is indistinguishable from the very source that generates it—namely, the individual's neural processing. The object of protection, therefore, can no longer be the information or data *as such*, but necessarily the mental activity that produces it.

This raises an additional concern: through new technologies, the real risk is that an individual's "thoughts"—in the form of brain waves—could be recorded, and potentially controlled, without the person's awareness, and thus without their knowledge, with the result that a mechanism such as prior consent could become obsolete and ineffective.

As for the problem of the impact on *mental integrity*, the right to it should, from this perspective, be rethought beyond its inscription within the logic of mental health protection, since it must be able to operate as a safeguard against any potential interventions by neurotechnologies that would alter a person's neural computation.

Finally, neuro-rights should also protect the way individuals perceive their own identity, shielding them from behavioural or self-perception changes that neurotechnologies might induce. Andorno and Ienca's proposal, therefore, is to achieve the creation of specific neuro-rights as a strategy to counter potential abuses of neurotechnologies and as a form of protection of the fundamental freedoms associated with individual decision-making in the context of neurotechnology use (Iena – Andorno 2017; Istace 2022; Blitz – Bublitz 2021; Lopez-Silva – Valera 2022).

The issue of neuro-rights has, unsurprisingly, been taken up and discussed in recent years in relation to the broader problem that concerns us here: the redefinition of subjective rights in response to the new protection needs linked to the development of neurotechnologies.

It is not possible here to cover the breadth of the debate or the range of positions that have emerged. Instead, we should aim to clarify some of the unresolved problems of a proposal—such as that of Andorno and Ienca—that is built on creating specific (neuro-)individual rights conceived as rights of the person (or of personality).

A first problem concerns the very *content* of such rights. If we look at the formulations usually given for neuro-rights, they are marked by a fundamental ambiguity. While they are often grounded on the principle of privacy, or sometimes on that of individual autonomy, it is clear that they cannot simply affirm the right of every individual to think autonomously, or to have control "over one's own mental states" (Lavazza 2018, p. 4).

Such formulations appear to lack any determinate semantic reference. They presuppose, in fact, that thought *as such* is something over which each of us exercises, in principle, a "mastery" or control, and that it would therefore always be possible to distinguish between thoughts or mental states of which I am the author and others that are somehow "imposed" on me from outside, by third parties. This, however, is debatable. Our thinking is always conditioned, influenced, if not determined, by others with whom we come into contact.

If we were to preserve its "autonomy," we would have to assume that the law should intervene, for example, to prohibit us from lying to others, giving them false information, or even exposing them to our opinions, desires, and so forth. As has been observed, strictly speaking, if someone lies to me, I could legitimately conclude—on the basis of the neuro-rights formulation under consideration—that they are altering my "mental integrity," since they are leading me "to believe, desire, act, or say things that I otherwise" would not (Rainey 2023, p. 10).

The likely reply would be that such neuro-rights should protect individuals from *unauthorized* intrusions, specifically from those technologies capable of directly accessing our thoughts by decoding our neural signals—something that another person could never do unaided. The problem, however, is that this presupposes we must always be able to ascertain whether and when such devices are accessing our "mind," whether they are merely recording data or actually altering our thoughts, and so on.

Yet the prospects of artificial intelligence challenge precisely this point. As Žižek has noted, since the BCI does not "read" my thoughts but their neuronal correlates—of which I am unaware—how could I know whether or not I am connected to such an interface?

The real risk, then, is not so much that technologies might interfere with our thoughts, but that they might do so without my ever perceiving such interference, while I continue to feel entirely free and self-determined in what I do. The fundamental problem, therefore, re-emerges: distinguishing between when I think "autonomously" and when my thoughts are conditioned, influenced, altered, or guided by a third party risks being empirically impossible.

## The Logic of Property

It is worth reflecting on a second issue. Neuro-rights have been conceived—primarily through the influence of Anglo-American jurisprudence—as rights of the person (or personality). Such a construction risks reproducing a set of problems, long discussed in legal scholarship, concerning the legal status of the body (Quigley 2018; Beaud 1993).

If neuro-rights are framed on the assumption that the brain (or the mind) *is* the person, this implies that any unlawful interference with, harm to, or intrusion into a person's mind constitutes an offence against the person themselves.

Let us consider an example—as simple as possible—partly adapted from Harari. Imagine that in the future we read books solely through devices—think of an advanced version of Kindle, enhanced with biometric sensors and scanners—that record how each sentence we read affects our heartbeat, blood pressure, and brain responses, thus enabling them to determine which pages made us laugh, bored us, interested us, and so on. The result would be that, after some time, such devices would be able to choose books for us better than we could ourselves.

The point that concerns us here, however, is another. Our right—as a right of the person—not to suffer unauthorized interference by the Kindle while reading would imply that, if it were to record our brainwaves without permission, it would be violating not a *res*—a thing—belonging to us, but our very person. The device, in this sense, could be held liable for personal injury and for infringing the person's right to self-determination and integrity.

The real question, however, is what happens to the heartbeat—or, in our case, to a "brainwave" or to certain neuronal discharges—once they become a "datum," that is, once the Kindle has recorded them and they have thus come to exist—as data, as information—separate and autonomous from the person themselves.

Who owns that data? Clearly, if it was identified with the person, it was in fact the person themselves. Only once recorded as "data" does it legally become a *res*. But a *res* without an owner, since it comes into existence as such only at the moment when it is no longer part of the person. This implies that the person can no longer claim ownership of it, as it will instead belong to whoever first takes possession of the "data."

In other words, if conceived as rights of the person, neuro-rights might allow us to punish—under civil or even criminal law—the conduct by which Kindle might appropriate our "data," but they would in no way allow what is truly essential: to ensure that such data remain our property, and that others cannot legitimately own it. Only the logic of property can guarantee this.

Legally, we are not our brain or our mind, but we own it—every part of it, every activity and every signal it generates.

As early as 1964, David Solomon had, I believe, clearly articulated the fundamental terms of the issue:

> Moreover, I believe that the astonishing human brain is man's most inalienable possession, his intellectual birthright. No person or institution has the moral right to muffle or inhibit its development. arrogate unto itself the consciousness to which No social authority can successfully right to dictate and fix the levels of men may aspire, whether those states are induced pharmacologically or otherwise. Die gedanken sind frei.
>
> (Solomon 1964, p. VIII)

The objection that this would "reify" the mind is meaningless and betrays a fundamental misunderstanding of the legal technology involved. To affirm that, legally, I am the owner of my body, and that the body is therefore a thing—a *res*—implies nothing as to what the body is on the "naturalistic," "ontological," or "philosophical" plane.

What the brain or mind *is* is irrelevant for the law; it is a question whose answer is of no intrinsic interest to it. What the law addresses is another problem altogether: what the mind must *be*, legally, so that it can be subjected to a set of protective mechanisms deemed appropriate.

The point is not, therefore, that "data" are mine because I own my mind; rather, it is that neuro-rights must be conceived as *property rights* so that it is possible for the "data" to remain mine even when taken from me.

Property rights, after all, remain the model of subjective right that offers the greatest protection for its holder: they are absolute rights, immediate, enforceable *erga omnes*, and inherent to their object. That they may concern a *res* which, as such, is not necessarily "corporeal" cannot, of course, be an objection.

Compared to the definition of neuro-rights as rights of the person, the construction of a property right over the brain (or mind) has a series of essential advantages, including:

(a) ensuring greater protection, particularly about the possibility that my "thoughts"—or, more precisely, their neuronal correlates—do not pass into someone else's "ownership" when extracted as "data."
(b) avoiding a multiplication of "rights" that would add to the already vast catalogue of "rights of the person," with content that is often indeterminate and vague.

The discussion would need to be developed in greater detail to arrive at a precise and analytical construction of a right over one's own brain, conceived in terms of property, rather than the logic of personality rights. What matters here, however, is the correct dogmatic framing of the issue.

## Philosophical Objection

The first issue is, we might say, "philosophical" in nature and, as already mentioned, must, in my view, be rejected.

When it is claimed that we do not own our data but *are* our data, and that therefore such data are not something the person "has" but rather something the person *is*, this assertion—even if correct at the descriptive level, concerning what data "really" are—provides no normative justification for maintaining that they should therefore be the object of a right of personality rather than a property right.

Whether "data" are or are not, legally speaking, things is not a problem that depends on what they "really" are, but solely on the consequences the law wishes to draw from classifying them in a certain way. It is meaningless to say that, by doing so, we "reify" the brain.

The point, rather, is to clarify that—at the level of legal logic—if one wishes to defend the idea, as we do here, that the data extracted, recorded, or generated through the "reading" of brain activity should remain under the power, control, and ownership of the individual, then legally this implies that the brain is the property of the subject (otherwise, as we have seen, the results of its activity would necessarily be *res nullius*).

Personally, I fail to understand why a property right over the mind would constitute an incorrect use of a legal category, as has sometimes been argued, preferring instead the idea of a "neuro-right" to cognitive liberty.

In my view, the opposite is true. Neuro-rights would not pertain to "things" but to the person themselves, in the sense that they would express the protection of "intrinsic" aspects or characteristics of the person, such as their "identity" or "mental integrity," and so forth. From the standpoint of legal science—its logic and dogmatic constructions—these are imprecise and poorly defined concepts.

Above all, they improperly conflate legal arguments with arguments concerning what things "naturally" are, with the description of facts, and so on. Consider, for example, the following passage from Jan-Christoph Bublitz, who argues against the idea that the "mind" could be the object of property:

> But even more constitutive of a subject, *cogito ergo sum*, is her mind. It is not just one aspect among many, but, arguably, it is what defines subjects, and hence, legal subjects. In fact, it is hard to conceive any conception of a legal subject in which the mind and mental capacities (e.g. acting from reasons, deliberation) are not among its necessary constitutive conditions. Thus, I submit, the claim "my mind is mine" is not based on property rights nor on a legally established relation of ownership between entities, not subject to distribution of meum and tuum, but rather the point from which any legal order originates, intrinsic to the very notion of legal subjects. The "innate"—not acquired—right of everyone is the right to one's person, to make use of one's mental and bodily powers and to remain free from interferences.
>
> (Bublitz 2013, pp. 242–243)

As can be seen, the author proceeds through a series of philosophical assumptions about what the mind is for the subject, and from there deduces what it legally ought to be:

(i) first, it is assumed that the mind is constitutive of the subject—that it is what defines subjectivity itself. In other words, there is no subject without a mind.

Bublitz explicitly takes up the Cartesian theme of the *cogito*. In itself, as already noted, the correctness of such reasoning can be questioned. In Descartes, it should be noted, the *cogito* is defined precisely where I arrive at knowing nothing else and having only the certainty that I am. But this means that the certainty of my being (*sum*) is separated from knowledge, from knowing (*cogito*).

The Cartesian dispositif, in other words, implies the opposite of what Bublitz assumes: it implies that "either I think or I am," that I have no choice but between being (and therefore not "thinking" in the proper sense—having nothing but the certainty of the fact that I am, without in any way knowing what I am) or thinking (but in that case losing all certainty about my being, about who I am). But the choice is always forced: the subject, in other words, is constituted as such precisely because it "chooses" being, and not thought.

For this reason, Lacan sees in the Cartesian *Cogito* the very moment of the constitution of the unconscious and, at the same time, of subjectivity, which implies that we "are" always where we do not think—that thought is by definition what separates the subject from itself (Ronen 2018; Žižek 1998). To claim, as Bublitz does, that the "mind" is constitutive of the subject in the sense that it would be defined above all by its "self-determination" through the mind, by its "being its own thought," is therefore, precisely from the philosophical standpoint, at least debatable.

In even simpler terms, one could say—as we shall see later—that neuroscience itself has by now demonstrated how our brain activity changes, adapts, and reorganizes constantly over time, and that therefore our brain changes in response to our experience. To use Nicholas Carr's words, we become, neurologically, what we think. This, however, means that it is the thoughts we have that determine our brain—what it is—and not vice versa. But that is not the point here.

(ii) rather, what is decisive is that from his premise Bublitz seeks to derive a precise normative consequence. Since the mind would be constitutive of what the subject "is," it would follow—so the argument goes—that the legal subject, too, could necessarily be defined only based on its being above all a mind, constituted by mental capacities, which as such could not "belong" to it as property, since they would be the constitutive conditions of its very identity. Here it is evident, I would say, that Bublitz adopts a naturalistic conception of the legal subject, as if it referred to the individual as they "are."

From a dogmatic and conceptual standpoint, this is an illegitimate assumption: "legal subject" is, in fact, any *x* that can be vested with the ownership of certain powers or capacities assigned by the legal order, regardless of whether it possesses them. Legally, it is perfectly coherent, for example, to attribute subjectivity to "things" or "entities" that do not even exist naturally, or that lack, for example, any "mind" or mental capacity.

In reality, Bublitz had offered a different argument a few lines before the passage cited, which can be reconstructed as follows: since the law recognizes subjects as responsible for their actions and thoughts, it always presupposes, implicitly, that they have the right to self-determination and, in particular, to "cognitive liberty," understood as a *right to mental self-determination*. Even in this case, however, the argument appears weak, for two reasons.

First, it is not true that if the law treats people as though their actions and thoughts were self-determined, it logically follows that it recognizes a right to such self-determination. It could simply assume self-determination as a fact—which, even if untrue in certain circumstances, the law might still choose to presume. In other words, there would be nothing logically contradictory in a legal system that treated subjects under the absolute presumption that they are always free to self-determine, even if this never actually occurred.

Secondly, the argument in itself risks missing the point. What must be protected is each individual's right to be the exclusive holder of, and to freely dispose of—not their "thoughts" (which would be meaningless)—but their brain and neural activities. The object of the right, in other words, is not the ability to "self-determine" concerning what one thinks. As noted above, a BCI does not read my "thoughts," but the neural processes correlated to them. It is therefore not some alleged "cognitive liberty" that must be protected (and, even before that, defined), but rather ownership of the brain and of neural processes. That such processes may be legally established as *things*, the object of a property right, does not appear to present any difficulty.

## Legal Objection

A second type of objection is specifically legal. Some concern the fact that, if the brain and its activity were the object of a property right, then they would be considered goods that, as such, could be transferred or "commodified."

In such cases, people could, through contract, transfer ownership of their brain or their brain activity. To this, one can respond that, in principle, nothing prevents the construction of an *inalienable* property right over the brain, while at the same time providing that it would always be possible to sell or transfer ownership of the "data" (Schwartz 2004).

This right to commercialize the results of recording one's brain activity could force companies to pay us for the extraction of the data they collect.

However, the more technically interesting objection is another one, which deserves consideration. It has rightly been pointed out that what devices actually operate upon are not our "thoughts," but neural processes, which such devices extract, decode, and transform into data (or, if one prefers here: thoughts, thought-contents). For this reason, it is objected that a property right could not possibly apply to such "data"—since they do not exist before and independently of their extraction/decoding—but only to brain activity, to a series of neural processes.

Moreover, a property right could never have as its object "data" or "information," but only and always "things." On this latter point, it should be stated from the outset that, legally speaking, a "thing" can be any *x* that the law institutes as such—even a human being can be a *res*, in certain contexts such as in Roman law, without this in any way denying that he or she remains, of course, a human being.

Strictly speaking, there is nothing, on a logical level, to prevent the law from making "thought-contents" extracted through brain-reading into *res* in the proper sense.[1]

In my view, it is precisely property rights that allow this issue to be resolved better than neuro-rights and, more generally, the logic of personal rights. It is to the latter that the objection should be directed.

Following that line of reasoning, I am not my "thoughts" (thought-contents)—since they properly do not exist except through their externalization/decoding; rather, I am my neural processes.

This implies, however, that if the illegitimate recording/decoding of these constitutes an offence or damage to my person, the thought/data, once created, comes into existence as a "thing" that could in no way be said to be mine—according to the logic already described: it would be a *res nullius*, over which the company would acquire ownership by virtue of possessing it.

The situation is quite different, however, if I own my brain. It is true that I am not the owner of the "data"—since it does not yet exist and only exists once externalized. But here, legally speaking, what happens is that the decoding comes to constitute an activity that, starting from the material I voluntarily provide (the brain and its neural processes), modifies/decodes what is my property to process it as "data" or "thought-content."

For me to acquire ownership of that data as well, it is sufficient to apply the legal principle—also the basis of what jurists have constructed as a specific mode of acquiring ownership, *accessio*—according to which "modifications of a thing do not affect the ownership of which the thing is the object" (Cannata 2001, p. 267).

1 The discussion, however, should be kept separate from that concerning the issue of ownership of personal data, given the fundamentally different nature of the matters at stake. In the case at hand, in fact, we are not dealing with "data" consisting of information *about* the individual—information that pre-exists its disclosure—but rather with data that constitute the "extraction" of cerebral activity. Such data might also be described as information *of* the individual, insofar as they are "rendered" by the very act of subjecting oneself to the process of reading. For the broader debate on personal data, reference may be made to Bergelson (2003) and Van Erp (2017).

The rules on property, in other words, make it possible to coherently structure my acceptance of the service that such devices offer in the following sense:

(a) what I transfer is certainly not ownership of my brain, nor—*a fortiori*—of the "data" (since these do not yet exist), but rather the right to use my brain as "material" from which the devices will be able to "translate" and extract "data," which are *res*.
(b) by the rule mentioned above, I therefore also become the owner of the data once they have been formed, which are thus by no means *res nullius* (as they would be if the logic of personal rights were applied).

It is clear that, in this context, the task of jurists will be to defend the idea that the product of the activity of such devices (the decoding that constitutes the "data") is not a new thing in relation to the original material (the brain and its neural signals), but only a new state of that material. The consequence is that the principle applicable is not that of occupation, but that of the persistence of the thing.

Moreover, this latter principle should not be confused with the rule of *accessio*, which jurists have historically derived from it as an original mode of acquiring property. The fact that digital "data" can be replicated, duplicated, and become ontologically "detached" from the brain matter from which they are derived is not a particularly relevant objection.

What occurs at the "ontological" level—the level of factual reality—is never binding for the law. If, in other words, digital data is legally instituted as a new state of the same material, the result is that ownership of the data necessarily follows ownership of the material itself. It is therefore in no way necessary, at the empirical level, for them to remain physically or functionally "united" to the principal good.

There is nothing "artificial" in this construction. On the contrary, it is entirely coherent and follows the proper logic of law. I will insist again on this point later because, in my view, jurists today tend instead to assume that legal reality can only be modelled based on what things empirically "are," with the consequence that, once they assume that on the neuroscientific level the individual "is" their thoughts, the law could have no alternative but to translate or convert that "truth" into a personal right.

## Brain and Technology

Suppose a neuropolitics aim is to promote the liberation of the brain from the relations of power in which it is enmeshed. In that case, its objectives must be to invent the techniques and practices through which such liberation can occur, starting from the fact that the brain, as a "human" brain—limited to the circuits identified by Leary—has always already been a product of those very relations of power.

This is what Leary had in mind when he wrote: "The paradox may be stated as follows: it becomes necessary for us to go out of our minds in order to use our heads" (Leary 1964, p. 12). The mind is already "programmed" to be limited. This is why Leary envisaged neurology and pharmacology as the domains that should provide "new means for altering consciousness, for propelling us into new dimensions of awareness"—in other words, as practices aimed at freeing the brain from the limits imposed upon it by our mental organization (the set of "consciousness–Ego–mind–adult personality" circuits).

If today such a transformation is possible not so much through psychedelic drugs, as Leary imagined, but through technology, then the task of neuropolitics will be to convert technological practices from "poison" into "remedy," to borrow Stiegler's expression—from what they currently are—mechanisms of dispossession of intelligence—into what they could be—techniques for liberating the brain (Stiegler 2010, p. 29).

It would be contradictory to seek to protect and shield the brain from the intrusions and influences of neurotechnologies while at the same time preventing it from freeing itself from the limits of "consciousness"—as Solomon calls them—which are always historically determined and thus the result of the technologies (and the power relations) that constitute it. Our "mental processes" are by no means given or natural.

We must avoid here any form of "biologism" and instead focus on the neuropolitical character of the functioning of organs and, in particular, of the brain. Stiegler spoke of a "transductive relationship" to refer to the fact that the brain does not evolve, develop, or change independently of the "technical" organs, of the symbolic supports (such as language), and of the social apparatuses with which it is in relation: the psychic apparatus is ceaselessly reconfigured by technical and technological apparatuses and, along with them, by social apparatuses:

> In psychosocial individuation, the pre-individual is individuated *simultaneously* socially and psychically, since the psychic and the social are two poles of a transductive relationship constructed as an always-differing resolution, but that is always productive of structures and stabilities occurring along with this differentiation-of the tensions concealing the pre-individual already-there that is common to them, translated through an inadequation at the core of the *I* and the *We*.
>
> (Stiegler 2011, p. 95)

For this reason, for Stiegler as well, the fundamental epistemological question of what we here call neuropolitics must be tied to the theory of the brain and its relationship to artificial organs, and thus from being a biological question it becomes also, and suddenly, a psycho-physiological question.

To clarify the point, we may follow—as Stiegler himself repeatedly does—the studies of neuroscientist Maryanne Wolf on how the human brain has been and continues to be transformed in correspondence with the techniques of writing that humans have historically adopted and with how reading practices have been taught and transmitted.

The invention of reading, Wolf observes, was made possible precisely because the human brain possesses "plasticity," the capacity to establish new connections among its pre-existing structures, and thus to be shaped by the kinds of experiences that techniques and technologies make possible. In other words, our brain learned—and continues to learn—to read because, with the invention of writing, neural circuits reorganized and adapted, forming new connections and new neural pathways (Wolf 2007).

That writing has altered our cognitive abilities, and thus also the functioning of the brain, was already the central thesis of works such as those of Havelock, Vygotsky, and Luria, for whom the more complex forms of thought do not originate in the brain as it is "naturally" given, but in the external conditions of life—namely, in the technologies, such as writing, that enable changes and new associations to be activated within it. Wolf, however, emphasizes that the very way one reads creates and constitutes different brains.

First, reading in one language or another produces different patterns of activation: those who read in Chinese activate neural pathways different from those of readers in English. Second, the brain changes, in its mode of reading, according to the technological supports that make reading possible (books, tablets, etc.). This is why Wolf has stressed that the transition from the "reading brain" to the "digital brain" is also the passage to new modifications of mental capacities, giving rise to a different way of learning and understanding. *Skimming*—or superficial reading—is thus not merely a mode of reading typical of digital texts, but a technique that modifies the brain itself.

The limit—or rather, the contradiction—of many discussions on neurorights lies in assuming that the brain should be protected against contemporary neurotechnologies as if the aim were to preserve its "innate" autonomy or some purported "biological nature."

The point, however, as Stiegler reminds us, is that the brain does not evolve independently of technical organs. It is, we might say, a product of technology (Bradley 2024; Wexler 2006; Bates 2024). To use his own terms, we could say that the exteriorization of the brain is the very condition of its constitution.

It is we—our technological and social structures—who make our brain, who shape it into what it is. But if that is the case, there is no reason why a neuropolitics should limit itself to defending, so to speak, the *status quo*—to defending that "mind," as Leary called it, which in reality is not genetically determined but is the outcome of a certain cerebral organization, historically shaped and thus already constituted by the techniques (and power relations) of a given society.

Following a conceptual strategy different from Stiegler's, and with different implications, Catherine Malabou has likewise emphasized the "historicity" of the brain, its always "unfinished" character, and its being essentially "free" in that, thanks to its plasticity, it is constantly capable of changing and readapting.

For this reason, Malabou claims not only that "Humans make their own brain, but they do not know they are doing so," but that we must finally "liberate" this freedom:

> It is thus not just a matter of uncovering, in the name of brain plasticity, a certain freedom of the brain but rather, starting from as precise a study as possible of the functioning of this plasticity, to free this freedom, to disengage it from a certain number of ideological presuppositions that implicitly govern the entire neuroscientific field and, by a mirror effect, the entire field of politics—and in this way to rescue philosophy from its irresponsible torpor. Philosophers, excepting "cognitive scientists," are not sufficiently interested in the problem, mostly misunderstand the cognitive sciences, and, in the end, are simply ignorant of the results of recent research on the brain. So they miss the ideological stakes as well. But *What should we do with our brain?* is not a question reserved for philosophers, for scientists, or for politicians—it is a question for everyone. It should allow us to understand why, given that the brain is plastic, free, we are still always and everywhere "in chains."
>
> (Malabou 2008, p. 11)

In other words, for Malabou, we must take seriously what neuroscience has discovered about the brain, recognizing that, so far, the "neuronal revolution" it has triggered has in reality led to nothing more than half-measures (improving "quality of life," achieving more effective treatment of disease, and so forth) and risks having no other outcome than ensuring that our brain serves merely to "move better, work better, feel better, or obey better."

More recently, Malabou has acknowledged that the "plasticity" of the brain is not what would distinguish it from cybernetic programmes and artificial intelligence, but rather what places brain and computer in a mirroring relationship, such that there is no "naturalness" to intelligence if by this we mean something distinct from what technological processes make possible (Malabou 2019).

Malabou, however, does not dwell on the problem we are outlining here—namely, the question of how the "liberation" of the brain does not simply involve shielding it from technological intrusions, but rather, on the contrary, entails a certain mode—yet to be invented and tested—of bringing them into relation. This is because, as has been said, the relationship between brain and technology is constitutive of the brain itself, of its current capacities as well as those it will develop or lose depending on the innovations that neurotechnologies will make possible.

When Solomon advocated a right over the brain, he assumed that psychedelic experimentation was the primary path for neuropolitics, understood here as a practice that frees the brain from the limits and the power relations within which it is forced to operate.

Today, the situation has changed in that it is neurotechnologies that will determine the path we take: either towards a progressive expropriation of the brain by them, or towards its use in ways that enable individuals to experiment with practices of liberation from its limits.

This, however, should not make us forget that Leary and other authors had, in fact, already closely linked psychedelia and cybernetics—to the point that the term *cyberdelia* has rightly been used in this regard (Hartogsohn 2022). Properly understood, the potential of lysergic drugs, from this point of view, has always been conceived as a technology whose very aim and function is the liberation of the brain.

It is, of course—it must be—possible to discuss how what is now called "cognitive enhancement" should be regulated and controlled, particularly concerning minors, to the most vulnerable segments of the population who could not afford to access new technological applications, or to the racial biases that neurotechnologies often reproduce (Benjamin 2019). But any such discussion—if it is to avoid falling into forms of paternalism or naturalism—should at least start from the following basic assumptions:

(a) there is no such thing as a "given" brain, or, if one prefers, no "natural" cerebral organization independent from the techniques and powers with which it is in relation. From this point of view—to take an example—one should ask why objections to cognitive enhancement through pharmaceuticals always seem to forget to note that we continue to raise children in social, economic, and cultural environments that mean that, already in preschool, there is a cognitive gap such that those from "culturally disadvantaged" families are exposed, on average, to 32 million fewer words than others. In other words, we are already "enhancing" certain brains and not others.

This is because no brain is not in a constitutive or transductive relationship with technologies that bring about its enhancement or its diminishment—and which, therefore, do not imply a "politics" of the brain that has always existed. From this perspective, when Harari hypothesizes a future in which brain enhancement will divide humanity into two castes, he should not forget that this division has always run through the history of the brain and its cognitive capacities—if only we think of how, in different eras, the separation was produced between brains capable of reading and writing and brains that were not.

(b) the claim that "we are our brains" thus risks becoming a double-edged sword. For if I am my brain—my "thoughts," my mental activity—this could imply that recognizing my right to modify or enhance it would mean recognizing that I can cease to be who I am, that I could "commercialize" my ceasing to be myself—with the result that, assuming that "identity" is also a right "to be myself," inalienable and unchangeable, the outcome would be to deny such recognition.

It is therefore necessary to defend the possibility of a legal construction of the brain as the exclusive property of the individual. Our brain is not what we are, but what we make of it. This does not mean, however, that the danger does not lie precisely here. What Rose and Abi-Rached call "neuro-ontology"—that is, the ontological implications and assumptions of modern neuroscientific theses—does not at all presuppose that we *are* our brains:

> In this emerging neuro-ontology, the claim is not that human beings are brains. The argument is different—that our selves are shaped by our brains but can also shape those brains. And indeed, references to neuroplasticity are used to underpin the argument that our brains are open to change across our lives, and that we can take conscious control of the ways in which those changes happen—we can learn the techniques to "rewire" our brains. It is in this form that neuroscientiic arguments are impacting upon conceptions of personhood and practices of self-fashioning. he pedagogies of "brain awareness" and the rise of practices and devices for working on the brain in the service of selfimprovement thus it comfortably with a more general array of techniques for working on the somatic self in the name of maximizing our well-being. In the name of improving the well-being of our societies, each of us is now urged to learn the techniques to manage our plastic, open, mutable brains in order to live a responsible life. The technologies of the neurobiological self it comfortably with contemporary beliefs that we can improve ourselves by knowing and managing our somatic, bodily, embrained selves.
>
> (Rose – Abi-Rached 2014, pp. 16–17)

As has already been emphasized, neuropolitics takes shape within a field in which technologies already operate on the assumption that the brain is something modifiable and "plastic." It is within this field that neuropolitics must work to find practices for the liberation of the brain capable of counteracting its expropriation by neurotechnologies in the service of cognitive capitalism.

## Legal Issues

If our "mind" is nothing other than what is constituted through its externalization—namely, through the techniques and technologies that support it—this means, as Stiegler has made clear, that there is no "mind," no "brain" that exists before such externalization, as some sort of "interiority" or pure substance that would subsequently make use of techniques—starting with language—as its "means" or "instruments."

But this implies that there is no "brain," before and independent of technologies, that we must or could defend and protect, as such, from "interference" or "intrusions." For this reason, "defensive" positions, such as those underlying the formulation of neuro-rights, risk being entirely inappropriate and failing to grasp the real problems to which neuropolitics must respond.

This does not mean that these problems are entirely new, or that they require new concepts to be addressed. But they must be tackled starting from the fact that they are not about dividing, separating, or fixing the boundaries of what would distinguish the "natural" (or biological) from the "technical" or the artificial. Rather, they are about deciding what kind of brain we are making for ourselves and, above all, for future generations—and therefore, what kind of "humans" we will be.

Neurotechnologies certainly raise new and specific issues when, as we have noted, it becomes possible to "read minds"—or more accurately: to decode mental states. Since our focus here is, first and foremost, on the law—the legal issues of neurotechnology—we can begin with an example concerning what has been hypothesized could occur in the judicial field.

Legal scholars such as Nita Farahany, as well as neuroscientists like Ienca, have suggested that the future application of brain-reading in criminal proceedings would entail the impossibility for the accused to lie or to refuse to answer—and thus the end of the *nemo tenetur se detegere* principle, that is, the right not to incriminate oneself.

The problem could be addressed by identifying the limits within which techniques for lie detection—and even more so for "brain reading"—could be admissible in court. But the example here serves to show how the very construction of the brain and its neuronal activities as "things" owned by the individual would suffice to eliminate the problem at its root, from a legal point of view.

The idea that the "reading of an accused's thoughts" could be relevant for his or her incrimination presupposes that what the accused "thinks," their thought (or rather: the decoding of their mental states), is legally the same as what they might "declare," that is, its manifestation. If, however, I have a property right over my brain and its neuronal activities, such an equivalence ceases to be possible or meaningful because:

(a) the content of thought is "mine" in the sense of a thing that is my property, whereas
(b) a declaration (that is: the manifestation of a thought, an intention, etc.) is legally "mine" in a different sense—namely, it is "mine" not because it is my property, but because I have authorship over it—that is, because the law treats me as the author of what I declare.

The idea that one can incriminate oneself through what one says is based on the logic that I must be held accountable, as author, for my words, for what I intended and chose to declare. It is by no means immediate or automatic that such logic should apply not to what I declare, but to contents of thought which, as such, are simply "present" in some way in my brain.

Even in this case, the logic of property works far more effectively and precisely than any neuro-right. Precisely because I do not "am" what I think, but rather own my "thoughts" (i.e., the neural signals and what results from their decoding), legally a thought would not be something for which I must or could be "held accountable" as its author, but only a *res* of which I am the owner.

This simple construction would, in itself, render pointless—without even the need to prohibit it—the use of neuro-polygraphs or other such devices in court or investigations. Suppose it is not, to my knowledge, generally considered. In that case, it is because even legal scholars tend to assume that a "thought," legally, cannot be anything other than what it "naturally" is (in this case also presupposing that what I think is, by definition, something that expresses what I "know," what I am, my "true" will, and so forth).

What they entirely overlook is that, once thoughts become possible objects of the law thanks to neurotechnology—which, until now, has historically been defined precisely by the irrelevance of thought as such and the relevance of its manifestation—the first question to ask is: what is (that is: how do we want to define) a thought, legally speaking?

The possible answers to such questions—this is the crucial point—must respond solely to the kinds of interests the law seeks to realize, and are in no way dependent on what a "thought" may be from an ontological, neuroscientific, or other standpoint. Introducing a legal discontinuity between what a thought is (a *res* in property) and what the manifestation of a thought is (a declaration by a subject who is accountable for it as its author) is not only perfectly possible. Still, it would, in itself, render meaningless, from a legal perspective, the possibility of self-incrimination for what one thinks. More generally, it would prevent "thoughts" from being legally treated as expressions of the subject's will, identity, or conduct.

A different matter is that of privacy risks or, more accurately, of the continuous extraction of surplus data by integrated devices that are already part of our daily lives. Even in the case of the "connected brain," we inevitably encounter the same issues that have already been studied concerning the extraction and analysis of data from our interactions with digital platforms. If I use Google, it appears that I am accessing a particular service free of charge.

In reality, I am providing Google with a series of pieces of information that I do not even know I am giving. As Zuboff has noted, in this new context, goods and services are merely in the service of surveillance. It's not the car that matters, but the behavioural data it generates; it's not the map that matters, but the behavioural data that comes from interacting with the map (Zuboff 2019).

The point, however, is that the devices we connect to extract far more information from us than we might expect. Again, Zuboff has stressed that the actual revenues of companies like Google come from the fact that more behavioural data is collected than is needed to improve the service, which is what enables algorithms to make predictions—predictions that are what have real economic value—about users' future behaviour.

It would be realistic to acknowledge that, in this field, the law has already lost its battle. No "declaration" of rights or constitutionalization of neurorights prevents or will prevent companies from extracting data from users without their even realizing it—and consent to such extraction risks having only a purely formal relevance, as already happens whenever we accept, without ever reading them, the terms and conditions of a given service.

From this perspective, not even a property right could guarantee that the extraction and expropriation of data will not occur. And yet, at least in theoretical terms, it may be able to frame the problem in more appropriate terms. If we hold fast to the logic defended in the preceding pages, it allows us to distinguish between:

(a) the consent I give for my brain activity to be used by devices to extract/decode a series of digital "data," which will come into existence only through my own use of the device.
(b) ownership of these data, which follows the rule of *accessio* or—more accurately—the legal principle that modifications of a material do not affect its ownership.

If, in other words, "digital data" are legally established as a new state that brain signals assume thanks to their decoding, then the "Sabinian" principle should apply: *quod ex eadem materia factum sit, dominus esset, quia sine materia nulla species effici possit* ("what is made from the same material belongs to the owner, for without the material no form can be produced"). If I make a vase from your metal, or a garment from your wool, you become the owner of these items, since they are not legally "new things," but modifications of the thing.

It is therefore by no means inevitable to configure the underlying legal relationship between the user and the company as if, by purchasing the service, I had ceded the right to such "data." This may, eventually, happen—but only insofar as these "data" have come into existence as a property of mine. Consider a BCI device that allows me—by wearing, for example, a helmet connected to a computer—to move the wheelchair I am confined to through "thought," that is, brain waves.

What I initially cede, strictly speaking, is the right to detect and record the brain signals I emit, to convert them into digital data, such that, for example, a certain recorded increase in some of my brain waves can be converted into the content of thought such as the command "turn right."

Now, this "data," which is a "content of thought," can be legally established as a *res*, as a thing that can be the object of property, as has been argued. And at that point, it comes into existence as my property by "accession," since it is merely the result of work carried out on my brain waves, of which I am the owner.

It is, of course, impossible—especially within the space of an essay—to offer a complete response to the questions posed by developments in neuroscience, even from a purely legal standpoint. The essential aim here has been to try to define a conceptual framework capable of linking, in broad outline, the issues of neuropolitics with how the law can help articulate them. From this point of view, the liberation of the brain will not be achieved, we believe, unless and until we are finally recognized as having exclusive ownership over it.

## Conclusions

The analysis of neuro-rights now allows us to draw a number of conclusions that will serve as a bridge to the next topic. What this discussion has sought to demonstrate is that law constitutes a specific and distinctive *technology*—one that operates by separating what things *naturally are* from the way in which they are constructed as what they *legally count as*.

More precisely, law functions so as to make it possible for individuals to enter into relations with themselves, with others, and with things through mechanisms that enable them to *subjectify* themselves in particular ways.

We have referred to this particular capacity of law as *cybernetic*, in order to indicate how it enables the "governance" of reality through the establishment of specific frameworks that allow individuals to equip themselves with artificial "organs"—as Savigny already observed—or, if one prefers, with forms of action that, by feeding back upon them, transform them and make them into new "subjects," different from what they "naturally" were (owners, fathers, and so on). This cybernetic dimension of law reveals its technical essence: a mechanism through which the normative order continuously produces and transforms its own subjects.

The example of the so-called *neuro-rights* helps, I believe, to clarify this point. The solution we propose is grounded precisely in the *cybernetic* capacities of law. To make the brain into a thing that can be owned—a *res*—constitutes a technical-legal operation that enables individuals to enter into a (non-"natural") relation with their own mental activity, thereby allowing for its effective legal protection.

In this sense, law proves not to be bound in any way—(a) either to what things *naturally are* (it is evident, after all, that neither the brain nor the mind is, strictly speaking, a "thing," and that there are good reasons to believe that a person does not *have* a mind but rather *is* their mind); (b) or to what things become as a result of technological innovations, such as those made possible by BCI mechanisms.

On the contrary, law exists precisely because, in order to realize the solutions it deems most just—that is, to bring about the configuration of interests it considers most appropriate—it operates by creating entirely autonomous forms of integration and connection between the individual and reality.

In our example, as noted above, the need to protect mental activity from external intrusion is ensured by making the brain a *res* in ownership—or, in other words, by establishing a human-brain relation such that the individual can relate to it *as if* it were a thing owned.

At this point, we may consider the first fundamental thesis of the book to be established: namely, that law constitutes an autonomous and distinctive *technique* and *technology*, one that differs from the algorithmic forms of "technological regulation" which have emerged in recent decades following the digital revolution and the advances in artificial intelligence.

This, however, makes it necessary to address a second issue: the nature of the relationship between the kind of regulation made possible by algorithmic technologies and the way in which law itself operates. It is in this context that we shall return to the debate on the "death of law," as developed by authors such as Brownsword and Lucy.

## References

Bates D.W. (2024), "Organologie et décision: sur la politique du cerveau." *Le portique. Revue de philosophie et de sciences humaine*, 2: 117–129.

Beaud O. (1993), *L'Affaire de la main volée*. Paris: Seuil

Benedikter R. – Giordano J. (2024), "Neuropolitics: New Frontiers for Policy and Strategy." In Giordano J. – Benedikter R. – Shook J.R. (eds.), *Neuroscience, Neuroculture, and Neuroethics. A Broad Overview*. Dordrecht: Springer, 65–74.

Benjamin R. (2019), *Race after Technology. Abolitionist Tools for the New Jim Code*. Cambridge: Polity.

Bergelson V. (2003), "It's Personal But Is It Mine? Toward Property Rights in Personal Information." *University of California Davis Law Review*, 3: 381–451.

Blitz M.J – Bublitz J.C. (2021), *The Law and Ethics of Freedom of Thought, 1. Neuroscience, Autonomy, and Individual Right*. London: Palgrave Macmillan.

Bradley J.P.N. (2024), *Critical Essays on Bernard Stiegler*. Cambridge: Cambridge Scholars.

Bublitz J.C. (2013), "My Mind is Mine!? Cognitive Liberty as a Legal Concept." In Hildt E. – Francke A. (eds.), *Cognitive Enhancement*. Berlin: Springer, 233–264.

Cannata C.A. (2001), *Corso di istituzioni di diritto romano*, I. Turin: Giappichelli.

Connolly W.E. (2002), *Neuropolitics. Thinking, Culture, Speed*. Minneapolis: University of Minnesota Press.

Devonis D.C. (2021), "Timothy Leay's Mid-carrer Shift: Clean Break or Inflection Point?" *Journal of the History of the Behavioral Sciences*, 1: 16–39.

Dyson F. (1999), *Back to the Future*. http://news.bbc.co.uk/hi/english/static/special_report/1999/12/99/back_to_the_future/freeman_dyson.stm.

Glen Boire R. (2003), "Mind Matters." *Journal of Cognitive Liberties*, 1: 7–10.

Hartogsohn I. (2022), "Cyberdelics in Context: On the Prospects and Challenges of Mind-manifesting Technologies." *Frontiers in Psychology*, 13: 1–8.

Ienca M. (2019), *Intelligenza2. Per un'unione di intelligenza naturale e artificiale-*. Turin: Rosenberg&Sellier.

Ienca M. – AndornoR. (2017), "Towards New Human Rights in the Age of Neuroscience and Neurotechnology." *Life Sciences, Society and Policy*, 13, 5: 1–27.

Istace T. (2022), "Neurorights: The Debate About New Legal Safeguards to Protect the Mind." *Issues in Law and Medicine*, 1: 95–114.

Krüger O. (2021), "'The Singularity is Near!' Visions of Artificial Intelligence in Posthumanism and Transhumanism." *International Journal of Interactive Multimedia and Artificial Intelligence*, 1: 16–23.

Kurzweil R. (2024), *The Singularity is near. When We Merge with AI*. London: Bodley Head.

Lavazza A. (2018), "Freedom of Thought and Mental Integrity: The Moral Requirements for Any Neural Prosthesis." *Frontiers in Neurosciences*, 12: 1–10.

Leary T. (1964), "Introduction." In Solomon D. (ed.), *Lsd: The Consciousness-Expanding Drug*. New York: Putnam's, 11–30.

Leary T. (1991), *Neuropolitique*. New Falcon: Scottsdale.

Lopez-Silva P. – Valera L. (2022), *Protecting the Mind. Challenges in Law, Neuroprotection and Neurorights*. Cham: Springer.

Malabou C. (2008), *What Should We Do with Out Brain?* New York: Fordham University Press.

Malabou C. (2019), *Morphing Intelligence*. New York: Columbia University Press.

Musk E. (2019), "An Integrated Brain-Machine Interface Platform with Thousands of Channels." *Journal of Medical Internet Research*, 10: 1–14.

Nicolelis M. (2012), *Beyond Boundaries. The New Neuroscience of Connecting Brains with Machines… and How It Will Change Our Lives*. New York: Griffin.

Quigley M. (2018), *Self-Ownership, Property Rights, and the Human Body: A Legal and Philosophical Analysis*. Cambridge: Cambridge University Press.

Qvortrup M. (2024), *The Political Brain. The Emergence of Neuropolitcs*. Budapest: Central European University Press.

Rainey S. (2023), "Neurorights as Hohfeldian Privileges." *Neuroethics*, 16, 9: 1–12.

Ronen R. (2018), "To Think or not to be. Lacan with Descartes." In Id., *Lacan with the Philosophers*. Toronto: Toronto University Press, 83–110.

Rose N. (2006), *The Politics of Life Itself. Biomedicine, Power, and Subjectivity in the Twenty-First Century*. Princeton: Princeton University Press.

Rose N. – Abi-Rached J. (2014), "Governing through the Brain: Neuropolitics, Neuroscience and Subjectivity." *The Cambridge Journal of Anthropology*, 1: 3–23.

Schreiber D. (2017), "Neuropolitics: Twenty Years Later." *Politics and the Life Sciences*, 2: 114–31.

Schwartz P.M. (2004), "Property, Privacy, and Personal Data." *Harvard Law Review*, 117: 2056–128.

Shanahan M. (2015), *The Technological Singularity*. Cambridge: MIT Press.

Solomon D. (1964), "Editor's Preface." In *Lsd: The Consciousness-Expanding Drug*. New York: Putnam's, VII–VIII.

Stiegler B. (2010), *Critique of Political Economy*. Cambridge: Polity.

Stiegler B. (2011), *Technics and Time, 3. Cinematic Time and the Question of Malaise*. Stanford: Stanford University Press.

Van Erp S. (2017), "Ownership of Data: The Numerus Clausus of Legal Objects." *Brigham-Kanner Property Rights Conference Journal*, 6: 235–57.

Wexler B. (2006), *Brain and Culture: Neurobiology, Ideology, and Social Change*. Cambridge: MIT Press.

Wolf M. (2007), *Proust and the Squid. The Story and Science of the Reading Brain*. New York: Harper.

Zizek S. (1998), *Cogito and the Uncounscious*. Durham – London: Duke University Press.

Zizek S. (2020), *Hegel in a Wired Brain*. London: Bloomsbury.

Zuboff S. (2019), *Surveillance Capitalism. The ht for a Human Future at the New Frontier of Power*. New York: PublicAffairs.

# 3 Technology and Legal Norms

## Introduction

As noted at the end of the previous chapter, the problem of the relationship between law and technology must now be addressed starting from the fact that, as has been shown, law itself is a particular kind of *technique*—one that must be distinguished from the way in which algorithmic technologies operate. These technologies have become a central topic of legal debate due to their potential applications in fields such as contractual enforcement, liability and tort law, evidence and judicial decision-making, as well as in the broader areas of data protection and criminal justice.

The thesis we shall put forward is that the distinction between law and algorithmic technologies must be sought in the very structure of their respective forms of normativity.

It is by now difficult to dispute—as previously suggested—that *Code is Law*: algorithmic technologies, many of which rely on data-driven processes or artificial intelligence, also generate their own form of *normativity*. They shape and direct human behaviour and decision-making, thereby functioning as genuine instruments of regulation.

For this reason, as we shall see, several authors now argue that what they call *technological regulation* is likely to replace—or at least to operate alongside—the more "traditional" forms of behavioural regulation achieved through legal norms.

The argument that will be developed in the following pages is that the *normativity* proper to law is fundamentally different in nature from the kind of *prescriptivity* embodied and conveyed by algorithmic rules, and therefore qualitatively distinct from the normative dimension of technological regulation.

This will require, however, a rather elaborate explanation and a set of assumptions that are, in many respects, counterintuitive. What I shall argue—drawing on certain insights that can be traced back to Kelsen—is that, in order to understand the kind of *normativity* that constitutes law, one must first challenge and abandon the idea that legal norms are "rules" of conduct, that is, prescriptions aimed at guiding behaviour.

DOI: 10.4324/9781003775850-4

It is important not to conflate the *ought-to-be*—which, I suggest, expresses the specific form of legal normativity—with the *duty to act* that a rule may prescribe. To say that legal norms consist in an *ought-to-be* does not mean that they simply require certain actions to be performed. Rather, it means that norms constitute their objects as entities whose existence depends on their being something that *ought to be*.

This definition, of course, requires further explanation, as it is not immediately intuitive, and the remainder of this analysis will be devoted to clarifying it. For now, however, a simple example may help to illustrate the essential point.

What I am suggesting is that legal norms are not so different from certain rules of chess—those that "regulate" what each piece can or cannot do. Consider, for instance, the rule stating that "the bishop must move diagonally." It would be entirely misleading to interpret this as prescribing that the bishop *ought to* move diagonally, as though it were being instructed to act. What the rule actually does is to *constitute* the bishop as a chess piece—namely, as that piece whose very existence consists in its *ought-to-be moved diagonally*.

In this sense, the bishop has no other "being" than its *ought-to-be* moved in that particular way. My claim is that legal norms must be understood in precisely this manner, and that only by doing so can we grasp the specific kind of normativity—distinct from mere prescriptivity—that defines law.

We should make clear, however, that this does not simply amount to endorsing the thesis—well defended by several scholars (in Italy, see for instance Carcaterra 1979)—that all legal norms are "constitutive" in nature. This is not because that thesis is mistaken in itself, but because our concern here is to clarify what *constitutivity* should mean within the framework of our argument, and how it *ought to be* understood in light of our conception of the *ought-to-be*. In this context, *constitutivity* does not merely refer to the creation of institutional facts—such as contracts, rights, or obligations—but to the establishment of entities whose very existence consists in their normative status.

Legal norms, in this sense, do not simply "constitute" social facts; they bring into being realities that exist as *ought-to-be*, rather than as mere empirical facts.

We must therefore examine in greater detail the argumentative steps that allow us to conceive a notion of legal normativity distinct from that which characterizes algorithmic mechanisms. In my view, this is the only path that can provide an effective response to those who, today, foresee the imminent "death of law," on the assumption that technological regulation will be able—whether partially or entirely—to replace the traditional ways in which law has operated until now.

## The Death of Law

In recent years, jurists have increasingly engaged in discussions about the impending "death of law," brought about by the gradual and inevitable replacement of legal rules with what is often referred to as "technological regulation" or "technological management." Roger Brownsword has argued that this will occur by 2061, while William Lucy has countered that we should be far less optimistic, since the death of law is likely to arrive much sooner (Brownsword 2015; Lucy 2022).

To clarify the matter, it is helpful to recall a distinction—now common among American legal scholars—between two regulatory "approaches," that is, two distinct logics concerning how conduct should be governed: the normative *East Coast* model, rooted in "classical" legal reasoning, and the *West Coast* model, which is based instead on the replacement of legal rules with "codes," characterized by a model of *technological management* (Brownsword 2005).

According to Brownsword, regulation following the normative logic typical of law begins with identifying the desired type of conduct (the *regulatory position*). Then it seeks to achieve it primarily by providing the addressees with reasons capable of motivating them to act accordingly.

By contrast, a technological management logic tends to disregard whatever reasons the addressees might—or might not—have for complying with the rule, moving instead in the opposite direction: designing mechanisms that eliminate the need for human participation or involvement in executing and enforcing the rules.

To take an example also discussed by William Lucy, consider how the two regulatory approaches distinguished by Brownsword would address a set of issues in road traffic, particularly the need to limit vehicle speed in specific stretches. Under the "legal rule" model, the authorities would likely: (a) set a maximum speed limit through a provision such as "Do not exceed 90 km/h;" (b) establish sanctions for non-compliance; and (c) ensure that such sanctions are generally and effectively applied. Variations in enforcement measures are, of course, possible.

Yet, according to Brownsword, such measures remain within a logic in which guiding conduct consists in providing reasons or motivations to induce the subjects concerned to behave in a certain way.

A "technological management" approach, by contrast, would proceed according to a different logic: it would seek to devise and implement measures ensuring that the desired outcome (not exceeding a given speed limit) is achieved regardless of the choices and actions of the individuals involved.

The most efficient solution, in this respect, would be to equip each vehicle with a speed limiter—such as the current Intelligent Speed Assistance system—making it impossible for the driver to exceed the prescribed limit on a given stretch of road. As Lucy observes, in such a case, the driver's decision to follow (or not follow) the "rule," and even their knowledge of the rule itself, becomes irrelevant and is eliminated:

> If we can ensure that it is impossible for cars to exceed the speed limit in this way, then that is a more direct and parsimonious way of achieving our regulatory goal than telling regulatees what is required of them, telling them what will follow if they behave contrary to the requirement and leaving them to decide whether or not to comply. Smart regulators—those looking for the cheapest and most effective means of achieving regulatory goals—must surely be technological managers.
>
> (Lucy 2022, p. 116)

Brownsword is perhaps the scholar who, in recent years, has most insisted on the transformations that blockchain technologies are already producing in the role that legal norms play, and will play, in society (Brownsword 2021).

The technological management model—non-normative by nature—will, he argues, coexist with the traditional model based on legal rules and will be adopted in contexts and for purposes as diverse as crime control, environmental protection, public health, and more. Ultimately, however, it will prevail. Inevitably. After all, as Lucy rightly asks, what objection could one raise to a mechanism that guarantees compliance with the law—say, with a rule imposing a speed limit—by making it impossible to violate?

On what grounds could one argue that it would be preferable to continue relying on the traditional regulatory mode, merely enacting a legal rule that states "do not exceed the speed limit"?

## Freedom of Choice

A likely objection to the regulatory technologies discussed above would be this: a "rule" that, by definition, cannot be violated—because it directly "causes" the desired behaviour—should not be regarded as a replacement for a legal norm.

Instead, it would constitute its negation, not a more effective equivalent. This objection assumes that the very concept of a legal norm entails the freedom of its addressee, insofar as it leaves to them the choice of whether to comply with the prescription. A norm prescribes what I must do, but precisely because it prescribes it, it necessarily leaves me to decide whether to do it or not.

This thesis, however, merits closer examination. We must first ask what is meant here by "freedom"—by the idea that the addressee is free to choose whether to fulfil the duty imposed by the norm (for example, not to exceed a specific speed limit).

This cannot mean that they have permission or a right not to comply—that they are entitled not to do what they must. One cannot at the same time have a duty to act and a right to decide whether to perform it, for the existence of the duty implies that acting otherwise constitutes a violation of the duty, not the exercise of a right.

Where one is recognized as having the right to choose arbitrarily between one course of action or another, there is no duty and no binding force at all.

The remaining possibility, then, is that "freedom" refers in fact to the empirical circumstance that the law can only entrust to the subject the choice whether or not to comply, since, *de facto*, it has no alternative.

Put differently, the legal norm cannot directly "cause" the desired conduct; it can only provide various kinds of reasons for the addressee to act accordingly. Individuals, therefore, would always remain, in the final analysis, "free" to determine themselves as they wish, since it is only through their own free will that the norm will—or will not—be carried into effect.

The problem, however, is that once we admit that this freedom is not a contradictory "right not to comply with the norm" but a mere fact, it becomes difficult to deny that, from the law's perspective, this fact is an empirical limit—an "obstacle" which the law, if it could, would gladly remove. Let us try to clarify the point further by adopting a different perspective.

## Guiding Conduct

Although jurists have given, and continue to give, highly diverse—and often conflicting—answers to the question "What is law?", they tend to agree on one point: law is, above all, a technique for regulating social behaviour and guiding conduct. In other words, the fundamental function of norms is to serve as "rules of human behaviour."

This conception treats the *ought-to-be* of norms as a duty that tends towards *being*: a duty that exists precisely insofar as it finds its realization in the conduct of its addressees. From this perspective, norms entirely lacking in effectiveness—norms never observed—would not truly be "rules of behaviour." Within this conception, it is therefore challenging to deny that a legal norm, by definition, is not merely that which obliges a certain conduct, but has as its intrinsic aim the actual performance of that conduct.

For this reason, jurists still speak of norms as "guides" to conduct, as possessing the capacity to "influence" human behaviour through various forms of "indirect coercion" and "pressure." The question, then, is whether such a characterization of law is genuinely compatible—and on what terms—with the idea that norms necessarily imply the freedom to violate them, the freedom to choose whether to comply.

A helpful distinction here is Frederick Schauer's distinction between *conformity* and *compliance*. To act in accordance with a norm is not the same as to comply with it. As Schauer notes, if I place a bowl of food before my dog and say "Eat!", she will, of course, begin to eat. Her conduct conforms to my command—she is doing exactly what I prescribed—but she is not "complying" in the sense of acting *because* I commanded it. For Schauer, to comply with a norm means at least that the norm itself is what causes us to behave in a certain way—i.e., it "causes" our conduct:

> It is the same with the commands of the law. When the law tells me to do what I would have done anyway, the law's commands are no more causally consequential than commanding my highly food-motivated dog to eat. Our interest in law is largely an interest in law insofar as it is causally consequential, and for that purpose the distinction between law that makes a difference to behavior and law that makes no difference is of central importance. If we are interested in obedience to law, we must focus on law's effect on people who, but for the law, would have done something other than what the law commands. Or, to put it differently, we are interested in the cases in which what the law commands differs from what a subject of the law would have done for law-independent reasons.
>
> (Schauer 2015, p. 49)

Why, then, do we say that norms influence or guide conduct rather than cause it? The reason is that the causal role of law in determining behaviour has traditionally been seen as problematic. Authors such as Schauer himself deny that law can exercise a genuinely causal influence on conduct. People may generally act in ways consistent with norms, but not because those norms provide their reasons for acting. They may do so out of fear, habit, or other motivations. The "cause" of their norm-conforming behaviour, then, is not the norm itself, but something else. In this sense, norms can never truly "determine" my will, and thus cannot be the "cause" of my behaviour.

We can now see in what sense norms imply the "freedom" to comply or not: it is the correlate of their lack of strictly causal power. I am "free" to violate the norm precisely because it cannot cause my behaviour, but only offer—directly or indirectly—reasons for it, which I may choose to follow or not.

Here we reach a turning point. To admit that norms cannot cause conduct is to assume that this is nevertheless their essential aim. When we say that norms have the function of "guiding" behaviour, we presuppose that the ideal norm would be one capable not merely of influencing but of directly causing the desired conduct. We then hasten to add that this is not empirically possible, thereby reintroducing the presupposition of individual freedom to choose whether to fulfil one's duties.

From this, a first set of observations can be drawn. In my view, to assume that law has the function of guiding conduct already necessarily denies that the concept of a legal norm logically entails the freedom to choose whether to comply. Quite the opposite: the thesis "a norm is such insofar as it can guide or influence behaviour" is nothing more than a reformulation of the thesis "a norm should be able to cause behaviour, but it cannot, because it cannot determine the will of its addressees, who remain free to choose whether to comply." If, in other words, a norm "influences" my conduct, it is only because it is not *de facto* able—though it is assumed it should be *de jure*—to cause it.

Until now, this has not posed particular problems for jurists, since it was taken for granted—too obvious to merit explicit mention—that prescriptive norms could not "cause" conduct, as this would require them to determine directly the will of their addressees. In this light, a norm such as "Do not exceed 90 km/h" could "cause" conduct only if it had the power to "cause" my will not to exceed that speed. But this seemed at best unprovable, and at worst impossible.

Even Georg Jellinek, who held that a norm is valid only when "it possesses the capacity to act in a motivating way, to determine the will" (*Eine Norm gilt dann, wenn sie die Fähigkeit hat, motivierend zu wirken, den Willen zu bestimmen*), had to concede that he could not explain how this could happen: "this capacity arises from the conviction—derivable from no further source—that we are obliged to follow it" (*Diese Fähigkeit entspringt aus der nicht weiter ableitbaren Überzeugung, daß wir verpflichtet sind, sie zu befolgen*) (Jellinek 1914, p. 333).

Matters change radically, however, once the norm becomes capable of causing what it prescribes—not because it acquires the power to coerce the will, but instead because it becomes capable of being applied without the cooperation of its "addressees."

## Technologies of Regulation

In what follows, I shall focus on how technology can "replace" the traditional function of legal norms as guides to conduct, precisely insofar as the conversion of a prescriptive norm into code enables it to "cause" directly the performance of the prescription itself.

Smart legal contracts—one of the best-known examples—execute a party's obligation directly through blockchain technology, making non-performance impossible (assuming there is no error in the code incorporating the parties' agreement) (Di Matteo– Cannarsa– Poncibò, 2019). Likewise, Digital Rights Management technologies can control the use of digital content, making it impossible to copy a given file or to play an illegally purchased DVD.

As has been observed, what disappears here is precisely the "freedom" to breach a norm or contractual provision—save for facing the consequences. In a system governed by self-executing smart contracts and other technical devices, the need for judicial intervention is significantly reduced because the very medium through which the rules are formulated—the code—is also the mechanism by which they are implemented (Becker 2022).

Over time, law and code could become so closely identified that the only way to break the law would be to sabotage or override the code itself:

> In a system regulated by self-enforcing smart contracts and other technical arrangements, there is less of a need for judicial enforcement, because the way in which the rules have been defined—*the code*—is the same mechanism by which they are enforced. Overtime, law and code may merge, so that the only way for people to infringe the law is to effectively break the code.
>
> (Wright – De Filippi 2015, p. 26)

In this regard, commentators have spoken of a shift from "Code is Law" to "Law is Code." (Hassan–De Filippi 2017; Oster 2021). The point is that the law increasingly defines rules based on code with a "guarantee of execution"—capable of being automatically enforced through algorithms and software—both in the private and public sectors. Compared to Lawrence Lessig's formulation of "Code is Law" (Lessig 1999), which stressed the capacity of algorithmic technologies and code to "govern" conduct as effectively, or even more effectively, than legal norms.

"Law is Code" marks the next step: legal rules are directly embedded within digital protocols rather than functioning as prescriptive provisions contained in legislative texts (Zavyalova–Shumskaia–Shumskaia 2019). Whether such self-executing mechanisms—capable of compelling performance and making violation practically impossible—constitute a substantial innovation or merely a refinement of existing legal devices is open to debate.

It is worth noting, however, that several authors have argued that there are fundamental incompatibilities between the logic of legal norms and that of "code."

On this view, a progressive replacement of the former by the latter would entail what we have called the "death" of law. In schematic form, the main differences on which attention has focused can be summarized as follows (see also: Amina – Martinez 2025):

These distinctions can, of course, be debated in detail; they are neither universally accepted nor necessarily well-founded. What matters for present purposes is the emphasis on the capacity to "force" performance of an obligation or compliance with a norm, in such a way as to render its breach or violation *de facto* impossible. As suggested earlier, this capacity appears to realize, rather than contradict, the very ideal towards which the legal norm tends, insofar as it is conceived as a "guide" to conduct.

The obituaries for the "end" of law, and the concerns about a "replacement" of legal normativity by technological regulation, thus risk missing

*Table 3.1* Key differences between regulation through legal norms and technological regulation

| *Regulation through legal norms* | *Technological regulation* |
|---|---|
| Publicity of the rule | Rule "embedded" in the thing |
| Transparency and knowability: the rule is normally written in a form that is clear and comprehensible to its addressee | Opacity of code: the user is generally unable to read or understand the embedded rule |
| Intrinsic ambiguity of norms → need for interpretation, capacity for contextual adaptation and flexibility. | Non-ambiguity of code → a language that requires no interpretation or contextual understanding |
| Presence of a symbolic function: norms often convey values, principles of justice, shared ideals | Absence of symbolic function: code regulates behaviour but does not convey values |
| Norms allow subjects to assume "responsibility" for their conduct as "authors" of it | Responsibility may become detached from the subject's ability to constitute themselves as "authors" of their conduct |
| Justiciability: norms are generally contestable, in various ways, before judicial authorities | Difficulty of contesting code, given the lack of access to it or the technical competence needed to understand its formulation and meaning |
| Enforcement is dependent on the "free" choice of agents | Automatic enforcement—impossibility of non-performance or non-compliance |

the point entirely. Rather than the end of law, technological devices seem to achieve the aim that law—conceived in a certain way—has always sought to attain.

Put differently: if law is nothing more than a set of statements expressing the demand or will that a particular conduct be performed, then technology can provide a more efficient, and probably less costly, means of securing that conduct than "rules" can. The temptation, as Lucy rightly observes, is irresistible, because on the prevailing conception of norms, there are no objective reasons to conclude that the "traditional" form of regulation is preferable to the technological one.

To clarify this further, let us return to the earlier example and suppose the legislature aims to ensure that drivers do not exceed 90 km/h. It is undoubtedly possible to proceed through the "legal" model of rules, prescribing that this limit shall not be exceeded. Yet if a technological device integrated into the vehicle can automatically cap the speed, one must ask: on what grounds should the legislature refuse to adopt it? Is this not precisely the "purpose" of the norm—to prevent speeds above 90 km/h?

Admittedly, this new mode of regulation excludes human agents from the implementation (and even the knowledge) of the norm, leaving them no possibility to "choose" whether to exceed the prescribed limit, whether to comply or not. But as argued above, this would contradict the normativity of norms only if norms were not identified—as they usually are—with prescriptions aimed at "influencing," "directing," or "regulating" conduct.

## Legal Norms and Causality

As we have seen, once the normativity of norms is interpreted as consisting in prescriptions aimed at "guiding" conduct, the idea that this necessarily requires agents to be able to choose whether to align their actions with the "duty" expressed by the norm does not follow from the concept of a norm itself. At most, it follows from a particular political choice underlying that concept—namely, the belief that it is "right" or "appropriate" to leave individuals the option of deciding what to do.

Strictly speaking, therefore, a device that makes it impossible to exceed 90 km/h would not render pointless the norm prescribing "Do not exceed 90 km/h."

On the contrary, it would ensure its full "implementation" and most perfect "realization." Thanks to such a device, the driver's behaviour would always comply with the norm; they would always perform the conduct the norm sought to "direct" (and which it can now directly determine or "cause").

Suppose we genuinely wish to question regulation through technology, and to demonstrate its differences—and incompatibility—with legal regulation. In that case, this is possible only if we can show that law, and norms in particular, do not consist in the "guidance" of conduct, which is merely a weaker way of conceiving causality.

Let us start again from here. Smart contracts, and indeed many technological devices designed to "regulate" behaviour, operate through an *if/then* mechanism: when a specific condition is met, a given effect automatically follows.

The conception of the legal norm as a guide to conduct, on closer inspection, answers to the same logic. When the norm is conceived as a hypothetical proposition of the form "If F, then C," jurists locate its "normativity" in the fact that it does not assert, as a matter of cause and effect, that if F then C will occur, but rather that C *ought to* occur. In other words, the norm does not describe the occurrence of the consequence but prescribes that it must happen.

The nature of this "prescription," however, remains ambiguous. On the one hand, it is admitted that the consequence does not follow automatically from the occurrence of the condition; on the other, it is assumed that in prescribing that it occur, the norm expresses the "will" that it should happen, advancing the claim that the commanded behaviour will follow.

If norms are understood as "prescriptions" and prescriptions, in turn, as speech acts intended to "guide, direct, influence, or modify human conduct," then a causal logic inevitably re-enters the definition of the legal norm.

For if it is true that the norm prescribes rather than "causes" the desired action, it is equally valid that it prescribes only since it expresses the intention to exert some influence over the conduct of its addressee. At this point, it is difficult to avoid the conclusion that the norm, in prescribing the duty to do *x*, also lays claim, as far as possible, to "causing" that behaviour.

This conclusion is, in fact, implicit in the way jurists usually understand the sense of *Sollen*—the "*ought-to-be*"—which they regard as characteristic of legal norms. They identify normativity with prescriptiveness, the *ought-to-be* with the legal duty to do something—i.e., the duty that a particular action should occur or be realized. This conception of prescriptiveness is common among jurists: "the prescriptive function [...] consists in giving commands, advice, recommendations, and warnings to influence and modify the behaviour of others—in short, to bring about action" (Bobbio 1993, p. 53).

In this way, the *Sollen* is identified with a duty that "tends" or is "directed" toward being; more than an "*ought-to-be*," it is an "*ought-to-become*"—a duty which, as such, is inherently oriented towards realizing itself in *being*. Within this conception, a prescription such as "You must pay taxes" expresses the idea that this duty becomes real, productive of effects, *wirklich*—only when the taxes are paid, when it is realized in the realm of being.

It is this logic that allows norms to be conceived as "guides" for action. To regard a norm as a prescription aimed at "influencing" behaviour is to consider the *ought-to-be* it expresses as "aiming at being," as achieving realization in being. Once the *ought-to-be* is conceived as a prescription, as a directive for behaviour (a duty "aimed" at being), the attempt to separate, at the level of logic, the duty to act from its actual performance proves illusory.

It is of no avail to assert that the norm does not "cause" X but merely prescribes the duty to do it. This distinction collapses as soon as "to prescribe" is identified with "to command that something be," and thus with a "directive" whose intrinsic purpose is to "influence" (and, ideally, to cause) the desired conduct. If the norm imposes a duty, it does so so that the duty should not exist merely as a duty, but as being, as the conduct that fulfils it.

The problem can now be restated. Within such a conception, technological modes of regulating conduct do not seem to "conflict" with the logic of law; on the contrary, they appear as what finally enables it to be fully realized. They make it possible for the duty prescribed by the norm actually to "cause" the desired behaviour and thus necessarily to realize itself in being.

True, one might reply that a norm posits a conduct as "owed" or obligatory, but need not do so to influence behaviour. Yet here a distinction must be drawn. One thing is to address the empirical question of what "intentions" a norm may express, and how it might seek to realize them. Another is to observe that if a norm is defined as a precept or directive concerning a behaviour qualified as "owed" or "obligatory," it necessarily and logically expresses the command that something be done, and thus that the conduct of the addressee be determined by it.

### *If/Then*

The thesis I wish to advance is that the conception of law as a rule of conduct rests on a mistaken interpretation of the meaning of the hypothetical form of the legal norm, understood as a proposition expressing a relation of *ought-to-be* (*Sollen*) between an antecedent and a consequent.

For the normative proposition "If A, then *ought-to-be* B" truly to express—as Kelsen maintained—a non-causal relation, it is not enough, as I have sought to show, merely to affirm that it prescribes the duty to do B or for B to occur. In my view, this is not the sense in which *Sollen* operates, even within Kelsen's doctrine.

First, it is necessary to take into account that the normative proposition, for Kelsen, involves a distinction between *Sollen* (*ought-to-be*) and *legal duty* (*Rechtspflicht*), or, if one prefers, action as owed. While the former refers to the consequent, *duty* refers instead to the conduct contrary to that envisaged by the norm as unlawful.

The *Sollen* thus qualifies the relation between the consequent and the antecedent, whereas *duty* qualifies a particular conduct. In the norm "If F, then C," the *Sollen* is referred to C, as consequent upon F, whereas the duty refers to ~F, i.e., to the conduct owed as the opposite of the unlawful act.

This distinction clarifies a first point: *Sollen* is not the action as owed. So much so that, for Kelsen, the norm "If F, then C" does indeed establish the duty of ~F, but not the duty of C.

Since, within this logic, an action can be said to be owed only insofar as its contrary is provided as the antecedent by the norm, it follows that the same norm cannot, at the same time, posit both the *Sollen* of a given consequence and its being owed. As Kelsen observes, "legal obligation is not, or not immediately, the behaviour that ought to be. Only the coercive act, functioning as a sanction, ought to be" (Kelsen 2005, p. 119).

| | |
|---|---|
| "If F, then ought-to-be C" | "If one steals, then one ought to be punished" |
| ↓ | C = *Ought-to-be* (*Sollen*) → punishment |
| F = Wrong (Delict) | ~F = Legal Duty (*Rechtspflicht*) → do not steal |

This means that *Sollen*—which is what the norm expresses—must be something different from prescribing a particular behaviour as owed. Thus, when Kelsen speaks of the *Sollen* of the sanction that follows the offence, he does not mean the duty to impose it, placed upon the judge.

Instead, *Sollen* is the meaning of a relation that constitutes, as such, the *relata*—that is, the terms it connects. To clarify, let us take as an example a norm of the form: "If X is theft, then it must be punished." A "theft," as such, is not something that exists independently of, and before, its *ought-to-be* punished.

Something counts as "theft" only insofar as a norm provides a specific penalty for it, thereby making it exist as an "offence." This is precisely what the normative proposition expresses: it makes theft nothing other than its *ought-to-be* punished.

Here, "*ought-to-be*" does not denote something that must happen, something "prescribed" to happen. Instead, it represents the very *being* of the terms it relates to. Juridically, theft has no other "being" than its *ought-to-be* punished, just as the sanction has no other "being" than the *ought-to-be* punished of theft.

This is the essential function of law, and of the legal norm in particular: to establish a reality that consists not in "being," but in "*ought-to-be*." This has nothing to do with "prescriptiveness." If, with "forbidden X," I simply prescribe not to perform a specific action X, I am presupposing that such action exists in reality, that it is something actual (or that will be actual) in the realm of being.

If, instead, in Kelsen's terms, the norm says "If X, then Y," what it does is to constitute X as something—a certain juridical reality—that has no "being" other than the *ought-to-be* of *Y*.

Of course, for Kelsen, the norm also fulfils a prescriptive function. But this is not the "normative" function expressed by the *Sollen*; rather, it concerns what is posited as a behaviour "owed" (in our example: not-X).

### ***Ways Out***

The foregoing considerations may allow us, at last, to identify the key point that separates the normativity proper to law from technological regulation. As we have seen, this is essential, since only by demonstrating a radical incompatibility between law and technology will it be possible to resist the *Law is Code* trajectory towards which our present is moving.

From this perspective, as I have sought to show, the strategy of defending the idea that legal norms logically entail the freedom of the addressee to choose whether to comply is bound to fail, for two reasons. The first, already noted, is that such a defence risks contradicting the very conception of norms usually presupposed—namely, that they are "prescriptions" aimed at "regulating" conduct. The second is that this strategy risks confusing a problem intrinsic to the logic of law with one concerning entirely different matters.

Consider a typical argument in favour of the idea that norms "imply" individual freedom. Paul Amselek, for example, defends this thesis in a fairly representative way. He first concedes that legal rules indeed have a "vocation to direct the conduct of citizens:" norms are "guides" to behaviour or, more precisely, directives aimed at influencing people's actions and obtaining—he writes—obedience.

He then imposes an immediate limit: "Man is not an automaton" (*l'homme n'est pas un automate*). No one, he argues, can be made directly to act; commands are always "mediated" by the individual's will, by their personal decision to carry them out or not. He adds: "Human relations of authority necessarily combine with this ontological particularity of the human being: man is a free being, that is, a being who guides himself, who possesses a power of self-determination over his movements" (Amselek 1993, p. 319).

For this reason, to "direct someone's conduct" can only mean to ask them to perform a specific behaviour. Legal rules, he concludes, have the function of directing the will of citizens, but they can do so only insofar as citizens voluntarily accept them.

Let's examine this line of reasoning closely. It becomes clear that Amselek, from the outset, refrains from claiming that the freedom to choose whether to comply follows logically from the very concept of a legal norm.

Instead, he presents this freedom as an "anthropological" or "ontological" fact: it is, *de facto*, impossible to "impose" a will on another, since man is not an automaton but, by definition, acts voluntarily and is therefore "self-determined."

But this risks being, at best, a restatement of the old logic *coactus voluit, sed tamen voluit*: I am always the one who, in acting, chooses to act—even when that "choice" is influenced or even coerced.

In truth, however, Amselek's claim is not simply descriptive. More than a "fact," the freedom of choice that law should always presuppose depends on an assumption of *legal policy*: namely, that law should never be able to "compel," but only to "request." (This, however, raises the question already noted: would this not imply the existence of a right, or at least a "permission," not to comply with the law?).

The point, then, is that many who maintain that law entails, by definition, the freedom to comply or not, do not notice that they are defending the quite different thesis that it would be proper for law to allow such a possibility.

Rather than describing law as it is, they are prescribing how it *ought to be*. In doing so, they shift the confrontation—and opposition—with technological regulation from the strictly juridical plane to that of political or legal policy. In contrast, it is first necessary to clarify how the very concept of the legal norm is incompatible with the "rules" of technology.

It is this second path, therefore, that I propose to follow. We can begin again with an example, distinguishing this time between three different strategies of "regulation." Suppose once more that we need to regulate the speed limit on a given road.

Adopting a "classical" legal model—based on the conception of the norm as a prescription—will mean introducing a rule such as "It is forbidden to exceed the limit" or "It is obligatory not to exceed the limit."

Such a rule prescribes a certain conduct while leaving the driver "free" to choose whether to comply. But, as already noted, this "freedom" is merely the consequence of the fact that, as a legal norm, the rule cannot "cause" what it nonetheless demands should happen.

Here, the duty not to exceed the limit is understood as a duty to perform a specific action, which is already—by definition—conceived as something realized in the moment it is performed. In other words, for the driver, observing the norm means acting not exceeding the limit, not merely recognizing that one *ought to* do so.

The second model, the technological one, represents the "completion" or "perfection" of the first. Here, by embedding the rule "Do not exceed the limit" in a device installed directly in the car, it becomes impossible for the driver to break the prohibition.

The device does not "negate" the prohibition; on the contrary, it automatically realizes it, guaranteeing its implementation. The duty not to exceed the limit is always, necessarily, converted into the fact that the limit will never be exceeded.

As I have argued, while this is perfectly coherent with the conception of duty already assumed in the first model, it reveals the contradiction inherent in that conception: an *ought-to-be* that exists only as *being* is no longer an *ought-to-be* but simply *being*. The first two models share the same (flawed) theoretical premise: identifying the *ought-to-be* with the duty-to-do, and with it, the idea that "observing" a duty means "performing" it.

The third model, which I would call the "normative" model, recovers Kelsen's distinction between *ought-to-be* and *duty*.

Here, the legal norm is not merely that which "prescribes" a certain conduct, but—above all—that which establishes a relation in which the antecedent consists in the *ought-to-be* of the consequent. The norm "If the speed limit is exceeded, then one must be punished" thus acquires a new meaning.

Suppose it prescribes not exceeding the limit (as a "duty" consisting in the conduct contrary to the unlawful act). In that case, it also does something else: through the sanction, it constitutes—makes exist—the "speed limit" as something that *ought-not-to-be* exceeded.

Consider this further. The norm's normativity lies in creating a new reality—a new "object," here the *speed limit*—which did not previously exist. It does not merely regulate it; it brings it into being.

The "speed limit" here does not indicate a fact or action to which a specific deontic qualification is assigned; it means, rather, an "object" whose being consists in its *ought-not-to-be* exceeded.

For example, the legal norm "Taxes must be paid" indeed prescribes the action of paying and attributes to that action the quality of "being obligatory." But, more fundamentally, it performs another operation—the properly normative one—by instituting taxes as that which *ought-to-be* paid.

This means that, juridically, a "speed limit" exists only insofar as it is something that *ought-not-to-be* exceeded. When technological regulation turns the speed limit into something that, by definition, is never exceeded, it does not "realize" the norm—it negates it, annulling it.

A speed limit that cannot be exceeded certainly "realizes" the duty not to exceed it. Still, it simultaneously abolishes its *ought-not-to-be* exceeded, making the object created by the norm cease to exist as such.

For this reason, technological devices cannot be regarded as substitutes for legal norms and their normativity, but only for prescriptions.

Consider an example from the context of smart contracts. Suppose Titius buys an item from Gaius for $100, with the parties agreeing to a right of withdrawal and refund within ten days. Through an escrow account, one could ensure that, upon withdrawal, the sum is automatically re-credited to Titius. Gaius would have no possibility of failing to return the price. Where is the problem?

Reconsider the reasoning. This type of contract is not incompatible with Gaius' duty to refund the sum; indeed, it ensures that such duty is always fulfilled.

But it is wholly incompatible with the norm's own normativity. To owe a sum, juridically, requires that the sum exist as something that *ought-to-be* returned. Otherwise, the very basis of the duty is removed. In our smart contract, while Gaius certainly has the duty to return the sum, the sum no longer exists as something that *ought-to-be* returned, since it is, from the outset, something that will necessarily be returned upon withdrawal.

In this sense, the purpose of norms is not primarily to prescribe behaviour or to impose duties, but to institute the reality of the *ought-to-be*. This does not mean making the duty's performance actual in reality, but making possible the experience of a reality whose being consists in its *ought-to-be*.

Returning to our example, the norm imposing a speed limit does not, in its primary function, aim to ensure that people in fact do not exceed a certain speed. Rather, it aims to make it possible to experience the speed limit as something that *ought-not-to-be* exceeded.

It is this possibility of experience that technology deprives us of: in a device that automatically controls speed, I cease to experience the "limit" as something that *ought-not-to-be* exceeded; it becomes nothing more than the speed that is never exceeded. In this sense, technological regulation is essentially anti-juridical and constitutes the negation of law, not its completion.

It abolishes the *de jure/de facto* distinction, which is a constitutive feature of the very concept of a norm. An Intelligent Speed Assistance system turns the *ought-not-to-be* exceeded into the *not-being* exceeded, thereby abolishing it.

The difficulty is that jurists themselves often fail to recognize this as the fundamental problem. By adopting a conception of the norm as a precept or prescription, rather than as a relation of *ought-to-be*, they are forced into a different defensive strategy—one that must not be confused with the one outlined here. Much of the literature insists that the problem lies in the fact that technological regulation eliminates the distinction between the existence of a rule and its execution/enforcement, since enforcement is no longer a separate and distinct moment from enactment (Hildebrandt 2020, p. 70).

However, this is not fully coherent with the idea that the rule is a "prescription," a precept that is such precisely because it claims to modify behaviour. On that assumption, it becomes difficult to maintain that validity (or the duty to observe the norm) is one thing and efficacy or effectiveness (its being observed) is another.

Although this distinction is drawn in theory, it remains entirely normal for jurists to assume that the duty to observe the norm finds its realization in the fact that the norm is observed. This weakens the sharp distinction between validity and efficacy, and with it the very objection to technological regulation.

## Open Questions

As already mentioned, the objections and criticisms that, in the current debate, address the model of technological regulation are varied and diverse. Many have pointed out, for example, the loss of transparency and procedures of democratic oversight; the risk that effective possibilities for ex post judicial intervention may disappear; that rules may cease to require mutual trust between the parties to function; that the interpretive activity of jurists and the contextual adaptation of norms become impossible; and so on.

These are, obviously, central issues. If we have set them aside here, it is because they do not go to the "heart" of the matter, which is to ask whether the concept of a legal norm—or, if one prefers, whether the normativity that characterizes legal norms—is or is not logically compatible with the type of regulation brought about by technology.

The answer suggested here is negative: "technological" normativity is the very negation of what is properly juridical. At the same time, however, we have tried to show that such an answer can be given if and only if one distances oneself from the conception of the norm, and of law in general, as a "guide" or "rule" of behaviour—so long as legal norms continue to be understood as propositions that prescribe a certain conduct to modify it.

Such a conception—if we may now summarize what has been argued—

(i) reintroduces into law the logic of causality, insofar as it identifies the function of the norm with its capacity to "influence" behaviour (where "influence," "guide," "regulate," "affect" are all expressions that refer back to the power to "cause" a certain effect).

(ii) identifies the *ought-to-be* in which norms consist with the duty to adopt a certain behaviour, a duty that, consequently, could be realized only by observing that very behaviour.
(iii) thereby makes it difficult to hold fast to the distinction—often presupposed and defended—between validity and effectiveness, between the duty to observe the norm and the fact of observing it.

Suppose failure to observe the norm, in certain circumstances, does not in itself render it invalid or non-existent. In that case, it is also true that, if the norm is understood as that which advances the claim to be observed, then effective observance—at least in generalized form—necessarily becomes a component of its very definition.

Conceived in these terms, legal normativity risks finding in technological normativity not, as has been said, its negation but, on the contrary, its completion—namely, that which succeeds in realizing the end toward which it is taken to tend. For this reason, other paths must be found to bring out the opposition between legal norm and technological rule, between law and code.

In the course of the chapter, we have therefore sought first to clarify why it is fundamentally illusory to imagine that "freedom"—that is, the idea that a norm would entail the voluntary choice of its addressee to follow it or not—could mark what is specific to law. This position is, in my view, exposed to at least three decisive objections:

(i) such "freedom" could not, in any case, be identified with a right to violate the norm and would therefore remain conceived as an empirical "fact."
(ii) the very nature of this "fact" would be problematic to define and ascertain—do individuals choose "voluntarily" and "freely" whether to respect and observe norms? How could this be shown?
(iii) finally, it is unclear why this "fact," as an empirical limit that the norm encounters in its claim to realize itself, should be defended on the juridical plane, when in reality it is assumed that the norm would be "perfect" precisely to the extent that it could directly "cause" the desired conduct.

The proposal formulated here takes a different route.

The irreducible difference between the normativity of law and technological regulation becomes clear when norms are rethought as propositions that make it possible to experience that upon which they bear—their "object"—as an *ought-to-be*.

To paraphrase, while giving it an entirely different sense, Searle's formula, we might say that norms operate above all according to the logic "X exists as an *ought-to-be* Y in context C."

Thus, the norm that regulates speed by limiting it to 90 km/h is not to be understood—except indirectly—as that which imposes on the driver the duty not to do something (such as exceed the limit), but rather as that which makes the object "90 km/h" exist as something that consists not in a being (the fact of maintaining a certain speed), but in its *ought-not-to-be* exceeded.

As we have seen, technological regulation operates oppositely: it reduces the *ought-to-be* to being, abolishing any distinction between what is and what *ought to be*. But to lose this capacity to institute a reality other than what is "in fact" is to lose law itself.

## *Ought-to-Be*

There is one point, above all, that requires further clarification in order to make the meaning of the argument advanced here more precise. What I maintain is that the specific normativity of law—that is, the normativity which makes a rule a *legal* norm—necessarily implies a distinction between what I have called the *ought-to-be* and the (legal) *duty*.

This distinction may be explained more precisely as follows:

(i) by *ought-to-be*, I refer to something analogous to what some authors describe as "*eidetic Ought*"—that is, a kind of duty that constitutes the very being of what it concerns (Di Lucia 2021). For example, in the rule of chess stating that "the bishop must move diagonally," the duty does not refer to an action that the bishop *ought* to perform, but rather defines what the bishop *is*. The bishop *is* its own being-obliged-to-move-diagonally.

(ii) by *duty*, in contrast, I refer to the circumstance in which a given action is qualified as due.

What I am arguing, then, is that every legal norm functions simultaneously on both of these levels. This allows me to respond more precisely to an objection already hinted at above. As we have seen, one might argue that ensuring the automatic execution of a contract (and thus making its non-performance impossible) does not negate the contract as such. On the contrary, execution would seem to be possible precisely *because* there exists a duty to perform the obligation, and the act of performance would merely realize that duty.

However, this objection fails to grasp the essential point. A contract is what it is only insofar as it *ought-to-be* performed. It is from this that the distinct *duty* of the parties to perform arises. If, by *contract*, we mean an agreement between two parties binding them to fulfil their respective obligations, then we must distinguish between the *ought-to-be* that constitutes the contract itself and the *ought* that refers to the conduct a party must adopt in order not to breach it.

In other words, the constitutive norm of the contract performs two simultaneous operations: (a) it institutes the contract as something that consists in its *ought-to-be* performed (its *eidetic Ought*), and (b) it prescribes, for that very reason, the duty of the parties to perform it.

In the case of *automatic execution*, however, at the very moment when the duty to perform is always and necessarily realized, what is denied—and thereby abolished—is the contract itself. The contract no longer consists in its *ought-to-be* performed, but rather in the mere *fact* of its constant execution. What disappears, therefore, is the *eidetic Ought*: the contract ceases to exist as something that *ought-to-be* performed, and hence ceases to exist *as a contract*. What remains is only the duty to perform—which, in this context, is reduced to a mere *command*. Indeed, even the duty to perform the contract is legally justified only insofar as the contract *is* something that *ought-to-be* performed—precisely what, in this case, is no longer true.

A further objection might be raised: legal scholarship has long regarded as fully *legal* certain contractual schemes that seem, in fact, to anticipate the very mechanisms of self-execution later introduced by technology. One example is what Italian jurist Natalino Irti has termed "exchanges without agreement"—cases such as the purchase of goods in supermarkets or department stores, where the contractual exchange takes place through a mechanical logic. The seller displays the goods; the buyer selects one; and there is no true *agreement* between the parties, only what Irti calls a mere "encounter in the thing," the "unilateral decisions to sell and to buy." (Irti 1998)

Irti himself recalls that the problem of the *automatic transaction* had already been noted by German scholars such as Auwers, Schiller, or Günther in the last decade of the nineteenth century and, in Italy, by Antonio Cicu as early as 1901, when discussing vending machines: the machine displays the product, and the insertion of a coin allows its purchase (Cicu 1901).

Yet, as has often been observed, this did not prevent the law from progressively adapting to such technologies, which were fully integrated into contract law without requiring any rethinking of its fundamental premises. The same, it could be argued, applies today: just as in the case of the vending machine—where "the contract is perfected when the coins are inserted into the machine"—the process of execution in *smart contracts* occurs without human intervention, but this does not, by itself, preclude the formation or existence of a "legally binding contract."

In my view, however, the analogy overlooks a crucial difference. In the case of the vending machine, one may indeed say that, by inserting the coin, I both conclude and perform the contract—precisely because that insertion constitutes the goods I intend to purchase *as something that ought to be paid for*.

Even in the vending machine, then, my duty to pay for the goods depends on, and finds its justification in, the fact that the goods enter into the contract *as something whose being consists in its ought-to-be paid for*—its *eidetic Ought*.

This is not the case, however, with the logic of smart contracts. Here, the purchased item *is always already paid for*—it exists from the outset as something that *will necessarily be paid*, not as something that *ought to be* paid.

There is, therefore, a decisive difference between (a) an *automatic transaction*, where the impossibility of non-performance depends on the mechanism by which it is through performance itself that the contract comes into being, and (b) a *self-executing transaction*, which *causes* its own performance always and necessarily. In the first case, the contract comes into existence as something that *ought-to-be* performed (even if it is the act of performance itself that brings it into being); in the second, the "contract" comes into existence as something that is *always necessarily performed*, thereby negating itself *as a contract*.

One might still reply, however, that when "code" makes it impossible not to pay for something—or impossible, say, to exceed a certain speed limit—it does so precisely because it *applies a norm* according to which that speed *ought not to be* exceeded, or that item *ought to be* paid for (in the sense of an *eidetic duty*). In other words, technology does not truly eliminate the *ought-to-be* of things, but merely "transfers" it, as it were, to the design phase of the code.

Even this, however, is not entirely correct. It is not true that code is *logically such* because it incorporates a norm. What code in fact incorporates is a rule of the type: "If the vehicle reaches 90 km/h, then the device reduces torque or fuel injection," thereby limiting engine thrust. This is indeed a *rule*, but it is not a *legal norm*: it merely establishes a causal link between the occurrence of an antecedent and the production of its effects.

It might be said, in response, that code behaves this way because it is itself mandated by a legal norm. Let us grant this. Such a norm would likely prescribe the obligation to design vehicles in a certain way—for example: "If X is a vehicle, then it must be equipped with an automatic speed limiter." Yet this norm does not institute the *speed limit* as something that *ought not to be exceeded*; rather, it constitutes *vehicles* as things that *ought to be equipped* with limiters. In other words, for a device to make it impossible to exceed a certain speed, it is not necessary that there exist a legal norm *instituting that limit* as something that *ought not to be* surpassed. Put more simply: it is not necessary that a norm prescribe a given limit as obligatory; it is sufficient that a norm prescribe the installation of automatic limiters in vehicles as obligatory.

At this point, however, further questions and clarifications are in order. The first concerns the meaning we ascribe here to *Sollen*, or *ought-to-be*. It could be objected that law, above all, is what attributes powers and faculties to subjects: to think of norms in terms of the *ought-to-be* they institute would prevent us from accounting for this dimension, from providing a framework capable of explaining how legal systems actually function. The example of the speed limit, in other words, might serve as a useful illustration, but could also risk being dismissed as a merely "textbook" case.

In my view, however, the matter is not so straightforward. The distinction between *duty* and *ought-to-be*—between what concerns the qualification of an action and what, by contrast, pertains to the mode of being of an object—allows us to clarify the point. It goes without saying that, when we refer to action, to the norm of conduct, the difference between permission and obligation, between the power to perform an act and the duty to do so, is fully relevant. To attribute a power is not the same as to impose a duty. This holds even in the case of the rules of a game: if duty refers to action—that is, to what *ought to occur*—it is evident that saying "the bishop must be moved diagonally" is by no means equivalent to saying "the bishop may be moved diagonally." In the first case, an action is posited as due; in the second, a possibility is attributed.

Things change, however, when we consider not the qualification of the action, but the *mode of being* of the bishop itself. When we speak of *Sollen*, this is what we mean: the circumstance in which the bishop *consists in* the very rule that concerns it. The bishop, in other words, is nothing other than the rule about the bishop. The bishop exists as that piece which *ought-to-be* what the rule says of it, in order to be what it is.

From this, it should be clear that one can speak of *ought-to-be* even in reference to norms that, at the level of conduct, confer powers rather than impose duties. Consider, for example, a norm such as "an obligation may be performed by a third party." Here, once again, we must distinguish between two operations proper to the norm:

(i) as a rule of action, it undoubtedly confers a *power*, understood as referring to the act of performance, and it does not impose a behaviour as due.
(ii) yet it is a rule of action precisely because—as we have seen—it *institutes* the obligation as something that *ought-to-be able* to be performed by a third party in order to be what it is, namely, an obligation.

From this perspective, *can-be* is a form of *ought-to-be*, whereas *power* is not a form of *duty*.

## Constitutive Rules

At this stage, it is worth considering certain objections raised by Andrea Romeo in a perceptive essay addressing the arguments of Brownsword and Lucy (Romeo 2024). While I broadly share the main lines of his critique, engaging with his remarks—which are grounded in a perspective close to Hart's—may help to clarify the specific conception of legal normativity that this book seeks to articulate. For reasons of space, I shall focus here on two of the points developed in his paper.

According to Romeo, technological regulation cannot dispense with the individual's cognitive access to the *internal aspect* of rules. His example is that of a control system which, upon automatically detecting that a driver's blood alcohol level exceeds a certain limit, prevents the vehicle from starting. Romeo argues that the driver must be able to recognize that the car's failure to start is not the result of a mechanical malfunction but a consequence imputable to his or her own behaviour. In order to act—that is, to know whether to repair the car or to wait until the blood alcohol level decreases—the driver must be able to refer to a *norm* explaining that the impossibility results from a specific action attributable to the driver. Moreover, the social-control function that such a device aims to secure would fail if drivers could not understand that the car does not start *because* they have drunk too much.

In my view, however, several distinct issues are conflated in this example. Let us proceed step by step. If we claim that the driver must be aware that the vehicle fails to start not due to a malfunction but due to the alcohol level, this, in itself, does not require the assumption of a legal norm. It is sufficient that the driver knows a *rule*—not a legal one—expressing a causal connection between behaviour and consequence, such as: "If the driver's blood alcohol level is X, then the engine will not start."

Romeo, however, adds a further requirement. He maintains that what is necessary is not only that the driver recognizes the causal link, but that the driver is able to perceive his conduct as a *reason* for being ascribed a given consequence—or, more generally, as a *reason for criticism* of the behaviour performed. This, he argues, can occur only if the impossibility is perceived as a *prohibition*—which would demonstrate that the device necessarily incorporates a legal norm.

Romeo thus correctly distinguishes between two different matters: on the one hand, the experience of a behaviour *causing* a certain consequence; on the other, the experience of that behaviour as the *reason* why a certain consequence is imputed to me. The first case does not, logically, presuppose the existence of a legal norm; the second, by contrast, does.

However, one might suggest that, in the example at hand, the "normativity" to which Romeo refers does not necessarily *precede* the causal mechanism, but may rather be *produced by it*. That is, individuals—precisely because they experience that the car fails to start *as a result* of their alcohol level—may come to *interpret* that experience in normative terms, concluding that the alcohol level is the *reason* why the device blocks the engine. In that case, what would be demonstrated is not that technological systems presuppose a given set of legal norms that they merely "apply" or incorporate, but rather that technological devices themselves are capable of *producing normative effects*.

What risks being shown, therefore, is the *normativity of code* ("Code is Law") rather than the necessity of a pre-existing legal norm enabling us to interpret the meaning of such codes.

At this point, I believe it is possible to clarify the fundamental issue at stake—namely, the very concept of normativity (or, more precisely, the *normativity of norms*, that which makes a norm what it is). Critiques of authors such as Brownsword typically proceed on the same terrain, assuming a conception of normativity—close to that of Hart—according to which it is understood in relation to *action*, to modes of acting, and thus concerns the way in which norms *guide behaviour*. In this sense, normativity is referred to the *guidance of action*: to the capacity of norms to influence the conduct of their addressees by providing them with reasons for acting in the way they prescribe (see Ződi– Bódig 2025).

As I have tried to show in the preceding pages, it is precisely this conception, in my view, that allows technology to *replace* law, since it realizes such "normativity" more efficiently. The line of inquiry I propose to pursue here, therefore, is that of an alternative conception of what we mean by normativity.

The second objection raised by Romeo—which in many respects converges with the arguments developed here—concerns the existence of *constitutive rules*. According to Romeo, technological systems could never replace the specific kind of normativity proper to law, since even the automatic regulation of a given behaviour would still presuppose reference to a rule that *constitutes* that very behaviour, following Searle's model of "X counts as Y in context C."

The only objection I would raise here is that it remains unclear what is meant by *constitutive* in this context. Let us try to clarify the point with an example. Consider once again a device that renders the non-performance of a contract impossible. Such a device certainly presupposes a constitutive rule of the type "X counts as a contract in context C." But what does *having value as a contract* actually mean here? What does it mean that X *counts as* a contract?

Assuming that a rule is (*eidetic*-)constitutive when it provides the condition of existence of the object to which it refers, the issue is that—as Conte had already noted (Conte 1995, pp. 332-334)—it may be either *ontic* or *deontic*. Adapting this distinction slightly to the perspective developed so far, we may say that:

(i) a rule such as "the bishop must move diagonally" is *deontic*, since it constitutes the bishop as that which *ought-to-be moved diagonally*—as something whose being consists in an *ought-to-be*.

(ii) a rule such as "in football, a team scores a point, called a goal, when the ball completely crosses the goal line between the posts," is certainly *constitutive* of the goal, but not *deontic*.

Conte observed that Searle's *constitutive rules* are in fact *non-deontic*; they are *ontic* in character. They certainly *constitute* a fact, but they do so as a *fact of reality*, as a way of *being*. Thought within Searle's model, the constitutive rule thus explains, returning to our example, when and under what conditions something—such as an agreement—*counts as* a contract, that is, when it *exists as* a contract; but it does not explain *what a contract consists in*.

The consequence is that, from the *ontic* rule "X counts as a contract," it does not follow that the parties have a duty to perform it, nor can such a duty, once posited as a prescription, be justified by that rule. Only *deontic constitutive rules*—in the sense developed above—are capable of accounting for what they prescribe. If players have the duty to move the bishop diagonally, it is because the bishop *is* its *ought-to-be moved diagonally*. Likewise, if the parties are obliged to perform the contract, it is because a contract *is* that which *ought-to-be performed*: its being consists in this *ought-to-be*.[1]

1 I shall leave aside here a possible objection that could, in principle, be raised—namely, that the conception of normativity proposed here entails a kind of *reification* of normative "objects." The objection might proceed as follows: to affirm that a contract is *that which ought-to-be performed* is merely another way of saying that there exists a duty to perform it. To distinguish between the two would imply assuming that there exists, in reality, an *object*—the contract—separate from the duty to perform it. Yet, following an approach such as Alf Ross', this kind of "entity" would be unnecessary: what *exists* is only the rule according to which, under certain conditions (for example, when the parties have concluded an agreement in a given form), certain effects follow—such as the parties' duty to perform the agreement. In my view, however, this objection is mistaken, because it can account only for the fact that the duty to perform has its *cause* in a certain antecedent fact (the conclusion of the agreement, and so on). But if one then asks, *why*—what is the *reason why*—the parties must perform the contract, the answer cannot be: because they have concluded a "contract." No *duty* ever follows from a *fact*. Nor is it sufficient to reply that the reason is that "the norm prescribes it," for this does not provide a reason but merely points to another *fact*. It says, in effect: one must perform the contract *because* the norm prescribes it. Yet the *normativity* of norms consists precisely in their ability to *give reason* for the duties they prescribe—otherwise, they would not be norms at all, but mere commands. If, then, it is the norm that introduces the duty to perform the contract, it does so because it *institutes* the contract as something that *ought-to-be performed*. To deny this is to deny normativity itself—that is, to reduce it to a *fact* or a set of facts. Any formulation of the type "one must perform the contract because" will ultimately explain the duty only by appealing to *facts* ("because the norm prescribes it;" "because the addressees recognize it as binding," and so on). Such explanations account only for a *fact*—why the duty exists, or why people recognize it as such—but they do not provide its *justification*. If we confine ourselves—as I have done so far—to considering a single norm, it is precisely the norm's function to go beyond merely prescribing or positing a duty to act: it *gives reason* for that duty by transforming the fact it takes as antecedent into an "object" whose very being consists in its *ought-to-be*. For this reason, the "intermediate step" that Ross criticized is, in fact, essential. The duty to perform the contract does not derive its reason from the fact that the parties have concluded an agreement under certain circumstances (that is the *cause*), but from the fact that the agreement they have concluded *is* something that *ought-to-be performed*—which we call a *contract*. It should now be clear that this does not imply any kind of hypostatization. "Contract" is not the name of a *real* entity in the empirically observable sense, but of the *agreement insofar as it ought-to-be performed*—that is, insofar as it is *instituted as a normative entity*.

It is from this point, I believe, that the fundamental antithesis between law and technology must be drawn. Even if it were true that a technological device ensures the performance of a contract precisely because it presupposes that the contract is something that *ought-to-be performed*, its operation nonetheless *produces* the contract as something that *is always performed*. In such a case, one could say that technology "replaces" and abolishes law—even when law is itself the condition for technology's own existence.

For this reason, I would argue that it is not sufficient to assume, as authors like Brownsword suggest, that the normativity proper to law will remain present even in a world of automated legal enforcement. Rather, it is necessary to maintain that, in order for normativity to be preserved, technologies must be configured in such a way that they allow for its preservation. This occurs if, and only if, the constitutive principle of legal reality continues to hold—namely, that it is a reality composed of *things* (subjects, objects, actions) whose *being* consists in their *ought-to-be*.

The irreducible difference between legal normativity and technological regulation thus becomes clear once norms are reconsidered as propositions through which we *experience* their objects as *ought-to-be*. To paraphrase—while giving it a wholly different meaning—Searle's formula, we might say that norms operate according to the logic: "X exists as an *ought-to-be* Y in context C."

Accordingly, a norm limiting speed to 90 km/h should not be understood—except indirectly—as imposing a *duty* on the driver not to exceed that limit. Rather, it is what allows the object "90 km/h" to exist as something that does not *consist in being* (in the factual sense of driving at a certain speed), but in its *not-ought-to-be-exceeded*. As we have seen, technological regulation works in precisely the opposite way: it reduces the *ought-to-be* to mere *being*, abolishing any distinction between what *is* and what *ought-to-be*. Yet to lose this capacity to institute a reality other than the one that simply *is*, is to losc law itself.

## References

Amina Y. – Martinez R. (2025), "Smart Contracts and Legal Enforceability: Decoding the Political Philosophy of Code as Law." *Interdisciplinary Studies in Society Law, and Politics*, 4, 2: 292–302.

Amselek P. (1993), "Les règles juridiques en tant qu'objets mentaux." *Archiv für Rechts- und Sozialphilosophie*, 79, 3: 311–320.

Becker K. (2022), "Blockchain Matters – Lex Crypytographia and the Displacement of Legal Symbolics and Imaginaries." *Law and Critique*, 33: 113–130.

Bobbio N. (1993), *Teoria generale del diritto*. Turin: Giappichelli.

Brownsword R. (2005), "Code, Control, and Choice: Why East is East and West is West." *Legal Studies*, 25: 1–21.

Brownsword R. (2015), "In the Year 2061: From Law to Technological Management." *Law, Innovation and Technology*, 7, 1: 1–51.

Brownsword R. (2021), *Law 3.0. Rules, Regulation and Technology*. Oxford: Routledge.
Carcaterra G. (1979), *La forza costitutiva delle norme*. Rome: Bulzoni.
Cicu A. (1901), "Gli automi nel diritto privato." *Il Filangieri*, 8: 561–659.
Conte A.G. (1995), *Filosofia del linguaggio normativo*, II. Turin: Giappichelli.
De Filippi P. – Wright A. (2018), *Blockchain and the Law: The Rule of Code*. Harvard: Harvard University Press.
Di Lucia P. (2021), "'Ought' is Spoken in Many Ways." In Di Lucia P. – Fittipaldi E. (eds.), *Revisiting Searle on Deriving "Ought" from "Is"*. London: Palgrave Macmillan, 177–188.
Di Matteo L.A. – Cannarsa M. – Poncibò C. (2019), *The Cambridge Handbook of Smart Contracts, Blockchain Technology and Digital Platforms*. Cambridge: Cambridge University Press.
Hassan S. – De FilippiP. (2017), "The Expansion of Algorithmic Governance: From Code is Law to Law is Code." *Field Actions Science Reports*, 17: 88–90.
Hildebrandt M. (2020), "Code-driven Law: Freezing the Future and Scaling the Past." In Markou C.Deakin S. – (eds.), *Is Law Computable? Critical Perspectives on Law and Artificial Intelligence*. Oxford: Hart, 67–83.
Irti N. (1998), "Scambi senza accordo." *Rivista trimestrale di diritto e procedura civile*, II: 347–364.
Jellinek G. (1914), *Allgemeine Staatslehre*. Berlin: Häring.
Kelsen H. (2005), *Pure Theory of Law*. Clark: The Lawbook Exchange.
Lessig L. (1999), *Code and Other Laws of Cyberspace*. New York: Basic Books.
Lucy W (2022), "The Death of Law: Another Obituary." *Cambridge Law Journal*, 81, 1: 109–138.
Oster J. (2021), "Code is Code and Law is Law—The Law of Digitalization and the Digitalization of Law." *International Journal of Law and Information Technology*, 29: 101–117.
Romeo A. (2024), "La norma non può morire. Brevi riflessioni critiche sugli ultimi necrologi del diritto e sulla gestione tecnologica della condotta umana." *Diritto&Questioni Pubbliche*, 1: 81–99.
Schauer F. (2015), *The Force of Law*. Cambridge, MA: Harvard University Press.
Wright A. – De FilippiP. (2015), *Decentralized Blockchain Technology and the rise of Lex Cryptographia*. https://ssrn.com/abstract=2580664.
Zavyalova E.B. – Shumskaia E.I. – Shumskaia A.I. (2019), "Transactions in the Digital Age: Blockchain Technology and Smart Contracts." *Journal of Law and Administration*, 15, 3: 32–38.
Ződi Z. – Bódig M. (2025), "Is Code Destined to Supersede Law? Behavioural Control by Code and Legal Theory." *The Cambridge Law Journal*, 1: 175–208.

# Conclusion

## Automata

The relationship between law and technology is today perhaps the most urgent and challenging issue confronting legal scholarship. This is true both because of the practical implications that the use of artificial intelligence, neurotechnologies, and digital devices is already producing within the legal domain—for instance, through the development of predictive algorithms in the field of justice—and because of the theoretical difficulties that arise when we ask what the very nature of law is, and what its normativity properly consists in.

In this sense, the situation has changed profoundly compared to a century ago, when the debate on "automata" in law briefly captured the attention of jurists. Scholars at the time had already intuited that what they called "automatic devices"—referring mainly to vending machines dispensing cigars, sweets, or tickets without human intervention—had little practical importance, yet a fundamental theoretical significance. These machines posed new legal questions concerning, for instance, contract formation and the performance of obligations.

For this reason, within the space of a single decade—roughly between 1890 and 1901—a remarkable number of studies appeared, particularly in German legal literature, addressing the problem of "automata" and their legal status. Alongside the Italian contribution of Cicu, one should at least recall earlier works such as Günther's *Das Automatenrecht* (1891) and Schiller's *Rechtsverhältnisse des Automaten* (1898).

A century later, the problem of automation in law has re-emerged with new intensity and urgency—so much so that several scholars have gone so far as to predict the impending "death of law," brought about by the rise of technological regulation.

We have therefore sought to offer some answers to this question—one that concerns, as should by now be clear, the very meaning of what law is, what its function consists of, and why we may still need it (or perhaps no longer do).

As already noted, the arguments advanced in this book can be summarized under two main theses. The first is intended to reconsider the way in which we traditionally speak of the relationship between "law and technology." It

DOI: 10.4324/9781003775850-5

argues that law is not something external or opposed to technology; rather, law itself constitutes a particular kind of technology (and, as noted in the Preface, of technique).

Its distinctive feature lies in the production of an "artificial" reality—one that stands apart from the natural world. Throughout the book, several examples have been offered to illustrate this point, leading us to describe law as a form of "cybernetics:" not a mere set of tools at the disposal of individuals, but a system of action that transforms the very subjects who employ it, making possible new forms of *subjectivation*.

From this follows the second thesis. We must ask, therefore: what distinguishes this particular technology that we call law? The answer we have proposed is that what defines law is the specific form of normativity that characterizes it.

This normativity, as we have argued, cannot be reduced to the prescriptive or behavioural logic that characterizes algorithmic regulation. Algorithms operate by transforming norms into causal mechanisms: they ensure that certain outcomes necessarily follow from given inputs, thereby collapsing the distinction between what *is* and what *ought to be*. Legal normativity, by contrast, depends precisely on maintaining that distinction. Law does not merely compel compliance or guide behaviour; it institutes a distinct normative reality—an order of things whose being consists in their *ought-to-be*.

This is why, even when technologies appear to replicate or automate legal functions, they cannot reproduce the normative dimension that makes law what it is. To automate the *ought* is, ultimately, to abolish it.

## Normativity and Prescriptivity

In the final chapter, I have sought to clarify what is properly meant by the notion of *"ought-to-be"* (*Sollen*), as distinct from the mere *duty* that refers to the action a norm prescribes as obligatory, permitted, or prohibited.

This is likely the point in the book that demands the greatest theoretical effort from the reader, not least because of its counterintuitive implications. On the one hand, this approach required shifting the discussion to a higher level of conceptual abstraction—one that is usually overlooked in debates on the impact of artificial intelligence on legal systems, the rise of predictive models of justice, or the automation of contractual relations. For this reason, the book deliberately refrains from addressing the specific technical issues that practitioners, judges, and legal officers already face, and will increasingly face, in their daily work. This choice rests on the conviction that only by understanding what constitutes the normativity of law can we begin to address the practical problems that will inevitably arise.

On the other hand, our proposal to distinguish legal normativity from the "prescriptive" nature of technological regulation requires moving beyond the idea that law is merely a *rule of conduct*—that is, an instrument designed to influence, direct, or guide human behaviour.

We are fully aware of how paradoxical this claim may sound. The idea that law exists to "regulate behaviour" appears not only intuitively obvious but is so widely shared among jurists that it is often taken for granted in both interpretation and argumentation.

After all, whether norms are defined as "reasons for action," as standards of conduct, as predictions of what judges will do, or in yet other ways, all these accounts still rest on the same underlying assumption: that normativity should be conceived primarily as a capacity to guide behaviour. Even Kelsen's own framework—on which our argument has partly relied—never entirely abandoned this idea.

Yet, as this book has sought to clarify, the distinction between normativity and prescriptivity entails precisely this. Although legal norms undoubtedly prescribe certain forms of behaviour, their normative character does not lie in the command itself, nor in the effects that such commands have on human conduct. What makes them *norms* is their capacity to constitute a specific kind of reality—one that exists as an *ought-to-be* rather than as a mere *being*.

In other words, law does not merely indicate how things *ought to* happen or to be; rather, it brings them into existence in such a way that they have no "being" other than their *ought-to-be*. Throughout the book, we have sought to clarify the meaning of this thesis through a series of examples. Let us return, for the last time, to one that has already been discussed: a norm N stating that "it is obligatory to pay taxes."

This norm certainly performs a prescriptive function: it makes a certain action—paying taxes—something that is *due*. In the terms used here, we might say that the prescription concerns a *duty*, a behaviour insofar as it is posited as *owed*. The crucial point, however, is that its normativity does not consist in, nor should it be confused with, this prescriptive aspect. The normativity of the norm lies instead in its instituting taxes as something that *ought to be paid*—as an "object" that has no being apart from its *ought-to-be* (for this reason, we have spoken of an "*eidetic Ought*").

An object such as "taxes" does not exist in nature, nor does it exist, even within the legal domain, as something that *is*. It exists only as something that *ought to be paid*.

What this book has sought to demonstrate is that technological regulation cannot reproduce—or even sustain—the separation between normativity and prescriptivity.

This means, however, that technological regulation does not merely contribute to the implementation, reinforcement, or greater efficiency of legal norms. Rather, it represents a distinct form of behavioural regulation which—precisely insofar as it makes it possible to embed and automatically execute the prescriptions derived from legal norms—at the same time nullifies and erases the very normativity that constitutes them.

Nothing prevents us, then, from imagining—or indeed from considering it plausible—that automation and, more broadly, algorithmic forms of regulation may eventually overlap with, and even replace, vast areas that have traditionally been governed by law. What must be emphasized, however, is that this would simply mean renouncing, in those domains, the law itself as a technology for constructing the reality in which we live—a *medium* through which we experience the world and our own action within it.

Is the death of law, then, near and about to come? We do not know. What we do believe, however, is that the way in which this question will eventually be answered depends largely on how we understand what law means—and what it would mean to renounce it.

As I have sought to show, as long as we continue to assume that law consists essentially in the regulation of conduct, objections to the use of algorithmic technologies will remain, at bottom, limited to a defence of the "human." This defence, in particular, takes the form of insisting that individuals must be directly involved in the process of applying norms and must remain free to choose whether or not to comply with them.

The book has sought to expose the limits of this conception. If what is ultimately at stake is the "freedom" of legal subjects, it is likely—as Lucy has also observed—that this freedom will gradually be sacrificed, perhaps without our even noticing it.

For this reason, we have proposed another path, suggesting that the truly central question lies elsewhere: whether—and under what conditions—it remains essential for us to experience the world, reality itself, through the *normativity of law*. The issue, in other words, is not whether it is worth ensuring that norms, as prescriptions, achieve greater "effectiveness," or even the possibility of automatic execution. It is, rather, whether we still wish to rely on a technology—such as law—capable of constructing a world in which things and our social relations appear to us not for what they *are*, but for what they *ought to be*.

## Law as a Technoscience

In this book, I propose to define law as a *technoscience*, in order to underline that it constitutes: (a) a *technique*, namely a set of procedures that generate objects that do not exist as such in reality and are not naturally "given" (what we might therefore call artefacts or artificial objects—such as property, marriage, obligations, and so on); and (b) a *technology*, that is, a body of knowledge concerning the processes required to produce these artefacts and the ways in which they operate and are coordinated.

I adopt the term *technoscience*, drawing on Don Ihde, because legal knowledge is "embedded," as it were, in the techniques through which it is expressed and which it employs: it does not pre-exist these techniques, but is formed together with them.

It is in this sense that law constitutes one of the essential forms through which individuals in our societies mediate their experience of reality. Law, that is, is one of the technologies through which our consciousness—or, more precisely, our way of perceiving and relating to ourselves and to others—is mediated.

We encountered several examples of this in the first chapter. We can now attempt to summarize and distinguish the different levels of this "mediality" that law constitutes. One may say that law *mediates* our experience of ourselves and of our relation to things and to others insofar as it is one of the technologies through which:

(a) we experience ourselves, and thus individuate ourselves, as subjects. The expression "legal subject," in its modern sense, can indeed also be understood in this way: it signals the fact that we relate to ourselves through the *medium* of legal concepts and legal technology.

Consider, for instance, the relation we have to our own body. We perceive and describe our body as "our own." It would be a mistake to overlook the fact that this occurs because our perception and experience of the body are already *mediated* by a technology that is (also) legal—one that enables us to conceive of belonging, of "mine" and "yours."

From this perspective, I would argue that the philosophical question of whether we *have* our body or *are* our body is secondary and derives from this prior structure. It is, in fact, the logic of law that provides the conditions of possibility for thinking these two alternatives with respect to how we understand "what is ours."

As Donellus already observed, something is said to be "ours" in the proper sense when it belongs to us in such a way that none of its constitutive elements depend on another: *proprie nostrum voco, quod ita nostrum factum est, ut nulla ex parte sit alterius*. But something is also said to be "ours" in a different sense—namely, in the sense of what is owed to us by another: *est etiam quod nobis debeatur*. It is legal "mediality," in other words, that makes possible two distinct logics of the "ours:" one in which it coincides with belonging as *being*, as *suitas*, and another in which it coincides with belonging as *having*, as *habere*.

For this reason, in the first chapter, I suggested that we might speak—drawing on Stiegler's insights—of an *organology* of legal technoscience as well. Law, after all, is what gives us a body. Or, more accurately: it *mediates* the experience we have of ourselves and of our body in such a way that this experience does not simply coincide with a supposedly "natural" perception of it. It has sometimes been argued that law constitutes a technique grounded in the *abstraction* of the body, an operation of "disincarnation" or "disappearance of the body" (Jean-Pierre Baud), allegedly underlying the Roman distinction between the individual and the person. To this I would respond that such an operation of "abstraction" should in fact be understood as an operation of *incarnation*, through which our body becomes something other than what it "naturally" is.

But we can speak of *organology*, in the strict sense, also in reference to the fact that law functions as something that—historically and variably—defines the ways in which we see or touch things.

The ways in which we interact with things, in other words, already incorporate legal technology, which transforms our very "organs." We might say that there exists a legal constitution of the hand insofar as it can grasp, and of the eye insofar as it can look.

And this proceeds according to rules different from those that govern "natural" perception—if such a thing exists at all, which it does not. It is in this light that Paul's famous passage must be read: *non est enim corpore et tactu necesse adprehendere possessionem, sed etiam oculis et affectu* (Paul. 54 ad ed. D. 41.2.1.214). The point is not to claim that possession may rest solely on "will," in the absence of any "material" grasp, but rather to make the eye—the gaze—into something that not only "sees" but in a strict sense *touches* and appropriates the thing.

The juridical elaboration of possession cannot be understood without recognizing this operation that separates possession from contact, from the "holding" proper to the hand: indeed, Paul states elsewhere that a sleeping person into whose hand an object is placed would in no way be said to possess it (*sicut si quis dormienti aliquid in manu ponat*).

Certainly, the strategy at work here is to give relevance to a certain kind of "conscious and voluntary" behaviour; but this cannot occur without a displacement of the senses which, on the one hand, deactivates the hand and, on the other, constitutes the eye as an organ that does not merely "see," but *touches*, retains, and holds something as mine.

What must be emphasized, then, is that in this context a specific solution to a juridical problem—namely, a certain "dematerialization" of *traditio*—could be formulated through an operation that assigns to the eye the function of "grasping:" the eye becomes that which can seize, apprehend, and retain the thing. In this case too, law constitutes the technology through which we may experience the eye not as an organ of vision, but as an organ of touch, of grasping.

(b) through law, we also experience our relation to things and to others. In this sense, law is the *medium* of collective individuation—of the way in which a "we" is constituted in relation to an "I" and a "you." At this level, it has already been emphasized that legal technology functions as the means through which our social relations are mediated. Property, kinship relations, and contract are techniques that define and transform the way we experience reality. This point was already underscored in the first chapter of the book.

Of course, the point is not to treat these legal conditions as "a priori." Precisely because they are historically determined, legal technologies lend themselves to different uses and acquire different meanings. As already noted, for example, the possibility of experiencing things in terms of *res*—as things-in-property or things capable of appropriation—can be articulated in multiple ways.

For this reason, we should avoid concluding that property always and necessarily entails a relation of domination between individuals and things, or that—inasmuch as it takes the form of private property—it is inherently tied to the development of forms of "possessive individualism." In fact, as we have seen by recalling Margarete Radin's theory, property schemes can also operate as techniques of individuation—that is, as modes through which the individual constitutes herself or himself.

When we refer to legal technology as one of the "modes of mediation" of collective individuation, we must also take into account the way in which law has constituted, in the modern West, the technology through which power—and the forms of domination that organize society—have been conceptualized. As Foucault already emphasized, law has historically been the technique through which modernity—at least since Hobbes—has given form to power, following the principle that law must be the very *form* of power, and that power must always be exercised in the *form* of law.

In Hobbes, all of this becomes particularly evident. This is not only because power does not exist, take shape, or operate except through legal technologies (the covenant, the obligation to obey, representation, law as command, and so forth).

More importantly, *ius* and *lex*—natural right and natural law—function in Hobbes as the very means through which individuals mediate their way of relating to themselves and to others. What Hobbes calls *jus*, by which he means "the Liberty each man hath, to use his own power, as he will himselfe, for the preservation of his own nature; that is to say, of his own life," must be understood in this precise sense.

It is what, already in the "state of nature," enables individuals to enter into relations with one another, to experience themselves and their relations to others in terms of claims they may advance. *Jus* does not simply *describe* the circumstance that human beings are *de facto* free to do as they wish in the natural condition. Rather, it states that this freedom is always already experienced as a rightful title to make claims, to assert demands, to lay hold of things.

A similar argument could be made for the notion of *lex naturalis*. Without being able, here, to examine the conceptual mechanisms of Hobbes' philosophy in detail, the point we wish to emphasize is that these considerations allow us to understand how the forms of *ius* and *lex* are, above all, forms of experience—that is, technologies that constitute part of our experience of reality.

## What Law Truly Is

At this point, we believe it becomes clearer in what sense law constitutes a technoscience that *mediates* our experience of reality.

Let us now add the final element, developed in the third chapter. Our claim is that this mediation is *normative*. Let me briefly summarize the point. I propose to define law as the technoscience through which we enter into relation with things, with others, and more generally with "reality," insofar as the latter appears to us not as what *is*, but as what *ought-to-be*.

We have devoted considerable effort to clarifying the distinction between what *ought-to-be* and what is *owed*, so as to avoid a range of ambiguities and misunderstandings that cannot be revisited here.

If we keep this distinction in mind, we may say that the specificity of law as a technoscience lies in the normativity through which it operates. Earlier, I suggested that law is the medium through which our consciousness is "mediated" in its experience of things, of itself, and of others—and that this mediation is normative.

This means, for instance, that in legal terms I do not experience something—say, this apple—except insofar as it has no other "being" than its *ought-to-be* delivered, returned, gifted, or, depending on the circumstances, its *may-be* claimed, and so forth. Likewise, as we have seen, law enables us to experience an action not simply as something *owed*, but as something whose very being consists in its *ought-to-be*.

This captures the basic logic through which legal technology operates. The concrete mechanisms that implement this logic—those that regulate conduct, structure legal relations, and so on—would naturally require further analysis.

Our concern here, however, has been to identify the distinctive form of normativity that defines law, so as to distinguish it—precisely, and without ambiguity—from the *prescriptive* logic that characterizes "technological regulation."

Likewise, as we have seen, law is what enables me to experience a given action not merely as something that is "owed," but as something whose very being consists in its *ought-to-be*.

Throughout the text, I have offered several examples to clarify this distinction. Let me add one more, which—at least in my view—helps to illuminate the point particularly well.

Suppose that my will provides a bequest to Titius on the condition that he refrain from performing a certain act—for example, that he never marry and remain unmarried for the rest of his life. This would imply, however, that since one can be certain that Titius has fulfilled the condition only at the moment of his death—when it is definitively established that he has not done what the testator prohibited—the condition could be considered satisfied only after the beneficiary's death. The consequence would be that Titius could inherit the property only once he is dead, and therefore never.

Let us examine this situation by contrasting different kinds of logic.

At the philosophical level—that is, within the discursive "technology" that philosophy constitutes—this is simply a self-invalidating paradox, in which the condition is such that its fulfilment renders impossible the very event it is meant to condition.

Or, if one prefers: a paradox in which the beneficiary can enjoy the bequest only from the moment at which he can no longer enjoy it. Philosophical *logos* cannot "resolve" such a case, which exists for philosophical reasoning solely in the logical form of a paradox.

Let us now ask how the same situation would be addressed within what we have called "technological regulation."

Here, we must assume that regulation is exclusively technological or algorithmic—meaning that it does *not* presuppose the existence of a juridical meta-rule such as: "If a suspensive condition is temporally impossible to verify during the beneficiary's lifetime, and the apparent intention is to reward a certain behaviour, then the condition must be reinterpreted as a resolutive one."[1]

Given this assumption, all that a technological device could do would be to guarantee, as a matter of fact, the impossibility for Titius to marry. To "solve" the problem at the level of *code*, we would therefore have to assume—admittedly, in a science-fiction scenario—that the condition may be deemed already fulfilled because technology makes it possible to establish with certainty that Titius will never be able to marry.

Technological regulation would operate here, as we have seen, by converting a duty *not* to perform an action into the impossibility of performing it.

Let us now turn to the juridical level. As is well known, it was the Roman jurists who introduced the rule—the so-called *cautio Muciana*—which allowed the beneficiary to acquire the legacy immediately, *as if* no condition were attached, provided that he undertook to return it should he subsequently

1 The point requires clarification. I am *not* assuming here that technology is incapable of expressing meta-rules or forms of normativity. It can, of course, be programmed—and instructed—to operate by applying legal rules. The issue lies elsewhere. The point is that an algorithmic or computational logic, *as such*, would never be able to resolve the problem raised by our example in the way that, as we shall see, the law in fact does.

perform the act prohibited by the testator. As Ulpian later commented, the usefulness of the *cautio* lies precisely in those conditions that are conceived as consisting in a non-action (*in non faciendo*).

What is crucial to observe is that only through that technoscience we call law can the series of operations be carried out that make it possible to transform reality in this way. Through the *cautio*, in fact: (a) the non-action, which is not an "event," is converted into a positive action that is empirically verifiable; (b) a suspensive condition is converted into an obligation to return the legacy (or, as in the current Italian Civil Code, into a resolutive condition).

Only law is capable of constructing a properly normative reality—one that is, as such, independent of what things "naturally" are.

What is "naturally" a non-action—something that, in itself, does not "exist," because it is not a verifiable fact or event—is, in law, reconstructed as a fact: namely, *entering into marriage*.

And it is precisely this transformation that allows the legacy to come properly into existence. As long as it is understood as something that must be transferred to Titius only if he does not marry, this "*ought-to-be*" is juridically impossible; indeed, it does not exist at all, insofar as it self-annihilates. Through the *cautio*, however, the legacy comes into existence as something that *ought-to-be* returned in the event of marriage.

This example enables us to understand more clearly in what sense law can be described as a technoscience, one that is fundamentally different from—and irreducible to—the kind of "technological regulation" expressed by the logic of *code is law*. But this becomes apparent, we should stress, only if one avoids conflating the operations of law with mere prescriptivity, that is, with the simple regulation of conduct.

The question of the "death of law," which some jurists have anticipated in light of technological innovations capable of replacing regulation through norms with regulation through algorithms, is closely tied to the considerations just outlined.

Only if legal scholarship continues to study and practise law as a technoscience—rather than merely as an instrument for regulating behaviour—will law remain irreplaceable. This, of course, does not mean that such an outcome is guaranteed.

After all, Pomponius' remark from the mid-second century CE still holds true: *constare non potest ius, nisi sit aliquis iuris peritus, per quem possit cotidie in melius produci* ("there can be no law unless there is a jurist through whom it may be improved day by day"). What we may say, in our terms, is this: there can be no law unless there is a legal science capable of understanding what law truly is—and thereby enabling it to remain alive.

# Index